1    25."
2    55.5"

# CLINICAL OPTICS

# CLINICAL OPTICS

Troy E. Fannin, O.D.
Theodore Grosvenor, O.D., Ph.D.

*College of Optometry*
*University of Houston*
*Houston, Texas*

**Butterworths**

Boston   London   Durban   Singapore   Sydney   Toronto   Wellington

**Library of Congress Cataloging-in-Publication Data**

Fannin, Troy E.
　Clinical optics.

　Includes bibliographies and index.
　1.　Optometry.　2.　Optics, Physiological.
I.　Grosvenor, Theodore P.　II.　Title.　[DNLM: 1.　Optics.
2.　Lenses.　3.　Refraction, Ocular.　WW　300　F213c]
RE951.F36　　1987　　　　617.7'5　　　　86-34288
ISBN 0–409–90060–5


Butterworth Publishers
80 Montvale Avenue
Stoneham, MA　　02180

10　9　8　7　6　5　4　3　2

Printed in the United States of America

# Contents

**CHAPTER SEVEN**
## Absorptive Lenses and Lens Coatings 179

**CHAPTER EIGHT**
## Multifocal Lenses 233

# Preface

This textbook is intended primarily for use by optometry students, but should prove useful also in programs for the training of optometric technicians and dispensing opticians.

In optometric educational programs, the subject of clinical optics (or ophthalmic optics, as it is sometimes called) is normally presented subsequent to courses in geometrical and physical optics. It is assumed, therefore, that the student taking up the subject of clinical optics will have an understanding of the following topics: the laws of reflection and refraction; the optical principles both of thin lenses and of thick lens systems; the optical principles of cylindrical and toric lenses; the optical principles of prisms; and the principles of chromatic dispersion, lens aberrations, interference, diffraction, and polarization.

In the past, a number of different "sign conventions" have been used by authors of optics textbooks. Nomenclature and notation (or symbols) have also varied from one textbook to another. Fortunately, in recent years most optometric textbooks in the English-speaking world have adopted the sign convention and nomenclature used by British authors including Emsley, Fincham and Freeman, and Bennett.

The introductory chapter will present the sign convention, nomenclature, and notation to be used throughout this textbook. For any terminology not included in the introductory chapter, the reader may wish to refer to W. A. H. Fincham and M. H. Freeman's *Optics,* 9th edition, London: Butterworths, 1980.

## ACKNOWLEDGMENTS

No textbook can be written as a result of the efforts of just one or two people. The authors would, therefore, like to acknowledge those individuals who helped make this book possible.

We want first to thank Dean William R. Baldwin and Associate Dean Jerald Strickland for their encouragement and support throughout the entire project. We would like to acknowledge the assistance of a large number of people who read, and critically evaluated, portions of the manuscript. These include Drs. William F. Long, Clifford Brooks, Michael P. Keating, Ralph P. Carifa, Henry B. Peters, William L. Brown, Robert B. Mandell, Robert Browning, Roger Boltz, and Jess B. Eskridge.

We would also like to thank Drs. Donald Pitts and Bernard Maslovitz for allowing us to use a large number of their transmission curves for absorptive lenses. And we particularly want to thank the college's Audio-Visual Department director, Mr. Jay MacMichael, and his staff for their many hours of diligent work in completing the several hundred illustrations for the book.

Last, but not least, we want to thank our wives for their patience during the many evenings and weekends spent in seclusion.

# INTRODUCTION

# Sign Convention, Nomenclature, and Notation

Sign convention, nomenclature, and notation establish the ground rules, or *standard of reference*, for all labeling and measurements. On the basis of this standard of reference, equations are developed that express the fundamental relationships of image formation by optical systems.

## 0.1. Sign Convention

Throughout this textbook the English system, as employed by Fincham and Freeman,[1] will be used. As illustrated in Figure 0-1, the elements of this sign convention are as follows:

1. Incident light will be considered as traveling *from left to right.* An arrowhead on a ray indicates the direction in which the light is traveling.
2. All distances (object distances, image distances, focal lengths, etc.) are measured *from the optical system.*
   a. If measured in the direction in which light is traveling (from left to right), distances will be considered as *positive.*
   b. If measured in the direction opposite to that in which light is traveling (from right to left), distances will be considered as *negative.*
3. Vertical distances measured from the optic axis to a point above the optic axis are *positive,* whereas vertical distances measured to a point below the optic axis are *negative.*
4. The angle between a ray and the optic axis is measured from the ray to the optic axis.
5. Angles of incidence, refraction, and reflection are measured from the normal to the ray.
6. Angles measured in a counterclockwise direction are considered as *positive*; angles measured in a clockwise direction are considered as *negative.*
7. An arrowhead on a line or a curve, fixing the limit of a given distance or angle, indicates the *direction* in which the distance or the angle is measured.
8. *Vergence* is defined as the curvature of a wavefront at a specific distance from the origin, or focus. If a wavefront is moving toward its focus (converging rays) the vergence is *positive,* as shown in Figure 0-2A; if a wavefront is emanating from its origin (diverging rays) the vergence is *negative,* as shown in Figure 0-2B. The unit used for specification of vergence is the *diopter.* For a given wavefront the vergence, in diopters, is given by the reciprocal of the distance (expressed in meters) from the wavefront to its center of curvature. Thus, the vergence of the converging wavefront shown in Figure 0-2A is +0.50 D, while the vergence of the diverging wavefront shown in Figure 0-2B is −0.25 D.

**Important Note.** Plus and minus signs are assigned to actual *values:* They are not assigned to symbols (i.e., letters) used in mathematical expressions, but to the numbers substituted for these symbols in solving problems.

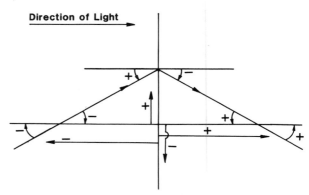

FIGURE 0-1. **Elements of the sign convention.**

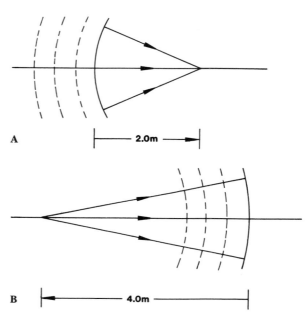

FIGURE 0-2. **Converging and diverging wavefronts.**

## 0.2. Nomenclature

A *ray* is a hypothetical line extending from the origin or focus of a wavefront, and is perpendicular to all wavefronts emanating from the origin or moving toward the focus of the wavefront. A ray represents the direction of propagation of the wavefront. Rays are considered to emanate from point sources or from any of an infinite number of points comprising an extended object, or a source of finite size.

A *pencil* is a bundle of rays emanating from a point source after passing through a limiting aperture.

The *chief ray* is the ray passing through the center of a limiting aperture.

A *beam* of light is a collection of pencils arising from an extended source, or source of finite size.

An *object* is the physical source of light (or of no light) that exists in object space. Objects, also called sources, may be either point sources or sources of finite size.

Objects may be either *real* or *virtual*. Real objects are those from which light actually radiates or from which it is reflected (Figure 0-3A). The object presented to the first source of an optical system is always real. Virtual objects are those toward which light is converging before interruption by a subsequent surface of an optical system (Figure 0-3B).

An *image*, produced by an optical system, is the optical counterpart of an object. It is formed by the light traveling from the object, in image space, after the optical system has acted upon it.

Images may be either *real* or *virtual*. A real image is one that is actually reached (and formed) by *converging* light rays (Figure 0-4A). A real image will be formed on a screen if the screen is placed in the image plane. A virtual image is formed by light that appears to be *diverging* from points in an optical system after treatment by the system (Figure 0-4B). Such an image will not be found on a screen placed in the image plane, but may be found by placing a convex lens in the optical system with the result that a real image will be formed. Due to the fact that the human eye is a converging optical system, a virtual image may be seen by an observer looking through the optical system.

*Object space* is the space in which light travels before encountering an optical system. Object space also encompasses the space beyond the optical system, in which light would have traveled had the optical system not interrupted it.

*Image space* is that space in which light travels after being acted upon by the optical system. Like object space, it encompasses the space on both sides of the system: the side on which it is actually traveling and the side from which it appears to have come.

The word *primary*, when used in conjunction with focal points, principal points, and nodal points, applies to points in object space, whereas the word *secondary* applies to points in image space.

## 0.3. Notation

The following general notation will be used in this text:

1. Lowercase letters are used to designate distances, angles, and constants such as an index of refraction.

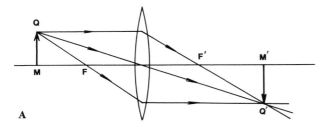

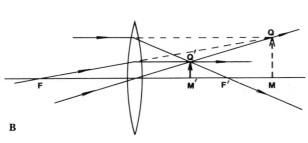

FIGURE 0-3. **Real and virtual objects: (A) MQ is a real object; (B) MQ is a virtual object.**

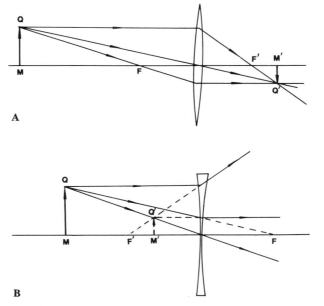

FIGURE 0-4. **Real and virtual images: (A) M'Q' is a real image; (B) M'Q' is a virtual image.**

2. Capital letters are used to denote points, reciprocals of distances, and complicated quantities involving a number of terms and variables.
3. Corresponding object and image quantities are differentiated by following the image quantity by a "prime" ('). For example, $f$ and $f'$ denote the primary and secondary focal lengths, respectively.
4. Subscript numerals are used to designate the order in which the surfaces of a lens or an element of a system are encountered by incident light. For example, $F_1$ and $F_2$ denote the front and back surfaces of a lens, respectively.

Commonly used notations include the following:

### Indices of Refraction

| | |
|---|---|
| $n, n'$ | Refractive indices of object and image space. |
| $n_1, n_2, \ldots$ | Index of refraction of the first medium, second medium, etc. |

### Object and Image Distances and Sizes

| | |
|---|---|
| $l, l'$ | Distance of object and image from the optical system. |
| $h, h'$ | Sizes of object and image. |

### Focal Lengths and Extra-focal Distances

| | |
|---|---|
| $f, f'$ | The primary and secondary focal lengths, measured from the optical system. |
| $x, x'$ | The extra-focal distances, or distances of the object and image from the primary and secondary focal points. |

### Radii and Curvatures of Surfaces

| | |
|---|---|
| $r_1, r_2, \ldots$ | Radius of the first surface, second surface, etc. |
| $C_1, C_2, \ldots$ | Center of curvature of the first surface, second surface, etc. |
| $R_1, R_2, \ldots$ | Curvature of the first surface, second surface, etc. |

### Angles

| | |
|---|---|
| $i, i'$ | Angle of incidence and angle of refraction, respectively. |
| $\alpha, \alpha'$ | Angular size of object and image, respectively. |

### Vergence

| | |
|---|---|
| $L$ | Vergence in object space ($n/l$). |
| $L'$ | Vergence in image space ($n'/l'$). |

### Refracting Power

| | |
|---|---|
| $F$ | Power of an optical system; ability to change vergence of light. |
| $F_1, F_2$ | Refracting power of front and back lens surfaces. |
| $F_n$ | Front vertex power (neutralizing power). |
| $F_v$ | Back vertex power. |
| $F_e$ | Equivalent power. |

### Focal Points, Principal Points, Nodal Points

| | |
|---|---|
| $F, F'$ | Primary and secondary focal points (or focal planes). |
| $P, P'$ | First and second principal points (or principal planes). |
| $N, N'$ | First and second nodal points (or nodal planes). |

### Magnification

| | |
|---|---|
| $M$ | Lateral magnification of a lens ($h'/h$, or $L/L'$). |
| $M$ | Angular magnification of a lens ($\tan \alpha'/\tan \alpha$, or $\alpha'/\alpha$). |

### Prism Notation

| | |
|---|---|
| $\beta$ | Refracting angle of a prism. |
| $\varepsilon$ | Refracting power of a prism; $P$ may also be used. |
| $\Delta$ | The prism diopter. |
| $d$ | Decentration. |

## Reference

1. Fincham, W. H. A., and Freeman, M. H. *Optics*, 9th edition. Butterworths, London, 1980.

## Bibliography

The student may find the following textbooks, on the subject of clinical optics and related areas, to be helpful.

Jalie, M. *The Principles of Ophthalmic Lenses*, 3rd edition. Association of Dispensing Opticians, London, 1977.

Morgan, M. W. *The Optics of Ophthalmic Lenses*. Professional Press, Chicago, 1978.

Brooks, C. W., and Borish, I. M. *System of Ophthalmic Dispensing*. Professional Press, Chicago, 1979.

Stimson, R. L. *Ophthalmic Dispensing*, 3rd edition. Charles C Thomas, Springfield, Illinois, 1979.

Tunnacliffe, A. H., and Hirst, J. G. *Optics*. Association of Dispensing Opticians, London, 1981.

Sasieni, L. S. *Principles and Practice of Optical Dispensing and Fitting*, 3rd edition. Butterworths, London and Boston, 1975.

Drew, R. *Professional Ophthalmic Dispensing*. Professional Press, Chicago, 1970.

# CHAPTER ONE

# Ophthalmic
# Lens Materials

Although lenses have been used as aids for vision in one form or another since ancient times, throughout most of the period of their use all such lenses were made of *glass*. Furthermore, from the eighteenth century until the period between World Wars I and II, virtually all single-vision lenses were made of a particular variety of glass known as *ophthalmic crown*.

Varieties of glass other than ophthalmic crown were first utilized for special purposes such as the reading segments of fused bifocal lenses, which were originally made of *flint* glass, having a higher index of refraction than crown glass. More recently, numerous "high-index" glasses have been developed, not only for bifocal segments but for higher powered single-vision lenses as well.

World War II served as a stimulus for the development of plastic materials of optical quality, and the development of plastic ophthalmic lenses proceeded rapidly once the war was over. With the advent of large lens sizes during the 1970s, plastic lenses, being only about half as heavy as glass lenses, have assumed an increasingly important role.

In this chapter both glass and plastic materials used for the manufacture of ophthalmic lenses will be discussed in terms of their development, their optical properties, and their impact resistance and other physical properties.

# GLASS

## 1.1. History of Glassmaking

Although glass is one of the oldest and most commonplace materials known to man, many questions about the exact nature of glass are still unanswered. Glass may be defined as an amorphous solid material which is obtained by cooling, without crystallization, an inorganic mixture that has been found to fuse at high temperatures. Most liquids change from a liquid to a solid abruptly and at a definite temperature, yet glasses behave in such a way that upon cooling the viscosity increases continuously, and there is no fixed temperature at which the material solidifies. For this reason, glass is often called a "supercooled" liquid.

Innumerable varieties of glasses are available today from the glass industry, formulated for specific and wide-ranging purposes: Hence, "glass" is a generic term like "metal," and when referring to different glass compositions it is appropriate to refer to "glasses" as we speak of "metals."

Although there is a large amount of information available today concerning the structure of glass, a generalized structural model of glass has yet to emerge. Due to the vast array of groups of materials capable of forming glass, a generally accepted structural model of glass is difficult to conceive.

The raw materials used to make modern glasses usually include quartz sand, soda (as sodium carbonate), and lime (as carbon oxide). The quartz sand provides silica, which is the principal ingredient of glass. Since the melting temperature of quartz is high (above 1700°C), soda is added as a flux to assist melting. Lime is added to increase its chemical stability and to decrease the solubility of the glass in water.

The first glass used by man probably came from volcanoes. Natural glass formed by volcanic action can be found in many parts of the world, the most common form being *obsidian*. This is a black, translucent material, and can be chipped and flaked to make tools, weapons, and ornaments. Other glassy materials, known as *fulgurites*, are long slender tubes of fused sand formed by lightning.

It is not known when glass was first manufactured. As noted by Gasson,[1] Pliny the Elder, a Roman historian, gave one account of the possible origin of the manufacture of glass. He described how a ship laden with *natron*, or soda blocks, was moved by Phoenician merchants on the banks of the river Belus near the Mediterranean Sea. The traders, while preparing a meal on the bank, used lumps of natron to support their pots. Upon being subjected to fire, the soda fused with the sand; to the Phoenicians' surprise, the natron melted, forming a liquid beneath the cooking pot that later cooled into a shining layer of what we now know as glass.

According to Pliny, cited by Obrig,[2] production of *permanent* glass involved the adding of magnesium limestone to the mixture of sand and sodium carbonate, and the craft of glassmaking was imported from Syria into Egypt. At what stage *glassblowing* was invented is a matter of dispute. Although Obrig[2] stated that the art of glassblowing was known in Egypt as early as 2000 B.C., other writers including Horne[3] and Gasson[4] gave the Syrians credit for inventing the technique toward the end of the first century B.C.

Regardless of where or when glassmaking began, the craft first flourished in Egypt early in the Eighteenth Dynasty (1580–1358 B.C.). Sophisticated and beautifully made glass appeared abruptly during this dynasty, suggesting that the Egyptians may have imported the skill from elsewhere.

In conquering Syria and Egypt during the first century of the Christian era, the Romans learned the glassmaking art and developed it into a large-scale industry. As the Roman Empire expanded, the art of glassmaking expanded to the boundaries of the empire; as the Roman Empire declined, the art of glassmaking declined also. With the fall of the Roman Empire the centers of glassmaking returned to the Middle East, and remained there until the occurrence of the Renaissance in the West.

The manufacture of glass during the Renaissance period has been traced by Horne.[3] By the year A.D. 1000, glass was being made in Rhineland forest factories, wood being used to fire the furnaces and the ashes being used for the potash flux needed for glassmaking. In the thirteenth century, glass furnaces were built in England (in Surrey and Sussex) by workers from Normandy, and the manufacture of windows and glass vessels was developed. By about A.D. 1550, a high-quality "crystal" glass had been developed in Venice and was exported to England. In an attempt to rival the Italian crystal, English glassmakers developed, in 1676, a form of crystal made of crushed flint stones (the first *flint* glass). By that time, glass was being produced in quantities in Italy, Germany, France, and England. As a result, a wide range of products made of glass was generally available.

## 1.2. The Development of Optical Glass

Glass has two properties that make it especially suitable for optical uses. It is transparent to the visible spectrum, and its surfaces may be worked so that they are also transparent and nonscattering.

One of the first references[4] to the use of the convex lens as a burning glass was made by the Greek playwright Aristophanes (450–388 B.C.) in his *Comedy of the Clouds*, written in 423 B.C.:

STREPSIADES: *You have seen at the druggist's that fine transparent stone by which fires are kindled?*
SOCRATES: *You mean glass, do you not?*
STREPSIADES: *Just so.*
SOCRATES: *Well, what will you do with that?*
STREPSIADES: *When a summons is sent to me, I will take this stone, and placing myself between it and the sun, I will, though at a distance, melt all the writing on the summons.* [Strepsiades was referring to the wax tablets used for writing at that time.]

The invention of spectacles (if we define "spectacles" as two lenses mounted in a frame, and used for the correction of ametropia) apparently occurred late in the thirteenth century. As noted by Hofstetter,[5] possible inventors of spectacles were the Chinese (as reported by Marco Polo in 1270), Roger Bacon, and Salvino degli Armati and Alesandro di Spina, both of Italy. These early lenses were apparently used only for the correction of *presbyopia*. However, during that period, good-quality homogeneous optical glass was not available. The invention of the movable-type printing press by Johannes Gutenberg in Germany, in 1440, greatly popularized reading and gave impetus to the craft of spectacle making, but the need for high-quality optical glass became obvious only with the invention of the telescope, by Galileo, in 1608.

The experiments of Fraunhofer on the analysis of the sun's spectrum, his specification of the refraction and dispersion of glass in terms of certain lines of the spectrum, and his interest in the color correction of telescope lenses all served to provide the stimulus for the Englishman John Dolland to construct in 1757 an *achromatic lens*, made of crown and flint glass components. Unfortunately the varieties of optical glass available at that time were not of sufficient homogeneity for use in telescope lenses.

The early optical glasses were known as either "crown" or "flint" glass. These terms originated in interesting ways. The adjective "crown" was originally applied to window glass used for seventeenth- and eighteenth-century houses in England. This glass was blown and whirled into a disk, having a knot left in the center by the working rod. Prior to the development of optical glasses especially made for lenses, pieces of crown glass were used for ophthalmic lenses and these lenses were known as "crown" lenses. Today crown refers to glass whose primary ingredients are silica, soda or potash, and lime.

In 1676 George Ravenscroft, an Englishman, used ground flints as a source of silica and incorporated a relatively large quantity of lead as one of the basic constituents of glass. The glass became known as "flint" glass. It was softer, heavier, clearer, and more brilliant than any other glass then available and was used for drinking vessels, bowls, and quality crystalware. Today, flint glasses are those that contain lead oxide.

One of the earliest makers of optical glass was P. L. Guinand of Switzerland, who discovered around 1814 that the homogeneity of glass could be greatly increased by *stirring* the glass, while in the crucible, in the molten state. In 1827 Michael Faraday began studies to improve the quality of optical glass in England. He developed methods of purifying the substances and constructed a *platinum* crucible that was remarkably resistant to attack by molten glass. However, this event went rather unnoticed and platinum did not come into use for nearly a century after Faraday's work. In 1839 the Chance brothers of England began the manufacture of a large range of optical glasses.

Beginning in 1876 Ernst Abbe and Otto Schott, of Jena, Germany, collaborated in experimenting with a large number of chemical oxides in the manufacture of glass, and developed a wide range of new glasses for optical purposes. Before 1880 the only types of glasses of optical quality were crown and flint. The addition of lead oxide to flint glass increased both the refractive index and the dispersive power as compared to crown glass. The invention of barium crown glass by Abbe in 1880 introduced a glass of high refractive index without an appreciable increase in dispersive power. As a result of the researches of Abbe and Schott, the firm of Carl Zeiss soon gained preeminence in the manufacture of optical glass.

In the United States, glassmaking was one of the first industries started in the Colony of Virginia, near Jamestown, in 1607. The Colony is said to have produced colored glass beads for use in trading with Indians. The first American attempt at producing optical glass was made by George Macbeth of the Macbeth–Evans Glass Co. in 1889 at Elwood, Indiana, but it soon was abandoned. The next attempt was made by Bausch and Lomb Optical Co. in 1912, and by 1915 Bausch and Lomb was producing substantial amounts of optical glass of good quality.

In the early months of 1917, when the imminence of the entry of the United States into World War I was apparent, the supply of optical glass was extremely critical. Imported stocks from Germany, Great Britain, and France were eliminated or were severely curtailed. In this emergency, the United States government set out to speed up the production of optical glass. With this stimulus, over 600,000 pounds of optical glass were produced during the remaining months of the war. Today, Bausch and Lomb, the Corning Glass Works, and the Pittsburgh Plate Glass Company comprise the major glass manufacturers in the United States.

## 1.3. The Manufacture of Optical Glass

The process of optical glassmaking has been described in some detail by Obrig,[2] and his description serves as a basis for the discussion in the following paragraphs.

The art of glassmaking consists of melting together the necessary ingredients at a temperature high enough to produce a liquid mass sufficiently fluid to permit any bubbles to escape; of mixing the molten mass thoroughly by stirring so that the composition is the same throughout; and of pouring and rolling the glass into a large sheet that is cooled slowly or letting the pot and the glass cool together at room temperature.

The ingredients are put into a *melting pot* 36 inches in diameter and 32 inches high, capable of making about 1000 pounds of glass. The ingredients making up the batch depend, of course, on the type of glass being made, but are principally oxides or salts of metals including silica, sodium and potassium, calcium, and aluminum. An additional ingredient, called *cullet*, is waste glass from previous melts. It is added to save valuable raw materials and to form a *glaze* on the surface of the pot, which reduces the corrosive action of the pot by the melting of the raw materials.

Before furnace operations are begun, the pot is gradually raised to a temperature of 800° to 1000°C and kept there for 3 to 5 days, following which the pot is glazed by the use of small pieces of cullet. The ingredients are then added, at intervals, until the temperature of the pot is raised to about 1400°C. During the melting process, the batch is full of bubbles of escaping gas. The escaping of the gas is

known as *fining*; toward the end of the melting and fining process (which requires several hours), scum, stones, and other materials rise to the top of the batch and are skimmed off. The melt is then stirred constantly, using long clay rods which may be moved either mechanically or by hand.

After completion of the melting, fining, and stirring processes, the molten glass has the consistency of heavy syrup. When cooled to a temperature of 1200°C, it is poured and rolled into sheets of various thicknesses, and each sheet of glass is placed in a heated annealing oven where it is gradually cooled to room temperature. After annealing, the glass is cut into small pieces, is reheated, and is then either pressed or molded into *rough blanks*. After inspection, the rough blanks are blocked on grinding shells and the first surface is ground and polished to the desired curvature. The blanks are then reblocked, and ground and polished on the second side.

The process just described, called the *batch process*, was formerly used for the manufacture of all optical glass. Now, it is used only for the production of relatively small quantities of glass, for example, special types of glass such as high-index glass and many varieties of colored glass. An automated method known as the *continuous-flow process* is now used for making large quantities of a particular type of glass. This process differs from the batch process in that the molten glass is not poured into sheets but is extruded, by means of a continuous process, and pressed into molds, making the rough blanks.

## 1.4. Varieties of Optical Glass

The critical properties of optical glass are the *index of refraction*, identified at the wavelength for the sodium D line, and stated as $n_d$, and the *dispersion*, defined as the variation of the index of refraction with wavelength. Dispersion is quantified by the Abbe number, $v_d$, known as the *nu* value.

The terms "crown" and "flint" applied to optical glass have lost their original meanings. Arbitrarily, glasses with nu values greater than 50 are called crown glasses, while glasses having nu values less than 50 are called flint glasses, although not all glasses are classified by this system. Many glasses are designated by descriptive names, for example, High-Lite, or abbreviations such as LHI.

The most meaningful classification is a six-digit designation whose first three digits are the decimal of the index of refraction and whose remaining three digits are the nu value. Thus, ophthalmic crown glass would be designated "523590."

The three main varieties of optical glass are ophthalmic crown glass, flint glasses, and barium crown glasses. *Ophthalmic crown* derives its name, as already discussed, from the sheets of glass originally blown for window glass. However, the term *crown* now is used to refer to the composition of the glass rather than to its method of manufacture. The ingredients of ophthalmic crown glass are silica (sand), 70%; sodium oxide (soda), 14–16%; calcium oxide (lime), 11–13%; and small percentages of potassium, borax, antimony, and arsenic. Ophthalmic crown glass is used for the great majority of single-vision glass lenses, and for the distance portion of most glass bifocal and trifocal lenses. As already noted, its index of refraction is 1.523 and its nu value is 59.

*Flint* glasses contain from 45% to 65% lead oxide, from 25% to 45% silica, and about 10% mixture of soda and potassium oxide. As compared to crown glass, flint glasses have a higher index of refraction (from 1.580 for light flint to 1.690 for dense flint) and a higher chromatic dispersion (a nu value of 30 to 40). They are used for bifocal segments for some *fused* bifocals (in which the segment must have a higher index of refraction than the major lens) and until recently for single-vision lenses of high power,

since due to the high index of refraction a *thinner* (although heavier) lens can be produced. Other high-index glasses have largely supplanted flint for this purpose.

*Barium crown* glasses contain from 25% to 40% barium oxide, which has the same effect as lead oxide in increasing the index of refraction, but without as great an increase in chromatic dispersion. These glasses have indices of refraction from 1.541 to 1.616, and nu values from 59 to 55 (comparable to that of crown glass). Their main use is for segments of the Nokrome series of fused bifocals. The optical properties of high-index glasses developed for fused bifocal segments, as compared to those of ophthalmic crown glass, are shown in Table 1-1.

In recent years, many specific varieties of glass, having special properties, have been developed. These include high-index glasses with refractive indices of 1.60, 1.70, and 1.80. These glasses have a high content of titanium oxide and are useful for reducing the thickness of high-powered lenses. Characteristics of some of these glasses are shown in Table 1-2.

Glass for *absorptive lenses* is made by the addition of metallic oxides to the raw materials in the batch. For example, the addition of cobalt results in a *blue* lens, chromium oxide in a *green* lens, magnesium in a *violet* lens, and uranium in a *yellow* lens. These and other metallic oxides may be used in combination to achieve the desired effect. In addition, cerium oxide and other materials have the property of absorbing *ultraviolet* radiation, and iron oxide has the property of absorbing *infrared* radiation.

*Photochromic* glass contains silver halide crystals, which under the influence of ultraviolet radiation separate into silver and halide ions. The silver and halide ions cluster together, and as the cluster becomes larger, it becomes opaque, causing the lenses

TABLE 1-1

**Optical Properties of High-Index Glasses Developed for Bifocal Segments, as Compared to Ophthalmic Crown Glass**

| Glass | Index of Refraction | Nu Value | Specific Gravity |
|---|---|---|---|
| Ophthalmic crown | 1.523 | 58.9 | 2.54 |
| Dense flint | 1.616 | 38.0 | 3.53 |
| Extra-dense flint | 1.690 | 30.7 | 4.02 |
| Barium crown | 1.701 | 31.0 | 2.99 |

TABLE 1-2

**Optical Properties of Currently Available High-Index Glasses, as Compared to Ophthalmic Crown Glass**

| Manufacturer | Glass | Index of Refraction | Nu Value | Specific Gravity |
|---|---|---|---|---|
| All manufacturers | Crown | 1.523 | 58.9 | 2.54 |
| Schott | 1.60 Crown | 1.60 | 40.7 | 2.62 |
| Chance-Pilkington | Slimline 640 | 1.60 | 41.0 | 2.58 |
| Schott | High-Lite | 1.70 | 31.0 | 2.99 |
| Chance-Pilkington | Slimline 730 | 1.70 | 31.0 | 2.99 |
| Chance-Pilkington | Slimline 750 | 1.70 | 51.0 | 3.38 |
| Hoya | LHI | 1.70 | 40.2 | 2.99 |
| Chance-Pilkington | Slimline 825 | 1.804 | 25.0 | 3.35 |
| Hoya | THI | 1.806 | 40.7 | 4.56 |

to darken in the presence of ultraviolet light. In the absence of ultraviolet radiation, the silver and halide crystals recombine, causing the lenses to fade. The main types of photochromic glasses are *boro-silicate* for Corning lenses and *alumino-phosphate* for Chance-Pilkington lenses.

Absorptive lenses (including photochromic lenses) will be discussed in greater detail in Chapter 7.

## 1.5. Desirable Characteristics and Defects of Optical Glass

Desirable characteristics of optical glass, described by Obrig,[2] are:

1. Homogeneity in both chemical composition and physical state.
2. Correct index of refraction and chromatic dispersion values.
3. Freedom from color.
4. A high degree of transparency.
5. A high degree of chemical and physical stability.

### Homogeneity

CHEMICAL COMPOSITION
Nonuniformity in the chemical composition of glass results in the following defects:

1. *Striae*—streaks or lines in the glass, caused by uneven mixing (also called *veins*), having a different index of refraction than the surrounding media.
2. *Bubbles*—usually formed during the melting and fining processes, occurring as a result of gaseous matter that fails to reach the surface before the glass cools (also referred to as seeds, air-bells, or boil).
3. *Inclusions*—consisting of stones and crystallites. Stones are undissolved particles of material in the finished glass. Crystallites are crystal bodies formed when the cooling process is too slow.
4. *Cloudiness*—caused by precipitated colloidal material (sulfates and chlorates) during the cooling period.

PHYSICAL STATE
Refractive index and structural characteristics of optical glass are strongly influenced by the thermal history of the material. Improper annealing procedures can produce (1) index differences throughout the bulk of the glass; (2) sufficient double refraction (birefringence) to cause image degradation; and (3)

residual stress that may cause fracture or distortion upon working the glass.

Defects in the homogeneity of glass are found upon final inspection of the glass or during the later stages of processing, either by a human inspector or by an optical inspection system.

### Index of Refraction and Dispersion Values

Index of refraction is defined as the ratio of velocity of light in a vacuum to the velocity in a given medium *at a given wavelength*. For ophthalmic crown glass, the index of refraction for the Fraunhofer D line (wavelength 589.3 nm)—but not for the C line (656.2 nm) or for the F line (486.1 nm)—is 1.523. Dispersive power is based on the change in index for different wavelengths (C and F, as compared to D): If the index of refraction of a substance is the same for *all* wavelengths, there will be no chromatic dispersion. Therefore, not only must the refractive index for the D line be correct, but the indices for the C and F lines must also be controlled, in order for the dispersive power (nu value) to be correct.

The index of refraction increases as the wavelength decreases. Dispersion will usually increase in glasses of high refractive index. As the density of a glass increases, the index of refraction usually (but not always) increases.

### Color and Transparency

Clear optical glass must have the highest transparency and transmission possible for light in the spectral region for which it is designed. Clear optical glass transmits only 92% of the incident light, due to a loss of approximately 4% at each of the surfaces due to reflections. Surface reflections decrease with increasing wavelength; therefore, an optical glass with no absorption would have a spectral variation in transmission that would change the color of the light passing through it.

High-index glasses transmit less than 92% of the incident light because of increased surface reflectance. If a tint is specified, the glass must have the correct transmission properties. Transmission is specified in terms of a lens thickness of 2 mm, and the thicker the lens the lower the transmission. Unwanted color in a batch of glass is due to the presence of impurities in the raw materials or is a result of material from the pot's walls entering into solution with the glass. Iron oxide is the most serious source of problems, causing an intense green or yellow discoloration.

*Chemical and Physical Stability*

Optical glass should be hard, durable, resistant to weather, and should not tarnish or stain. Flint and barium crown glasses are softer than ophthalmic crown glass, and are less chemically tolerant. Materials used for fused bifocals must be physically stable, the major lens and segment glasses having equal expansion rates so that stress does not occur along the line of fusion.

## PLASTIC MATERIALS

## 1.6. Introduction

A plastic material is defined as "a polymeric material (usually organic) of large molecular weight which can be shaped by flow; usually refers to the final product with fillers, plasticizers, pigments and stabilizers included (versus the resin, the homogeneous starting material)."[6]

While terminology often changes with technology, the adjective "plastic" is still widely used to describe ophthalmic lenses made from one of these materials. In the early years of development, plastics had a reputation of being cheap substitutes for traditional materials, a reputation often abetted by improper processing or by incorrect use of materials. However, at the present time, if plastics are processed properly and tailored for appropriate applications, they may offer many advantages as compared to the materials they replace.

Because of the negative connotation of the term "plastic," ophthalmic lens manufacturers have used other terms to describe lenses made of plastic materials, such as "organic" lenses (because the lenses are made by processes involving organic chemistry, as compared to the "mineral" nature of glass lenses, which are manufactured from mineral materials). Other adjectives used to describe plastic lenses are "resin" and "hard resin." No term has yet been universally adopted.

Most plastics are synthetic materials formed by combining various organic ingredients with inorganic materials such as carbon, hydrogen, oxygen, nitrogen, chlorine, and sulfur. Some raw materials may come directly from plant and animal sources such as wood and cotton (cellulose) and milk (casein). More commonly, the raw materials are derivatives of fossil-formed products including oil, coal, and natural gas.

Although billions of pounds of plastics are produced each year, the amount used for optical purposes is only a very small fraction of the total. Most of the research efforts concerning the optical uses of plastic materials have been financed by private industry with the goal of developing new optical products with attractive properties and with desirable indices of refraction and dispersive powers. In many instances the amount of information available is scanty, because the material formulation and processing technology are proprietary secrets.

## 1.7. Manufacturing Processes

Plastic materials can be classified into two main groups, based on the physical properties of the finished product: *thermoplastic* materials, which soften when heated and therefore can be remolded, and *thermosetting* materials, which once hardened cannot be softened, even at high temperatures.

*Thermoplastic* materials have their molecules arranged in long chains, and such materials are usually supplied in pellet, granular, or sheet form. The material softens when heated and can be stretched, pressed, or molded into complex shapes with no appreciable change in its chemical structure. When cooled, the material hardens and shrinks, having the same configuration as the mold in which it was heated. This process is known as *injection molding*. Since no chemical change transpires, the softening and hardening cycle may be repeated indefinitely. Thermoplastic materials are generally less dimensionally stable than thermosetting materials (to be described below) and they can withstand less heat without deformation.

Examples of thermoplastic materials are acrylates (including Plexiglas, Lucite, and the familiar polymethyl methacrylate used for the manufacture of "hard" contact lenses), cellulose acetate and cellulose nitrate (zylonite and celluloid, both of which are used for the manufacture of spectacle frames), polycarbonate, polystyrene, the nylons, and the vinyls.

*Thermosetting* materials are usually supplied in liquid monomer form and cast in molds. Plasticizers, filters, dyes, binding modifiers, other monomers, and catalysts may be added to the basic chemical compound. After the catalyst is added and the material is heated, *polymerization* occurs. This is the linking together of the molecules to form longer molecular chains, the molecules forming a three-dimensional, cross-linked, lattice pattern rather than the two-dimensional pattern of the thermoplastic material.

This process transforms a liquid monomer first into a syrup and then into a gel, and finally into a solid. The relationship between the time and temperature required in the hardening process is known as the *curing cycle*. Once hardened, the material cannot be softened (even at high temperature). The process is irreversible, and it is this characteristic that distinguishes this material from the thermoplastic materials. If the thermosetting materials are subjected to a high temperature, the material decomposes without melting or substantially softening.

Thermosetting materials have good dimensional stability, are rigid, are relatively insensitive to heat, are flame resistant, can withstand solvents, and have relatively hard surfaces. Examples of thermosetting materials are allyl diglycol carbonate (Columbia Resin 39, or CR-39, used for the majority of plastic lenses), epoxies, phenolics (Bakelite), and melamine compounds (used for tableware).

When two monomers are mixed the resultant material, called a *copolymer*, possesses some of the characteristics of each of the constituent materials. The two monomers need not originate from the same group: That is, one can be thermosetting and the other thermoplastic.

## 1.8. Development of Optical Plastics

Although plastic materials have been available for many years, plastic lenses are relatively new compared to glass lenses, and provided little competition to glass lenses until the decade of the 1970s: With FDA requirements of testing lenses for impact resistance, and the increasing popularity of larger lens sizes together with the availability of an easily used dyeing system for creating "fashion tints," plastic lenses were suddenly in great demand.

Just as World War I served as the impetus for the development of the optical glass industry in the United States, World War II served as the impetus for the development of the plastics industry. One of the plastic materials developed during World War II was *polymethyl methacrylate*, a synthetic thermoplastic resin developed for aircraft windshields. This material has been known in the United States as *Lucite* or *Plexiglas*, and in England as *Perspex*, and has been used for a variety of products from plumbing fixtures to the well-known PMMA contact lenses. This material was also used for the manufacture of spectacle lenses (such as the *Igard* lens, made in Great Britain), and although it had the advantage of being much more shatterproof than nontempered glass, it had the unfortunate disadvantage of *scratching* more easily.

Another plastic material developed during World War II, also for use in military applications, was the material *allyl diglycol carbonate*, known as *Columbia Resin 39*, or simply *CR-39*. Just before 1940, the Columbian Southern Division of Pittsburgh Plate Glass Industries was asked to develop a lightweight transparent plastic material for use in military aircraft. Concentrating on thermosetting rather than on thermoplastic materials, a series of 170 clear allylic resins were compounded and tested. The thirty-ninth compound, designated as CR-39, was the allyl diglycol carbonate monomer, and it was selected as having the optimum properties for further development for use in military aircraft.

In 1947 Robert Graham, an optometrist, formed the Armorlite Lens Company in Pasadena, California, and later described[7] the first ophthalmic lenses made in the United States from CR-39. The Armorlite Lens Company subsequently manufactured CR-39 lenses in commercial quantities for the optical professions. This material, although scratching more easily than glass, is much more scratch resistant than polymethyl methacrylate. The CR-39 lens, now over 35 years old, currently dominates the plastic lens market.

In 1957 General Electric developed a new plastic material, a *polycarbonate resin*, called *Lexan*. This material, of great mechanical strength and high service temperatures, was first produced in ophthalmic lenses in 1978 by Gentex Corporation.

In 1982 Corning Glass Works announced the development of a lens called *Corlon*, a two-layer ophthalmic composite material consisting of a glass lens backed by a very thin layer of polyurethane film, which is bonded to the back surface of the glass lens. Under a licensing agreement with Coburn Optical Industries, the lens material is being made available to optical laboratories in stock single-vision lenses under the trade name *C-Lite*.

## 1.9. Manufacture of Plastic Lenses

*CR-39 Lenses*

Ophthalmic lenses made of CR-39 are cast from allyl diglycol carbonate monomer. Currently the only American source of allyl diglycol carbonate monomer is the Columbia Division of Pittsburgh Plate Glass Industries, which supplies it to qualified casters (ophthalmic and nonophthalmic) as a yellowish viscous liquid that produces CR-39 in a variety of forms and sizes. Although the basic process of making CR-39 lenses is common to most manufacturers, some add a copolymer or other additives, such as antiyellowing agents, ultraviolet absorbers, and mold releasers. While copolymers and additives may reduce lens shrinkage, make casting easier, and make lenses lighter and tougher, the lenses are usually impaired in other ways, such as being less scratch resistant.

After adding the catalyst and other ingredients, the liquid resin is poured into a mold. The mold consists of two glass surfaces, one concave and the other convex, assembled with a gasket of predetermined thickness to control center and edge thickness of the lens. The inside surfaces of the mold, which impart the curvature to the lens, are made with great precision and highly polished in order to produce finished lens surfaces of high quality. The glass molds are chemically tempered to resist the flexing action due to the shrinkage of the CR-39 material during polymerization. The shrinkage is about 14%. The entire mold is then placed in an oven and subjected to a controlled time/temperature relationship called the *cure cycle*. After completion of the cure cycle, the molds are removed from the oven, dismantled, and separated from the finished lenses.

The preparation of the monomer in combination with the cure cycle used will determine the physical properties of the finished lens. Lower temperatures and longer curing times produce lenses that have superior rigidity, dimensional stability, impact resistance, and scratch resistance.

*Polycarbonate Lenses*

Polycarbonate is a thermoplastic material. It begins as a solid, is melted down, and then injected into a mold at a temperature of about 320°C. In the injection process the polycarbonate will conform, under pressure, to the highly polished surfaces of the injection mold. A device will squeeze the lens, to prevent shrinkage and ensure optical accuracy of the surfaces. Each lens-forming cycle requires approximately 90 to 130 seconds. After removal from the molds, the lenses are inspected and processed through a coating machine. Because the surface hardness of polycarbonate is much softer than CR-39, *all* polycarbonate lenses receive a hard coating to increase scratch resistance and chemical protection. Once the coating process is completed, a heat-curing process produces polymerization and cross-linking of the coating. The coating solvents evaporate, leaving behind a silica-based solid coating approximately 4–6 μm thick.

## 1.10. Optical and Physical Properties of Plastic Lenses

*CR-39*

This material has an index of refraction of 1.498, a nu value of 58, and a specific gravity of 1.32. Its desirable characteristics include the following:

1. *Lightness.* Because of its low specific gravity, a CR-39 plastic lens weighs approximately half (1.32/2.54 = 0.52) that of an ophthalmic crown glass lens of identical size and prescription. For a strong prescription requiring thick lenses or the use of a frame with a large eye size, the weight difference is significant.

2. *Impact Resistance.* Without the need for special treatment, CR-39 lenses are sufficiently impact resistant to pass the FDA impact-resistance tests.

3. *Chemical Inertness.* CR-39 lenses are resistant to virtually all solvents including acetone, benzene, and gasoline, and to most chemicals other than highly oxidizing acids.

4. *Resistance to Pitting.* CR-39 material is more resistant than glass to pitting from small hot particles such as welding spatter and particles thrown off grinding wheels.

5. *Resistance to Fogging.* CR-39 material has a much lower thermal conductivity than glass, and hence fogging and misting do not occur as readily as a result of sudden temperature changes as with glass.

6. *Tintability.* Almost any tint or color can be applied to the surface of a CR-39 lens. A uniform tint over the entire surface or a gradient tint of varying intensity or color may be applied. If desired, the tint can be removed and the lens dyed with a different color.

7. *Versatility in Optical Design.* Repletion casting techniques permit optical designers more latitude in incorporating desirable optical features that traditionally have been prohibitive in cost and hence largely unavailable. Examples include aspheric surfaces for cataract lenses, magnifiers with aspheric surfaces, high-powered lenticular lenses, occupational trifocals, progressive addition multifocal lenses, and reverse slab-off construction—all of which are now available at a reasonable cost.

These lenses, however, are not without some disadvantages, which include:

1. *Surface Abrasion.* The resistance to surface abrasion is appreciably lower in CR-39 material than in ophthalmic crown glass. The surfaces must be worked carefully and patients must be instructed properly in their care. However, manufacturers have responded to this problem by developing durable hard coating systems that greatly improve scratch resistance. The coating should preferably be placed on *both* surfaces.

2. *Warpage upon Glazing.* Targrove et al.[8] reported that after a 12-month evaluation significant amounts of lens warpage were almost universally revealed by the use of a Geneva lens gauge. Such warping may go undetected on a lensometer if the front and back surface warpages are essentially equal. The base curves provided in a particular lens series for a given prescription are selected by the lens designer to minimize oblique astigmatism and mean power error. Warpage changes the initial form of the lens and affects its performance for these two aberrations, as well as the magnification and distortion properties of the lens. Smith and Wientzen[9] studied the theoretical performance changes due to warpage and concluded that, if the manufacturer adheres to the 1 D tolerance for induced warpage as set forth in the ANSI Z80.1-1979 Standard,[10] the changes were not likely to cause difficulties except for adaptation and annoying differences between one pair of glasses and another.

3. *Increased Thickness.* As a result of its relatively low index of refraction, a CR-39 lens of a given power will have a greater thickness difference between the center and the edge than will a glass lens. The thickness difference, of course, is increased with large eye sizes, and may elicit patient complaints based on cosmetic appearance.

4. *Poor Photochromic Properties.* As yet a plastic photochromic lens, in which the darkening and fading cycles are permanent, has not been developed.

The American Optical *Photolite* lens, a CR-39 photochromic lens, loses 50% of its photochromic properties in approximately 2 years.

### Polycarbonate

The unique characteristic of polycarbonate is that it is vastly superior to both ophthalmic glass and CR-39 plastic in terms of impact resistance. Polycarbonate differs from CR-39 in having an index of refraction of 1.586, which is higher (rather than lower) than that of glass. Accordingly, the difference between center thickness and edge thickness is less than that for glass, so for a minus lens having the same center thickness as a glass lens this material will result in a thinner edge. The specific gravity of polycarbonate material is less than that of CR-39 (1.20 as compared with 1.32), resulting in a lens of lighter weight. For high-powered lenses, the decreased weight is enhanced by decreased edge thickness (minus lenses) or decreased center thickness (plus lenses).

An abrasion-resistant coating containing an ultraviolet absorber is applied to both the front and back of the lens. The lens may be tinted by an immersion procedure.

An optical disadvantage of polycarbonate, as compared to either CR-39 or glass, is its low nu value (of only 30), resulting in a larger amount of chromatic dispersion. This chromatic dispersion usually manifests itself by the creation of color fringes around light sources when viewed directly through high-powered lenses.

Surfaces of polycarbonate lenses are difficult to mold in such a way that the lenses are free of surface waves or other noticeable blemishes. The material is also difficult to work with: edging and beveling require special tools and techniques.

Data concerning indices of refraction, nu values, and densities for various plastic materials, as compared to ophthalmic crown glass, are summarized in Table 1-3.

TABLE 1-3
**Optical Properties of Currently Available Plastic Lens Materials, as Compared to Ophthalmic Crown Glass**

| Material | Index of Refraction | Nu Value | Specific Gravity |
|---|---|---|---|
| Ophthalmic crown | 1.523 | 58.9 | 2.54 |
| Polymethyl methacrylate | 1.490 | 57.2 | 1.19 |
| CR-39 | 1.498 | 58.0 | 1.32 |
| Polycarbonate | 1.586 | 30.0 | 1.20 |

*Corlon*

Lenses made of this material are described by Corning as *bonded* lenses, having a front layer of glass and a back layer of polyurethane. The manufacturer states that, for minus lenses, Corlon lenses are up to 25% thinner than traditional lenses, and that the lenses are up to 25% lighter than all-glass lenses.

The glass portion of the Corlon lens is available in untempered single-vision white crown glass (approximately 1.3 mm thick) and in Photogray Extra (1.5 mm thick). The polyurethane layer is 0.4 mm thick for both glasses. The polyurethane layer may be tinted to solid colors or gradient tints, using water-based dyes specifically developed for use with the Corlon lens.

Because the lens is laminated, individual drop-ball testing (see Section 1.11) is not required at the fabrication and dispensing level. However, the lens manufacturer is required to conduct "statistical" drop-ball testing, and to certify compliance with this requirement to the fabricating laboratories and to dispensers.

Coburn Optical Industries, Inc., under exclusive license from Corning Glass Works, manufactures the Corlon lens as the *C-Lite* lens. Coburn provides a custom bonding service that bonds the polyurethane-layer to single-vision lenses outside the stock range, to most multifocal styles (not including the Executive style), and to previously edged and tempered lenses.

---

## THE STRENGTH OF LENS MATERIALS

### 1.11. FDA Policies

Prior to 1971, whether or not a spectacle wearer's lenses were rendered impact resistant was left to the discretion of the practitioner or the dispenser. However, on May 15, 1971, the U.S. Food and Drug Administration published, in the *Federal Register*, a statement of policy concerning impact-resistant lenses. This policy stated that all lenses had to be impact resistant by February 1972, unless the practitioner judged otherwise. The statement of policy began with the following paragraphs:

> Examination of the data available on the frequency of eye injuries resulting from the shattering of ordinary crown glass lenses indicates that the use of these lenses constitutes an avoidable hazard to the eyes of the wearer.
>
> The consensus of the ophthalmic community is that the number of eye injuries would be substantially reduced by the use of eyeglasses and sunglasses of either plastic lenses, heat-treated crown glass lenses, or lenses made impact resistant by other methods.
>
> To protect the public more adequately from potential eye injury, eyeglasses and sunglasses must be fitted with impact resistant lenses, except in those cases where the physician or optometrist finds that such lenses will not fulfill the visual requirements of the particular patient, directs in writing the use of other lenses, and gives written notification to the patient.

Since impact-resistant lenses are available in both heat-tempered and chemically tempered crown glass, chemically tempered high-index glass, and both CR-39 and polycarbonate plastic materials (both of which are inherently impact resistant), it is unlikely that an occasion would arise in which a patient's visual requirements could *not* be met by the use of impact-resistant lenses. However, if such an occasion should arise, the practitioner must give the patient written notice that the lenses are not impact resistant. Unless such written notice is given, it is the *practitioner's* responsibility to make sure that the patient's prescription is filled with impact-resistant lenses.

The FDA policy gives the practitioner the option of ordering "heat-treated glass lenses, plastic lenses, laminated glass lenses, or glass lenses made impact resistant by other means," and states that all such lenses must be capable of withstanding an impact test in which "a ⅝ inch steel ball weighing approximately 0.56 ounce is dropped from a height of 50 inches upon the horizontal upper surface of the lens. The ball should strike within a ⅝ inch diameter circle located at the geometrical center of the lens." In order to pass the test, the lens must not fracture. The term "fracture" means that the lens cracks through its entire thickness across an entire diameter, into two or more pieces, or that lens material visible to the naked eye becomes detached from the ocular surface.

Each individual lens must be submitted to the drop-ball test, with the exception of certain categories of lenses that the manufacturer must "subject to the impact test a statistically significant sampling of lenses from each production batch." These categories include (1) raised-edge multifocal lenses, such as Executive-style bifocal or trifocal lenses; (2) laminated glass lenses, which are likely to crack as a result of drop-ball testing; (3) plastic lenses; and (4) nonprescription spectacle lenses ("over-the-counter lenses"). The manufacturer is responsible for deciding what constitutes a "batch," and what constitutes a statistically significant sample of a batch.

The FDA policy does not specify a minimum thickness for a lens in order for it to be considered impact resistant, and the ANSI Z80.1-1979 *Recommendation for Prescription Ophthalmic Lenses*[10] contains no minimum thickness requirement. However, the tolerance for thickness, when specified, is ±0.3 mm.

In describing to a patient lenses that meet the FDA requirements, the practitioner should make sure to use the term *impact resistant*. Terms such as "unbreakable" and "shatterproof" are misleading and should not be used.

## 1.12. The Strength of Glass

Glass has an extremely high theoretical tensile strength, in excess of 4,000,000 pounds per square inch.[11] However, annealed glass breaks at 2,000 to 10,000 pounds per square inch, only a fraction of the theoretical strength, and is considered to be a weak material. Unlike metals, glass does not exhibit *ductility*, being so brittle that it displays virtually no plastic deformation prior to fracturing.

The difference between the theoretical and actual tensile strengths of glass is thought to be due to *flaws* in glass materials. The flaws may be either intrinsic or acquired. Intrinsic flaws may result from slight material inhomogeneities at the time of manufacture or from residual stress due to improper annealing processes; acquired flaws occur in the form of minute surface and edge imperfections, often known as "Griffith flaws," resulting from the grinding, polishing, or edging operations or from handling damage while in use. The surface and edge flaws are the primary contributors to reduced impact resistance of ophthalmic lens materials. Hence, surface and edge quality of ophthalmic lens materials largely control impact resistance.

When a stress field is applied to a lens, those flaws act as stress concentrators: Locally the stress is many times the nominal stress.[12] The flaws may widen under applied stress with the result that the strength of the lens is further reduced. Fracture propagation and failure occur when the stress at the flaw exceeds the intrinsic strength of the glass. A flaw or crack in a lens enlarges only as a result of the tensile stress on the lens surface: A crack will have a tendency to close by compressive forces across the crack. If the surface is placed under compression, tensile stresses occur in the interior of the lens to compensate for the compressive stress. The compression makes the lens stronger because any applied tensile strength must overcome the built-in compression before the surfaces can be put under tension.

Brandt[13] lists four mechanisms by which a lens may fail under impact:

1. Fracture of front surface origin, due to simple elastic denting of the surface only (Figure 1-1). This occurs primarily when the lens is hit by a small, lightweight, high-velocity missile.

2. Fracture of rear surface origin, due to compound flexure of the lens (Figure 1-2). This fracture

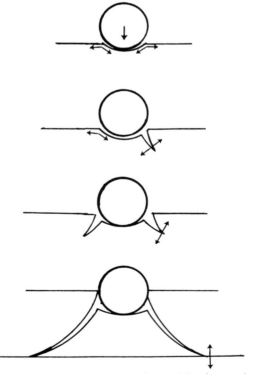

FIGURE 1-1. **Fracture of front surface origin, due to simple elastic denting of the surface. (From N.M. Brandt, *Am. J. Optom. Physiol. Opt.,* Vol 51, p. 985, © The American Academy of Optometry, 1974.)**

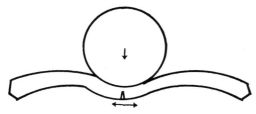

FIGURE 1-2. **Fracture of rear surface origin, due to compound flexure of the lens. (From N.M. Brandt, *Am. J. Optom. Physiol. Opt.*, Vol 51, p. 985, © The American Academy of Optometry, 1974.)**

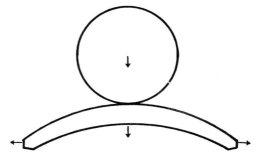

FIGURE 1-3. **Edge fracture, due to simple flexure or flattening of lens. (From N.M. Brandt, *Am. J. Optom. Physiol. Opt.*, Vol 51, p. 985, © The American Academy of Optometry, 1974.)**

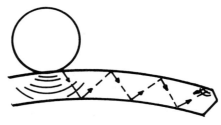

FIGURE 1-4. **Edge fracture and failure due to elastic wave reflection (From N.M. Brandt, *Am. J. Optom. Physiol. Opt.*, Vol 51, p. 985, © The American Academy of Optometry, 1974.)**

is usually caused by a missile of moderate mass and moderate velocity, striking a negative-powered lens whose center is thinner than the edge.

3. Edge fracture, due to simple flexure or flattening of the lens (Figure 1-3). This usually occurs with a heavy, low-velocity missile striking a positive lens whose edge is thinner than the center.

4. Edge fracture and failure, due to elastic wave reflection (Figure 1-4). An elastic wave whose origin is the site of impact on the lens spreads to the edge which contains a flaw large enough to originate fracture.

## 1.13. Methods of Tempering Glass Lenses

Thermal tempering was the classic (and only) method of rendering ophthalmic lenses impact resistant until relatively recently, when the chemical tempering method was developed.

### Thermal or Air Tempering

Any treatment that places the surface of a lens under compression will provide the lens with greater impact resistance than that possessed by annealed glass. Thermal (air) tempering is such a method. The process of thermal tempering of glass is more than 100 years old, one of the first patents having been taken out by François de la Bastie in 1874.[14] However, de la Bastie and other early inventors of thermal tempering processes failed to understand the nature of the process, believing that they were *hardening* the surface of the glass.

In thermal tempering, the finished lens (without the frame) must be heated to a temperature near the softening point (about 650°C) and then rapidly cooled in a blast of air on both surfaces. The heating process causes the glass to expand, and the subsequent rapid cooling causes the surface to quickly become rigid while the interior mass of the lens is still slightly plastic. As the interior mass of the lens cools, it contracts and exerts tension on the resisting rigid surface volume. When reaching room temperature the surface is under compression and the central portion of the lens (i.e., below the surface) is under tension.

The layer of compression at the surface of the lens may be compared to the rim of a *bicycle wheel*.[14] Each of the spokes of the wheel has to be tightened, so as to place the spoke in just the right degree of tension. This tension exerts forces on the rim, keeping the rim in a state of compression.

The compression is not always uniform across the lens because the cooling rate varies with the volume of air reaching different parts of the lens surface and with the thickness of the lens. Careful control of heating time and temperature is required for air tempering. Proper heating produces a compression layer, minimally distorting the lens surfaces. Overheating will cause warping of the surfaces, and underheating will lead to insufficient compression.

Compression of the surface with resulting tension of the internal portions produces *birefringence*, or double refraction, of the lens. If a heat-tempered lens is placed between two Polaroid filters oriented at right angles to each other (in an instrument called a *polariscope*) the birefringence patterns are readily visible. The birefringence patterns are usually characterized as "Maltese cross" patterns, but an infinite number of patterns are possible. According to Wigglesworth,[15] lenses cannot reliably be assessed for impact resistance on the basis of their displayed birefringence patterns: Such a pattern indicates that the lens has been tempered, but unfortunately it doesn't indicate whether or not the process was done satisfactorily.

### Chemical Tempering

Although applied to ophthalmic lenses only relatively recently, techniques for chemical strengthening of glass are not new. In 1890, Tegetmeier[16] observed that large stresses could be generated in the surface of a glass by exposure to molten potassium or lithium salts. In 1965, 75 years after Tegetmeier's observation, Weber[17] was issued a patent for chemical strengthening of lenses by *ion-exchange injection*, in which large ions are exchanged for small ones below the strain temperature.

In 1971, Corning Glass Works adapted the ion-exchange process of chemical strengthening to ophthalmic lenses, and it was introduced to the ophthalmic community in 1972 when the FDA impact-resistance test requirements came into effect. In this process, the surface of the lens is put into compression, as with thermal tempering, but at a considerably lower temperature. The compression layer is created by exchanging small ions present in the glass with larger ions from a molten salt bath in which the lens is processed, as shown in Figure 1-5. The crowding of a large ion into a network site formerly occupied by a smaller ion causes a stress on the adjacent network, if the network is stable. This "stuffing" action and resultant stress on the rigid network produces the compressive stress built into the surface of the glass. Since the exchange occurs at a temperature well below the strain point of the glass, the stresses are not relieved by network movement. Tensile stresses develop in the interior of the glass, to compensate for the surface compression.

Using this process, a batch of finished lenses is "cooked" in a molten bath for a period of about 16 hours during which the large ions in the bath replace the smaller ions in the glass. The amount of ion exchange is proportional to the square root of the time, and depends also upon the temperature.

There are two ion-exchange systems used for ophthalmic lens tempering, and the composition of the glass determines the system used. For crown glass lenses, both clear and tinted, the bath consists of 100% $KNO_3$, and sodium ions in the lens are replaced by larger potassium ions present in the bath. However, for Corning photochromic lenses, the bath consists of 40% (by weight) $NaNO_3$ and 60% $KNO_3$, and the process involves two steps: (1) Lithium ions in the lens are first replaced by larger sodium ions present in the bath, and (2) sodium ions in the lens are then replaced by still larger potassium ions in the bath.

The required temperature varies, depending upon the variety of glass. For clear crown glass, the bath requires a temperature within ±5° of 470°C, whereas for tinted crown glass the required temperature is within ±5° of 440°C. If clear and tinted crown glass are tempered in the same batch, a temperature with ±5° of 450°C is used. For photochromic lenses, the temperature is somewhat lower, being within ±5° of 400°C.

Glass manufacturers have experimented with, and changed to some extent, the composition of ophthalmic crown glass, in order to improve ion exchange. As a result, older lenses do not obtain the strength and compression levels of the newer formulations.

Due to the lower temperature (as compared to the temperature required for heat tempering), there is less chance of lens warpage, and irregularities do not occur as with heat tempering. Resurfacing and reedging are possible with chemically tempered lenses, but not with heat-tempered lenses. After resurfacing and reedging, however, the lenses should once again

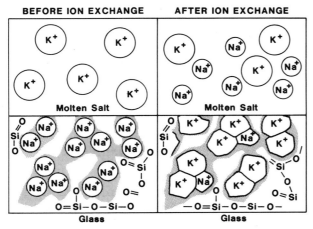

FIGURE 1-5. **The ion exchange process used in chemical tempering. Courtesy of Corning Glass Works, Corning, NY.**

be submitted to the tempering process. Since the compressive stress developed in chemically strengthened lenses is much greater than that of heat-tempered lenses, these lenses are from two to three times more impact resistant.

Some antireflective and tint *coatings* may be damaged or discolored at the temperature of the chemical bath, so, when chemically tempered lenses are to be coated, the coating operation should be done *after* the tempering operation. Lenses that are to be coated should not be drop-ball tested until after the coating operation: This procedure will eliminate any lenses which may have been weakened by the coating process.

To visually identify a chemically tempered lens, the lens (removed from the frame) should be immersed in glycerine between crossed polarizing plates. When viewed through the surfaces, a narrow band is visible around the periphery of the lens. Commercial devices such as the American Optical *Chem-Check* and the Kirk Optical *Tempr-A-Scope* are available for the visual identification of chemically tempered lenses.

One disadvantage of chemical tempering is the fact that it requires a 16-hour (overnight) period, thus adding one day to the time required to receive the finished lens from the laboratory. However, this time delay is to some extent offset by the fact that the lenses are tempered in batches, as opposed to thermal tempering in which lenses must be processed one at a time.

Recently, equipment has been developed that ultrasonically stimulates the chem-tempering process.[18] The ultrasonic stimulation of a chem-tempering bath appears to result in higher strength lenses than conventional processing, and requires a treatment time of only 2½ to 4 hours. The ultrasonic stimulation appears to achieve this effect by improving the efficiency of transfer of potassium ions into the glass.

## 1.14. Impact Resistance of Plastic Lenses

Both CR-39 and polycarbonate lenses, as already stated, are intrinsically impact resistant. However, their great advantage, as compared to lenses made of air-tempered or chemically tempered glass, is their *light weight*. Although CR-39 lenses are considerably lighter than glass lenses, for a lens of a given prescription and center thickness a polycarbonate lens (having an index of refraction of 1.568 and a specific gravity of 1.20) will be lighter in weight than any

other available lens. Both CR-39 and polycarbonate material have the disadvantage of being softer than glass and therefore being more easily scratched: The scratch resistance of CR-39 lenses can be increased by the use of a scratch-resistant coating, but polycarbonate material is so soft that such a coating is mandatory.

When the impact resistance of CR-39 lenses has been compared to that of air-tempered and chemically tempered glass lenses, conflicting (and interesting) results have been reported. Renaldo, Keeney, and Duerson[19] found chemically tempered glass lenses to be 100%–350% more impact resistant than air-tempered glass lenses, using low-velocity drop-ball tests. Wigglesworth[20] reported that 3-mm-thick air-tempered glass lenses have a higher impact resistance than 3-mm CR-39 plastic lenses when tested with a 1-inch steel ball on the drop-ball test. However, with ballistic tests using a ⅛ inch steel ball the positions were reversed, 3-mm CR-39 lenses being more impact resistant than air-tempered lenses. Welsh et al.[21] tested untreated glass, air-tempered glass, chemically tempered glass, and CR-39 plastic lenses with a ballistic device using a 4.76-mm (³⁄₁₆ inch) missile, and found the CR-39 lens to have the greatest impact resistance and the untreated glass lens to have the least impact resistance. The differences between impact resistances of air-tempered and chemically tempered glass lenses were slight.

Goldsmith and Gonden[22] applied strain to chemically tempered glass lenses and CR-39 plastic lenses with steel balls, aluminum cylinders, and pine boards at defined velocities. Fracture of chemically tempered glass lenses occurred in the range of 30,765 to 161,920 pounds per square inch while CR-39 plastic lenses survived in the 32,384 to 40,480 pounds per square inch range but failed beyond that range.

Many other studies have been reported, comparing impact resistance of CR-39 lenses to that of chemically tempered glass lenses. Analysis of the results of these studies indicates that chemically tempered glass lenses tend to be superior for large-mass, low-velocity objects (such as a large stone or a baseball), whereas CR-39 plastic lenses are likely to be superior for small, high-velocity, sharply pointed objects.

When plastic lenses fracture, they tend to break into a smaller number of pieces than in the case of glass lenses, and the edges of the plastic fragments are not as sharp as the edges of glass fragments. Also, there are fewer small, fine particles originating at the line of fracture (i.e., flaking) of a plastic lens.

Surface scratching and pitting tend to reduce the survivability of all lenses. However, they compromise the strength of glass lenses more than that of plastic lenses, because a glass lens depends, for its strength, upon a compressive layer at the surface: Scratching the surface of a glass lens has the effect of reducing the surface compression and therefore reducing the impact resistance.

## 1.15.  Lenses for Occupational and Educational Use

The standards for lenses and frames for occupational and educational use have been set forth by the American National Standards Institute, in ANSI Z87.1-1979, *Practice for Occupational and Educational Eye and Face Protection.*[23] This standard applies to all occupational and educational operations or processes excluding X rays, gamma rays, high-energy particulate radiation, lasers, and masers.

Spectacles designed for industrial exposure must pass a more stringent drop-ball test than that required by the FDA for impact-resistant lenses. The lens must be removed from the frame and subjected to a test in which a 1-inch steel ball, weighing approximately 2.4 ounces, is dropped from a height of 50 inches onto the convex surface of the lens. Minimum thickness for a prescription lens is 3.0 mm except for high plus lenses, for which the edge thickness may be reduced to 2.5 mm provided the lens passes the impact-resistance test as specified.

*Tinted* lenses for industrial use should not be worn indoors unless called for by the nature of the particular occupation or when prescribed for an individual by an ophthalmic practitioner. *Absorptive* lenses (shade numbers 1.5 through 3.0) must be supplied in pairs, and lenses for the two eyes must have equal luminous transmission values, within 10%. *Filter* lenses, which are much darker than absorptive lenses (shade numbers 4.0 through 14.0) must be supplied in pairs, and luminous transmission values of the two lenses must agree within 20%. To qualify as a *clear* lens, the luminous transmission must be not less than 89%. Variable tint (i.e., photochromic) lenses are not allowed for indoor occupations and are allowable only for those outdoor tasks that do not involve hazardous ultraviolet or infrared radiation. Absorptive lenses, both for general wear and for occupational and educational use, are discussed in more detail in Chapter 7.

Spectacles for industrial use require not only unique lenses, but *special frames* as well. The use of frames not meeting the ANSI standard is definitely in violation of the standard, even though the lenses are in compliance.

## References

1. Gasson, W. The Early Story of Glass, Part I. *The Optician*, Vol. 156, No. 4051, Nov. 22, 1968.
2. Obrig, T. E. *Modern Ophthalmic Lenses and Optical Glass.* Chilton Co., Philadelphia, 1935.
3. Horne, D. F. *Spectacle Glass Technology*, p. 1. Adam Hilger, Bristol, 1978.
4. Gasson, W. The Early Story of Glass, Part II. *The Optician*, Vol. 156, No. 4052, Nov. 29. 1968.
5. Hofstetter, H. W. *Optometry: Professional, Economic and Legal Aspects.* C. V. Mosby Co., St. Louis, 1948.
6. *Dictionary of Scientific and Technical Terms*, 2nd ed. McGraw-Hill, New York, 1978.
7. Graham, R. A New, Light Weight and Unbreakable Plastic Ophthalmic Lens. *Amer. J. Optom. Arch. Amer. Acad. Optom.*, Vol. 24, pp. 495–497, 1947.
8. Targrove, B. D., Miller, J. W., Tredici, T. J., Kislin, B., Rahe, A. J., and Provines, W. F. Glass vs. Plastic Lenses—An Air Force Replacement and Durability Study. *Amer. J. Optom. Arch. Amer. Acad. Optom.*, Vol. 49, pp. 320–329, 1972.
9. Smith, F. D., and Wientzen, R. V. Prediction of Visual Effects from the Warpage of Spectacle Lenses. *Amer. J. Optom. Arch. Amer. Acad. Optom.*, Vol. 50, pp. 616–631, 1973.
10. *American National Standard Recommendations for Prescription Ophthalmic Lenses*, Z80.1-1979. American National Standards Institute, Inc., New York, 1979.
11. Chase, G. A., Kozlowski, T. P., and Krause, R. P. Chemical Strengthening of Ophthalmic Lenses. *Amer. J. Optom. Arch. Amer. Acad. Optom.*, Vol. 50, pp. 470–476, 1973.
12. Berger, R. E. Impact Testing of Ophthalmic Lenses: Stress Distribution and "Search" Theory. *J. Amer. Optom. Assoc.*, Vol. 47, pp. 86–92, 1976.
13. Brandt, N. M. The Anatomy and Autopsy of an "Impact Resistant Lens." *Amer. J. Optom. Physiol. Opt.*, Vol. 51, pp. 982–986, 1974.
14. Lueck, I. B. *Toughened Safety Lenses.* Scientific and Technical Publication No. 23. Bausch and Lomb, Inc., Rochester, New York, 1961.
15. Wigglesworth, E. C. The Birefringence Fallacy. *Amer. J. Optom. Physiol. Opt.*, Vol. 52, pp. 320–327, 1975.
16. Tegetmeier, F. Electrical Conductivity of Glass and Rock Crystal. *Ann. Phys. Chem.*, Vol. 41, No. 18, p. 18, 1890.
17. Weber, N. U.S. Patent No. 3,218,220, 1965.
18. Duckworth, W. H., and Rosenfield, A. R. Strength of Glass Lenses Processed in Ultrasonically Stimulated Chemtempering Bath. *Amer. J. Optom. Physiol. Opt.*, Vol. 61, pp. 48–53, 1984.
19. Renaldo, D. P., Keeney, A. H., and Duerson, H. L., Jr. Ion Exchange Tempering of Glass Ophthalmic Lenses. *Amer. J. Ophthal.*, Vol. 80, pp. 291–295, 1975.
20. Wigglesworth, E. C. The Comparative Performance of Eye Protection Materials. *Austr J Optom*, Vol. 55, pp. 461–469, 1972.
21. Welsh, K. W., Miller, J. U., Kislin, B., Tredici, T. J., and Rahe, A. J. Ballistic Impact Testing of Scratched and Un-

scratched Ophthalmic Lenses. *Amer. J. Optom. Physiol. Opt.*, Vol. 51, pp. 304–311, 1974.

22. Goldsmith, W., and Gonden, D. Fracture Resistance of Ophthalmic Lenses. *Amer. J. Optom. Physiol. Opt.*, Vol. 60, pp. 914–919, 1983.

23. *American National Standard Practice for Occupational Eye and Face Protection*, Z87.1-1979. American National Standards Institute, Inc., New York, 1979.

## Questions

1. Describe the "nature" of glass.

2. List the principal ingredients of ophthalmic glass.

3. Discuss the origin of the terms "crown" and "flint" used to describe glass.

4. Describe the batch process of ophthalmic glass-making.

5. List the desirable characteristics of optical glass.

6. Describe the common defects found in the chemical composition of glass.

7. Distinguish between thermoplastic and thermo-setting plastic materials.

8. Compare the characteristics of CR-39 and poly-carbonate plastic lenses.

9. Describe the construction of the Corlon lens.

10. What is meant by the term "impact resistance"?

11. Describe how glass lenses are strengthened by thermal tempering.

12. Discuss the significance of birefringence patterns in thermally toughened lenses.

13. Describe the process of chemical tempering of glass lenses.

14. Discuss the relative impact resistance of CR-39 and chemically tempered lenses.

15. How are lenses designed for occupational use tested for resistance to breaking?

# CHAPTER TWO

# Characteristics of Ophthalmic Lenses

In this chapter the fundamental characteristics of ophthalmic lenses will be presented, in terms of both their physical and optical characteristics. Under the heading *physical characteristics*, topics to be discussed include the concept of curvature, surfaces of revolution, the relationship between curvature and refracting power, the lens measure (or lens "clock"), lens form, lens blanks and base curves, specification of cylinder axes, prescription writing, and transposition.

Topics to be discussed in the second part of the chapter, under the heading *optical characteristics*, include basic terminology, image formation by spherical lenses, image formation by cylindrical and toric lenses, the concept of the conoid of Sturm, the power in an oblique meridian of a cylindrical or toric lens, obliquely crossed cylinders, astigmatism due to lens tilt, and the optical principles of the Maddox rod.

The discussion of optical characteristics of ophthalmic lenses will be continued in the two chapters to follow, which will be concerned with power specification and measurement, and ophthalmic prisms and decentration.

# PHYSICAL CHARACTERISTICS

## 2.1. Curvature

The curvature of a surface is defined as the angle through which the surface turns in a unit length of arc. Referring to Figure 2-1, an object traveling in a circular arc, AB, when at point A will be traveling in the direction AA', and when at point B will be traveling in the direction BB'. The angle between the two directions of travel will be θ', which is equal to the angle θ subtended by the two radii CA and CB. Since AA' and BB' are tangents to the curve, we may define curvature as follows:

$$\text{Curvature} = \frac{\theta}{\text{arc AB}}.$$

Since θ, in radians, $= \dfrac{\text{arc AB}}{r}$,

$$\text{curvature } (R) = \frac{1}{r}.$$

There are several methods of specifying curvature. The simplest method is to use the relationship between curvature (R) and radius (r) as illustrated in the above definition. Since curvature is inversely proportional to radius, it frequently is convenient to use the radius for the specification of curvature. As shown in Figure 2-2, the smaller circle, having a radius of curvature of 0.50 m, obviously has a greater curvature (a greater change in angle per unit arc length) than does the larger circle, having a radius of curvature of 1 m. Note that, if the radius is used to specify curvature, a large degree of curvature will be represented by a small quantity and a small degree of curvature will be represented by a large quantity. To avoid this confusion, it is recommended that curvature be specified in terms of the reciprocal (R) of the radius, rather than in terms of the radius itself (r).

### Units of Curvature

As Bennett[1] has pointed out, there is no universally accepted name for a unit of curvature. When the radius is specified in meters, the unit *reciprocal meters* ($m^{-1}$) may be used.[2] Morgan[3] uses the *diopter* as a unit of curvature when the radius is expressed in meters. However, the term *diopter* has become associated with a large number of useful applications, including surface power of an ophthalmic lens (Section 2.3); the curvature of a wavefront at a specified distance from its source, that is, *vergence* (Section 3.1); reduced vergence, that is, $l/n$ and $l'/n'$; and focal power of a lens or of an optical system (Section 3.1). The use of the diopter as a unit of curvature of an *optical surface* especially causes confusion with diopters of *surface power*. It is best, therefore, to restrict the use of the term diopter to vergence, reduced vergence, surface power, and focal power of ophthalmic lenses.

### The Sagitta as a Unit of Curvature

Another method of specifying curvature is to use the *sagitta*, which in Latin means *arrow*. As shown in Figure 2-3, the sagitta, *s*, is the distance between a point on the circle and the midpoint of a chord of the circle. The relationship between the sagitta, the chord, and the radius of a circle can be derived from

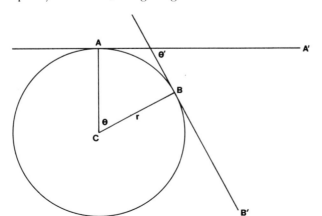

FIGURE 2-1. **Curvature (see text for explanation).**

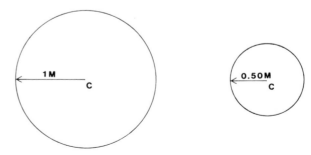

FIGURE 2-2. **The smaller circle has the greater curvature. (C = center.)**

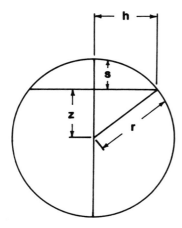

FIGURE 2-3. **The sagitta as unit of curvature.**

this figure. Using the relationship between the sides of a right triangle,

$$r^2 = h^2 + z^2$$

$$z^2 = r^2 - h^2$$

$$z = \pm\sqrt{r^2 - h^2}.$$

Since $s = r - z$,

$$s = r \pm \sqrt{r^2 - h^2}. \qquad (2.1)$$

In practice, the *minus* root of this expression is used, because the plus root results in the diameter minus the sagitta.

The above expression is the *exact* formula for the value of the sagitta, $s$. However, it is also possible to derive an *approximate* formula:

$$r^2 = h^2 + (r - s)^2$$

$$r^2 = h^2 + r^2 - 2r + s^2.$$

When the quantity $s$ is small in relation to $r$, as in the case of an ophthalmic lens, the quantity $s^2$ may be dropped from the formula. By transposition, we have

$$2\,rs = h^2$$

or

$$s = \frac{h^2}{2r}. \qquad (2.2)$$

### EXAMPLE
Calculate the sagitta of a glass surface having a refracting power of 15.00 D, 42 mm in diameter, and having an index of refraction of 1.523, first by using the exact formula and then by using the approximate formula.

1. By using the exact formula, $s = r - \sqrt{r^2 - h^2}$. Since

$$F = \frac{n' - n}{r} = \frac{1.523 - 1}{r} = \frac{0.523}{r},$$

$$r = \frac{0.523}{F} = \frac{0.523}{15} = 0.0349 \text{ m}.$$

$$h = 0.021 \text{ m},$$

that is, one-half the chord, and

$$s = 0.0349 - \sqrt{(0.0349)^2 - (0.021)^2}$$

$$= 0.0349 - \sqrt{0.00122 - 0.00044}$$

$$= 0.0349 - \sqrt{0.00078}$$

$$= 0.0349 - 0.0279$$

$$= 0.0070 \text{ m} = 7.0 \text{ mm}.$$

2. By using the approximate formula, $s = h^2/2r$,

$$s = \frac{(0.021)^2}{2(0.0349)} = \frac{0.00044}{0.0698}$$

$$= 0.0063 \text{ m} = 6.3 \text{ mm}.$$

When working with highly curved surfaces, such as those found in contact lenses, the *exact* expression for calculating $s$ must be used, since $s$ is not small in relation to $r$ (that is, the sagittae of the surfaces are not small in relation to their radii).

When considering the curvature of one of the surfaces of an ophthalmic lens, the chord length (equal to $2h$) is considered as the *aperture*, or width,

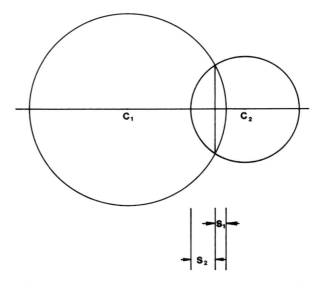

FIGURE 2-4. **When two lenses have the same aperture, the circle having the steeper curvature has the longer sagitta. (C = center.)**

of the lens. As shown in Figure 2-4, if two lens surfaces have the same aperture, the surface having the steeper curvature will have the longer sagitta.

## 2.2. Surfaces of Revolution

Most surfaces placed on ophthalmic lenses are surfaces of revolution, which are formed by rotating a plane curve about an axis within its plane. A *spherical surface* is generated by rotating a circle or an arc about one of its diameters, as shown in Figure 2-5. The intersection of any plane passing through a spherical surface will always be a circle. A *great circle* is formed on a spherical surface when it is intersected by a plane passing through the center of the sphere, and the curvature along any great circle is equal to the curvature of the generating circle. The shortest distance between two points on a spherical surface is the flattest arc connecting them. This distance lies along a great circle through the two points. This concept is useful for navigational purposes, when plotting the course between two points on the surface of the earth.

A *plane surface* has zero curvature and may be considered as a special case of a spherical surface, having an infinite radius.

A *cylindrical surface* is generated by rotating a straight line about another straight line that is parallel to it. As illustrated in Figure 2-6, this is the simplest surface of revolution. The curvature of a cylinder is zero along any line parallel to the axis of

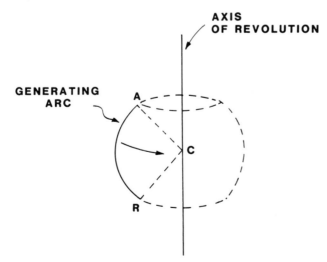

FIGURE 2-5. **A spherical surface can be generated by the rotation of an arc, AR, about an axis of revolution that passes through the center of curvature.**

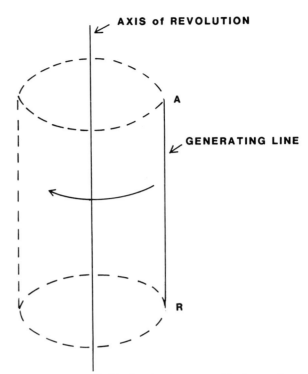

FIGURE 2-6. **A cylindrical surface is generated by the rotation of a line about an axis of revolution that passes through the center of curvature.**

revolution, and a maximum along the line of intersection of the surface by a plane that is perpendicular to the axis of revolution. The axis meridian (the meridian of zero curvature) and the meridian of maximum curvature are always at right angles to each other and are known as the *principal meridians*.

A *toric surface* is generated by rotating a circle or an arc about an axis that lies in the same plane but does not pass through the center of curvature of the arc. This is illustrated in Figure 2-7. The curvature of a toric surface varies from a minimum in one *principal meridian* to a maximum in the other, the two principal meridians being 90° apart. If the axis about which the toric surface is generated is beyond the center of curvature of the arc, as shown in Figure 2-7A, a surface having the shape of a *donut*, or *tire*, is generated. For this surface, the shorter radius is in the vertical meridian and the longer radius goes around it. If the axis is closer than the center of curvature of the circle or arc, as shown in Figure 2-7B, a surface having the shape of a *barrel* is generated, having the shorter radius of curvature in the horizontal meridian. If the arc or circle is oriented concave toward the axis, a *capstan* surface is generated, as shown in Figure 2-7C. It is of interest that a sphere is a special case

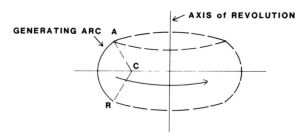

**A    DONUT OR TIRE TORIC SURFACE FORMATION**

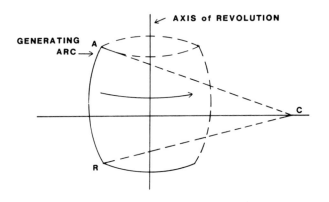

**B    BARREL TORIC SURFACE FORMATION**

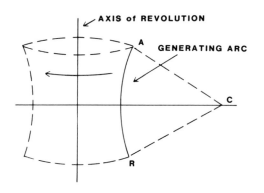

**C    CAPSTAN TORIC SURFACE FORMATION**

FIGURE 2-7. **Toric surface formation. A, a donut, or tire, surface; B, a barrel surface; C, a capstan surface.**

of a toric surface, in which the axis passes through the center of curvature of the arc or circle.

Referring again to the diagrams in Figure 2-7 depicting the three types of toric surfaces, it should be understood that the *outer* surface of each of these diagrams demonstrates the *positive* form of the toric

surface, whereas the *inner* surface of each diagram portrays the *negative* form of the toric surface. The tire formation (Figure 2-7A) is the one that is the most commonly used in ophthalmic lens construction: For it and the barrel form (Figure 2-7B), the radii of curvature in the two principal meridians are *both positive* (for the outer surface) or *both negative* (for the inner surface). However, in the case of the capstan form (Figure 2-7C), for either the outer or inner surface, the radii of curvature in the two principal meridans have *opposite* signs. This form is rarely used in ophthalmic lens construction.

## 2.3. Relationship between Curvature and Refracting Power of a Surface

Although the form of a surface is defined by specifying its curvature, the refracting power of a surface depends upon both the curvature and the index of refraction of the material forming the surface. In geometrical optics the relationship between surface power, curvature, and index of refraction is given by the equation

$$F = \frac{n' - n}{r}, \qquad (2.3)$$

where $F$ is the refracting power in diopters, $r$ is the radius of curvature of the surface in meters, $n$ is the index of refraction of the medium through which light passes before reaching the surface, and $n'$ is the index of refraction of the medium on the emergent side of the surface. In clinical optics we deal almost entirely with air/glass or air/plastic surfaces, and since the index of refraction of air is unity it is customary to define $F$ by the relationship

$$F = \frac{n - 1}{r}, \qquad (2.4)$$

where $n$ is the index of refraction of the glass or plastic and the light is moving from the air to the surface of the lens.

## 2.4. The Lens Measure

The refracting power of a surface may be determined by an instrument called the *lens measure*, lens clock, or lens gauge. As shown in Figure 2-8, the lens measure has two fixed pins and a central, spring-loaded, movable pin. The position of the movable

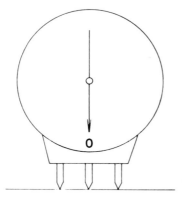

FIGURE 2-8. **The lens measure (on a flat surface).**

pin in relation to the fixed pins is indicated by a pointer that is activated by a system of gears. If the instrument is placed on a flat surface, the protrusion of the central pin will be equal to that of the fixed pins with the result that the scale reading will be zero. If placed on a convex surface, the protrusion of the central pin will be less than that of the fixed pins, but if placed on a concave surface the protrusion of the central pin will be greater. As shown in Figure 2-9, since the chord length (the distance between the two outer pins) has a constant value for the instrument, the position of the central pin indicates the sagitta, $s$, of the surface.

For the lens measure to provide a direct reading of diopters of refracting power of a surface, it is necessary to state the relationship between refracting power, index of refraction, sagitta, and chord length ($h$ being equal to one-half the chord length). This can be done by rewriting Eq. (2.2) and (2.4) in terms of the value of $r$ and by solving for the value of $F$.

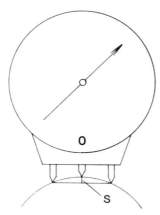

FIGURE 2-9. **The lens measure on a convex surface. The position of the central pin measures the sagitta of the surface.**

Rearranging the terms in Eq. (2.2) we have

$$r = \frac{h^2}{2s},$$

and rearranging the terms in Eq. (2.4) we have

$$r = \frac{n-1}{F}.$$

Substituting for $r$ in both equations, we get

$$\frac{h^2}{2s} = \frac{n-1}{F},$$

which we solve for $F$:

$$F = \frac{2(n-1)s}{h^2}. \qquad (2.5)$$

If for a given lens measure the distance, $h$ (one-half the chord length), is equal to 1 cm and the instrument is calibrated for an index of refraction of 1.523, the movement of the central pin for each diopter of refracting power of the surface can be determined by solving Eq. (2.5) for $s$ and substituting the values of $h$ and $n$:

$$s = \frac{h^2 F}{2(n-1)}$$

$$= \frac{(0.01)^2}{2(1.523 - 1)}$$

$$= \frac{0.0001}{1.046}$$

$$= 0.0000956 \text{ m, or } 0.0956 \text{ mm.}$$

Some lens measures are calibrated for an index of refraction of 1.53 rather than 1.523, with the result that the reading is in error, being slightly high.

Another instrument used in ophthalmic laboratories to measure curvature is the *curve gauge*. A curve gauge is a brass template having curved edges with curvatures corresponding to dioptric values marked on the gauge (see Figure 2-10). These gauges are used for checking the curvature of surfacing tools, and the dioptric value marked on the gauge represents the refracting power for glass of index 1.53 with a curvature equal to that of the gauge.

When either a lens measure or a curve gauge is calibrated for an index of refraction other than that of the material being used, the correct power of the surface can be determined by the use of a conversion factor that is derived as follows:

$$F_C = \frac{n_c - 1}{r},$$

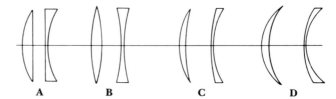

FIGURE 2-11. **Spherical lens forms: A, plano-convex and plano-concave; b, bi-convex and bi-concave; C, periscopic; D, meniscus.**

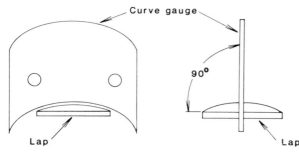

FIGURE 2-10. **The curve gauge, used to check the curvature of a surfacing lap.**

where $F_C$ = the surface power for which the curve gauge is calibrated, and $n_c$ = the index of refraction for which the curve gauge is calibrated.

$$F_T = \frac{n_t - 1}{r},$$

where $F_T$ = the true surface power, and $n_t$ = the index of refraction of the material being used. Setting both of the above equations equal to $r$, and then equal to each other, we have

$$\frac{n_t - 1}{F_T} = \frac{n_c - 1}{F_C}$$

and

$$F_T = F_C \frac{n_t - 1}{n_c - 1}. \qquad (2.6)$$

For example, when a lens measure or curve gauge calibrated for an index of refraction of 1.53 is used for a material having an index of refraction of 1.523, the conversion factor is

$$\frac{1.523 - 1}{1.53 - 1} = 0.986,$$

with the result that a reading of +5.00 D would indicate that the true power of the surface is

$$+5.00 \, (0.986) = +4.93 \text{ D}.$$

If CR-39 plastic, having an index of refraction of 1.498, is the material, the conversion factor will be

$$\frac{1.498 - 1}{1.53 - 1} = 0.940.$$

## 2.5. Lens Form: Spherical Lenses

The term *form*, when applied to an ophthalmic lens, refers to the relationship between the front and back surface curvatures of a lens. For a lens of a given power, an infinite number of forms are possible. Some of these are shown in Figure 2-11. The simplest form for a lens is *plano-convex* or *plano-concave* (Figure 2-11A), in which one surface is flat and the other is either convex or concave. Another simple lens form is the *bi-convex* or *bi-concave* (Figure 2-11B), having half the power on the front surface and half on the back surface.

With few exceptions, modern ophthalmic lenses have convex front surfaces and concave back surfaces (Figures 2-11C and D). Such lenses are sometimes referred to as "bent" lenses. The earliest bent lens, introduced early in this century, was the *periscopic* lens. This lens had a back surface power of $-1.25$ D (for plus lenses) or a front surface power of $+1.25$ D (for minus lenses). A later development was the *meniscus* lens, having a $-6.00$ D back surface power (for plus lenses) or a front surface power of $+6.00$ D (for minus lenses).

## 2.6. Lens Form: Cylindrical and Toric Lenses

Cylindrical and toric lenses, used for the correction of astigmatism (or, more often, for the correction of astigmatism combined with myopia or hyperopia), may also exist in an infinite number of forms. Some of these forms are shown in Figure 2-12. For a patient who has astigmatism but no myopia or hyperopia, the simplest form of lens is the *plano-cylinder*, having one flat surface (either front or back) and one cylindrical surface. The cylindrical surface has plano power in one principal meridian and plus or minus power in the opposite principal meridian (plus if the cylinder is on the front surface, and minus if it is on the back surface). Diagrams of plus and minus cylindrical lenses are shown in Figure 2-13.

For a patient who has either myopia or hyperopia in addition to astigmatism (the usual situation) a spherical surface will be ground on one side of the lens and a toric surface on the other. Traditionally, the toric surface has been ground on the front surface of the lens. However, since a toric back surface decreases the amount of meridional magnification (a difference in the induced size of the retinal image in the two principal meridians) as compared to that obtained by a toric front surface, the standard procedure in recent years has been to grind the toric surface on the back and the spherical surface on the front. Diagrams of toric surfaces are shown in Figure 2-14.

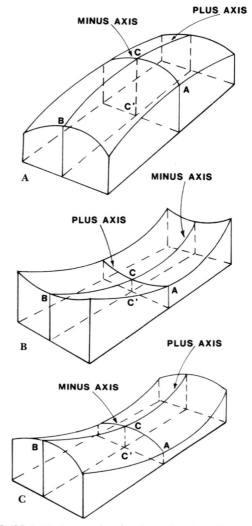

FIGURE 2-14. Perspective drawings of toric surfaces: A, plus power in both principal meridians; B, minus power in both principal meridians; C, plus power in one principal meridian and minus power in the other.

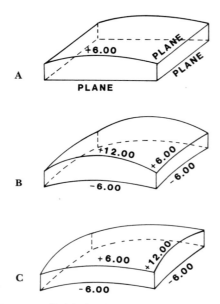

FIGURE 2-12. Cylindrical and toric lens forms: A, plus planocylinder front surface and flat back surface; B, plus toric front surface and concave spherical back surface with net power in the vertical meridian of zero; C, plus toric surface and spherical back surface with net power in the horizontal meridian of zero.

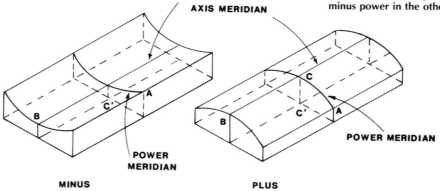

FIGURE 2-13. Perspective drawings of minus and plus cylinders.

## 2.7. Lens Blanks and Base Curves

A lens manufacturer makes (or obtains) optical glass in the form of *rough blanks*. A rough blank is a relatively thick, molded piece of optical glass that must be ground and polished on both sides. Once one of the surfaces has been ground and polished, it is called a *semifinished blank*. When both the front and back surfaces have been ground and polished, it is ready to be cut, edged, and mounted in a frame, and is called an *uncut lens*.

In discussing the physical form of a lens, the term *base curve* is widely used. The word *curve* is commonly used, perhaps erroneously, in the sense of surface power. The term *base curve* has several nuances of meaning, but refers to the standardized curvature ground onto the lens by the manufacturer for a grouping of lens powers. It is a common manufacturing practice to group lenses into ranges of powers and to make all lenses within a particular group with the same curvature on one surface.

The selection of base curve is made when the lens is in the semifinished stage, in which case the base curve will always be on the *finished* side of the lens. For example, for a spherical meniscus lens in plus power, the back surface of the lens will have been surfaced with a −6.00 D spherical curve, and the front surface will remain unfinished. When toric surfaces were routinely ground on the front surface of the lens, the flatter of the two curvatures was considered to be the base curve. However, for the recently introduced negative toric lenses, the manufacturer continued to finish the front side of the lens, in this case using a spherical surface for the base curve.

The monochromatic lens aberrations (particularly oblique astigmatism and curvature of image) can be minimized by the careful selection of base curves. A lens designed for the purpose of minimizing these aberrations is called a *corrected curve*, or *best-form* lens. The first corrected-curve lenses to be introduced in this country were the *Orthogon* lens (by Bausch and Lomb) and the *Tillyer* lens (by American Optical). Lens design and corrected-curve lenses will be discussed in Chapter 5.

## 2.8. Specification of Cylinder Axes

The axis of a cylindrical or toric lens is specified in such a way that when you face the patient (or the front surfaces of a pair of ophthalmic lenses) the axis increases in a *counterclockwise* direction, from 1° to

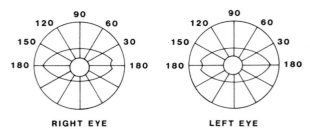

FIGURE 2-15. **Specification of cylinder axes (facing the patient).**

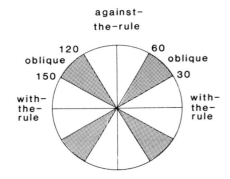

FIGURE 2-16. **With-the-rule, against-the-rule, and oblique astigmatism for minus-cylinder notation.**

180°, as shown in Figure 2-15. Astigmatism is said to be *with the rule* if the minus axis of the correcting cylinder is at 180° or within 30° of 180°, and is said to be *against the rule* if the minus axis of the correcting cylinder is at 90° or within 30° of 90°. It is said to be *oblique* if the minus axis of the correcting cylinder is between 30° and 60° or between 120° and 150°. With-the-rule, against-the-rule, and oblique astigmatism are illustrated in Figure 2-16.

## 2.9. Prescription Writing and Transposition

When writing the prescription for an ophthalmic lens, the spherical power of the lens is given first, followed by the cylindrical power and the cylinder axis. For example, the prescription

$$-2.00 \text{ DS } -0.50 \text{ DC axis } 180$$

would indicate that the spherical component of the prescription is −2.00 D, the cylindrical component is −0.50 D, and the cylinder axis is 180°. In practice, *DS* and *DC* are omitted, and the letter *x* is used to specify *axis*:

$$-2.00 \ -0.50 \text{ x } 180.$$

Since a toric surface may be ground either on the front or on the back of the lens, the prescription for a lens may be written in either the *plus-cylinder* form or the *minus-cylinder* form. The above prescription is obviously written in the minus-cylinder form. However, if the cylinder were to be ground on the front of the lens, the prescription would be written as follows:

$$-2.50 +0.50 \text{ x } 90.$$

The relationship between the spherical and cylindrical components of an ophthalmic lens can be visualized by the use of *optical crosses*, also known as *power diagrams*. For a given lens, three crosses are drawn, the two limbs of each cross representing the principal meridians of the lens. For the cross used to represent the spherical component, the power is the same in both principal meridians. For the cross used to represent the cylindrical component, the maximum power of the cylinder is specified in the meridian 90° from the axis meridian, while the power in the axis meridian is specified as zero (since there is *no* power in the axis meridian of a cylindrical lens).

The optical crosses shown in Figure 2-17A represent the lens prescription in minus-cylinder form $(-2.00 -0.50 \text{ x } 180)$, whereas those in Figure 2-17B represent the prescription for the same lens in plus-cylinder form $(-2.50 +0.50 \text{ x } 90)$. Note that, for both sets of optical crosses, the *total* power in the horizontal meridian (combining spherical and cylindrical components) is $-2.00$ D, while the *total* power in the vertical meridian is $-2.50$ D. Note also that, for the minus-cylinder form, the total power in the *most plus* meridian is selected as the spherical power, while for the plus-cylinder form the total power in the *least plus* meridian is selected as the spherical power.

Considering another example, the lens prescription for a patient having hyperopia and astigmatism is

$$+1.00 -0.75 \text{ x } 30.$$

It will be noted that this prescription is written in the minus-cylinder form. However, if the lens is to be manufactured in the plus-cylinder form, the laboratory technician will have to transpose the prescription to the plus-cylinder form. Referring to the optical crosses shown in Figure 2-18A, representing the prescription as written in the minus-cylinder form, note that the total power in the 30° meridian is $+1.00$ D, while the total power in the 120° meridian is $+0.25$ D. To write a prescription for this lens in the plus-cylinder form, the power in the meridian of least plus power is selected as the spherical power, and the difference between that power and the power in the most plus meridian is selected as the cylindrical power. Thus, the prescription in plus-cylinder form is

$$+0.25 +0.75 \text{ x } 120,$$

as shown in Figure 2-18B.

There are many instances in which it is necessary to *transpose* a lens prescription from minus-cylinder form to plus-cylinder form, or vice versa. For example, optometrists routinely use minus-cylinder refractors or phoroptors and therefore write lens prescriptions in the minus-cylinder form, so when a lens is to be made in the form of a front surface cylinder it is necessary for the laboratory to transpose the prescription into plus-cylinder form (as indicated in the above example). Conversely, ophthalmologists routinely use plus-cylinder refractors or phoroptors and

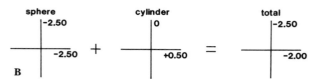

FIGURE 2-17. **Optical crosses (power diagrams): A,** for the prescription $-2.00 -0.50 \text{ x } 180$; **B,** for the prescription $-2.50 +0.50 \text{ x } 90$.

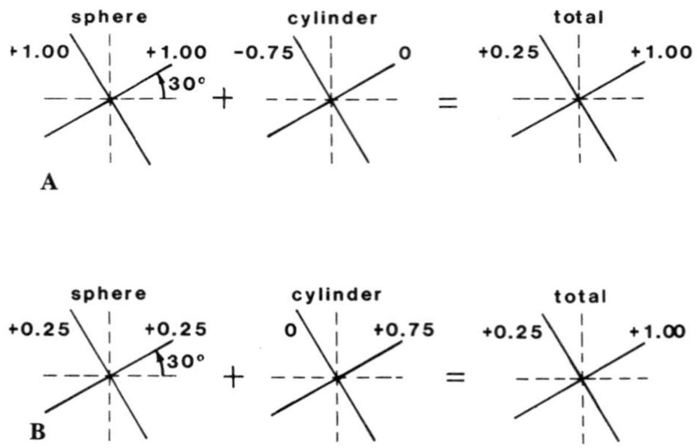

FIGURE 2-18. **Optical Crosses (power diagrams): A,** for the prescription $+1.00 -0.75 \text{ x } 30$; **B,** for the prescription $+0.25 +0.75 \text{ x } 120$.

therefore write prescriptions in plus-cylinder form, so when a lens is to be made in the form of a back surface cylinder it is necessary for the laboratory to transpose the prescription into minus-cylinder form.

*Three-Step Rule for Transposition*

To transpose a sphero-cylindrical prescription from minus-cylinder form to plus-cylinder form or vice versa, the following three-step rule may be used:

1. Add the sphere power and the cylinder power algebraically, to obtain the new spherical power.
2. Change the sign of the cylinder (from minus to plus, or from plus to minus).
3. Rotate the cylinder axis 90°.

Inspection of the examples given in the optical crosses shown in Figures 2-17 and 2-18 will show that, by using the three-step rule, the prescription −2.00 −0.50 x 180 may be transposed to −2.50 +0.50 x 90, and the prescription +1.00 −0.75 x 30 may be transposed to +0.25 +0.75 x 120. When using the three-step rule, the novice should check his work by means of optical cross diagrams.

*The Crossed-Cylinder Form*

A "crossed-cylinder" lens is one having a plus cylinder ground on the front surface and a minus cylinder ground on the back surface, with the axes of the two cylinders being 90° apart. An example of a crossed-cylinder lens is

+0.50 DC x 90 combined with −0.50 DC x 180.

Optical crosses representing the prescription for this lens are shown in Figure 2-19. Note that the total power of this lens in the horizontal meridian is +0.50 D, while the total power in the vertical meridian is

−0.50 D. In optometric practice, crossed cylinders are used in refining the axis and power of the patient's cylindrical correction, and are also used for near-point testing (for example, to determine the power of a tentative bifocal addition).

The following rule may be used to transpose a sphero-cylindrical prescription to a crossed-cylinder prescription:

1. Find both the plus-cylinder and minus-cylinder forms of the prescription.
2. Connect the extremes.

For example, if a lens prescription, written first in the minus-cylinder form and then in the plus-cylinder form, is

+0.50 DS −1.00 DC x 180

−0.50 DS +1.00 DC x 90 ,

connecting the extremes will result in the prescription

+0.50 DC x 90 combined with −0.50 DC x 180.

Note that this is the prescription given above, as an example of a crossed-cylinder lens.

When a prescription has been written in the crossed-cylinder form, the following rule may be used to transpose it into the sphero-cylindrical form:

1. Use the cylindrical power you encounter first as the spherical power.
2. For the cylindrical power, change the sign of the new spherical power and add it algebraically to the second cylinder power.
3. For the cylinder axis, use the axis of the second cylinder.

Continuing with the above example, for the lens whose crossed-cylinder prescription is

+0.50 DC x 90 combined with −0.50 DC x 180,

when we apply this rule we obtain the sphero-cylindrical prescription

+0.50 DS −1.00 DC x 180,

and, using the original three-step rule, this can be transposed into the plus-cylinder prescription

−0.50 DS + 1.00 DC x 90.

It will be noted that each of these sphero-cylindrical prescriptions is similar to the corresponding sphero-cylindrical prescription (for the same lens) given above, thus corroborating the correctness of the re-

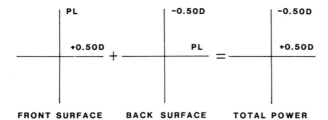

FIGURE 2-19. **Power diagrams for a +0.50 D/−0.50 D cross cylinder.**

sult. The result may also be checked by means of the optical cross diagrams shown in Figure 2-19.

Since transposition between minus-cylinder and plus-cylinder forms may be encountered routinely in optometric practice, the student is advised to memorize the three-step rule for sphero-cylindrical trans-position. However, since the transposition into (and out of) the crossed-cylinder form is not encountered routinely, if rules for these transpositions are forgotten one can easily resort to the use of optical crosses to make the transposition.

## OPTICAL CHARACTERISTICS

### 2.10.    Basic Terminology

**Optic Axis.**  The optic axis of a lens is an imaginary line connecting the centers of curvature of the two surfaces of the lens, which is also normal to both of the surfaces, as shown in Figure 2-20. Although the great majority of ophthalmic lenses possess only one line that is normal to both surfaces, and therefore have a true optic axis, there are a number of possible lens forms that have more than one line normal to both surfaces and for this reason do not possess a true optic axis. In all of these special cases, the two surfaces are concentric in one or both of the principal meridians.

As shown in Figure 2-21, lens forms not possessing a true optic axis include (A) a lens having plane parallel surfaces and (B) a lens having concentric parallel spherical surfaces. Figure 2-22 illustrates other lens forms that have no natural optic axis: (A) a plano-cylindrical lens, in which all lines that are normal to both surfaces are confined to a single plane through which the cylinder passes; (B) a lens having a cylindrical surface on one side and a spherical surface on the other, for which all lines that are normal to both surfaces pass through the center of curvature of the spherical surface; (C) a lens having a toric surface on one side and a spherical surface on the other, for which all lines normal to both surfaces pass through the center of curvature of the spherical surface; and (D) a lens having a toric surface on one side and a spherical surface on the other, for which all lines that are normal to both surfaces pass through the center of curvature of the arc that generates the toric surface.

**Optical Center.**  The term optical center is used in geometrical optics in connection with spherical lenses, to designate the point on the optic axis intersected by the path of a ray of light between the two surfaces that, after refraction by the second surface, is parallel to its path prior to incidence at the first surface (see Figures 2-23A and B). Although for the

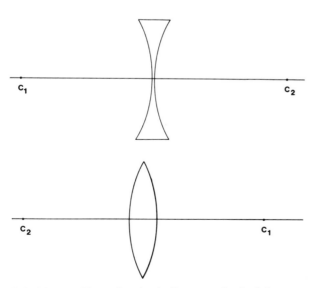

FIGURE 2-20.  **The optic axis (the line normal to both lens surfaces, passing through the centers of curvature of the surfaces).**

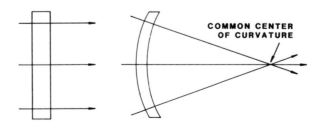

FIGURE 2-21.  **Lenses that have no natural optic axis. A line normal to one surface at any point is also normal to the other surface: A, lens with plane parallel surfaces; (B), lens with concentric spherical surfaces.**

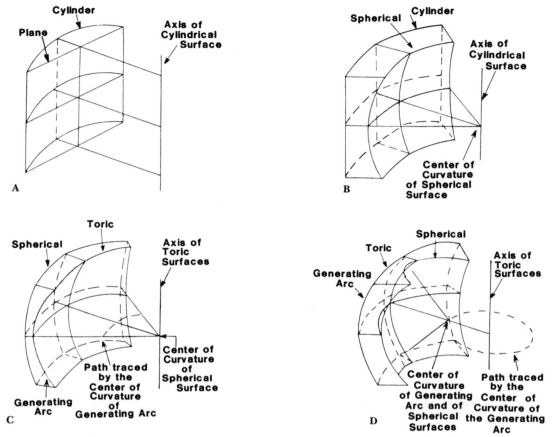

FIGURE 2-22.   Lenses in which the two surfaces are concentric in one of the two principal meridians. Any line in the meridian section that is normal to one surface is normal to the other. See text for explanation.

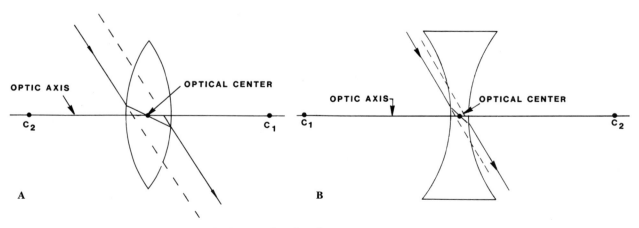

FIGURE 2-23.   The optical center: A, of a plus lens; B of a minus lens.

bi-convex and bi-concave lenses shown in Figure 2-23A and B the optical center is located within the lens, for any *bent* ophthalmic lens the optical center falls outside of the lens.

The optical center of a lens can be found by using the following method of construction (see Figures 2-24A and B).

1. Draw a line from the center of curvature of one surface to the surface, and let the point of intersection of this line with the surface be called A.
2. Draw another line parallel to the first line, passing through the other center of curvature and intersecting the other surface at a point designated as B.
3. Draw a line connecting points A and B. The point at which this line or its extension crosses the optic axis is the optical center of the lens. Any ray directed toward this point will, after passing through the lens, continue in the same direction as it had before entering the lens.

It is also possible to find the position of the optical center mathematically, by using the formula

$$A_1O = \frac{r_1(d)}{r_1 - r_2} = \frac{F_1(d)}{F_1 + F_2}, \qquad (2.7)$$

where $A_1$ = the front vertex of the lens, $O$ = the optical center, $d$ = the lens thickness, $r_1$ and $r_2$ =

radii of front and back surfaces, and $F_1$ and $F_2$ = refracting power of front and back surfaces.

### EXAMPLE 1
If we have a lens in which $F_1 = F_2 = +4.00$ D, having a thickness of 4 mm, then

$$A_1O = \frac{4(0.004)}{4 + 4} = \frac{0.016}{8} = 0.002 \text{ m.}$$

$$A_1O = 2 \text{ mm.}$$

Therefore, the optical center $O$ lies midway between the front and back surfaces of the lens.

### EXAMPLE 2
If the front surface power is +4.00 D, and the back surface power is plano, with a thickness of 4 mm, then

$$A_1O = \frac{0(0.004)}{4 + 0} = 0.$$

This indicates that the optical center is at the front vertex, or apex, of the lens.

### EXAMPLE 3
If the front surface power is +9.00 D and the back surface power is −5.00 D, with a thickness of 4 mm, then

$$A_1O = \frac{-5(0.004)}{9 - 5} = \frac{-0.02}{4} = -0.005 \text{ m.}$$

$$= -5.0 \text{ mm.}$$

The minus sign indicates that the optical center is 5.0 mm anterior to the front surface of the lens.

**Vertices or Poles.** The points on the front and back surface of a lens that are penetrated by the optic axis are called the front and back *vertices*, or *poles*, of the lens. When ophthalmic lenses were made only in flat, "unbent" forms (plano-convex, plano-concave, bi-convex, and bi-concave), the optical center was either on one surface of the lens (coinciding with the front or back pole or vertex) or within the lens. Although the optical center of a modern ophthalmic lens is usually not located on or within the lens (as already discussed) it is a common, although incorrect, practice to use the term optical center, rather than the term pole or vertex, when referring to the point on the lens penetrated by the optic axis.

**Major Reference Point.** When an ophthalmic lens is decentered in order to create a prismatic effect, it is necessary to specify the point on the lens for which the prismatic effect corresponds to that called for in the prescription. This point is known as the

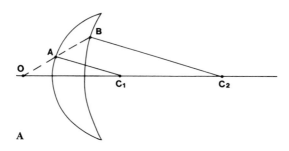

A

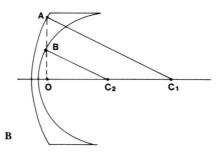

B

FIGURE 2-24. **Graphical method of locating the optical center of a lens.** See text for explanation.

*major reference point.* If the prescription does not contain a prismatic effect, the major reference point coincides with the front pole or vertex of the lens. Decentration and prismatic effects will be discussed in Chapter 4.

## 2.11. Image Formation by a Spherical Lens

From geometrical optics we know that, when parallel rays of light from a distant object point are incident upon a convex spherical lens, a point image will be formed on a screen placed in the secondary focal plane of the lens. As shown in Figure 2-25, if the screen is now placed some distance in front of or behind the secondary focal plane of the lens, a *blur circle* will be formed on the screen. As shown in this figure, the size of the blur circle increases with the distance between the screen and the secondary focal plane of the lens. The size of the blur circle also depends upon the diameter of the aperture of the lens, increasing as the size of the aperture increases. It should be understood that a blur circle will be formed only when the aperture is *circular* in form: For an elliptical or square aperture, the "blur circle" will be elliptical or square.

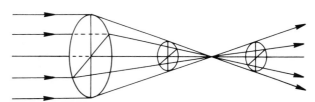

FIGURE 2-25. **Image formation by a spherical lens. A point image is formed at the secondary focal plane of the lens, but at any other plane a blur circle will be formed.**

## 2.12. Image Formation by Cylindrical and Toric Lenses

If the convex spherical lens shown in Figure 2-25 is replaced by a convex plano-cylindrical lens, the point image for a point object will be replaced by a *focal line,* as·shown in Figure 2-26. The axis of the cylindrical lens shown in this diagram is located in the vertical meridian, with the result that the focal line (located in the vertical meridian) is formed by refraction of rays in the horizontal meridian. Since no refraction occurs in the vertical meridian, the length of the focal line (for rays coming from a distant axial

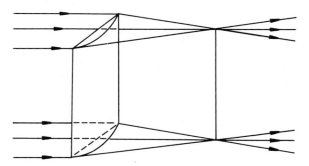

FIGURE 2-26. **Image formation by plano plus cylinder. The point image is replaced by a focal line located in the vertical meridian, formed by the refraction of rays in the horizontal meridian.**

point) is equal to the diameter of the aperture. If the screen is placed in front of or behind the position of the focal line, a *blur ellipse,* rather than a blur circle, will be formed. For any position of the screen, the extent of the ellipse in the axis meridian will be equal to the length of the focal line (and also equal to the diameter of the aperture), but the extent of the ellipse in the power meridian of the lens will increase with increasing distance from the plane of the focal line.

If the convex plano-cylindrical lens is now replaced by a convex *toric* lens, not one, but two focal lines will be formed (Figure 2-27). In this diagram, the horizontal meridian is the meridian of greatest refraction, with the result that the vertical focal line, V (formed by refraction of rays in the horizontal meridian), is formed closer to the lens than the horizontal focal line, H (formed by refraction of rays in the vertical meridian). If the screen is placed at a short distance in front of or behind either of the two focal lines, a blur ellipse will be formed, whose major axis will be parallel to the focal line in question. If the screen is placed at the point along the optic axis of the lens where the vergence of the refracted rays is midway between that of the two focal lines, the *circle of least confusion,* C, will be formed.

## 2.13. Terminology and Basic Concepts Regarding Image Formation by a Lens

The following terminology is of importance in understanding image formation by spherical, cylindrical, and toric lenses.

**Homocentric.** A homocentric bundle of rays is one in which all rays meet, after refraction by the lens, at a common point. This will occur when a

FIGURE 2-27.  **Image formation by a sphero-cylindrical lens having its greatest refraction in the horizontal meridian. Note that the refraction in the vertical meridian forms the horizontal focal line, and the refraction in the horizontal meridian forms the vertical focal line. The circle of least confusion is located at such a point that its dioptric power is midway between that of the two focal lines.**

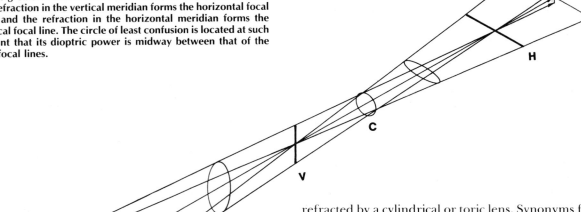

bundle of rays from a single object point is refracted by a spherical lens, if aberrations are neglected (Figure 2-25). Synonyms for homocentric are *monocentric* and *stigmatic* (the root *stigmatic* meaning *point*).

**Astigmatic.** An astigmatic bundle of rays is one in which all rays from a single object point fail to meet, after refraction by the lens, at a common point, but form two line images perpendicular to one another (Figure 2-27). This will occur when a bundle of rays is

refracted by a cylindrical or toric lens. Synonyms for astigmatic are *nonhomocentric* and *nonmonocentric*.

**Conoid of Sturm.** The conoid of Sturm is the entire three-dimensional figure formed by the bundle of rays refracted by a toric lens, as shown in Figure 2-28. This includes, for various positions of the screen, the vertical (V) and horizontal (H) focal lines, blur ellipses, and the circle of least confusion (C). For the lens shown in Figure 2-28, having a power of +2.00 DS − 1.00 DC x 180, the vertical focal line (V) is located 50 cm from the lens, the horizontal focal line (H) is located 100 cm from the lens, and the circle of least confusion (C) is located 67 cm from the lens.

**Interval of Sturm.** The interval of Sturm is the dioptric or physical distance between the two focal lines. The focal line nearest the lens is called the *primary* focal line, whereas that farthest from the lens is called the *secondary* focal line.

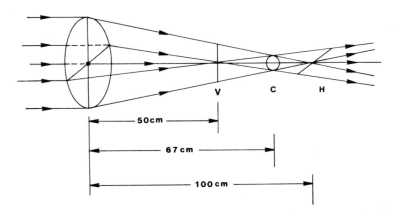

FIGURE 2-28.  **The conoid of Sturm. For a lens having the prescription +2.00 −1.00 x 180, the vertical focal line is located 50 cm from the lens, the horizontal focal line is located 100 cm from the lens, and the circle of least confusion is located 67 cm from the lens.**

## 2.14. Dimensional Aspects of the Conoid of Sturm

The positions or locations of each of the focal lines may be found by applying the familiar thin lens formula,

$$F = L' - L, \qquad (2.8)$$

to each principal meridian. In this formula, $F$ refers to the refracting power in the meridian under consideration, $L'$ refers to the image vergence, and $L$ refers to the object vergence. If the physical distance of each focal line from the lens is desired, it may be found by solving the thin lens formula for image vergence and then taking the reciprocal of the image vergence.

Expressions for lengths of the primary and secondary focal lines and for the diameter of the circle of least confusion can be derived with the aid of Figures 2-29 and 2-30. These diagrams represent the refraction of an astigmatic lens with its principal meridians located in the horizontal and vertical meridians. Plus power exists in both meridians, with the greatest amount of plus power in the horizontal meridian. For an object point on the optic axis, a vertical focal line (V'V") is located at a distance $l_1'$ from the lens and a horizontal focal line (H'H") is located at a distance $l_2'$ from the lens. The circle of least confusion (C'C") is located at a distance $l_c'$ from the lens. Figure 2-30 shows the top and side views of the astigmatic imagery shown in Figure 2-29.

From Figures 2-29 and 2-30, we may derive formulas for the lengths of the focal lines, using similar triangles. In Figure 2-30B, triangles MNH and V'V"H are similar. Therefore,

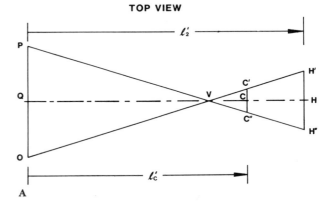

**TOP VIEW**

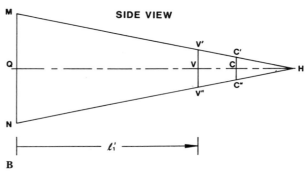

**SIDE VIEW**

FIGURE 2-30. **Diagram showing top and side views of the astigmatic imagery shown in Figure 2-29.**

$$\frac{V'V''}{MN} = \frac{VH}{QH} = \frac{l_2' - l_1'}{l_2'}$$

$$V'V'' = \frac{l_2' - l_1'}{l_2'} (MN)$$

$$V'V'' = \frac{\dfrac{1}{L_2'} - \dfrac{1}{L_1'}}{\dfrac{1}{L_2'}} (MN)$$

$$V'V'' = \frac{L_1' - L_2'}{L_1'} (MN).$$

Since the distance MN is equal to the aperture size, the length of the primary focal line is given by the expression

$$\frac{L_1' - L_2'}{L_1'} \text{ (aperture size)}. \qquad (2.9)$$

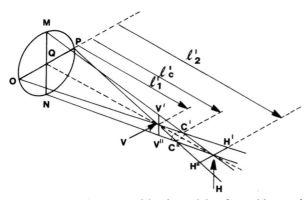

FIGURE 2-29. **Diagram used for determining the positions and lengths of the focal lines and the position and size of the circle of least confusion.**

In Figure 2-30A, triangles POV and H'H"V are similar. Therefore,

$$\frac{H'H''}{OP} = \frac{VH}{QH} = \frac{l_2' - l_1'}{l_1'}$$

$$H'H'' = \frac{l_2' - l_1'}{l_1'} (OP)$$

$$H'H'' = \frac{\dfrac{1}{L_2'} - \dfrac{1}{L_1'}}{\dfrac{1}{L_1'}} (OP)$$

$$H'H'' = \frac{L_1' - L_2'}{L_2'} (OP).$$

Since OP represents the aperture size, the length of the secondary focal line is given by the expression

$$\frac{L_1' - L_2'}{L_2'} \text{ (aperture size)}. \qquad (2.10)$$

Note that, for a circular aperture, OP = MN, but, for a noncircular aperture, OP ≠ MN. Note also that the circle of least confusion (discussed below) is a circle *only* for a circular aperture.

The position ($l_c'$) and length (C'C") of the circle of least confusion can also be found by similar triangles. In Figure 2-30A, triangles POV and C'C"V are similar, and in Figure 2-30B triangles MNH and C'C"H are similar. Therefore,

$$\frac{C'C''}{PO} = \frac{C'C''}{MN} = \frac{l_c' - l_1'}{l_1'} = \frac{l_2' - l_c'}{l_2'}$$

or

$$(l_c')(l_2') - (l_1')(l_2') = (l_1')(l_2') - (l_c')(l_1')$$

$$(l_c')(l_1') + (l_c)(l_2') = (l_1')(l_2') + (l_1')(l_2')$$

$$l_c'(l_1' + l_2') = 2l_1'l_2'$$

$$l_c' = \frac{2(l_1')(l_2')}{l_1' + l_2'} \qquad (2.11)$$

$$l_c' = \frac{2\dfrac{1}{L_1' + L_2'}}{\dfrac{1}{L_1'} + \dfrac{1}{L_2'}}$$

$$l_c' = \frac{2}{L_1' + L_2'}, \qquad (2.12)$$

which provides the location of the circle of least confusion.

The dioptric value of the circle of least confusion is therefore

$$L_c' = \frac{L_1' + L_2'}{2}. \qquad (2.13)$$

Since

$$\frac{C'C''}{MN} = \frac{l_c' - l_1'}{l_1'},$$

then

$$C'C'' = \frac{l_c' - l_1'}{l_1'} (MN).$$

Substituting the value of $l_c'$ in Eq. (2.12),

$$C'C'' = \frac{\dfrac{2}{L_1' + L_2'} - l_1'}{l_1'} (MN)$$

$$C'C'' = \frac{\dfrac{2}{L_1' + L_2'} - \dfrac{1}{L_1'}}{\dfrac{1}{L_1'}} (MN)$$

$$C'C'' = \frac{2L_1' - L_1' + L_2'}{L_1'(L_1' + L_2')}\left(\frac{L_1'}{1}\right) (MN)$$

$$C'C'' = \frac{L_1' - L_2'}{L_1' + L_2'} (MN).$$

Therefore, the diameter of the circle of least confusion may be found by the expression

$$\frac{L_1' - L_2'}{L_1' + L_2'} \text{ (aperture size)}. \qquad (2.14)$$

### EXAMPLE
Given a lens, +3.50 −0.75 x 140, having a diameter of 50 mm and a point object at a distance of 1 m in front of the lens, find the following:

1. The location of the line foci.
2. The lengths of the focal lines.
3. The location of the circle of least confusion.
4. The diameter of the circle of least confusion.

1a. Location of the primary focal line (formed by the meridian with +3.50 D power):

$$L_1' = L + F_1$$

$$= -1.00 + 3.50 = +2.50 \text{ D.}$$

$$l_1' = \frac{1}{L_1'} = \frac{1}{+2.50} = 0.4 \text{ m}$$

$$= 40 \text{ cm behind the lens.}$$

1b.  Location of the second focal line (formed by the meridian with +2.75 D power):

$$L_2' = L + F_2$$

$$= -1.00 + 2.75 = +1.75 \text{ D.}$$

$$l_1' = \frac{1}{L_2'} = \frac{1}{+1.75} = 0.555 \text{ m}$$

$$= 55.5 \text{ cm behind the lens.}$$

2a.  Length of the first focal line:

$$\frac{L_1' - L_2'}{L_1'} \text{ (aperture size)} = \frac{(+2.50 - 1.75)50}{+2.50}$$

$$= \frac{(0.75)50}{+2.50} = 15 \text{ mm.}$$

2b.  Length of the second focal line:

$$\frac{L_1' - L_2'}{L_2'} \text{ (aperture size)} = \frac{(2.50 - 1.75)50}{+1.75}$$

$$= \frac{(0.75)50}{+1.75}$$

$$= 21.4 \text{ mm.}$$

3.  Location of the circle of least confusion:

$$L_c' = \frac{L_1' + L_2'}{2} = \frac{2.50 + 1.75}{2}$$

$$= \frac{+4.25}{2} = +2.125 \text{ D.}$$

$$l_c' = \frac{1}{L_c'} = \frac{1}{+2.125} = 0.471 \text{ m}$$

$$= 47.1 \text{ cm behind the lens.}$$

4.  Diameter of the circle of least confusion:

$$\frac{L_1' - L_2'}{L_1' + L_2'} \text{ (aperture)} = \frac{(2.50 - 1.75)50}{(2.50 + 1.75)}$$

$$= \frac{.75 \times 50}{+4.25} = 8.8 \text{ mm.}$$

It should be understood that in any plane perpendicular to the optic axis of a lens, other than the circle of least confusion, the rays of light are spread over a relatively large area (as shown in Figure 2-27), whereas they are most highly concentrated at the circle of least confusion. For this reason, when presented with a distant object, a patient who has uncorrected hyperopic astigmatism can maximize his visual acuity by accommodating in such a way that the circle of least confusion (for each point on the object) is located on the retina. This is analogous to the situation occurring when light is refracted by a spherical lens, in which the greatest concentration of light (and the best visual acuity) occurs when the image for each point on the object is placed on the retina. For a single line object, depending on its orientation, the patient may be able to maximize visual acuity by placing one of the line foci of the interval of Sturm on the retina. For a more complex configuration, the patient can accommodate in such a way as to manipulate the interval of Sturm on the retina in order to maximize visual acuity. The manipulation of accommodation may be a source of asthenopia (eyestrain) for a person with uncorrected astigmatism.

## 2.15. Misconceptions Concerning Astigmatic Image Formation

The student can easily be misled into believing that the entire conoid of Sturm is formed simultaneously, and that some of the light makes up each focal line, some makes up the circle of least confusion, and some makes up the blur ellipses. However, the conoid of Sturm should be thought of in terms of the *progression of the bundle of rays*, with time, after refraction by the lens. As the bundle of rays progresses with time (very rapidly indeed!), the first focal line, blur ellipses, the circle of least confusion, additional blur ellipses, and the second focal line are all formed in succession (assuming that one places a screen at each of these points, oriented perpendicular to the optic axis of the lens). This is analogous to what happens when light is refracted by a spherical lens. As the bundle of rays progresses, with time, blur circles of decreasing size, then the point image, then blur circles of increasing size are sequentially formed (again, assuming that someone has placed a screen at each position where an image is desired).

Another misconception is that, somehow, the image of the *whole object* consists of the two focal lines and the circle of least confusion. It should be understood that the conoid of Sturm is formed for a *point object*, just as the point image for a spherical lens is formed for a point object. Since, for a spherical lens, each point on the surface of an object forms a point image or a blur circle, it follows that for an astigmatic lens each point forms a conoid of Sturm with two line foci, blur ellipses, or the circle of least confusion, as the case may be: The resultant image of a whole object is made up of the totality of the images formed for all object points (for a specified image plane).

The *clock dial* (or *sunburst dial*), used in the clinical testing of astigmatism, serves as a good example of the concept that an object can be considered as being made up of an infinite number of points, each of which may form a point image or a blur circle (in the case of a spherical optical system) or a focal line or any other configuration in the conoid of Sturm (in the case of a cylindrical optical system).

The clock dial test for astigmatism consists of a chart having radiating lines representing the numbers on the face of a clock. Prior to beginning the test the patient's eye is "fogged": That is, sufficient plus lens power is placed before the eye to make sure that the entire interval of Sturm (if astigmatism exists) is located in front of the retina. If the patient has *no astigmatism*, each object point will form a point image in front of the retina, with the result that all of the lines on the clock dial will appear to be equally clear. If the patient *does* have astigmatism, the line (on the clock dial) that will appear to be the most distinct will be the one parallel to the principal meridian of the eye having the greatest refractive power. The minus-cylinder lens correcting the eye's astigmatism will have its axis position perpendicular to the direction of the clearest line on the retina. Minus-cylinder power is increased until the patient reports that all of the radial lines are equally distinct. This technique provides the cylindrical correction, i.e., the correction for the astigmatic portion of the refractive error.

Figure 2-31A is a simplified clock dial with the lines represented as a series of point objects. For an astigmatic optical system, each point object will form a conoid of Sturm. If the principal meridians are horizontal and vertical, Figure 2-31B represents the astigmatic imagery at the vertical line focus and Figure 2-31C represents the astigmatic imagery at the horizontal line focus. It is obvious that, if the power in the vertical meridian is greater than the power in the horizontal meridian (with-the-rule astigmatism), the clearest line on the clock dial will be the *vertical* line; if the power in the horizontal meridian is greater than that in the vertical meridian (against-the-rule astigmatism), the clearest line on the clock dial will be the *horizontal* line.

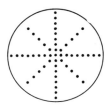

**A. OBJECT (SIMPLIFIED CLOCK DIAL)**

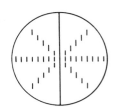

**B. IMAGE AT VERTICAL LINE FOCUS**

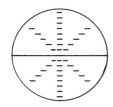

**C. IMAGE AT HORIZONTAL LINE FOCUS**

FIGURE 2-31. **A simplified clock dial, illustrating image formation by a sphero-cylindrical lens.**

### 2.16. The Spherical Equivalent

Spherical equivalent power is defined as the average (or *nominal*) spherical power of a cylindrical or sphero-cylindrical lens. It is determined, for a sphero-cylindrical lens, by algebraically combining *one-half* of the cylindrical power with the spherical power.

There are a number of situations in optometric practice in which it is helpful to convert a lens formula or prescription into the spherical equivalent form. For example, how can the following lens formulas be compared, to determine which has the most overall plus power?

(A)    +2.00 DS −1.00 DC x 180,

(B)    +3.00 DS −2.00 DC x 180.

Algebraically combining one-half of the cylindrical power with the spherical power, for each of the two formulas, we have

(A)      +2.00 DS −0.50 DS = +1.50 DS,

(B)      +3.00 DS −1.00 DS = +2.00 DS.

The second formula, as shown here, has 0.50 D more plus spherical equivalent power than the first formula. As shown in Figure 2-28, the spherical equivalent power of the lens is represented by the *dioptric value of the circle of least confusion*.

One situation in which the spherical equivalent concept is used is when, for one reason or another, a practitioner decides to ignore or to reduce the cylindrical component of the patient's refraction, prescribing the spherical equivalent in order to position the circle of least confusion on the retina. For example, if a patient's refractive finding is −2.00 DS −0.50 DC x 180 and if the practitioner decides not to prescribe the cylinder power, he would very likely prescribe the spherical equivalent, which would be −2.25 DS. Even though the patient has 0.50 D of uncorrected astigmatism, the circle of least confusion would fall on the retina (with accommodation relaxed).

Occasionally a practitioner will find that a patient has a large amount of previously uncorrected astigmatism, and will decide to *reduce* (but not completely ignore) the cylindrical correction. This may be done particularly if the cylinder has an oblique axis. For example, if the refractive finding is +2.00 DS −3.00 DC x 45, the practitioner may decide to prescribe only −2.00 D of cylinder power. In this case, the lens to prescribe, in order to place the circle of least confusion on the retina, is found by algebraically adding one-half of the *cylinder reduction* to the original sphere power, to obtain the new sphere power. The prescribed lens power will therefore be +1.50 DS −2.00 DC x 45. It should be understood that the original lens formula, +2.00 DS −3.00 DC x 45, and the modified prescription both have a spherical equivalence of +0.50 D.

The Jackson cross cylinder, which is frequently used in refraction for determining the axis and power of the cylindrical component, is an example of a lens that has zero spherical equivalent power. A "+0.50 D/−0.50 D" Jackson cross cylinder (+0.50 D cylinder on the front surface and −0.50 D cylinder on the back surface, with the axes 90° apart), when transposed to the sphero-cylindrical form, can be expressed as either

−0.50 DS +1.00 DC

or

+0.50 DS −1.00 DC.

In either case, using the rule for determining spherical equivalent power, the cross cylinder will be found to have a spherical equivalent power of *zero*. The placement of a Jackson cross cylinder before the eye, therefore, allows us to change the power in each of the principal meridians without changing the average, or spherical equivalent, power.

The concept of spherical equivalent power is also useful when we wish to compare refractive findings obtained by different forms of instrumentation or by more than one examiner. For example, when a newly developed automatic refraction system is being evaluated, it is customary to compare the results obtained in routine clinical refraction in terms of the spherical component (that is, the spherical portion of the lens formula), the spherical equivalent power, the cylinder power, and the cylinder axis.

### 2.17. Power in an Oblique Meridian of a Cylindrical Lens

Since a cylindrical lens has its maximum refracting power in one meridian (the power meridian) and has zero power in the meridian at right angles (the axis meridian), it is reasonable to assume that in any oblique meridian the refracting power will lie somewhere between zero and that of the power meridian. However, it will be recalled that a cylindrical lens forms a clear image only for each of the two focal lines, each line being perpendicular to the principal meridian that forms the line. The fact that we have only two focal lines can be accounted for if it is understood that two incident rays from a given object point will reunite after refraction *only* if they lie in a principal meridian: Any two rays, from a single object point, that lie in an *oblique* meridian will fail to reunite after refraction. Such rays are said to exhibit "skew" convergence or divergence.

The following question arises: Is it possible to talk of refracting power in a given oblique meridian when there is no clear image formed by the lens in that meridian? Since the diopter is the reciprocal of focal length, does it follow that diopters of power exist only when a clear image is present?

Using the theorems developed by the Swiss mathematician Leonard Euler (1707−1783) concerning the curvatures of two normal sections intersecting each other at right angles at a point of the curved surface, Southall[4] derived an equation for the curvature along a meridian of a cylindrical surface other than a principal meridian, such that

$$R_\alpha = R_c \sin^2 \alpha , \qquad (2.15)$$

where $R_c$ is the maximum curvature and $\alpha$ is the angle from the axis meridian to the unknown oblique meridian. It should be understood that this equation results in only an approximation, since the curvature $R_\alpha$ is not constant along the entire oblique meridian. For a normal section along an oblique meridian of a cylindrical surface, the resulting curve is *elliptical*. If only a small, central portion of the curve in an oblique meridian is considered, we may assume that the curvature of this small portion is *circular*. This assumption is valid, for ophthalmic lenses, because the area of the cylindrical surface used is only a small portion of the complete cylinder from which it is taken.

Since curvature is related to surface power, we can derive a relationship for $F_\alpha$ as follows:

$$F = (n' - n) R$$

or

$$R = \frac{F}{n' - n}$$

$$R_\alpha = \frac{F_\alpha}{n' - n} = R_C \sin^2 \alpha$$

$$\frac{F_\alpha}{n' - n} = \frac{F_c \sin^2 \alpha}{n' - n}$$

or

$$F_\alpha = F_C \sin^2 \alpha, \qquad (2.16)$$

where $F_C$ is the cylinder power and $\alpha$ is the angle from the axis to the meridian in question.

For any two meridians on a cylindrical surface of power $F_C$ that are 90° apart, that is, meridians that are $\alpha$ and $\alpha + 90$ from the cylinder axis,

$$F_\alpha = F_C \sin^2 \alpha$$

$$F_{\alpha + 90} = F_C \sin^2(\alpha + 90) = F_C \cos^2 \alpha .$$

Therefore, the algebraic sum of the powers in the two perpendicular meridians would be

$$F_\alpha + F_{\alpha + 90} = F_C \sin^2 \alpha + F_C \cos^2 \alpha.$$

$$= F_C(\sin^2 \alpha + \cos^2 \alpha)$$

$$= F_C(1)$$

$$F_\alpha + F_{\alpha + 90} = F_C.$$

Thus, the algebraic sum of the powers in any two meridians perpendicular to each other on a cylindri-

cal surface is constant and equal to the power of the cylinder.

Although, in the strictest sense, power, as traditionally defined, does not exist in an oblique meridian of a cylindrical surface, the sine square law has many useful applications. Bennett and Rabbetts[5] have used the term "notational power" to denote the power in the oblique meridian of a cylindrical surface, calculated by the sine square equation. This equation is sufficiently accurate for determining sagittal depths and thicknesses of oblique meridians for ophthalmic lenses of average dimension. Keating and Carroll[6] have shown that the sine square law is related to *minimizing the blur* or *maximizing the modulation transfer function* for meridional refraction of a line object.

Long and Haine[7] have shown that the sine square law is applicable to laser refraction: It is possible to refract oblique meridians and reconstruct the refraction in conventional notation (sphere, cylinder, and axis) using the sine square law. While meridional refraction does not provide a clear image, it *does* provide the position of "best focus" in each of several preselected meridians.

Keating[8] states that dioptric power exists in an off-axis meridian, this dioptric power having a *curvature* component and a *torsional* component. The torsional component can be demonstrated when working with oblique meridians in a number of routine optometric procedures: These include the "skewed reflex" seen in retinoscopy, misalignment of the mire images in keratometry, and the scissors motion occurring in lensometry.

The references cited here indicate that the approximations arrived at by the sine square law provide accurate numerical results under the conditions for which it is applied. One should, however, recognize the unique nature of "oblique power," as well as its limitations. Specifically, the prismatic effect at any point along an oblique meridian of a spherocylindrical lens cannot be calculated using the power determined by the sine square formula. This will be discussed further in Chapter 4.

In the above paragraphs, Southall's derivation $(R_\alpha = R_c \sin^2 \alpha)$ was used to derive the formula for the power in the oblique meridian of a cylindrical lens. This formula will now be derived from first principals, on the basis of the diagram shown in Figure 2-32. Referring to this diagram, the line segment CB represents the axis meridian (the meridian of zero power) of a cylindrical surface: CA represents the power meridian, and CP represents an oblique meridian at an angle $\alpha$ from the axis meridian. If $F_P$ is defined as the refractive power in the oblique merid-

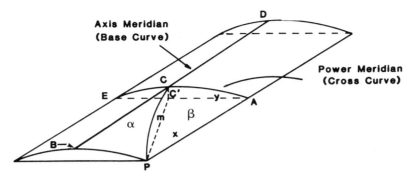

FIGURE 2-32. **Diagram for derivation of the power in an oblique meridian of a plano-cylinder.**

ian, CP, and if $F_C$ is the power in the power meridian, CA, then it follows from Eq. (2.4) that

$$F_P = 2(n - 1) \frac{CC'}{m^2}$$

and

$$F_C = 2(n - 1) \frac{CC'}{y^2}.$$

By solving the latter equation for CC' and substituting the result in the former,

$$F_P = F_C \frac{y^2}{m^2},$$

but $y^2 = m^2 \sin^2\alpha$; therefore,

$$F_P = \frac{F_C m^2 \sin^2 \alpha}{m^2}$$

$$F_P = F_C \sin^2\alpha. \qquad (2.17)$$

In calculating the power in an oblique meridian of a cylindrical surface, it is convenient to use a table of values of $\sin^2\alpha$, such as may be found in any book of mathematical tables, or to use a pocket calculator.

TABLE 2-1.
**Values of Sin$^2$ α for Various Angles, Making Use of the Approximation Sin$^2$ α = α/90**

| Angle | Sin$^2$ |
|---|---|
| 0° | 0.00 |
| 30° | 0.25 |
| 45° | 0.50 |
| 60° | 0.75 |
| 90° | 1.00 |

However, if one simply remembers representative values of $\sin^2 \alpha$, a graph can be sketched showing the sine square values in all meridians of a cylindrical surface, as shown in Figure 2-33.

Approximate values of $\sin^2 \alpha$ can be obtained by using $\alpha/90$ instead of $\sin^2 \alpha$ (see Table 2-1).

## 2.18. Power in an Oblique Meridian of a Toric Lens

As with a cylindrical lens, the concept of the power in an oblique meridian of a toric lens has a limited meaning, but the concept of surface power in an oblique meridian of a toric surface is useful in calculating the difference between center thickness and edge thickness. Referring to Figure 2-34, CA and CB represent the principal meridians of a toric surface. CC'' represents a line perpendicular to the surface and is the vertex depth for the arc CP. C'C'' is the

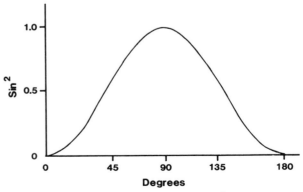

FIGURE 2-33. **Graph showing values of sin$^2$ from 0 to 180 degrees.**

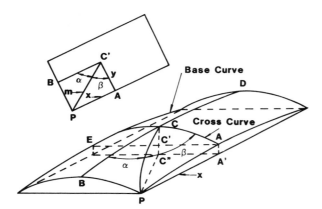

FIGURE 2-34. **Diagram for derivation of the power in an oblique meridian of a sphero-cylindrical lens.**

vertex depth of the arc AP, and CC' is the vertex depth of the arc CA. The plane PAA' is parallel to the plane CC''B containing arc CB, and the curvature of the arc AP is assumed to be the same as that of arc CB. Using Eq. (2.4) again,

$$F_A = 2(n - 1) \frac{CC'}{y^2},$$

$$F_B = 2(n - 1) \frac{C'C''}{x^2},$$

and

$$F_P = 2(n - 1) \frac{CC''}{m^2}.$$

Since $CC'' = CC' + C'C''$, one can solve the first two equations for $CC'$ and $C'C''$, and substitute in the

third equation. Thus,

$$F_P = \frac{y^2 F_A + x^2 F_B}{m^2}.$$

But $y^2 = m^2 \sin^2 \alpha$ and $x^2 = m^2 \sin^2 \beta$; therefore,

$$F_P = F_A \sin^2 \alpha + F_B \sin^2 \beta. \qquad (2.18)$$

Since $\sin^2 \beta = 1 - \cos^2 \beta = 1 - \sin^2 \alpha$,

$$F_P = F_B + (F_A - F_B)\sin^2 \alpha. \qquad (2.19)$$

Equation (2.19) may be arrived at empirically by solving Eq. (2.17) a number of times. $F_B$ is sometimes referred to as the spherical component, and $(F_A - F_B)$ as the cylindrical component of the toric surface. By thus analyzing the toric surface into spherical and cylindrical components one may restate Eq. (2.19):

$$F_P = F_S + F_C \sin^2 \alpha, \qquad (2.20)$$

where $F_S$ and $F_C$ are the powers of the spherical and cylindrical components, respectively.

Applications of Eqs. (2.17) and (2.20) in calculating lens thickness will be discussed in Chapter 3.

## 2.19. Obliquely Crossed Cylinders

If two or more cylindrical lenses are placed in apposition, they may be resolved into a single spherocylindrical lens whose principal meridians are 90° apart. The easiest way to visualize what takes place is to plot sine square curves for each of the cylindrical lenses on a graph, and then summate the two curves. This is shown in Figure 2-35, for the special case of two plano-cylinders of the same power, having their

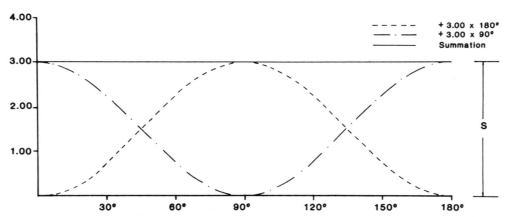

FIGURE 2-35. **Summation of sine square curves for two plus cylinders of equal power whose axes are 90 degrees apart.**

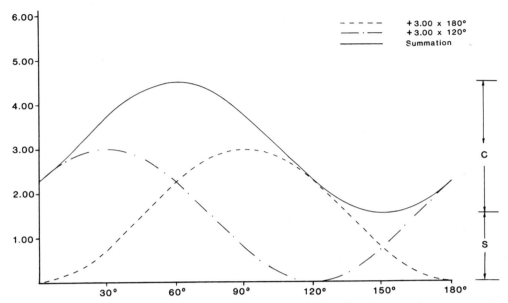

FIGURE 2-36. **Summation of sine square curves for two plus cylinders of equal power whose axes are not 90 degrees apart.**

plus axes 90° apart (+3.00 DC x 180 and +3.00 DC x 90). When the sine square curves for these two cylinders are summated, the result is a *straight line*, representing a spherical value of +3.00 D. However, when we plot sine square curves for two cylinders of equal power *not* having their plus axes 90° apart (+3.00 DC x 180 and +3.00 DC x 120), as shown in Figure 2-36, summation of the two curves results in a third sine square curve representing a spherocylindrical power, as nearly as can be determined by inspection, of +1.00 DS +3.00 DC x 150.

The minimum and maximum positions of the resultant sine square curve lie at approximately 150° and 60°: These positions represent the principal meridians for the resultant sphero-cylindrical lens.

*Vector Analysis Method*

A more precise method of resolving obliquely crossed cylinders involves the use of *vector analysis*. However, the method to be described differs from ordinary vector analysis in that all angles are *doubled*. Suppose that we have two cylindrical lenses, having powers $F_1$ and $F_2$, whose axes are separated by the angle gamma (γ), with gamma defined as the *acute* angle between the two axes, measured in a counterclockwise direction from $F_1$ to $F_2$. In plotting the angle 2γ for vector analysis, it is always laid out in a counter-

clockwise direction from the axis of cylinder $F_1$ to the axis of cylinder $F_2$.

As shown in Figure 2-37, the axes of cylinders $F_1$ and $F_2$ are chosen in such a way that the angle γ between the axes of the two cylinders will be the *acute* angle (rather than the obtuse angle) between the two axes. The power of cylinder $F_1$ is then plotted along the x-axis to the right of the origin, and the power of cylinder $F_2$ is plotted at an angle 2γ (twice the angle between the two cylinder axes). Construction lines are then drawn, parallel to $F_1$ and $F_2$, and a diagonal line is drawn from the origin. The length of the diagonal line will be equal to the power of the resultant cylinder, and the angle between the diagonal line and the line representing the axis of cylinder $F_1$ (which we will call 2θ) is equal to *twice* the angle between the axis of the resultant cylinder and the axis of the cylinder $F_1$.

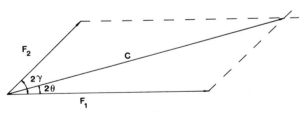

FIGURE 2-37. **Diagram for using the vector method of obtaining the angle 2θ and the cylinder power.**

(Note that, when using this system, both cylinders must have the *same sign*, in order for the resultant cylinder axis to fall between the axes of the cylinders $F_1$ and $F_2$. If the cylinders both have the same power, the resultant cylinder axis will be halfway between the two original axes; if one cylinder is stronger than the other, the resultant axis will fall closer to the axis of the stronger cylinder.)

Using the vector analysis method, we may determine the power and axis of the resultant cylinder, but the spherical component must be determined mathematically. The formula for this purpose is derived in the following discussion.

### Mathematical Method

We will now make use of vector analysis to derive mathematical formulas for the cylinder axis, cylinder power, and spherical power. By dropping a perpendicular from the end of the vector representing the resultant cylinder to the *x*-axis (see Figure 2-38), it is possible to derive an expression that makes it possible to find angle $2\theta$. Referring to Figure 2-38, where $a = OP$ and $b = NO$, then, in triangle MOP,

$$\tan 2\theta = \frac{a}{F_1 + b},$$

in triangle NOP,

$$\sin 2\gamma = \frac{a}{F_2},$$

and

$$a = F_2 \sin 2\gamma.$$

In triangle NOP,

$$\cos 2\gamma = \frac{b}{F_2},$$

and

$$b = F_2 \cos 2\gamma,$$

with the result that

$$\tan 2\theta = \frac{F_2 \sin 2\gamma}{F_1 + F_2 \cos 2\gamma}. \quad (2.21)$$

The angle $\theta$ is the resultant cylinder axis, measured in a counterclockwise direction from the axis of cylinder $F_1$.

A formula for determining the power of the resultant cylinder, $C$, may be derived as follows: In triangle MOP,

$$\sin 2\theta = \frac{a}{C},$$

and

$$C = \frac{a}{\sin 2\theta},$$

Since $a = F_2 \sin 2\gamma$, substituting,

$$C = \frac{F_2 \sin 2\gamma}{\sin 2\theta}. \quad (2.22)$$

Now that we have expressions for determining the power and axis of the resultant cylinder, it is necessary only to determine the *spherical* component of the lens formula. The final spherical component will include the spherical components of the two original sphero-cylindrical lenses, as well as the spherical power due to the resultant sphero-cylindrical lens.

An expression for the resultant spherical power can be derived with the help of Figure 2-39. In this diagram, the dashed lines (which can be labeled $S_1$ or

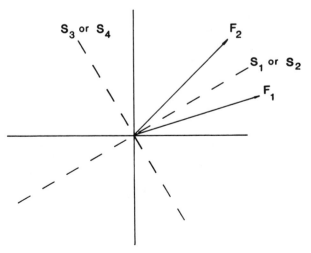

FIGURE 2-39. **Diagram for the derivation of a mathematical expression for the resultant spherical power.**

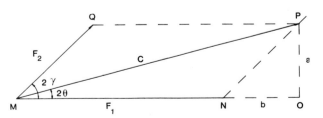

FIGURE 2-38. **Diagram for derivation of formulas for cylinder axis and power.**

$S_2$ and $S_3$ or $S_4$) are perpendicular to one another and represent the principal meridians of the resultant sphero-cylindrical lens. As derived in Section 2.17, for a cylinder of any power, the power in any two oblique meridians lying at right angles to each other will add up to the total power of the cylinder. If we forget about $F_2$ for the moment, considering only the powers $F_1$, $S_1$, and $S_3$,

$$S_1 + S_3 = F_1,$$

and if we forget about $F_1$, considering only the powers $F_2$, $S_3$, and $S_4$,

$$S_2 + S_4 = F_2,$$

it then follows that

$$S_1 + S_3 + S_2 + S_4 = F_1 + F_2.$$

Since $S_1$ and $S_2$ will combine to make up the resultant sphere power, $S$, and since $S_3$ and $S_4$ will combine to make up the power of the resultant sphere plus the power of the cylinder, $S + C$,

$$S_1 + S_2 = S$$

and

$$S_3 + S_4 = S + C.$$

Substituting,

$$S + (S + C) = F_1 + F_2$$

$$2S + C = F_1 + F_2$$

$$S = \frac{F_1 + F_2 - C}{2}. \qquad (2.23)$$

To the resultant sphere power, we must add the spherical components of the original lens formulas.

The procedure for combining obliquely crossed cylinders will be illustrated by an example, using first the vector analysis method and then the mathematical method.

### EXAMPLE

Given the obliquely crossed cylinders, plano + 2.50 DC x 30 and plano −2.00 DC x 140, find the resultant sphero-cylindrical lens.

The first procedure is to put both cylinders in the same form, either plus or minus. The most common procedure is to put them both into plus-cylinder form:

> plano + 2.50 DC x 30,

and

> −2.00 DS +2.00 DC x 50.

The two cylinders are plotted in the form of a vector diagram (see Figure 2-40) as follows:

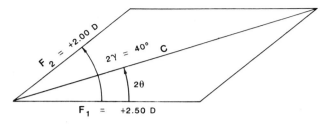

FIGURE 2-40. **Two cylinders in plus cylinder form, +2.50 DC x 30 and +2.00 DC x 50, plotted in the form of a vector diagram.**

1. $F_1 = +2.50$ D, $F_2 = +2.00$ D, $\gamma = 50° − 30° = 20°$, and $2\gamma = 40°$.
2. Measuring counterclockwise, $F_1$ is plotted (2.5 units long) along the x-axis, to the right of the origin, and $F_2$ is plotted (2.0 units long) at an angle of 40°.
3. The angle $\theta$ may be found by *direct measurement* from the vector diagram, using a protractor. Using the method of direct measurement, the angle $2\theta$ is found to be approximately 18° with the result that $\theta$ is approximately 9°.
4. Since the angle $\theta$ is defined as the angle between the axis of cylinder $F_1$ and the axis of the resultant cylinder, the axis of the resultant cylinder is equal to 30° +9°, or 39°.
5. The power, $C$, of the resultant cylinder may be found by direct measurement from the vector diagram. By direct measurement, it is found to be approximately +4.25 D.
6. The sphere resulting from the two obliquely crossed cylinders is found by Eq. (2.23).

$$S = \frac{F_1 + F_2 - C}{2}$$

$$= \frac{+2.50 +2.00 -4.23}{2}$$

$$= +0.135 \text{ D}.$$

7. Adding the −2.00 DS from one of the original lens formulas (the other had plano sphere power), the formula for the resultant lens is approximately

> (−2.00 DS + 0.135 DS) +4.25 DC x 39°

or

> −1.865 DS +4.25 DC x 39°.

Rounding off, we get

> −1.87 DS +4.25 DC x 39°.

The sphero-cylindrical result of obliquely crossed cylinders can also be determined by calculation.

1. When the angle $\theta$ is found by calculation, Eq. (2.21) is used:

$$\tan 2\theta = \frac{F_2 \sin 2\gamma}{F_1 + F_2 \cos 2\gamma}$$

$$= \frac{2 \sin 40}{2.5 + 2 \cos 40}$$

$$= \frac{1.286}{4.032} = 0.319$$

$$2\theta = 17.69°$$

$$\theta = 8.85°.$$

Hence, the resultant plus-cylinder axis is $30° + 8.85° = 38.85°$.

2. To find the power of the resultant cylinder by calculation, Eq. (2.22) is used:

$$C = \frac{F_2 \sin 2\gamma}{\sin 2\theta}$$

$$= \frac{1.286}{0.304} = +4.23 \text{ D.}$$

3. The sphere resulting from the two obliquely crossed cylinders is found by the use of Eq. (2.23):

$$S = \frac{F_1 + F_2 - C}{2}$$

$$= \frac{2.50 + 2.00 - 4.23}{2}$$

$$= +0.135 \text{ D.}$$

4. When the $+0.135$ DS is added to the $-2.00$ DS from the original lens formula, the total sphere power is $-1.865$ D.
5. The total resultant sphero-cylinder has a power of $-1.865$ DS $+4.23$ DC x 38.85, or approximately $-1.87$ DS $+4.25$ DC x 039.

It will be recalled that, when both cylinders have the same sign, the resulting axis will be between the two original axes, and that if the two cylinders are of equal power the resultant axis will be halfway between the two; but if they are of unequal powers the resultant axis will be shifted toward the axis of the stronger of the two cylinders. In our examples, the axis of the resultant cylinder was closer to $F_1$ than to $F_2$: This was expected, since the cylinder $F_1$ was slightly stronger than $F_2$ (Figure 2-41).

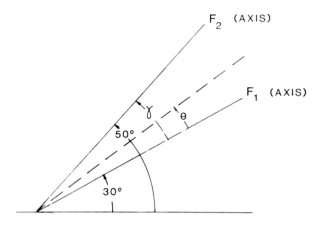

FIGURE 2-41. **Diagram showing the position of the resultant axis for the two cylinders plotted in Figure 2-40.**

### Polar Coordinate Method

In 1950, American Optical Company[9] described a vector analysis system for resolving obliquely crossed cylinders by using a polar coordinate plot in which the labeled meridians were only half as large as the corresponding real angles. For an angle of 30° (labeled on the polar plot) the actual angle would be 60°. Using this system, it is not necessary to multiply angle $\gamma$ by 2 or to divide angle $2\theta$ by 2, as done with the vector analysis method illustrated by Figure 2-37.

Humphrey[10] reintroduced this type of coordinate system in 1976, to explain how the astigmatic properties of an eye can be measured by means of the Humphrey Vision Analyzer without requiring an azimuthal rotation of any optical elements (note that an "azimuthal rotation" is a rotation like that of the hands of a clock). The Humphrey Vision Analyzer is a computer-assisted instrument for subjective refraction, in which any cylindrical correction can be created by a combination of two crossed cylinders, one having axes oriented at 45° and 135° and the other having axes oriented at 90° and 180°: During the testing procedure, the axes of the two crossed cylinders remain in these positions (even though their powers may be altered).

A vector, when confined to a given plane, can be represented by two numbers. These two numbers can be either (1) the length or magnitude of the vector, and its angle with respect to some axis in the given plane, or (2) the components of the vector along two directions in the plane, usually the horizontal ($x$) and the vertical ($y$) axes. Upon first glance, cylindrical lenses appear to be vectors since they can be described in terms of both a magnitude and a direction. But upon closer examination it becomes obvious that they are not vectors—they do not obey the vector addition law. For example, the combination of the two cylinders +1.00 DC axis 180 and +1.00 DC axis 90, having their axes 90° apart, results in *no cylindrical power* rather than the expected result (by vector analysis) of 1.41 D.

Humphrey's method permits cylindrical measurements to behave like simple vectors so that they can be added in the ordinary manner. The only difference between the Humphrey diagram (Figure 2-42) and a conventional polar coordinate plot is that the angular scale indicates angles that are only *one-half* the angles one would measure with a protractor: An angle that we would measure as 90° with a protractor would be labeled 45°.

For example, suppose we wish to combine the following two lenses:

(A) +5.00 DS −3.00 DC x 030,

(B) −3.00 DS +2.00 DC x 080.

Cylinders may be stated in either plus or minus cylinder form, but both cylinders must have the *same sign*. Therefore, transposing the formula for lens A,

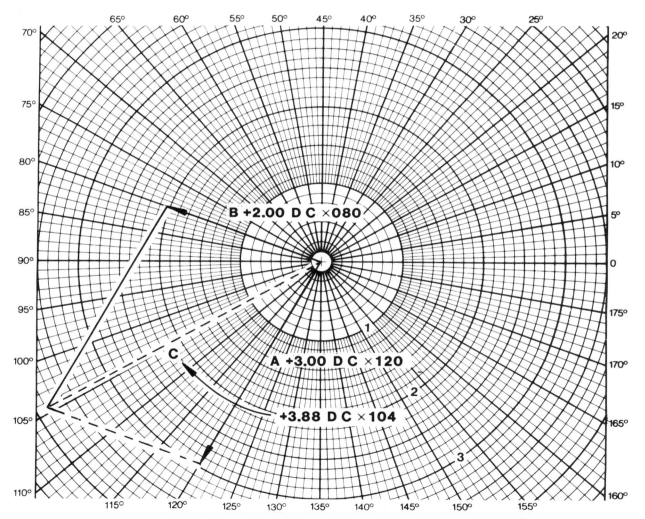

FIGURE 2-42. **Humphrey's polar coordinate diagram for resolving obliquely crossed cylinders. From Humphrey Instruments, San Leandro, CA.**

we have

$$(A) \quad +2.00 \text{ DS} +3.00 \text{ DC} \times 120,$$

$$(B) \quad -3.00 \text{ DS} +2.00 \text{ DC} \times 080.$$

*Step 1.* Construct a box format, as shown in Figure 2-43A.

*Step 2.* Write the two lens powers in the box, as shown in Figure 2-43B.

*Step 3.* Compute the equivalent sphere for each lens and enter it in the appropriate box (Figure 2-43C). For this example, the equivalent sphere for lens A is

$$+2.00 +3.00/2 = +3.50 \text{ D},$$

and the equivalent sphere for lens B is

$$-3.00 +2.00/2 = -2.00 \text{ D}.$$

*Step 4.* Add the equivalent spheres for lenses A and B in the appropriate box (Figure 2-43D).

*Step 5.* Plot the value of the two cylinders on the Humphrey diagram (Figure 2-42), plotting each cylinder at an axis that is actually *twice* the real angle.

*Step 6.* Complete the vector parallelogram. The length and direction of the diagonal, $C$, represent the resultant cylinder power and axis. Enter the values of the resultant cylinder power and axis in the appropriate boxes (Figure 2-43E). In this example, the vector sum is

$$+3.88 \text{ DC} \times 104$$

*Step 7.* Compute the resultant sphere value, using the following formula:

$$S_R + C_R/2 = SE_{SC},$$

where $S_R$ = the resultant sphere power, $C_R$ = the resultant cylinder power, and $SE_{SC}$ = the spherical equivalent of the resultant sphero-cylinder power. Transposing, we have

$$S_R = SE_{SC} - C_R/2$$

In this example,

$$S_R = +1.50 - 3.88/2$$

$$= +1.50 - 1.94$$

$$= -0.44 \text{ D}$$

|   | SPH. | CYL. | AXIS | EQUIV. SPH. |
|---|---|---|---|---|
| A |   |   |   |   |
| B |   |   |   |   |
| RES. |   |   |   |   |

A

|   | SPH. | CYL. | AXIS | EQUIV. SPH. |
|---|---|---|---|---|
| A | +2.00 | +3.00 | 120 |   |
| B | -3.00 | +2.00 | 080 |   |
| RES. |   |   |   |   |

B

|   | SPH. | CYL. | AXIS | EQUIV. SPH. |
|---|---|---|---|---|
| A | +2.00 | +3.00 | 120 | +3.50 |
| B | -3.00 | +2.00 | 080 | -2.00 |
| RES. |   |   |   |   |

C

|   | SPH. | CYL. | AXIS | EQUIV. SPH. |
|---|---|---|---|---|
| A | +2.00 | +3.00 | 120 | +3.50 |
| B | -3.00 | +2.00 | 080 | -2.00 |
| RES. |   |   |   | +1.50 |

D

|   | SPH. | CYL. | AXIS | EQUIV. SPH. |
|---|---|---|---|---|
| A | +2.00 | +3.00 | 120 | +3.50 |
| B | -3.00 | +2.00 | 080 | -2.00 |
| RES. |   | +3.88 | 104 | +1.50 |

E

|   | SPH. | CYL. | AXIS | EQUIV. SPH. |
|---|---|---|---|---|
| A | +2.00 | +3.00 | 120 | +3.50 |
| B | -3.00 | +2.00 | 080 | -2.00 |
| RES. | -0.44 | +3.88 | 104 | +1.50 |

F

FIGURE 2-43. **Steps required for resolving obliquely crossed cylinders using the Humphrey diagram. From Humphrey Instruments, San Leandro, CA.**

This value is placed in the appropriate box (Figure 2-43F). The formula of the resultant spherocylinder is therefore

$$-0.44 \text{ DS} +3.88 \text{ DC} \times 104.$$

### 2.20. Astigmatism due to Lens Tilt

When a homocentric pencil of light is incident normally along the optical axis of a spherical lens, the refracted pencil of light is stigmatic (Figure 2-44A). However, if the lens is pantoscopically tilted so that the pencil of light is obliquely incident upon the optical axis of the lens at its front pole, as shown in Figure 2-44B, the refracted pencil will be astigmatic. The astigmatism occurs simply as a result of the obliquity of the incident light. Not only is a cylindrical component introduced, but the spherical power changes also.

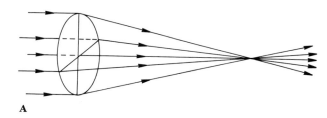

**A**

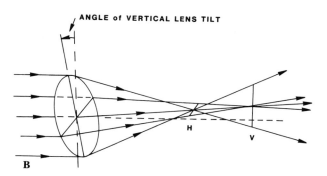

**B**

FIGURE 2-44. **Astigmatism due to lens tilt: A, a homocentric pencil of light incident normally along the optic axis of a spherical lens; B, the lens is rotated around a horizontal meridian with the result that the lens is tilted in the vertical meridian with respect to the homocentric bundle of incident light. The obliquity of incidence in the vertical meridian increases the power of the vertical meridian as compared to that of the horizontal meridian, with the result that the refracted pencil is astigmatic. The refracted rays in the vertical meridian form a horizontal focal line, H, whereas the refracted rays in the horizontal focal line form a vertical focal line, V.**

The spherical refraction produced by the lens increases in power, and the cylinder that is generated has the same sign as that of the original sphere. The axis of the resulting cylinder is parallel to the axis of rotation of the lens. For example, if a plus lens is "pantoscopically" tilted (tilted inward toward the cheekbone) so that the wearer looks through the pole of the lens obliquely, plus cylinder is induced, and since the axis of rotation is horizontal, the axis of the induced plus cylinder is horizontal. Likewise, if a minus lens is pantoscopically tilted, a minus cylinder is induced, having its axis horizontal.

For small values of angle $\alpha$, the power of the new sphere is given by the expression

$$S' = S_o \left( 1 + \frac{\sin^2 \alpha}{2n} \right), \qquad (2.24)$$

where $S'$ = the power of the new sphere, $S_o$ = the power of the original sphere, and $\alpha$ = the angle of obliquity. If the index of refraction of the glass or plastic is assumed to be 1.50, the above equation then becomes

$$S' = S_o \left( 1 + \frac{\sin^2 \alpha}{3} \right). \qquad (2.25)$$

The induced cylinder is given by the expression

$$C = S' \tan^2 \alpha. \qquad (2.26)$$

#### EXAMPLE

For a +10.00 D spherical lens, whose index of refraction is 1.50, tilted 15° around the horizontal axis, the power of the new sphere will be

$$S' = 10 \left( 1 + \frac{\sin^2 15}{3} \right)$$

$$= 10 \left( 1 + \frac{0.2588^2}{3} \right)$$

$$= 10(1 + 0.067/3) = 10(1 + 0.022)$$

$$= +10.22 \text{ D},$$

and the power of the cylinder will be

$$C = S' \tan^2 \alpha$$

$$= 10.22 \tan^2 15$$

$$= 10.22(0.2679)^2$$

$$= 10.22(0.072)$$

$$= +0.74 \text{ D}.$$

This lens behaves (as shown in the power diagram in Figure 2-45) as if it had the formula +10.22 DS +0.74 DC x 180.

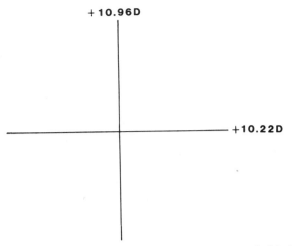

**+10.96D**

**+10.22D**

FIGURE 2-45. **Power diagram indicating the sphero-cylindrical power that results when a +10.00 D spherical lens is tilted 15 degrees around the horizontal meridian.**

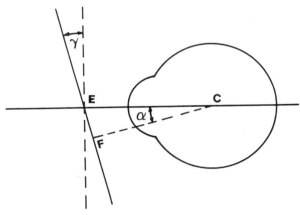

FIGURE 2-46. **Diagram used to establish the relationship between the amount of pantoscopic tilt and the level of the optical center, in order for the optic axis of the lens to pass through the center of rotation of the eye.**

It should be understood that pantoscopic tilt results in a change in power (spherical as well as cylindrical) *only* when the optic axis of the lens fails to go through the center of rotation of the eye (which is considered to be 14 mm behind the corneal apex).

By making certain approximations, a very simple relationship can be demonstrated between the pantoscopic tilt of a lens and the level at which the center must be placed in order to cause the optic axis to pass through the center of rotation of the eye. Referring to Figure 2-46, the point E lies directly in front of the pupil when the eyes are in the primary position, and the line CE lies in the primary horizontal plane of the head. When the angle α is expressed in degrees,

$$\sin \alpha = \sin \gamma = \frac{EF}{EC}$$

$$EF = EC(\sin\gamma).$$

Since $\sin 1° = 0.0175$,

$$EF = EC (\gamma) (0.0175),$$

and if EC is assumed to have a value of 27 mm,

$$EF = 27 (0.0175)$$

$$= 0.4725$$

$$= \text{approximately } 0.5. \quad (2.27)$$

Therefore, in order to make the optic axis of the lens pass through the center of rotation of the eye, the amount the optical center must be lowered from the pupillary level, in millimeters, is approximately equal to *one-half the pantoscopic tilt, in degrees.*

The distance EC will vary somewhat from one patient to another, but from a practical point of view, there is little to be gained by treating this distance as a variable.

## 2.21. The Maddox Rod

The Maddox rod is actually a series of glass or plastic rods (as shown in Figure 2-47) designed for use in phoria measurement. As the patient views a small, intense source of light (called a "muscle light") with the Maddox rod placed in front of one eye, she sees a "streak," or line, oriented in the direction perpendicular to the orientation of the rods (Figure 2-48). The same effect could be produced by the use of a single rod, making multiple rods unnecessary.

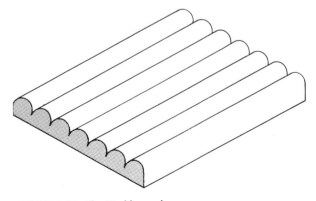

FIGURE 2-47. **The Maddox rod.**

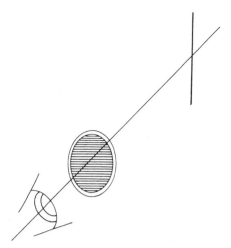

FIGURE 2-48. **When a patient views an intense source of light through a Maddox rod, she sees a "streak" of light, oriented in the direction perpendicular to the direction of the rods.**

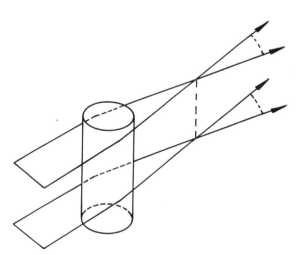

FIGURE 2-49. **A single Maddox rod, acting as a plano-cylindrical lens. Light is not refracted in the direction parallel to the axis of the rod (the cylinder axis meridian) but *is* refracted in the direction perpendicular to the axis of the rod (the power meridian), forming a focal line parallel to the axis of the rod. A second focal line is formed at *infinity*.**

However, the use of a single rod would require very precise positioning of the patient's eye, whereas with multiple rods the patient's eye may be located behind any of the rods. The rods are usually *red* in color, in order to encourage dissociation of the two eyes and to make the streak of light more noticeable.

Each of the individual rods acts as a plano-cylindrical lens. As shown in Figure 2-49, if a single Maddox rod has its axis oriented vertically, rays of

light *will not* be refracted in the direction parallel to the axis of the rod (the axis meridian of the cylinder), but *will* be refracted in the meridian perpendicular to the axis of the rod (the power meridian of the cylinder), forming a focal line in the vertical meridian (parallel to the axis of the rod). This being the case, one might expect the streak produced by the Maddox rod to be located parallel to the axis of the rod. However, this is not the case: As already noted, the streak is oriented in the meridian *perpendicular* to the direction of the orientation of the rods. How can we account for this?

Although we tend to think of a plano-cylindrical lens as forming only one focal line, it actually forms *two* focal lines, However, the focal line parallel to the power meridian of the lens is formed at infinity, just as the image formed by a plano-spherical lens, for parallel incident rays, is formed at infinity. The focal line formed at infinity can be seen only if it is focused onto a screen by a converging optical system: The *patient's eye* acts as the converging system that focuses the focal line on the retina (the screen).

The above explanation accounts for the appearance of the streak, seen perpendicular to the direction of orientation of the Maddox rod or rods, but what about the *other* focal line? Why don't we see it? The reason we don't see it is that the image formed by the power meridian of the rod is so close to the eye that the resultant vergence brought about by the rod

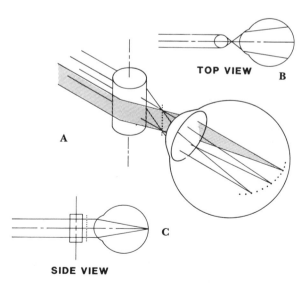

FIGURE 2-50. **When a single Maddox rod is placed in front of the eye, the first focal line is too close to the eye to be focused by the eye's optical system, and is therefore "wasted"; however, parallel rays of light in the meridian parallel to the axis of the rod *are* focused, forming a focal line perpendicular to the axis of the rod.**

(having a radius of curvature of only 2 or 3 mm) is so great that the eye is completely incapable of focusing it on the retina. This focal line is therefore "wasted," and the patient sees only the line that is formed at infinity.

As a schematic illustration of how the Maddox rod functions, consider a single Maddox rod having a power of +40.00 DC axis 90, placed 3 cm in front of the patient's cornea (see Figure 2-50). A bundle of rays from a distant object point, incident upon the Maddox rod, will form a vertical line focus 5 mm in front of the cornea.

With the cornea located so close to the vertical focal line, what effect does the optical system of the eye have on the vergence in the horizontal meridian? The eye's optical system will have little or no effect. There is nearly *infinite* vergence of rays in the horizontal meridian in the corneal plane, since it is so close to the vertical focal plane of the cylindrical lens. Compared to the nearly infinite vergence for incident rays, the 60 D change in vergence brought about by the optical system of the eye is insignificant. However, in the vertical meridian the situation is entirely different: The eye's optical system refracts the parallel rays, in this meridian, forming a horizontal focal line on the retina.

## References

1. Bennett, A. G. *Ophthalmic Lenses*, p. 12. The Hatton Press, London, 1968.
2. Fincham, W.A.H., and Freeman, M.H. *Optics*, 9th edition, p. 62. Butterworths, London, 1980.
3. Morgan, M.W. *The Optics of Ophthalmic Lenses*, p. 14. Professional Press, Chicago, 1978.
4. Southall, J.P.C. *Mirrors, Prisms and Lenses*, 3rd edition, pp. 300–306. Macmillan, New York, 1933.
5. Bennett, A.G., and Rabbetts, P.B. Refraction in Oblique Meridians of the Astigmatic Eye. *Brit. J. Physiol. Opt.*, Vol. 32, pp. 59–77, 1976.
6. Keating, M.P., and Carroll, J.P. Blurred Imagery and the Sine Squared Law. *Amer. J. Optom. Physiol. Opt.*, Vol. 53, pp. 66–69, 1976.
7. Long, W.F., and Haine, C.L. The Endpoint of the Laser Speckle Pattern in Meridional Refraction. *Amer. J. Optom. Physiol. Opt.*, Vol. 52, pp. 582–586, 1975.
8. Keating, M.P. Dioptric Power in an Off-Axis Meridian: The Torsional Component. *Amer. J. Optom. Physiol. Opt.*, Vol. 63, pp. 830–838, 1986.
9. *Transposition of Obliquely Crossed Cylinders*. Publication of the Bureau of Visual Science. American Optical Co., Southbridge, Mass., 1950.
10. *Adding an Over-refraction to a Prescription Using the Humphrey Diagram*. Operator's Manual Application Note 017.0, Humphrey Instruments, San Leandro, Cal., 1976.

## Questions

1. What is the sagittal depth of a surface of −5.00 D that has a 50-mm diameter? Index of refraction is 1.523.

2. The radius of curvature of a surface is 5.25 cm. If the refracting power is 13.30 D, what is the index of refraction?

3. What would be the reading of a lens measure calibrated for 1.530 when "clocking" a surface whose refracting power is +5.00 D, if the index of refraction of the glass is 1.523?

4. A fused bifocal has a front surface power of +13.00 D as measured by a lens measure calibrated for a refractive index of 1.53. The index of refraction of the major portion is 1.523 and the index of refraction of the segment is 1.69. What is the actual front surface power of the major portion and of the segment?

5. A lens measure graduated for spectacle crown glass with an index of refraction of 1.53 is placed on the surface of a steel sphere 40 cm in diameter. What reading will it give?

6. A lens gauge is used on a surface of −4.00 D having an index of 1.50. The lens gauge reading on the surface is −5.00 D. For what index of refraction is the lens gauge calibrated?

7. A lens gauge is calibrated for a glass with an index of refraction of 1.50. An infinitely thin bi-convex lens of +7.00 measures only +2.50 on each surface when measured with the lens gauge. Find the index of refraction of the glass.

8. The curved surface of a convex plano-cylindrical lens mates exactly with the curved surface of a −3.00 D plano-cylindrical lens made with glass of index 1.45. When placed together with their axes parallel the combination has +1.00 D along the power meridian. What is the index of refraction of the glass of the convex plano-cylindrical lens?

9. If the index of the material from which a contact lens is made is 1.42, and the radius of curvature of the front surface is 6 mm, what is the refracting power of this surface?

10. The back surface of a lens is −3.00 D. The front surface measures +6.50 D in the vertical and +9.00 D in the horizontal. Write the prescription of the lens.

11. (a) Calculate the power in the meridian 30 degrees from the axis of a −6.00 D cylinder.
    (b) What is the power 30 degrees from the power meridian?

12. What is the power in the 0, 30, 45, 60, 90, 120, 150, and 180 meridians of the following lenses?
    (a) −2.00 DS +4.00 DC x 120
    (b) +4.00 DC x 180/−2.00 DC x 60

13. A lens gauge calibrated for a refractive index of 1.523 is placed on a cylindrical surface with a radius of curvature of 10 cm, such that the line connecting the three legs makes an angle of 30° with the cylinder axis. What would be the reading of the lens gauge?

14. Locate the focal lines and circle of least confusion for the following lenses:
    (a) +10.00 DS −5.00 DC x 90
    (b) +5.00 DS −10.00 DC x 90
    (c) +5.00 DS −5.00 DC x 90
    (d) −5.00 DS −5.00 DC x 90
    (e) +3.00 DS −6.00 DC x 90

15. For a point object at infinity, an unknown lens has a vertical line focus at $33\frac{1}{3}$ cm and a horizontal line focus at 50 cm behind the lens. What is the prescription?

16. The meridian of least refractive power of an astigmatic eye is 30 degrees. When the eye fogged, what line will appear clearest on a clock dial?

17. A parallel pencil of light falls axially on a lens having a power of +3.00 DS −2.00 DC axis 90. What is the power of the lens that, placed at the circle of least confusion, would restore the emergent light to parallelism in both meridians? (Assume both lenses to be infinitely thin.)

18. Find the sphero-cylindrical resultant equivalent of the following combination of obliquely crossed cylinders:
    (a) +4.00 DC x 20/−2.75 DC x 65
    (b) +2.00 DC x 20/+3.00 DC x 70
    (c) −1.75 DC x 120/+1.25 DC x 135
    (d) +4.00 DC x 80/−2.00 DC x 135
    (e) +2.25 DC x 40/−4.00 DC x 115

19. With +3.00 D in the vertical meridian and +1.00 D in the horizontal meridian, what is the prescription written in the cross-cylinder form?

20. A lens with a power of +7.00 DS +2.50 DC x 50 has a −3.00 D surface on the back side. What is the lowest surface power on the front surface?

21. The fixed legs of a lens measure calibrated for spectacle crown glass, $n = 1.523$, are separated by 22 mm when the instrument is new, but, after wear, become separated by 24 mm. What must be the index of refraction of a glass in order for the instrument to give the correct surface power?

22. A −10.00 D spherical lens with an index of 1.50 has pantoscopic tilt of 10°. The pole of the lens lies directly in front of the pupil when the head is in the primary position. Through the polar area, what is the effective power of the lens?

23. What is the spherical equivalence of a lens that has a power of +5.00 DS −3.00 DC x 90?

# CHAPTER THREE

# Power Specification and Measurement

There is a number of ways of specifying the refracting power of an ophthalmic lens. These include approximate power (also called nominal power), according to which the power of a lens is specified in terms of its front and back surface powers without regard to thickness; back vertex power and front vertex power, according to which the power of a lens is specified in terms of the refracting power for emergent rays at its back surface or front surface; equivalent power, according to which the power of a "thick" lens or optical system is specified in terms of the power of a single thin lens; and effective power, according to which the power of a lens is dependent upon its distance from the wearer's eye. Of all of these methods of power specification, only *back vertex power* is used routinely by optical laboratories and practitioners.

Although the practitioner can determine approximate power by the use of the lens gauge and can determine front vertex power by means of hand neutralization with trial lenses, methods routinely employed in modern optometric practice all involve the measurement of back vertex power. Instrumentation for the measurement of back vertex power includes the lensometer or vertometer, the projection lensometer, and the automatic (computerized) lensometer.

# POWER SPECIFICATION

## 3.1. Introduction

Although designers of lenses used in cameras and other optical instruments have always described these lenses in terms of *focal length*, ophthalmic lenses are described in terms of *refracting power*. Refracting power is defined as the change in vergence that occurs when light passes through a lens.

Light consists of waves which, in a medium such as air or glass, advance perpendicularly to their own wavefronts. If a wavefront is flat (or plane), the paths of the waves are parallel. However, if the wavefront has a convex curvature, as when light emanates radially from a point source, the wave paths are divergent; and if the wavefront has a concave surface, the wave paths will converge to a focus. All that any lens can do is to produce a change in vergence on the surface of an incident wave. The power of a lens describes the extent of this change.

The unit used for specifying the power of a spectacle lens is the *diopter*, abbreviated by the letter D. As noted by Levene,[1] a French ophthalmologist, F. Monoyer, suggested the term in 1872 at an ophthalmological congress in Heidelberg. Originally the term diopter described the refracting power of a lens in air with a focal length of 1 m. As mentioned in Section 2.1, the term diopter has also been used to describe the curvature of a surface, the diopter of curvature being equal to the reciprocal of the radius of the surface in meters. Thus, if the radius of a spherical refracting surface is 0.5 m, its curvature is 2 D.

The term diopter is also used to describe the curvature of a wavefront at a specified distance from its focus or its source. The reciprocal of this distance expressed in meters is commonly called *vergence*, with the unit of vergence being the diopter. The term *reduced distance*, either in object space or in image space, is the actual distance divided by the index of refraction of the medium, such as

$$\frac{l}{n} \quad \text{or} \quad \frac{l'}{n'}.$$

*Reduced vergence* is the reciprocal of reduced distance. That is,

$$\frac{1}{l/n} \quad \text{or} \quad \frac{1}{l'/n'},$$

which can be written as

$$\frac{n}{l} \quad \text{or} \quad \frac{n'}{l'}.$$

If we let

$$L = \frac{n}{l} \text{ and } L' = \frac{n'}{l'},$$

then $L$ and $L'$ are examples of reduced vergence, expressed in terms of the diopter. The fundamental expression for the refractive power, $F$, of an optical system is that it is equal to the reduced image vergence, $L'$, minus the reduced object vergence, $L$, or

$$F = L' - L. \tag{3.1}$$

Hence, refractive power is the ability to produce a *change in reduced vergence*.

Expression (3.1) is very useful in dealing with thin lens object-image relationships. When applied to single spherical refracting surfaces, the expression for power becomes

$$F = \frac{n'}{l'} - \frac{n}{l} = \frac{n' - n}{r}, \tag{3.2}$$

and since, for an object at infinity,

$$l' = f'$$

and

$$F = \frac{n'}{f'} = \frac{n' - n}{r},$$

and since, for an image at infinity,

$$l = f$$

and

$$F = \frac{-n}{f} = \frac{n' - n}{r},$$

Therefore,

$$F = \frac{n'}{f'} = \frac{-n}{f}. \tag{3.3}$$

Many methods of specifying power of ophthalmic lenses have been used. The following constitute the most common methods.

## 3.2. Approximate Power

The approximate power of an ophthalmic lens is given by the simple formula

$$F_A = F_1 + F_2, \qquad (3.4)$$

where $F_1$ and $F_2$ are the front and back surface powers, as measured by the lens measure (lens "clock") as shown in Figure 3-1. Recall (see Section 2.4) that the lens measure indicates the correct surface power *only* if the lens is made of the same material for which the lens measure has been calibrated. Significant errors will occur if a lens measure calibrated for an index of refraction of 1.53 or 1.523 is used (without using a correcting factor) for measuring the surface power of a CR-39 plastic lens ($n = 1.498$) or a High-Lite glass lens ($n = 1.70$).

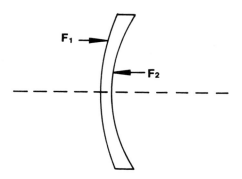

FIGURE 3-1. **Front and back surface powers, as measured by the lens "clock."**

Note that *thickness* doesn't enter into the approximate power formula: This is because this formula results in the correct power only for a lens of zero thickness. Since most ophthalmic lenses cannot be considered to be infinitely thin, a more accurate expression for lens power is required. Such expressions include back vertex power, front vertex power, and equivalent power.

## 3.3. Back Vertex Power

When an ophthalmic lens is ordered from a laboratory, the power of the lens is specified as *back vertex power*. The back vertex power of a lens is defined as the reciprocal of the reduced distance from the back pole of the lens to the secondary focal point (the reduced distance, it will be recalled, is the actual distance divided by the index of refraction).

An expression for the back vertex power of a lens in air may be derived with the aid of Figure 3-2. Using this diagram, back vertex power as defined above is the reciprocal of the distance $A_2F'$. By using the general equation for refraction at a single spherical surface,

$$\frac{n'}{l'} = \frac{n}{l} + \frac{n' - n}{r},$$

for the second surface of the lens (where $n$ is the index of refraction of the lens and 1 is the index of refraction of the surrounding air), we obtain the

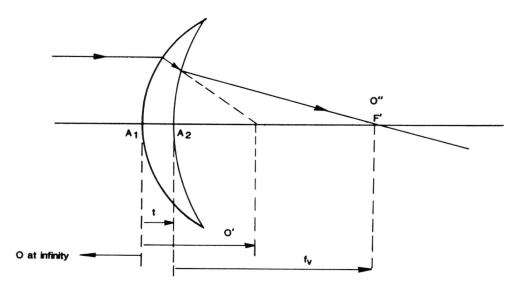

FIGURE 3-2. **Diagram for derivation of the expression for the back vertex power of a lens.**

equation

$$\frac{1}{f_v} = \frac{n}{O' - t} + \frac{1 - n}{r_2} = F_V$$

$$\frac{1}{f_v} = \frac{n}{O' - t} + F_2. \tag{3.5}$$

When applied to the first surface, we obtain the equation

$$\frac{n}{O'} = \frac{1}{\infty} + \frac{n - 1}{r_1}$$

$$\frac{n}{O'} = F_1. \tag{3.6}$$

Solving Eq. (3.6) for $O'$ (note that $O' = n/F_1$), and substituting in Eq. (3.5),

$$F_V = \frac{1}{f_v}$$

$$= \frac{n}{\dfrac{n}{F_1} - t} + F_2 = \frac{\dfrac{F_1}{n}(n)}{\dfrac{F_1}{n}\left(\dfrac{n}{F_1} - t\right)}$$

$$F_V = \frac{F_1}{1 - \dfrac{t}{n}F_1} + F_2. \tag{3.7}$$

However,

$$\frac{F_1}{1 - F_1\left(\dfrac{t}{n}\right)} = F_1 + F_1^2\left(\frac{t}{n}\right) + F_1^3\left(\frac{t}{n}\right)^2 + \dots$$

This is an infinite series in which each term gets smaller and smaller, and in ophthalmic optics only the first two terms of the series are used. The back vertex power of a lens, then, can be stated either in terms of the exact formula (3.7) or in terms of the approximate formula,

$$F_V = F_1 + F_2 + F_1^2\,\frac{t}{n}. \tag{3.8}$$

Either the exact formula or the approximate formula may be used if $F_1$ and $F_2$ are given and it is desired to Find $F_v$. However, the exact formula is easier to use if $F_v$ and $F_2$ are known and it is desired to find $F_1$, since the use of the approximate formula in this situation would require solving a quadratic equation.

Although the power of a lens may be stated in terms of either back vertex power, front vertex power, or equivalent power, there are two important reasons for using *back vertex power* for specifying the power of an ophthalmic lens:

1. Since the back vertex power is measured from the back pole of the lens (an easy-to-locate reference point) to the secondary focal point, this allows us to place the lens in any position we wish, as long as we choose a suitable power to place the secondary focal point of the lens at the far point of the eye (the concept of the far point of the eye will be discussed in Chapter 5). Whether the lens is placed in the "spectacle plane" or is placed on the cornea (a contact lens), we can specify its back vertex power to make it have the desired effect.

2. Use of back vertex power allows unlimited freedom in terms of lens form—the "bend," or cross-section shape of the lens. Back vertex power allows us to use any form of lens both in examining and in fitting the patient, and all we have to do is to make the secondary focal point of the lens coincide with the far point of the eye.

In order to measure back vertex power, an instrument known as the *lensometer*, or *vertometer*, is used. The use of this instrument will be discussed in a later section of this chapter.

### 3.4. Front Vertex Power, or Neutralizing Power

One method of measuring the power of an ophthalmic lens is by "neutralizing" it with a lens from a trial case. Two lenses are said to neutralize each other if, when placed in contact, their total refracting power is zero. Strictly speaking, when one lens is neutralized by another it is implied not only that their focal lengths are equal in magnitude, but that the secondary focal point of the known lens is coincident with the primary focal point of the unknown lens. These conditions are realized in the case of a plano-convex and a plano-concave lens placed with their surfaces in contact (see Figure 3-3).

The commonly used procedure of placing a trial case lens (calibrated in terms of back vertex power) with its back pole in contact with the front pole of a spectacle lens, and "neutralizing" the lens, measures the *front vertex power* of the spectacle lens. Front vertex power (also called *neutralizing power*) is defined as the negative reciprocal of the reduced distance from the front pole of the lens to its primary focal point.

An expression for neutralizing power can be derived in a manner similar to that for back vertex power. As defined above, neutralizing power is the negative reciprocal of the distance $A_1F$ in Figure 3-4. By using the general equation for refraction at a spherical surface, the refraction at the first surface is

given by

$$\frac{n}{0'} = \frac{1}{f_N} + \frac{n-1}{r_1}$$

$$\frac{1}{-f_N} = F_1 - \frac{n}{0'} = F_N \qquad (3.9)$$

and the refraction at the second surface is given by

$$\frac{1}{\infty} = \frac{n}{0' - t} + \frac{1-n}{r_2}$$

$$\frac{1}{\infty} = \frac{n}{0' - t} + F_2. \qquad (3.10)$$

Solving Eq. (3.9) for $0'$ (note that $0' = t - n/F_2$), and substituting in Eq. (3.9),

$$F_N = \frac{1}{-f_N} = F_1 - \frac{n}{0'}$$

$$= F_1 - \frac{n}{t - \dfrac{n}{F_2}} = F_1 - \frac{n\left(\dfrac{F_2}{n}\right)}{\left(t - \dfrac{n}{F_2}\right)\dfrac{F_2}{n}}$$

$$= F_1 - \frac{F_2}{\left(t - \dfrac{n}{F_2}\right)\left(\dfrac{F_2}{n}\right)}$$

$$F_N = F_1 + \frac{F_2}{1 - \dfrac{t}{n}F_2}. \qquad (3.11)$$

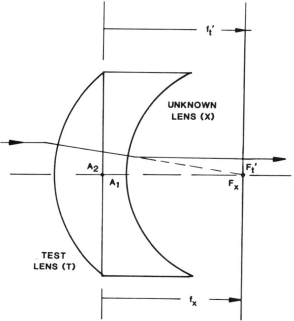

FIGURE 3-3. **Neutralizing power: The secondary focal point of the known lens is coincident with the primary focal point of the unknown lens.**

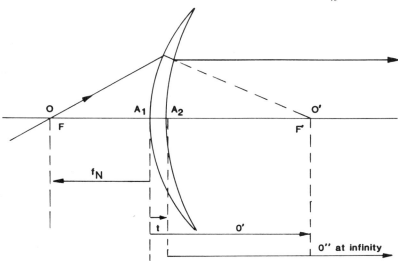

FIGURE 3-4. **Diagram for derivation of the expression for the front vertex power (neutralizing power) of a lens.**

However,

$$\frac{F_2}{1 - \frac{t}{n} F_2} = F_2 + F_2{}^2 \frac{t}{n} + F_2{}^3 \left(\frac{t}{n}\right)^2 + \ldots \, .$$

Dropping all but the first two terms of the series, the neutralizing power of a lens can be stated in terms of the approximate formula,

$$F_N = F_1 + F_2 + F_2{}^2 \frac{t}{n}. \qquad (3.12)$$

*Comparison of Approximate Power, Back Vertex Power, and Neutralizing Power*

The following examples will serve to illustrate the differences between approximate power, back vertex power, and neutralizing power.

### EXAMPLE 1

Given a lens for which $F_1 = +6.00$D, $F_2 = -7.00$ D, $t = 2.0$ mm, and $n = 1.523$. Find $F_A$, $F_V$, and $F_N$.

For approximate power,
$$F_A = +6.00 - 7.00 = -1.00 \text{ D}.$$

For back vertex power,
$$F_V = +6.00 - 7.00 + 36\frac{0.002}{1.523}$$
$$= +6.00 - 7.00 + 0.047$$
$$= -0.95 \text{ D}.$$

For neutralizing power,
$$F_N = +6.00 - 7.00 + 49\frac{0.002}{1.523}$$
$$= +6.00 - 7.00 + 0.06$$
$$= -0.94 \text{ D}.$$

### EXAMPLE 2

Given a lens for which $F_1 = +16.00$ D, $F_2 = -4.00$ D, $t = 7.5$ mm, and $n = 1.523$. Find $F_A$, $F_V$, and $F_N$.

For approximate power,
$$F_A = +16.00 - 4.00 = +12.00 \text{ D}.$$

For back vertex power,
$$F_V = +16.00 - 4.00 + 256\frac{0.0075}{1.523}$$
$$= +16.00 - 4.00 + 1.26$$
$$= +13.26 \text{ D}.$$

For neutralizing power,
$$F_N = +16.00 - 4.00 + 16\frac{0.0075}{1.523}$$
$$= +16.00 - 4.00 + 0.08$$
$$= +12.08 \text{ D}.$$

As shown by these examples, for low-powered, relatively thin lenses, the differences between approximate power, back vertex power, and neutralizing power are relatively small; but for thick, high-powered lenses, the differences can be large. The error involved in using hand neutralization for the lens in Example 2, rather than using the lensometer or vertometer, would be 1.18 D. For meniscus ophthalmic lenses, the front vertex power is less than the back vertex power.

## 3.5. Equivalent Power

Many optical devices are constructed with a series of lenses separated by air or arranged with a series of curved surfaces separated by media having different refractive indices. Most of these complex systems are *symmetrical*; that is, the centers of curvature of the surfaces all fall on a common optical axis.

On occasion it is conceptually and mathematically convenient to express the result of a complex optical system in terms of an imaginary single, thin lens that would produce an image of a distant object of the same size and at the same position as produced by the system. Replacing an optical system with an imaginary single thin lens allows the application of simple object-image relationships to the equivalent lens.

The focal length of the thin lens that will produce an image size and an image position similar to those produced by the system is called *equivalent focal length*. The reciprocal of the equivalent focal length in meters is defined as the *equivalent power*.

To determine the position of the thin equivalent lens with respect to the system, the location of the principal planes of the system must be known. In a symmetrical optical system there is only a single pair of planes which have the property that, in these planes, the magnification is positive unity (+1); that is, the object and the image of the object are the same size, and the image is erect. These planes are called the *principal planes*, and the points of intersection of the optical axis with these planes form the *principal points* of an optical system.

The plane associated with the object space is called the *primary principal plane*, and the plane associated with the image space is called the *secondary principal plane*. The distance from the primary principal point (P) of the system to the primary focal point (F) is known as the *primary equivalent focal length*, and the distance from the secondary principal point (P') to the secondary focal point (F') is called the *secondary equivalent focal length*. The reciprocal of the secondary equivalent focal length is known as the *equivalent power* of the system.

Thus, if a single thin lens with a secondary focal length of P'F' is situated at P', it will achieve the effect of the lens system (see Figure 3-5). It is obvious, then, that the secondary focal length as measured from this single thin lens is equal to the secondary equivalent focal length of the system as measured from the secondary focal point. Equivalent power is frequently referred to as *principal plane refraction*.

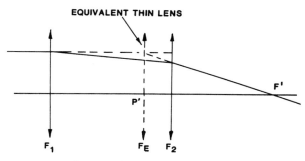

FIGURE 3-5. **The equivalent power of a lens system is equal to the power of a single thin lens placed at the secondary principal plane of the system.**

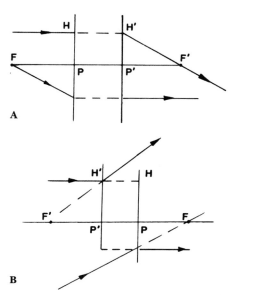

FIGURE 3-6. **The principal planes and focal points of an optical system. A, a converging system; B, a diverging system.**

The following is a useful method of conceptualizing the principal planes of an optical system:

When any ray in either object space or image space goes through the focal point associated with that respective space, in the opposite space there will be a ray parallel to the optical axis of the system. The rectilinear portions of the ray in these two spaces will intersect in a point on the principal plane with which the respective focal point is associated. Therefore, as shown in Figure 3-6, the ray going through F will intersect the parallel ray in the image space on the primary principal plane (H), and the ray going through F' will intersect the parallel ray in the object space of the secondary principal plane (H').

The formula for calculating the equivalent power of any two optical elements may be derived from Figure 3-7. The lens shown in this diagram is assumed to be in air. In the similar triangles $JA_1F_1'$ and $KA_2F_1'$:

$$\frac{h_1}{h_2} = \frac{f_1'}{f_1' - t}. \tag{a}$$

In the similar triangles $OP'F_s'$ and $KA_2F_s'$:

$$\frac{h_1}{h_2} = \frac{f_E'}{l_2'}. \tag{b}$$

Therefore,

$$\frac{f_E'}{l_2'} = \frac{f_1'}{f_1' - t}. \tag{c}$$

Refraction at the second surface may be expressed as

$$\frac{1}{l_2'} - \frac{n}{f_1' - t} = \frac{1}{f_2'}$$

or

$$\frac{1}{l_2'} = \frac{1}{f_2'} + \frac{n}{f_1' - t}. \tag{d}$$

If equation (d) is divided by equation (c), we obtain

$$\frac{1}{f_E'} = \frac{1}{f_2'} + \frac{n}{f_1'} - \frac{1(t)}{f_1'(f_2')}.$$

Multiplying the last term in the above expression by $n/n$, we have

$$\frac{1}{f_E'} = \frac{1}{f_2'} + \frac{n}{f_1'} - \left(\frac{1(t)(n)}{f_1'(f_2')(n)}\right).$$

Rearranging the last term,

$$\frac{1}{f_E'} = \frac{1}{f_2'} + \frac{n}{f_1'} - \left(\frac{1}{f_2'}\right)\left(\frac{n}{f_1'}\right)\left(\frac{t}{n}\right).$$

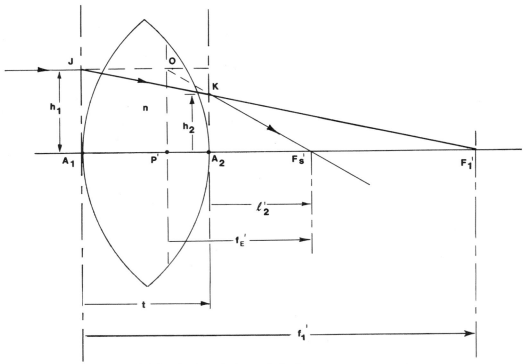

FIGURE 3-7. **Diagram for deriving the formula for equivalent power.**

The above equation may be restated as

$$F_E = F_1 + F_2 - cF_1F_2 \qquad (3.13)$$

and

$$F_E = \frac{-n}{f_E} = \frac{n'}{f_E{}'}, \qquad (3.14)$$

where $F_1$ = the power of the first element; $F_2$ = the power of the second element; $c$ = the reduced distance from the secondary principal point of the first element to the primary principal point of the second element, that is,

$$c = \frac{P_1{}'(P_2)}{n_2}, \qquad (3.15)$$

where $n_2$ is the index of refraction of the intervening medium; $f_E$ = the primary equivalent focal length, that is, $PF$; and $f_E{}'$ = the secondary equivalent focal length, or $P'F'$.

When formula (3.13) is used for two thin lenses in air, $F_1$ and $F_2$ represent the powers of each of the two lenses, and $c$ represents the distance between the lenses; but when it is used for a thick lens, $F_1$ and $F_2$ represent the surface powers of the lens and $c$ represents the reduced thickness, $t/n$, of the lens.

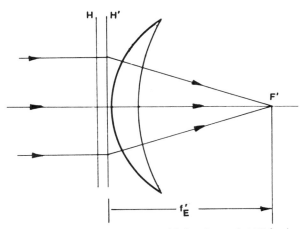

FIGURE 3-8. **Equivalent power: A thin lens located at H', having a focal length of $f_e'$, would have the same effect as the thick lens.**

When the equivalent power formula is used to express the power of an ophthalmic lens, the resulting infinitely thin lens, with the power $F_E$, is located at the secondary principal plane (H') of the lens (see Figure 3-8). Unfortunately there is no easy way to determine the location of the secondary principal plane of a lens. An additional problem is the fact that the principal planes move backward or forward as the *form* of the lens is changed. For these reasons the concept of equivalent power is seldom used for ophthalmic lenses, but it *is* used in the case of most complex optical systems, such as those used in low-vision aids. The various methods of specifying focal lengths are shown in Figure 3-9.

The relationships between equivalent power, back vertex power, and front vertex power are illustrated in the following example.

### EXAMPLE

Given a +5.00 D lens and a +8.00 D lens, separated (in air) by a distance of 12.5 mm. Find (1) equivalent power; (2) back vertex power; (3) front vertex power; (4) the positions of the principal planes.

1. To find the equivalent power of the system,

$$F_E = F_1 + F_2 - cF_1F_2$$

$$= +5.00 + 8.00 - \frac{0.0125}{1}(5.00)(8.00)$$

$$= +13.00 - 0.50 = +12.50 \text{ D.}$$

2. To find the back vertex power of the system,

$$F_V = \frac{F_1}{1 - \frac{t}{n}F_1} + F_2$$

$$= \frac{5}{1 - \frac{0.0125}{1}(5.00)} + 8.00$$

$$= \frac{5.00}{0.9375} + 8.00$$

$$= 5.33 + 8.00 = +13.33 \text{ D.}$$

3. To find the front vertex power of the system,

$$F_N = \frac{F_2}{1 - \frac{t}{n}F_2} + F_1$$

$$= \frac{8.00}{1 - \frac{0.0125}{1}(8.00)} + 5.00$$

$$= \frac{8.00}{0.9} + 5.00$$

$$= 8.89 + 5.00$$

$$= +13.89 \text{ D.}$$

4. To find the positions of the principal planes of the system,

$$F_E = \frac{1}{f_E'}$$

$$f_E' = \frac{1}{F_E} = \frac{1}{12.5} = 0.08 \text{ m} = 8 \text{ cm}$$

$$f_E = -\frac{1}{F_E} = -\frac{1}{12.5} = -0.08 \text{ m} = -8 \text{ cm}$$

$$F_V = \frac{1}{f_V}$$

$$f_V = \frac{1}{F_V} = \frac{1}{+13.33}$$

$$= 0.075 \text{ m} = 7.5 \text{ cm}$$

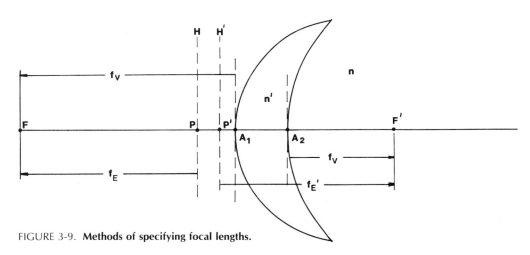

FIGURE 3-9. **Methods of specifying focal lengths.**

$$F_N = -\frac{1}{f_N}$$

$$f_N = -\frac{1}{+13.89}$$

$$= -0.072 \text{ m} = -7.2 \text{ cm}.$$

If $e$ = distance from the first lens to $P$, and if $e'$ = distance from the second lens to $P'$, then

$$e = f_N - f_E$$

$$= -7.2 - (-8) = 0.8 \text{ cm},$$

and

$$e' = f_V - f_E'$$

$$= 7.5 - 8$$

$$= -0.5 \text{ cm}.$$

These results indicate that the primary principal point lies 0.8 cm to the right of the +5.00 D lens, and the secondary principal point lies 0.5 cm to the left of the +8.00 D lens.

## 3.6. Effective Power

The effective power of a lens may be defined as the ability of the lens to focus parallel rays of light at a given plane. Although the term has several slightly different uses, its major use is to indicate the change in lens power required if a lens is moved from one position to another in front of the eye. It is well known that plus lenses are "more effective" (that is, bring about more change in vergence than needed) and that minus lenses are "less effective" (bringing about less change in vergence than needed) as they are moved farther from the eyes.

Suppose, for example, that a test lens located at a distance of 15 mm in front of the cornea produces the desired optical effect and has a back vertex power of +10.00 D. Parallel rays of light will be focused 10 cm behind the lens, as illustrated by lens A in Figure 3-10. If, after selecting a frame for the patient, it is found that the spectacle lens will be located at B, say, 10 mm in front of the cornea, parallel rays no longer will be focused at plane F by lens A. A lens whose back

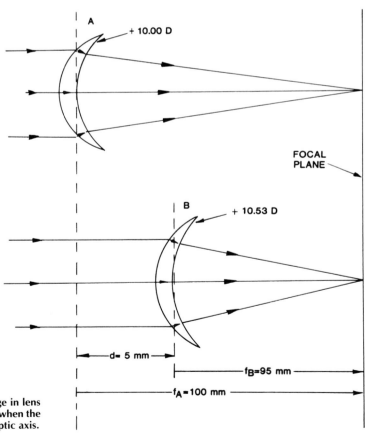

FIGURE 3-10. **Diagram illustrating the required change in lens power to cause parallel rays to focus in the same plane when the lens is moved in the fore and aft direction, along the optic axis.**

vertex power is equal to that of lens B must now be used to focus parallel rays at F. The back focal length $f_B$ of lens B is equal to the back focal length $f_A$ minus the distance $d$. In the case being considered, $f_A = 0.10$ m and $d = 0.005$ m, so

$$f_B = F_A - d$$
$$= 0.10 - 0.005 = 0.095 \text{ m}$$

and

$$F_B = \frac{1}{0.095} = +10.53 \text{ D.}$$

In other words, the "effective power" of a lens with reference to position B (that is, its ability to focus parallel rays of light in plane F) is $+10.53$ D.

With the aid of the relationships shown in Figure 3-10, we may derive a formula for use in determining the required power of a lens, when the lens is moved from its original position, A, to a second position, B.

$$f_B = f_A - d$$
$$= \frac{1}{f_A - d}$$

$$= \frac{1}{\frac{1}{F_A} - d}$$

$$F_B = \frac{F_A}{1 - dF_A} . \qquad (3.16)$$

In this formula, $d$ is given a plus sign if the lens is moved toward the eye (as in the above example), and is given a minus sign if the lens is moved away.

Using formula (3.16) to solve our example, we have

$$F_B = \frac{+10}{1 - 0.005\,(+10)} = +10.53 \text{ D.}$$

It can now be seen that back vertex power as defined above is simply the ability of a lens to focus *parallel rays* with reference to its back pole. Thus the term "effective power" is frequently used synonymously with "back vertex power." It is convenient to use back vertex power because ophthalmic lenses are placed in the "spectacle plane," at a fixed distance from the cornea, and back vertex power gives the effective power of the lens in the spectacle plane.

---

## POWER MEASUREMENT

---

### 3.7. Hand Neutralization

As stated in Section 3.4, two lenses are said to neutralize each other if, when placed in contact, their combined power is zero. Prior to the invention of the lensometer, optometrists routinely made use of hand neutralization in order to verify the powers of prescription lenses received from the laboratory. Even after the lensometer became available, for some years this instrument was used mainly by laboratories, many practitioners preferring to use hand neutralization.

With practice, hand neutralization gives fairly accurate findings on minus lenses and low-powered plus lenses, but it becomes less accurate with moderate and high plus lenses. Despite its limitations, hand neutralization is a highly useful procedure for use in estimating the power of a lens quickly when a lensometer is not readily available.

In performing hand neutralization, a lens from the trial set (calibrated in back vertex power) is used to neutralize the unknown lens by placing its *back* pole in contact with the *front* pole of the unknown lens, giving rise to the term *front vertex power*. Due to the fact that ophthalmic lenses typically have concave back surfaces, it is in most instances impossible to place the neutralizing lens in contact with the back surface of the unknown lens. Trial case lenses, used for hand neutralization, usually have less bend, or flexure, than ophthalmic lenses, with the result that it is usually possible to place the back surface of the neutralizing lens in contact with the front surface of the unknown lens.

*Hand Neutralization of Spherical Lenses*

To neutralize an unknown spherical lens, the following procedure is used:

1. The lens is held at arm's length, back surface facing the observer, while viewing a target at a dis-

tance of approximately 6 m from the lens. The target must have both vertical and horizontal contours: a large cross, a large square, or a 20/400 letter E are adequate targets. To minimize the effects of aberrations, attention should be directed principally to the *transverse* motion of the image seen through the central zone of the lens. Care should be exercised that the contact does not scratch the lenses, especially when *plastic* lenses are neutralized.

2. The observer slowly moves the lens, both vertically and horizontally, noting whether the transverse movement of the target appears to be in the same direction as that of the movement of the lens, or in the opposite direction. Movement in the *same* direction (*with* motion) indicates that the lens is a *minus* lens (see Figure 3-11); movement in the *opposite* direction (*against* motion) indicates that the lens is a *plus* lens (see Figure 3-12) as long as the distance from the lens to the observer's eye is less than the focal length of the lens. If the distance from the lens to the observer's eye is *longer* than the focal length of the lens, *with* motion will be seen, but the image will be inverted.

3. A lens of the opposite power is selected, and is placed with its back surface in contact with the front surface of the unknown lens. With experience, it is possible to estimate very closely the required power of the neutralizing lens. Both lenses are slowly moved, vertically as well as horizontally, and the judgment of *with* or *against* motion is again made. Referring to the situation in Figure 3-3 in which a minus

lens is being neutralized: If the neutralizing lens is of insufficient power, *with* motion will be seen; if the neutralizing lens is too great in power, *against* motion will be seen; and if the neutralizing lens is of the correct power, no motion will be seen and the lens will have been neutralized. If, for example, it is found that no motion is seen when a +4.00 DS is used as the neutralizing lens, the power of the unknown lens is −4.00 DS.

It will be noted that the stronger the lens to be neutralized, the slower the speed of the motion; and the closer the power of the neutralizing lens to that of the unknown lens, the faster the speed of the motion.

Since it was known in advance that the lens to be neutralized was a *spherical* lens, it was expected that the same speed (and direction) of motion would be seen in both the vertical and horizontal meridians. However, if the unknown lens is a cylindrical or sphero-cylindrical lens, the speed (and possibly also the direction) of motion will differ in different meridians.

*Hand Neutralization of Cylindrical and Sphero-cylindrical Lenses*

To neutralize a cylindrical or a sphero-cylindrical lens, the following procedure is used:

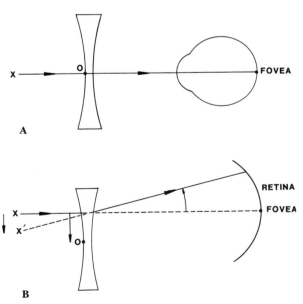

FIGURE 3-11. **Hand neutralization: *with* motion, for a minus lens.**

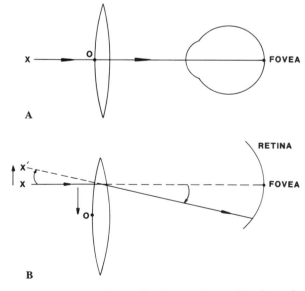

FIGURE 3-12. **Hand neutralization: *against* motion, for a plus lens.**

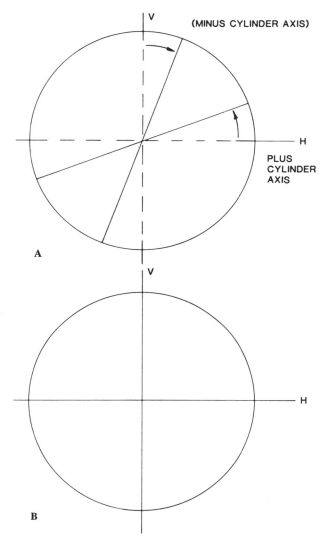

FIGURE 3-13. **Hand neutralization of a lens containing cylindrical power: A, off axis, showing scissors motion; B, on axis.**

appear to be parallel to (and continuous with) the limbs that can be seen outside the lens (Figure 3-13B).

3. Once the orientation shown in Figure 3-13B has been found, in which both limbs of the cross are continuous inside and outside the lens, further rotation of the lens will show that the scissors motion itself possesses either *with* or *against* motion: If the line seen inside the lens rotates *with* the direction of the rotation, the line target is parallel to the minus cylinder axis (Figure 3-14), whereas if the line inside the lens rotates *against* the direction of rotation, the line target is parallel to the plus cylinder axis (Figure 3-15). Since ophthalmic lens prescriptions are routinely specified in terms of minus cylinders, we are interested in finding the position of the *minus* cylinder axis.

4. Once the two principal meridians have been located, the power in each meridian is neutralized separately, using spherical lenses. Once the power of the lens in each principal meridian has been found (remembering that the power is the *opposite* to that of the neutralizing lens), the prescription of the lens is written in minus cylinder form. For example, if the lens is neutralized in the horizontal meridian with a −3.00 DS and in the vertical meridian with a −2.00 DS (indicating that the power of the unknown lens is +3.00 D in the horizontal meridian and +2.00 D in the vertical meridian), the prescription of the unknown lens would be written in the form

$$+3.00 \text{ DS} - 1.00 \text{ DC} \times 180.$$

A recommended procedure is always to neutralize the *minus axis* meridian first. This will mean that this meridian will always have the most plus (or least minus) power, and that the difference in power between the two principal meridians then becomes the cylindrical power of the lens. The power diagrams shown in Figure 3-16, based on this example, should aid in understanding this procedure.

5. The major difficulty involved in hand neutralization is the determination of *cylinder axis*. This may be done by drawing a line on the back surface of the lens with a grease pencil or a felt pen, while firmly holding the lens in the position in which the line of the cross representing the axis is continuous within and outside the lens. Following this, the axis is measured by means of a lens protractor. This procedure involves an inherent error of approximately ±5°.

6. An additional difficulty involved in hand neutralization is that of marking the optical center, or pole, of the lens. This is done by placing a small dot on the lens at the point where the two lines of the cross meet inside the lens (when the lens is held at such an orientation that the two lines are continuous

1. As with the spherical lens, the lens is held at arm's length, back surface facing the observer, while viewing a target having vertical and horizontal contours, through the lens, at a distance of approximately 6 m.

2. The observer then rotates the lens, and *scissors motion* is seen. As shown in Figure 3-13A, the vertical and horizontal lines of a cross target will be seen within the lens as being displaced from their positions as viewed outside the lens; and when the lens is oriented in such a manner that the two limbs of the cross are parallel to the principal meridians of the lens, the limbs of the cross seen within the lens will

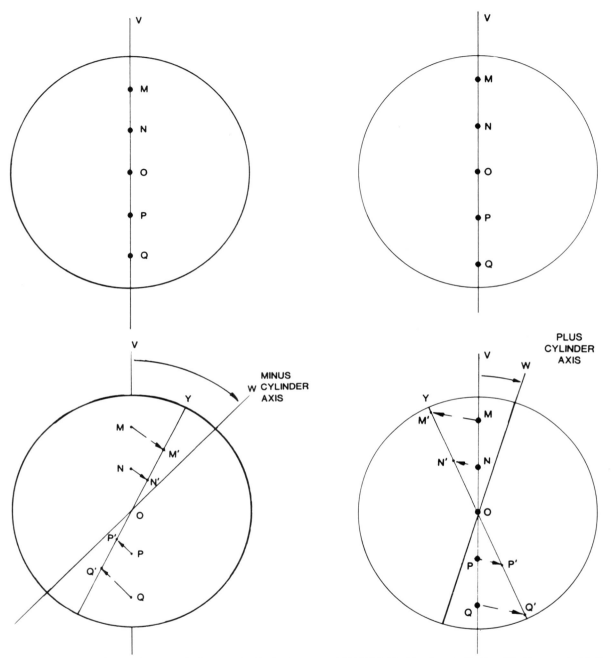

FIGURE 3-14. **Hand neutralization: The line seen inside the lens rotates *with* the direction of rotation of the lens, and the line target is parallel to the minus cylinder axis.**

FIGURE 3-15. **Hand neutralization: The line seen inside the lens rotates *against* the direction of rotation of the lens, and the line target is parallel to the plus cylinder axis.**

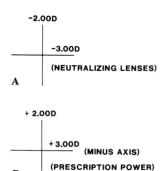

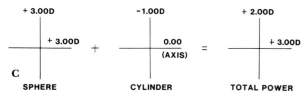

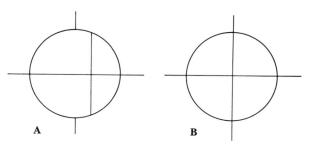

FIGURE 3-17. **Hand neutralization of a lens containing prism: A, vertical line is displaced to the right (when looking through the ocular side of the right lens), indicating base-in prism; B, the base-in power is neutralized by means of a trial case prism having base-out prism power.**

FIGURE 3-16. **Power diagrams illustrating the process of hand neutralization: A, power of neutralizing lenses; B, power of lens being neutralized; C, power diagrams for sphere power, cylinder power, and total power of lens being neutralized. Using this procedure, the lens formula can easily be written in minus cylinder form.**

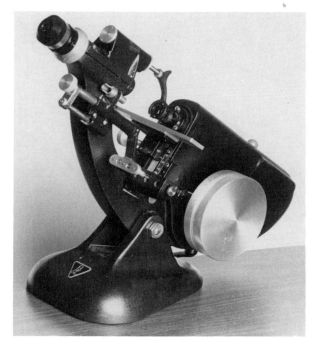

FIGURE 3-18. **The lensometer.**

within and outside the lens). Again, this requires a steady hand and involves a certain amount of error. However, simple "spotting" or "centering" machines are available for marking the cylinder axis and the pole of the lens: These instruments were the precursors of the modern lensometer.

If a lens (spherical, cylindrical, or sphero-cylindrical) contains *prism*, the lines of the cross seen through the lens (when the principal meridians of the lens have been found) will be displaced in one direction or the other, perhaps even outside the lens. The direction of the base of the prism in the unknown lens can be determined by remembering that, when viewing through a prism, the image of a given object is displaced *toward the apex* of the prism. For example, if the vertical line is seen to be displaced toward the right (looking through the ocular side of the lens), while neutralizing the right lens of a pair of glasses, the prism in the unknown lens would be *base-in* prism (Figure 3-17A). To measure the amount of base-in prism, *base-out* neutralizing prisms are used, finding the amount of base-out prism power necessary to cause the vertical line of the cross to be undeviated by the lens (Figure 3-17B).

## 3.8. The Lensometer

The *lensometer*, or *vertometer*, shown in Figure 3-18, is an optical instrument designed for the purpose of measuring the vertex power of a spectacle lens. As shown in Figure 3-19, the basic elements of the lensometer include a *focusing system* and an *observation system*. The focusing system consists of a light source, a movable target, a *standard lens* (collimating lens), and a lens stop with a small central aperture. Depending upon the particular instrument, the target

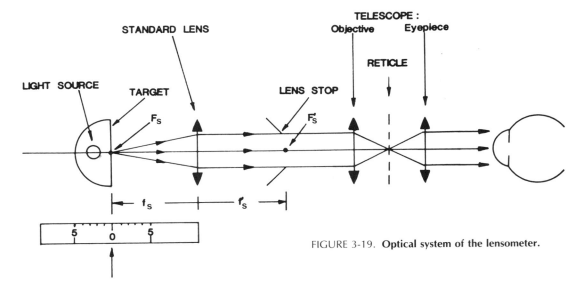

FIGURE 3-19. **Optical system of the lensometer.**

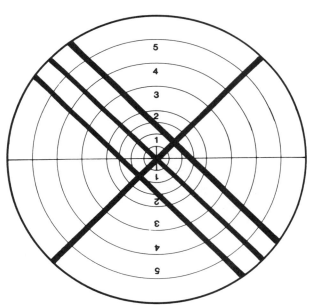

FIGURE 3-20. **Lensometer target having a cross pattern, with single and triple lines.**

may be in the form of a cross (usually having a single line oriented in one direction and triple lines oriented in a direction 90° away), as shown in Figure 3-20, or it may have a series of small dots or points arranged in a circular pattern, as shown in Figure 3-21.

When using the lensometer with the target consisting of two perpendicular lines, the target must be rotated until the perpendicular lines are parallel to the two principal meridians of the lens being neutralized. When the target is in focus for one principal meridian, the image of the line (or lines) seen clearly will be oriented *perpendicularly* to that principal meridian. The cylinder axis can then be read directly from the cylinder axis scale. The cylinder power is simply the dioptric difference between the power readings found for the two line images. For example, if we have a lens whose power is +1.00 DS −2.00 DC x 120, Figure 3-22A shows the appearance of the lensometer target for the "most plus" meridian (note that the scale reading shows +1.00 D and the single line in the 30° meridian is in focus), whereas Figure 3-22B shows the appearance of the target in the "least plus" meridian (the scale reading now being −1.00 D and the triple lines in the 120° meridian being in focus).

When using the lensometer with a target formed of dots arranged in a circle, the target itself is not rotated. When cylindrical power is present, each dot source forms its own conoid of Sturm with two line foci, each focal line being oriented perpendicularly to the principal meridian for which the target is in focus. The power wheel of the lensometer can be adjusted, sequentially, until each of the line images is clearly focused. The cylinder axis is then determined by projecting the orientation of the line images on a cylinder axis scale. As with the lensometer having

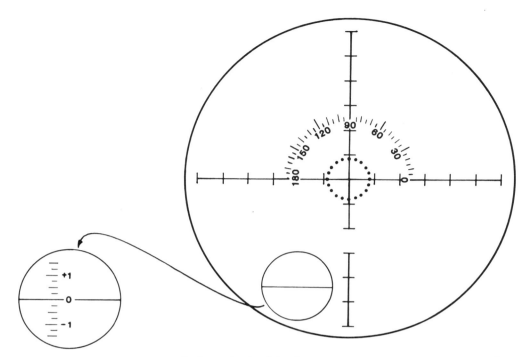

FIGURE 3-21. **Lensometer target having a circular dot pattern.**

perpendicular line targets, the cylinder power is the dioptric difference between the power readings found for the two perpendicular line foci. For the lens described above, having a power of +1.00 DS −2.00 DC x 120, Figure 3-23A shows the appearance of the lensometer target for the "most plus" meridian (note that the scale reading shows +1.00 D and that each dot is elongated to form a line in the *30°* meridian), whereas Figure 3-23B shows the appearance of the target in the "least plus" meridian (the scale reading being −1.00 D and each dot forming a line parallel to the 120° meridian).

The power of the *standard lens* must be greater than that of any lens to be measured in the instrument, and usually is from +20.00 D to +25.00 D. The lens stop or lens rest is designed to hold the unknown lens with its back surface against the stop, thus making the measurement of *back vertex power* possible. The lens stop is located at the secondary focal plane of the standard lens. By placing the lens stop in this position, the movement of the target from its zero position is now linearly related to the *back vertex power* of the lens being neutralized.

With the lens stop located at the secondary focal plane of the standard lens, the apparent size of the

image of the target remains constant. This phenomenon occurs because the equivalent power of the combination of the standard lens and the spectacle lens, acting together as a thick lens, remains equal to the equivalent power of the standard lens without any lens before the lens stop.

The observation system consists of a low-powered Keplerian telescope with an objective lens, a reticle located in the secondary focal plane of the objective lens, and an adjustable eyepiece. The reticle (Figure 3-24) usually contains a set of cross hairs for focusing, a prism scale, and a cylinder axis scale (the latter only when a dot target is used).

Without any lens before the lens stop, the zero position of the target is the anterior focal plane of the standard lens. In this position, light emerging from the standard lens is parallel, and it enters the objective lens of the telescope parallel, resulting in a clear image of the target on the reticle. If the eyepiece is properly adjusted, a clear image will be seen by the observer.

As mentioned earlier, the movement of the target from its zero position is directly proportional to the back vertex power of the lens being tested. This can be demonstrated by the application of the Newtonian

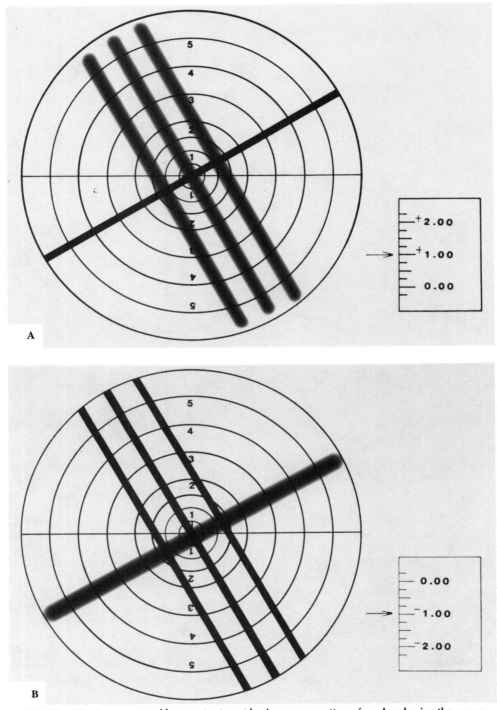

FIGURE 3-22. **Appearance of lensometer target having a cross pattern, for a lens having the power +1.00 DS −2.00 DC x 120: A, for the "most plus" meridian; B, for the "least plus" meridian.**

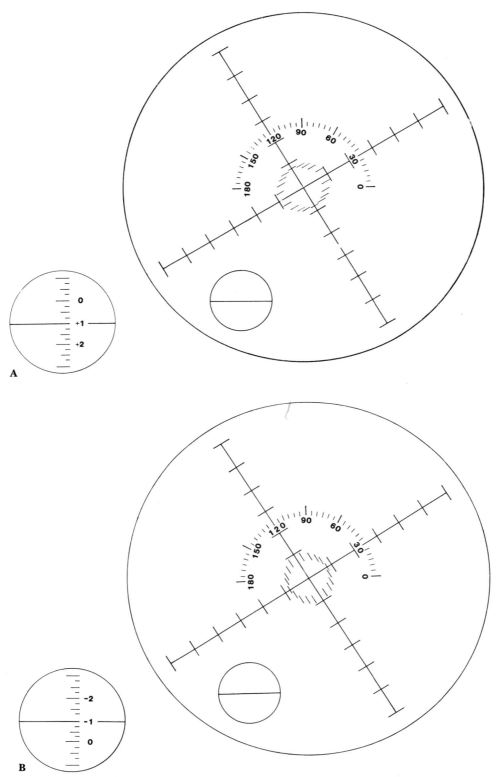

FIGURE 3-23. **Appearance of lensometer target having a circular dot pattern, for a lens having the power +1.00 DS −2.00 DC x 120: A, for the "most plus" meridian; B, for the "least plus" meridian.**

equation involving the extra-focal distances $x$ and $x'$, in which $xx' = ff'$. In the lensometer the displacement of the target from the anterior focal point $(F_S)$ of the standard lens is equal to the distance $x_S$. The displacement of the image from the secondary focal point $(F_S')$ of the standard lens is equal to the distance $x_S'$. When a lens is neutralized in a lensometer, the secondary focal length of the lens under test is then equal to the distance $x_S'$.

Under all conditions, in order to have a clearly focused image on the reticle, the rays forming the image must be parallel when they impinge upon the objective lens. In order for the rays to emerge paral-

lel from a test lens, the rays must leave the secondary focal point of the test lens (for plus lenses) or be headed for the secondary focal point of the test lens (for minus lenses).

When a concave lens is placed in the instrument (Figure 3-25), it causes light rays from the target to diverge, so in order for the target to be seen clearly it is necessary to move the target farther away from the standard lens (by means of a drum provided for this purpose). The distance through which the lens is moved, $x_S$, is such that the distance $x_S'$ (from the lens stop to the image position) is equal to the secondary focal length of the lens $(f_T')$ being tested (as the lens will be worn before the eye).

On the other hand, when a convex lens is placed in the instrument (Figure 3-26), it causes rays from the target to converge, with the result that the target must be moved closer to the standard lens. The distance through which the target must be moved is such that the distance, $x_S'$, from the lens stop to the new position of the image of the target is now equal to the secondary focal length, $f_T'$, of the lens being tested (as the lens will be worn before the eye).

For both the concave test lens and the convex test lens, the distance shown as $x_S'$ may appear to be equal to the *primary* focal length, $f_T$. However, it is actually equal to the *secondary* focal length, $f_T'$, when the lens is worn before the eye. This occurs because, when measuring a lens in the lensometer, the light travels from the back of the lens to the front.

*Calibration of the Lensometer*

The calibration of the lensometer is based upon the Newtonian formula,

$$xx' = ff',$$

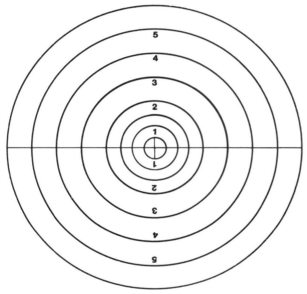

FIGURE 3-24. **Appearance of the reticle target in the telescope eyepiece.**

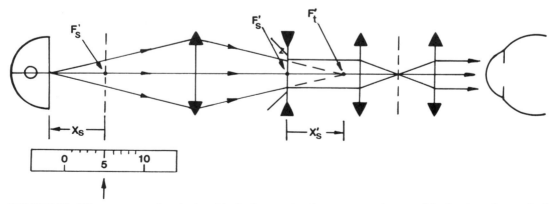

FIGURE 3-25. **When a concave lens is placed in the lensometer, the target must be moved farther from the standard lens.**

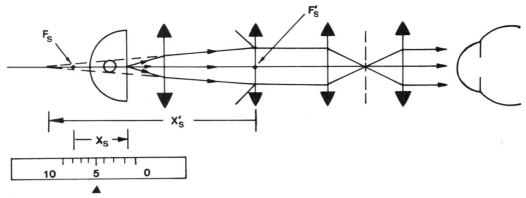

FIGURE 3-26. **When a convex lens is placed in the lensometer, the target must be moved closer to the standard lens.**

where, as already discussed, $x$ is the distance from the primary focal plane of the lens to the object, and $x'$ is the distance from the secondary focal plane of the lens to the image. When this formula is applied to the standard lens, we have

$$x_S x_S' = f_S f_S'.$$

Since, for the lensometer, $f_S' = -f_S$,

$$x_S x_S' = -f_S{}^2$$

and

$$x_S = -\frac{f_S{}^2}{x_S'}.$$

As the lens is worn before the eye,

$$x_S' = -f_T'.$$

We now substitute,

$$x_S = \frac{f_S{}^2}{f_T'}.$$

Since $1/f_T' = $ back vertex power, we have

$$x_S = f_S{}^2 F_V. \qquad (3.17)$$

Note, in the above formula, that since $f_S{}^2$ is a constant, $F_V$ changes with $x_S$ in a *linear* manner, as already discussed: A given amount of rotation of the measuring drum brings about an equal change, in diopters, over the whole range of the instrument, with the result that the numbers on the measuring drum are *equally spaced*.

It should be noted that, if the unknown lens is placed with its *front* surface against the lens stop, *front* vertex power can be measured. When this is done, although the power readings can be taken directly from the power scale, the cylinder axis readings must be modified by subtracting the axis reading from 180° (unless the principal meridians are 180° and

90°). For example, if a lens having a back vertex power of +2.00 DS −1.00 DC x 120 is placed on a lensometer with its front surface against the lens stop, the lens reading will now be +2.00 DS −1.00 DC x 60. If we subtract 60 from 180, we get 120°, which was the original axis.

A displacement of the image of the target from the center of the reticle indicates the power and direction of the prismatic effect of the lens at its position in front of the lens stop. It will be recalled that the telescope through which the target image is viewed is a Keplerian telescope. This telescope inverts the image of the target, with the result that, whereas base-up prism normally displaces the image downward, toward the apex of the prism, the inversion of the image of the target by the telescope causes the image to be displaced in the *upward* direction. In other words, the displacement of the image is in the *same* direction as that of the base of the prism.

### Extent of Target Movement

The extent of the target movement required to measure a lens of a given power depends upon the power of the standard lens. This will be illustrated by the following examples:

**EXAMPLE 1**

For a lensometer fitted with a standard lens having a power of +20.00 D, what is the extent of the target movement in measuring a lens having a power of +1.00 D?

$$x_S = f_S{}^2 (F_V)$$
$$= (0.05^2)(1)$$
$$= (0.0025)(1)$$
$$= 0.0025 \text{ m} = 2.5 \text{ mm}.$$

Since the amount of target movement is the same for a diopter on *any* part of the power scale, this example shows that the amount of target movement for a lensometer with a +20.00 D standard lens is 2.5 mm per diopter of power of the unknown lens.

### EXAMPLE 2

If the above lensometer is moved from a +20.00 D reading to a −20.00 D reading, how far does the target move?

$$x_S = (0.05^2)\,(40)$$
$$= (0.0025)\,(40)$$
$$= 0.1 \text{ m} = 10 \text{ cm}.$$

The total extent of target movement, therefore, for a lensometer having a +20.00 D standard lens is 10 cm, or 100 mm.

### EXAMPLE 3

Given a lens having the prescription

$$+3.00 \text{ DS} -6.00 \text{ DC x } 180,$$

how far does the target move in measuring the power in the two principal meridians with a lensometer that is fitted with a standard lens of +25.00 D?

Note that this lens has a refracting power of +3.00 D in the horizontal meridian and −3.00 D in the vertical meridian, the difference in power between the two principal meridians being 6.00 D. Therefore, the amount of movement will be

$$x_S = f_S^2 F_V$$
$$= (0.04)^2\,(6)$$
$$= 0.0016\,(6)$$
$$= 0.0096 \text{ m} = 9.6 \text{ mm}.$$

## Operation of the Lensometer

Operation of the lensometer, in verifying the prescription of a pair of glasses, involves the following steps:

1. The first procedure, before placing the glasses in the instrument, is that of focusing the eyepiece. Beginning with the eyepiece all the way out (turned all the way to the left), the eyepiece is slowly turned inward as the operator views the cross hairs until they are in sharp focus. The focus may then be verified by turning the power drum (usually located on the right-hand side of the instrument) until the image of the cross target or the dot target is in sharp focus. When this occurs, the power drum reading should be 0.00

D, and both the image of the target and the cross hairs should be sharply focused.

2. The glasses are then placed in the instrument with the right lens in the lens stop, with its *back* surface against the stop. The spring-loaded clamp is activated, holding the lens in place, and the lens is centered (approximately) in the lens stop, by moving the glasses from side to side and moving the stage (upon which the glasses rest) up or down, as required.

3. If the instrument is equipped with a target having a single line in one direction and triple lines in the direction 90° away, the power drum is turned until the target is in sharp focus. (Whether the drum is to be turned in the minus or plus direction will have already been determined by noting the direction of motion of a distant object seen through the lens, as in hand neutralization.) When both sets of lines are in sharp focus for the same setting of the power drum, as shown in Figure 3-27 (or when the two sets of lines are equally blurred, as shown in Figure 3-28), the lens is a *spherical* lens.

4. If either the single line or the triple line target comes into focus separately, the power drum is turned to the setting where the line or lines in the *most plus* meridian are in focus. At this point the axis scale is rotated, if necessary, until the *single line* is in sharp focus (Figure 3-22A). The power drum reading at this point will indicate the spherical power of the lens. The power drum is then turned in the minus direction until the triple lines are in focus (Figure 3-22B). The difference between this reading and the first reading of the power drum indicates the power of the cylinder, and the axis scale provides the cylinder axis. The lens prescription is then recorded in the minus cylinder form.

5. During the process of adjusting the power drum for each of the two principal meridians, the operator should *center* the lens, so that the target is centered with respect to the reticle scale. Once the lens has been centered, the lens is "spotted," using the spring-loaded, inked, spotting device. This will enable the operator to measure the distance between the centers of the two lenses, once the prescription for both lenses has been determined.

6. If the instrument is equipped with a target in the form of a circular array of dots, all of the dots will come into sharp focus simultaneously if the lens is a spherical lens. However, if the lens contains cylindrical power, there will be a power drum setting at which each dot is elongated in one meridian, and another setting at which each dot is elongated in the opposite meridian. With this instrument, the procedure is to first find the *most plus* position of the power drum in which the dots are in focus (although elon-

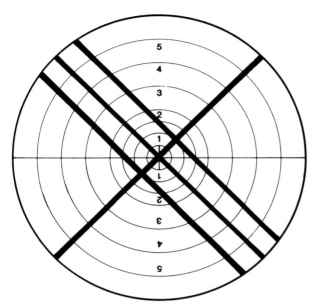

FIGURE 3-27. **Lensometer target with the two sets of lines in sharp focus.**

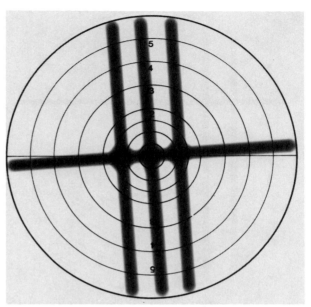

FIGURE 3-28. **Lensometer target with the two sets of lines equally blurred.**

gated), as shown in Figure 3-23A. The power drum reading at this point is the spherical power. The drum is then moved in the minus direction until the dots are in focus but elongated in the opposite direction, as shown in Figure 3-23B. The axis scale is then rotated until the axis indicator is parallel to the direction of elongation of the dots. The difference between the two readings is the cylinder power, and the axis scale reading provides the minus cylinder axis.

7. The lens clamp is now loosened, the glasses are moved along the stage until the left lens is in front of the lens stop (*without changing the height of the stage*), and procedures 2 through 5 are repeated. Since the height of the stage has not been changed, it will be possible to determine the presence of a vertical prismatic effect. If, when the target is in focus and centered in the horizontal direction, it is found that the center of the target is above or below the center of the reticle scale, vertical prism is present. By using the prism scale in the reticle, the amount and direction of the base of the prism can be determined. If the center of the target is displaced upward with regard to the center of the prism scale, base-up prism is present; if the target is displaced downward, base-down prism is present. The amount of prism may be read off the prism scale, each unit on the scale usually representing 1 prism diopter.

If *prism* has been ground into the lens being tested, it may be impossible to center the lens in the lens-

ometer (that is, so that the perpendicular line target or the dot target will be centered with respect to the reticle scale in the lensometer eyepiece). In such a case, the center of the perpendicular line target or dot target will be displaced to one side or the other (for horizontal prism), upward or downward (for vertical prism), or obliquely (for oblique prism). When this occurs, the eyepiece of the instrument must be rotated so that the reticle scale intersects the center of the perpendicular line or dot target (as shown in Figure 3-29). When this has been done, the reticle scale is coincident with the base-apex line of the prism. The *amount* of prism may be determined by reading it off the concentric circles in the reticle scale, and the *direction* of the prism base is given by the direction of the center of the lensometer target with respect to the center of the reticle scale. For example, if for the *right* eye the lensometer target is displaced temporally, resulting in a scale reading of 2 prism diopters (as shown in Figure 3-30), the result would be 2 prism diopters *base out.*

## 3.9. Lensometer Calibration, Alignment, and Measurement Errors

In the daily use of a lensometer, the operator should be cognizant of the following points, which are of

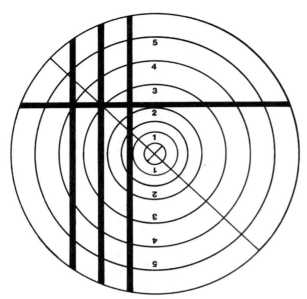

FIGURE 3-29. **Measuring a lens containing oblique prism. The eyepiece of the lensometer is rotated so that the reticle scale is coincident with the base – apex line of the prism, and the amount of the prism and the direction of the base are then read and recorded.**

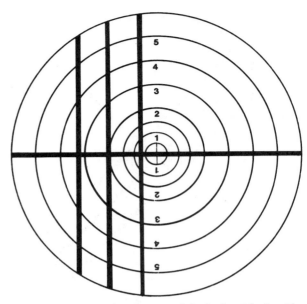

FIGURE 3-30. **Measuring a lens containing horizontal prism. The eyepiece is rotated so that the reticle scale is coincident with the base – apex line of the prism, and the amount of the prism and the direction of the base are then read and recorded.**

importance in ensuring that the instrument is functioning properly:

1. *Adjustments and Calibration.* It is imperative that the *eyepiece* of the instrument be adjusted to the eye of the individual operator, using the procedure described in the previous section. This is particularly important when more than one person is routinely using the instrument. The *zero setting* of the lensometer is also an important consideration. When the eyepiece is correctly adjusted and the target is clearly focused, the dioptric scale reading should be zero with no test lens in the instrument. If the dioptric scale reading is *not* zero, the scale setting is incorrect. With many instruments it is possible to reset the scale to the correct reading.

A properly calibrated lensometer should not only read zero with no lens in place, but should read correctly for high-powered (i.e., over 10.00 D) plus and minus lenses. Precision test lenses are available from some of the major lens manufacturers, for the purpose of checking the accuracy of the lensometer. An error that is always in the same direction, that is, either too much plus or too much minus (for strong plus lenses and minus lenses), indicates that the *lens stop* is in an incorrect position. If both strong plus lenses and strong minus lenses read too high or too low, the power of the standard lens is likely to be in-

correct. Cylinder axis errors and misalignment of the marker device or the lens platform can be checked by the use of cylindrical test lenses.

The *centration of the reticle* may be checked by rotating the reticle. If the reticle is properly centered, the center of the reticle should remain stationary during rotation. To check for *centration of the target*, the center of the target image, as the target is rotated, should coincide with the center of the reticle. If the target is not centered, errors in prism measurement will occur.

2. *Measurement Problems.* Measurement of the power of a *contact lens* may be a problem. If the lens stop is large and the back surface of the contact lens is steeply curved, the back pole of the contact lens will not lie in the plane of the lens stop, but will lie *in front* of the stop. An error is thus induced in the measurement of back vertex power.

If the front surface of the contact lens is placed against the lens stop, *front vertex power* will be measured. With the front surface of the contact lens against the lens stop, the front pole of the contact lens lies *behind* the plane of the lens stop. In most cases the distance from the plane of the lens stop is smaller when the front surface is placed against the stop than when the back surface is placed against the stop: Therefore, the error is greater for *back* vertex power than for *front* vertex power (Figure 3-31). Although

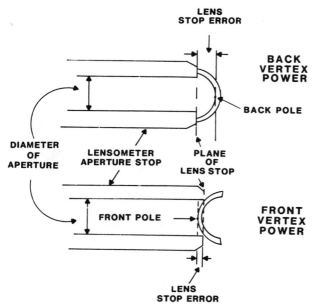

FIGURE 3-31. **Measuring the power of a contact lens. Because of the small radius of curvature, the back pole of the contact lens cannot be placed in the plane of the lens stop.**

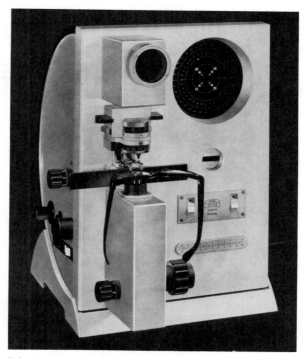

FIGURE 3-32. **A projection lensometer.**

the measurement error is smaller for front vertex power than for back vertex power, it is important to understand that, since contact lens powers are specified in back vertex power, they must be *measured* in terms of back vertex power also. Several aperture devices have been manufactured to assist in placing the back pole of the contact lens in the plane of the lens stop. Some lensometers are equipped with special lens holders which assist in making measurements of back vertex power.

The measurement of *bifocal* or *trifocal additions* also presents a problem, owing to the fact that the lensometer is designed to measure lenses that are intended for distance vision, whereas a bifocal addition is intended for near vision. This will be discussed in the chapter on multifocal lenses, Chapter 8.

## 3.10. Projection Lensometers

With the projection lensometer (Figure 3-32) the operator views the target on a screen, rather than through a telescope. In addition to the fact that focusing is not necessary, this arrangement has the advantage that two or more people may look at the target at the same time. The glasses to be verified are placed in a horizontal (rather than a vertical)

position. In all other respects, the operation of the instrument is similar to the operation of a standard lensometer.

The projection lensometer is particularly convenient for verifying *contact lens* power, since the contact lens is simply placed (back surface downward) on the lens stop. On the other hand, with a standard lensometer a contact lens must be placed in a vertical position against the lens stop: This procedure is not only inconvenient but may result in pressure being put on the contact lens, which tends to cause flexure of the lens while the reading is taken.

## 3.11. Automatic Lensometers

A number of completely automatic lensometers have recently been introduced (Figure 3-33). Such an instrument contains a microcomputer that processes the data, displays the result, and provides a printout of the measurement including sphere, cylinder, axis, and prism power. Each of these instruments employs a different principle in determining the measurements. They are designed to rapidly measure the back vertex power in all meridians, and to select the meridians of maximum and minimum power (the principal meridians).

The process of checking and analyzing ophthalmic lenses occurs automatically and rapidly, with a substantial saving of time and effort. Apart from the saving of time and effort, the automatic lensometer eliminates the human errors involved in reading the scale and the arithmetical errors that could occur in subtracting the two readings in order to obtain the cylinder power. In addition, any controversy over the variability of measurements obtained by different operators can be resolved.

These automatic instruments are expensive and since an optometrist uses a lensometer only intermittently during the working day, a decision whether or not to purchase an automatic instrument must involve weighing the cost against the saving in time and the possibly greater accuracy. However, a high-volume optical laboratory routinely makes use of one or more lensometers continuously throughout the day; and since the verification of spectacles is a demanding task, requiring a high level of concentration, the use of automatic instruments in a laboratory can be justified on the basis of improvement in the efficiency of the verification process.

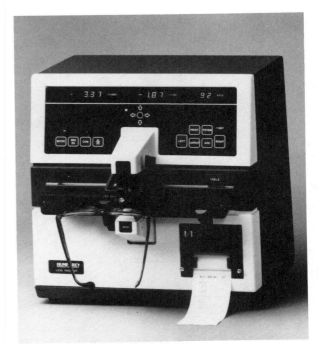

FIGURE 3-33. **An automatic lensometer.**

# RELATIONSHIPS BETWEEN LENS POWER AND LENS THICKNESS

Lens thickness is often an important consideration in the selection of the materials used in designing a patient's eyewear. Since a minus lens is thicker in the periphery than at the center, the dispenser will often want to exert some control over the *edge thickness* of the lens; and since a plus lens is thicker at the center than in the periphery, the dispenser may want to exert some control over the *center thickness* of a plus lens. Thickness usually becomes an important factor for lens prescriptions in the neighborhood of about ±4.00 D or more. However, different problems are presented by minus and plus lenses: For a minus lens, an edge thickness that is greater than necessary will detract from the cosmetic effect of the lenses, but will usually not create a problem of excess weight unless the lens is very large, whereas a center thickness that is greater than necessary, for a plus lens, will not only detract from the cosmetic effect of the glasses but will often cause the glasses to be unduly heavy. The problem of lens weight can, of course, be controlled to a great extent by using *plastic* lenses which, as pointed out in Chapter 1, have a specific gravity of a little more than half that of glass lenses.

### 3.12. The Sagitta Formula

It will be recalled from section 2.1 (and with reference to Figure 2-3) that the sagitta, $s$, of a lens surface having a radius of curvature $r$ and a chord length $h$ is given by the formula

$$s = \frac{h^2}{2r}.$$

By making use of the relationship between the radius of curvature and the refracting power of a surface,

$$F = \frac{n-1}{r},$$

we have

$$s = \frac{Fh^2}{2(n-1)}$$

and

$$F = \frac{2(n-1)s}{h^2}. \tag{3.18}$$

In this formula, the distance $s$ is measured from the arc to the chord, and is positive if measured toward the right and negative if measured toward the left.

If we are given the sagitta of each of the surfaces of a lens, together with the chord length and the index of refraction of the material, we will be able to determine the refracting power of each of the lens surfaces and, by algebraically adding the two surface powers, to determine the approximate power of the lens.

### EXAMPLE

Given a round lens (Figure 3-34) having the sagittae $s_1 = 4.8$ mm and $s_2 = 2.8$ mm, a diameter (chord length) of 40 mm, and an index of refraction of 1.50, find the approximate power of the lens.

For the front surface of the lens,

$$F_1 = \frac{2(n-1)s_1}{h^2}$$

$$= \frac{2(1.50-1)(0.0048)}{(0.02)^2}$$

$$= +12.00 \text{ D},$$

and for the back surface,

$$F_2 = \frac{2(1-1.50)(0.0028)}{(0.02)^2}$$

$$= -7.00 \text{ D}.$$

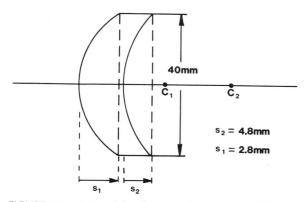

FIGURE 3-34. **Determining the approximate power of a lens on the basis of the sagittae and chord length.**

We find the approximate power by adding the two surface powers,

$$F_A = +12.00 - 7.00 = +5.00 \text{ D}.$$

It will be recalled that the *lens measure*, used to determine the refracting power of a lens surface, is constructed so that it has a constant chord length, $2h$, and is calibrated for a specific index of refraction. The above example, therefore, could have been solved by the use of the lens measure.

### 3.13. Formula Relating Power to Center Thickness and Edge Thickness

An approximate formula relating the power of an ophthalmic lens to the difference between center thickness and edge thickness may be derived with the help of Figure 3-35. As shown in this figure, the center thickness (also called polar thickness), $t_C$, and the edge thickness, $t_P$, can be related to the sagittae, $s_1$ and $s_2$, by the formula

$$t_C = s_1 + t_P - s_2,$$

or

$$t_C - t_P = s_1 - s_2, \tag{3.19}$$

remembering that $s$ is negative when measured (from arc to chord) toward the left. Solving Eq. (3.18) for $s$, we have

$$s = \frac{Fh^2}{2(n-1)},$$

and substituting this value for $s_1$ and $s_2$, we have

$$t_C - t_P = \frac{F_1 h^2}{2(n-1)} - \frac{F_2 h^2}{2(1-n)}$$

$$= \frac{F_1 h^2}{2(n-1)} + \frac{F_2 h^2}{2(n-1)}$$

$$= \frac{(F_1 + F_2)h^2}{2(n-1)}.$$

Since $F_1 + F_2$ is equal to approximate power, $F_A$, we have

$$t_C - t_P = \frac{F_A h^2}{2(n-1)}. \tag{3.20}$$

The use of this formula will be demonstrated by means of an example.

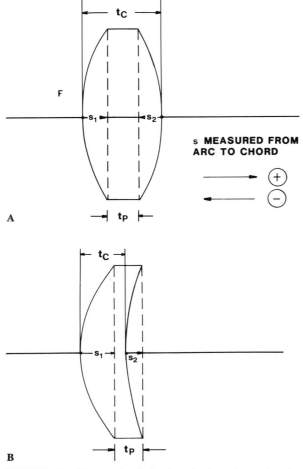

**A**

**B**

FIGURE 3-35. **Diagram for derivation of a formula relating the power of an opthalmic lens to the difference between the center thickness and the edge thickness.**

s **MEASURED FROM ARC TO CHORD**

$\oplus$

$\ominus$

*EXAMPLE*
Given a round lens, 50 mm in diameter, having a power of +1.00 D and an index of refraction of 1.523. What is the difference between the center thickness and edge thickness ($t_C - t_P$) of this lens?

$$t_C - t_P = \frac{+1\,(0.025)^2}{2\,(0.523)}$$

$$= 0.0005975 \text{ m}$$

$$= \text{approx. } 0.6 \text{ mm.}$$

On the basis of this result we can state the following *rule of thumb*: For every diopter of re-

fracting power, using a 50-mm lens blank, there will be a change in thickness of 0.6 mm between the center and the edge. Although this rule of thumb will not always give a precise result, it will provide an approximate result for both plus and minus lenses, for predicting an increase or a decrease in thickness.

*The Exact Sagitta Formula*

When $s$ is not small in relation to $r$, that is, when dealing with a *highly curved surface*, the approximate sagitta formula,

$$s = \frac{Fh^2}{2(n-1)},$$

will give a result that is considerably less than the true value. In such a case, the sagittal values of $s_1$ and $s_2$ in the formula

$$t_C - t_P = s_1 - s_2$$

must be calculated by the exact formula,

$$s = r - \sqrt{r^2 - h^2}.$$

An example will illustrate the difference in results when the approximate formula is used, rather than exact formula, in calculating sagittae and lens thickness.

*EXAMPLE*
A lens has the following parameters: $F_1 =$ +12.00 D, $F_2 = -6.00$ D, $n = 1.523$, edge thickness ($t_P$) = 1.0 mm, and lens diameter = 60 mm. What will be the center thickness, $t_C$, (1) using the approximate sagitta formula and (2) using the exact sagitta formula?

1. Using the approximate sagitta formula,

$$s_1 = \frac{Fh^2}{2(n-1)}$$

$$= \frac{+12\,(0.03)^2}{2(1.523 - 1)}$$

$$= \frac{+12(0.0009)}{1.046}$$

$$= 0.0103 \text{ m}$$

$$= 10.3 \text{ mm.}$$

$$s_2 = \frac{Fh^2}{2(1-n)}$$

$$= \frac{-6(0.03)^2}{2(1-1.523)}$$

$$= \frac{-6(0.0009)}{1.046}$$

$$= 0.00516 \text{ m } = 5.16 \text{ mm.}$$

$$t_C - t_P = s_1 - s_2$$

$$t_C = s_1 - s_2 + t_P$$

$$= 10.3 - 5.16 + 1.0$$

$$= 6.14 \text{ mm.}$$

2. Using the exact sagitta formula,

$$s_1 = r - \sqrt{r^2 - h^2}.$$

$$= \frac{1.523 - 1}{F} - \sqrt{\left(\frac{1.523-1}{F}\right)^2 - h^2}$$

$$= \frac{0.523}{12} - \sqrt{\left(\frac{0.523}{12}\right)^2 - (0.03)^2}$$

$$= 0.0436 - \sqrt{0.001899 - 0.0009}$$

$$= 0.0436 - \sqrt{0.001}$$

$$= 0.0436 - 0.03162$$

$$= 0.01198 \text{ m } = 11.98 \text{ mm.}$$

$$s_2 = r - \sqrt{r^2 - h^2}.$$

$$= \frac{1.523 - 1}{F} - \sqrt{\left(\frac{1.523-1}{F}\right)^2 - h^2}$$

$$= \frac{-0.523}{-6} - \sqrt{\left(\frac{-0.523}{-6}\right)^2 - (0.03)^2}$$

$$= 0.0872 - \sqrt{0.007598 - 0.0009}$$

$$= 0.0872 - \sqrt{0.0067}$$

$$= 0.0872 - 0.08185$$

$$= 0.00535 \text{ m } = 5.35 \text{ mm.}$$

$$t_C - t_P = s_1 - s_2$$

$$t_C = s_1 - s_2 + t_P$$

$$= 11.98 - 5.35 + 1.0$$

$$= 7.63 \text{ mm.}$$

By comparing the approximate value (6.14 mm) and the exact value (7.63 mm), we can deter-

mine the percentage error resulting from the use of the approximate formula:

$$\text{Percent error} = \frac{(7.63 - 6.14)100}{7.63}$$

$$= 19.51\%.$$

This demonstrates that, for highly curved surfaces, when lens thickness is determined by the use of the approximate sagitta formula, a large error results.

## 3.14. Thickness Calculations for Cylindrical and Sphero-Cylindrical Lenses

For a plano-cylindrical lens, Figure 3-36 shows that along the *axis* meridian thickness has a constant value, whereas along the *power* meridian thickness changes in the same manner that it does for a spherical lens. The thickness at any point, P, lying in an oblique meridian, OP, may be determined by either of two methods. For each of these methods, either the center thickness or the thickness of a specific peripheral point on the lens must be known. For example, if the edge thickness is known, the center thickness may be found by applying Eq. (3.20) to the power meridian of the lens.

1. The first method uses the perpendicular distance from the peripheral point, P, to the cylinder axis. If the peripheral point is described in relation to a specified point, O, on the cylinder axis along an oblique meridian at an angle, α, to the cylinder axis,

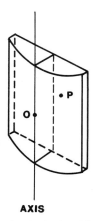

**AXIS**

FIGURE 3-36. **For a plano-cylindrical lens, thickness has a constant value along the axis meridian, but changes in the same manner as for a spherical lens in the power meridian.**

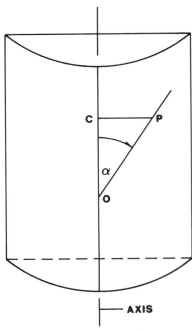

FIGURE 3-37. **Determining the perpendicular distance from a peripheral point, P, to a point, C, on the cylinder axis.**

as shown in Figure 3-37, the perpendicular distance from point P to point C on the cylinder axis is found by the use of equation

$$CP = OP \sin \alpha$$

The value of CP is then substituted for $h$, and the full power of the cylinder is used for $F$ in Eq. (3.20).

2. Using the second method, the power of the cylinder in the oblique meridian is calculated from equation

$$F_\alpha = F_C \sin^2 \alpha,$$

and the distance OP is substituted for $h$, while the value of $F_\alpha$ is substituted for $F$ in Eq. (3.20).

The statement was made in Section 2.14 that the concept of refracting power in an oblique meridian of a cylindrical lens is controversial, since a cylindrical lens has refracting power only in the *power meridian* (having zero power in the axis meridian). However, the concept of refracting power in an oblique meridian is useful—and completely valid—for the purpose of calculating the *thickness* at a point in an oblique meridian of a lens.

Both of the above methods for determining the thickness in an oblique meridian of a cylindrical lens will be illustrated by the use of an example.

*EXAMPLE*
Given a 46-mm round lens, centered in the eyewire, having a power of +5.00 DC axis 90, an index of refraction of 1.523, and an edge thickness of 1.0 mm at the *thinnest point* along the edge. Find the thickness at a point, P, located at the edge of the lens and such that a line drawn from this point to the center of the lens subtends an angle ($\alpha$) of 30° from the cylinder axis (see Figure 3-38).

For either method of solving the problem, we must first find the center thickness, $t_C$. Note that for a plus cylinder the thinnest point along the edge of the lens will be the point where the cylinder power meridian meets the edge of the lens, specified as point A. Therefore, for the power meridian (the 180° meridian),

$$t_C - t_P = \frac{h^2 F}{2(n-1)}$$

$$
\begin{aligned}
t_C &= \frac{h^2 F}{2(n-1)} + t_P \\[6pt]
&= \frac{(0.023)^2(+5)}{2(1.523 - 1)} + 0.001 \\[6pt]
&= \frac{0.002645}{1.046} + 0.001 \\[6pt]
&= 0.00253 + 0.001 \\[6pt]
&= 0.00353 \text{ m} = 3.53 \text{ mm.}
\end{aligned}
$$

Using the first method, the distance CP, or $h$, is found as follows:

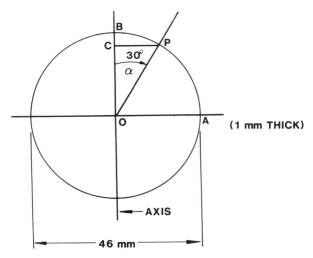

FIGURE 3-38. **Diagram used to find the thickness at point P (see text for explanation).**

$$CP = OP \sin \alpha$$

$$= OP \sin 30$$

$$= 23(0.5) = 11.5 \text{ mm.}$$

Since all points along the axis have the same thickness, that is,

$$t_O = t_C = t_B \ ,$$

we can find the thickness at point $t_P$ as follows:

$$t_C - t_P = \frac{h^2 F}{2(n-1)}$$

$$t_P = t_C - \frac{h^2 F}{2(n-1)}$$

$$= 0.00353 - \frac{(0.0115)^2(+5)}{1.046}$$

$$= 0.00353 - 0.00063$$

$$= 0.0029 \text{ m} = 2.9 \text{ mm.}$$

Using the second method, the power along OP is found as follows:

$$F = F_C \sin^2 \alpha = (+5) \sin^2 30°$$

$$= (+5)\,(0.25) = +1.25 \text{ D.}$$

Since the distance OP is 23 mm, we can find the thickness at P by

$$t_P = t_C - \frac{h^2 F}{2(n-1)}$$

$$= 0.00353 - \frac{(0.023)^2(+1.25)}{1.046}$$

$$= 0.00353 - 0.00063$$

$$= 0.0029 \text{ m} = 2.9 \text{ mm,}$$

which gives the same result as the first method.

Either of these two methods may be used to calculate thickness changes in cylindrical lenses. Remember that if the perpendicular distance from the peripheral point is found, the full power of the cylinder must be used in Eq. (3.20). However, if the oblique distance from a perpendicular point is used, the power along the oblique meridian must be used in Eq. (3.20).

To determine the thickness at a point in the periphery of a *sphero-cylindrical* lens, it is necessary to consider both the spherical and cylindrical components. Using $F_S$ to designate the spherical power and $F_C$ to designate the cylinder power, the thickness due to both the spherical and cylindrical components, at a peripheral point P located at a distance $h$ from the center of the lens along a meridian oriented at $\alpha$ degrees from the cylinder axis, may be found by the use of the formula

$$t_C - t_P = \frac{h^2(F_S + F_C \sin^2 \alpha)}{2(n-1)} \qquad (3.21)$$

## Reference

1. Levene, J. R. *Clinical Refraction and Visual Science*, p. 45. Butterworths, London, 1977.

## Questions

1. A ray incident upon a refracting surface of index 1.50 is headed toward a point on the optic axis 20 cm behind the surface. After refraction by the surface, it emerges parallel to the optic axis. What is the refracting power of the surface?

2. If a 7-mm-thick lens has a +16.00 D front surface and a −6.00 D back surface and if it is made out of glass of index of refraction 1.523, what is its back vertex power?

3. A lens has back vertex power of +4.00 D and an ocular surface of −6.00 D. If the lens is 3.00 mm thick and the index of refraction is 1.50, what is the front surface power?

4. If the power of the front surface of a lens with an index of refraction of 1.523 is +12.00 D and its center thickness is 10 mm, what must be the power of the back surface to make the lens have:
   (a) zero approximate power?
   (b) zero vertex power?
   (c) zero neutralizing power?
   (d) +10.00 D vertex power?

5. A −10.00 D lens gives the desired optical effects when in a given position in front of the eye. What new lens 8 mm nearer the eye is necessary to give the same result?

6. A +13.00 DS −3.00 DC x 180 lens gives the desired correction when in a given position in front of the eye. If the lens is to be worn 5 mm closer to the eye, what lens power is necessary to give the same result?

7. The equivalent power of a thick lens is +12.50 D. The back vertex power is +15.00 D. What is the distance of the second principal plane from the back vertex of the lens?

8. By hand neutralization, a single-vision lens is neutralized along the 15° meridian by a +2.25 D trial lens and along the 105° meridian by a +0.50 D trial lens. What is the prescription of the lens being neutralized?

9. A certain lensometer manufactured to have a standard lens of +20.00 D was accidentally fitted with a +25.00 D standard lens. With the target in focus the wheel was then adjusted to read zero power when no lens was before the lens stop. If a +12.50 D lens is placed before the lens stop, what will be the reading on the wheel with the target again in focus?

10. A lensometer with a standard lens of +20.00 D is incorrectly constructed in that the lens rest is 1 cm too close to the standard lens. With no lens before the lens stop, the wheel is adjusted to read zero. If a +20.00 D lens is placed before the lens rest, what will be the reading when the target is again brought into focus?

11. In using a lensometer-type instrument, it is found that the target moves 8 mm when focusing between the two principal meridians of a +3.00 DS +5.00 DC x 90 lens. What is the power of the "standard lens" in the instrument?

12. The standard lens of a lensometer is +20.00 D. The single line (sphere) of the target is in focus 10 mm behind the plano position. The triple or cylinder line is in focus 5 mm in front of the plano position. If the axis indicator is 90°, what is the lens prescription?

13. A lens of unknown power is placed in a vertometer and the target is moved until the image of the target is seen clearly on the reticle. The distance from the standard lens to the target is now 60 mm. The power of the standard lens is +20.00 D. What is the power of the unknown lens?

14. A front surface bifocal lens is placed with the ocular surface against the lens stop of a lensometer. Through the distance portion of the lens the readings are as follows: single line reading is +2.00 D; triple line reading is +3.25 D; axis dial reading is 15; through the segment portion the single line reading is +4.25 D. With the lens reversed (the front surface against the lens stop) the readings are as follows: single line reading through the distance portion is +2.25 D; single line reading through the segment portion is +4.25 D. What is the prescription of this lens?

15. A lens with a center thickness of 3 mm is measured with calipers and the minimum edge thickness is found to be 2 mm in the 30° meridian and the maximum edge thickness is 5 mm in the 120° meridian. The lens is 40 mm in diameter and has an index of 1.50. Write the prescription for this lens.

16. What is the edge thickness of a 42-mm, −3.75 D lens with a 0.7-mm center thickness? The index of refraction is 1.523.

17. What is the center thickness of a 50-mm, +4.00 D lens with a 2-mm edge thickness? The index of refraction is 1.523.

18. What is the highest minus power possible with a 40-mm-diameter flat blank which is 5 mm thick? The center thickness of the finished lens is to be 1 mm thick. The index of refraction is 1.523.

19. If the vertex power required is +4.00 D and on request the center thickness was increased from 5 mm to 8 mm without consideration for this change, what will be the resultant change in power? The back surface power is −10.00 D and the index of refraction is 1.523.

20. A meniscus lens has surface powers of +10.00 D and −6.00 D. It is circular and its diameter is 40 mm and its edge thickness is 0.5 mm. The index of refraction is 1.5. Find the thickness of the thinnest flat plate of glass from which the lens can be ground.

# CHAPTER FOUR

# Ophthalmic Prisms and Decentration

*[handwritten notes:]*

RELATIVE DISPERSION

$$V = \frac{n_d - 1}{n_f - n_c}$$

DISPERSION

$$\frac{n_f - n_c}{n_d - 1}$$

In geometrical optics we learned that a prism has the property of changing the direction of a beam of incident light without changing the vergence of the light.

Although the primary function of an ophthalmic lens is to compensate for a refractive anomaly by changing the vergence of incident light, it is sometimes necessary to incorporate a prismatic component in the lens, in order to compensate for an anomaly of binocular vision. The prismatic component may be placed with its base-apex meridian in any desired direction: in the horizontal meridian (as base-in or base-out prism), in the vertical meridian (as base-up or base-down prism), or in any oblique meridian.

An obvious method of incorporating prismatic power into an ophthalmic lens is to grind the prism into the lens, in the surfacing process. However, even a spherical or cylindrical lens will induce prismatic power whenever the visual axis passes through any point in the lens other than its optical center: Thus, a second method of providing prismatic power in an ophthalmic lens is to "decenter" the lens.

In this chapter, the terminology and optical properties of prisms will first be presented. This will be followed by discussions of units of measurement; the effects of prisms on the eyes; prismatic effects of spherical and cylindrical lenses; the resolving of oblique prisms; prescribing prism by decentration; the effects of prism power on the thickness of a lens; and the use of Fresnel Press-on prisms.

## 4.1. Terminology

A *prism* is defined as a portion of a transparent isotropic substance included between two polished, nonparallel surfaces. Although we usually think of a prism as having flat surfaces, ophthalmic prisms usually have curved, rather than flat, surfaces, and may be used in conjunction with spherical, cylindrical, or sphero-cylindrical power.

A prism has a thick portion, called the *base*, and a thin portion, called the *apex*. The straight line at the apex, in which the two faces of the prism meet, is known as the *edge* of the prism. The included angle between the two surfaces forming the edge is known as the *refracting angle*, and is denoted by the Greek letter β (beta). A *principal section* of a prism is a section made by a plane perpendicular to the edge of the prism. In Figure 4-1, the plane of the paper is a principal section, and the edge of the prism is a line extending perpendicularly in front of (and behind) the paper, at the apex. Figure 4-2 is a perspective view of a plano-prism with plane surfaces, indicating the refracting angle, β, the edge, and the base-apex

direction. In ophthalmic optics we are concerned only with rays of light traveling in principal sections: Other rays are called *skewed* rays, and are of no interest to us.

## 4.2. Refracting Power of a Prism

The angle of deviation of light brought about by a prism is denoted by the Greek letter ε (epsilon). As shown in Figure 4-3, rays of light are always deviated toward the *base* of the prism. However, for an observer viewing an object through the prism, the object appears to be displaced toward the *apex* of the prism.

With the aid of Figure 4-4, it is possible to derive the formula for the refracting power of a thin prism. When rays of light strike the first face, AB, of the prism, there is a change in the speed of light, but since the rays of light strike the prism at an angle of 90°, there is no deviation of the light. However, when light strikes the second face, BC, of the prism, there is

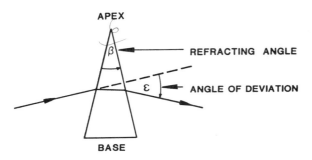

FIGURE 4-1. **A principal section of a prism. The edge of the prism is perpendicular to the plane of the paper, at the apex.**

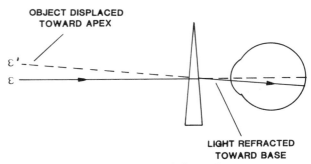

FIGURE 4-3. **The action of an opthalmic prism.**

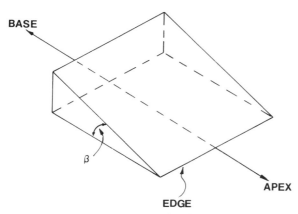

FIGURE 4-2. **A perspective view of a plano-prism.**

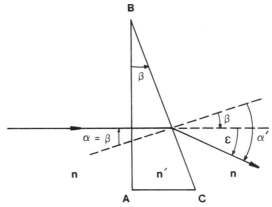

FIGURE 4-4. **Diagram for deriving the formula for refraction by a thin prism.**

a deviation of the light, since it does not strike the face at a 90° angle. Constructing a perpendicular to the surface BC, we can indicate the refracted ray, which makes an angle, ε, from the incident ray. By inspection of the diagram, we note that we can find two additional angles that are equal to the refracting angle, β. We also note that the angle of refraction at the surface BC is equal to angle β plus angle ε. Applying Snell's law,

$$n \sin \alpha = n' \sin \alpha'.$$

Since, in this diagram, α is equal to β and α' is equal to β + ε, we have

$$n' \sin \beta = n \sin(\beta + \epsilon).$$

If we now limit our consideration to thin prisms, having refracting angles of about 10° or less, we can make the approximation that an angle (in radians) is equal to its sine, with the result that

$$n'\beta = n\beta + n\epsilon$$

For a thin prism in air, $n = 1$, and the index of refraction of the material from which the prism is made can be represented by $n$ (rather than $n'$). The above equation then becomes

$$n\beta = \beta + \epsilon$$

and

$$\epsilon = \beta(n - 1). \tag{4.1}$$

In this formula, the angles ε and β may be expressed in either radians or degrees. For a medium having an index of refraction of 1.50, in air, this formula reduces to

$$\epsilon = \beta(1.5 - 1) = \frac{\beta}{2},$$

indicating that the angle of deviation of an ophthalmic prism is equal to half the refracting angle of the prism.

## 4.3. Specification of the Power of an Ophthalmic Prism

Although the refracting angle, β, of a prism is always specified in *degrees*, the angle of deviation, ε, or *refracting power* of an ophthalmic prism, is usually specified in *prism diopters*.

A prism diopter is defined as a deviation of 1 unit at a distance of 100 units. Although any unit can be used, it is convenient to think of the deviation in *centimeters*, in which case the prism diopter can be defined as a deviation of 1 cm at a distance of 1 m.

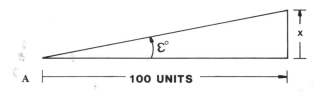

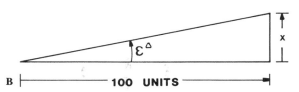

FIGURE 4-5. **The prism diopter: A, angle of deviation given in degrees; B, angle of deviation given in prism diopters.**

To derive an expression for the deviation brought about by a prism, in prism diopters, we begin with Eq. (4.1), which states that the deviation in degrees is equal to the refracting angle of the prism times the index difference,

$$\epsilon° = \beta°(n - 1).$$

Referring now to Figure 4-5A, we have an angle of deviation, ε°, such that the displacement is $x$ units for a distance of 100 units. Inspection of the diagram shows that

$$\tan \epsilon° = \frac{x}{100}$$

or

$$x = 100 \tan \epsilon°.$$

If we now state the angle of deviation in terms of prism diopters, as shown in Figure 4-5B, we can substitute ε for $x$ (since we have defined a prism diopter as a deviation of 1 unit at a distance of 100 units), with the result that

$$\epsilon^\Delta = 100 \tan \epsilon°, \tag{4.2}$$

$$\epsilon^\Delta = 100 \tan \beta°(n - 1), \tag{4.3}$$

and

$$\tan \beta° = \frac{\epsilon^\Delta}{100(n - 1)}. \tag{4.4}$$

Equation (4.4) is not an exact expression but, as will be shown later, involves very little error for small angles. It should be understood that, whereas the angle ε can be stated in either degrees or prism diopters, the refracting angle of the prism, β, is always stated in degrees.

It should also be understood that the prism diopter differs from the degree or the radian in being a unit of *tangent measurement*, whereas the degree and the radian are units of *arc measurement*. As illustrated in Figure 4-6, a radian is the angle subtended by an arc that is equal to the radius of a circle. In this diagram, the length of the arc BC is equal to either of the radii of the circle, AB or AC. Since there are $2\pi$ radians in a circle, we can say that

$$1 \text{ radian} = \frac{180}{\pi} = 57.5°$$

and that

$$1° = \frac{\pi}{180} = 0.0175 \text{ radian}.$$

Another unit of arc measurement is the *centrad*, which is defined as the angle subtended by an arc equal to 1/100th of the radius of a circle (or, simply, equal to 1/100th of a radian). The relationship between the

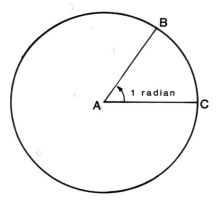

FIGURE 4-6. **An angle of 1 radian. The length of the arc BC is equal to the length of the radius, AB or AC.**

centrad and the prism diopter is illustrated in Figure 4-7. It can be seen that, for small angles, the centrad and the prism diopter are essentially equal.

*Relationship between Degrees and Prism Diopters*

For small angles, an angle specified in degrees may be converted to prism diopters, or vice versa, with very little error. In the case of the conversion of degrees to prism diopters,

$$1° = 0.0175 \text{ radian}$$
$$= 1.75 \text{ centrads}$$

and

$$1° = 1.75 \text{ prism diopters.} \qquad (4.5)$$

In the case of conversion from prism diopters to degrees,

$$1 \text{ prism diopter} = 1 \text{ centrad}$$
$$= \frac{1}{1.75°}$$

and    *degree (.57)*

$$1 \text{ prism diopter} = 0.57° \qquad (4.6)$$

As shown in Figure 4-8, as an angle increases in size, the tangent of the angle (which is analogous to prism diopters) increases more rapidly than the angle itself (the arc angle, expressed in degrees, centrads, or radians). We may now ask: How large must an angle be before it is no longer accurate to assume that prism diopters and centrads are equal?

Reference to Table 4-1 indicates that the error involved in assuming that prism diopters and centrads are equal increases from −0.1% for an angle of 10° to −8.3% for an angle of 30° and to −55.8% for an angle of 57.3° (which is equal to 100 centrads, or 1

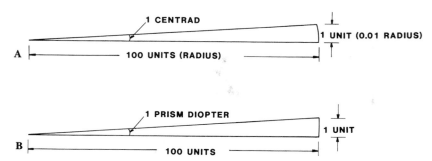

FIGURE 4-7. **The relationship between (A) the centrad (arc measurement) and (B) the prism diopter (tangent measurement). Not to scale.**

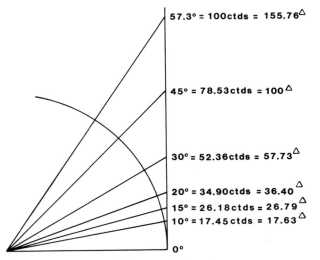

FIGURE 4-8. **As an angle increases in size, the tangent measurement (prism diopters) increases more rapidly than the arc measurement (degrees, radians, or centrads).**

TABLE 4—1
**Comparison of Degrees, Centrads, and Prism Diopters[a]**

| Degrees | Centrads | Prism Diopters | Percentage Error[b] |
|---|---|---|---|
| 10 | 17.45 | 17.63 | −0.1 |
| 15 | 26.18 | 26.79 | −0.8 |
| 20 | 34.90 | 36.40 | −1.4 |
| 30 | 52.36 | 57.73 | −8.3 |
| 45 | 78.53 | 100.00 | −27.3 |
| 57.3 | 100.00 | 155.76 | −55.8 |

[a]Where centrads = degrees/0.573, and prism diopters = 100 tan ε°.
[b]If it is assumed that prism diopters are equal to centrads.

radian). Since ophthalmic prisms are almost never prescribed in powers greater than 5 prism diopters (2.9°) for each eye or a total of 10 prism diopters (5.7°) for both eyes, the error involved in making the conversion is not significant.

The measurement of phorias and fusional vergences, however, involves the use of prisms in powers as high as 25–40 prism diopters (14.04° to 21.8°), with the result that an error on the order of 1 to 1.5% is involved in these measurements if one makes the assumption that prism diopters are equal to centrads. The matter is further complicated by the fact that rotary prisms used in phoroptors or refractors are located at a great enough distance from the eyes to bring about a decrease in the *effectivity* of such prisms when used for near testing distances (this will be discussed in Section 4.17).

## 4.4. Relationship between Refracting Angle and Angle of Deviation

As stated in Eq. (4.1), the angle of deviation of a prism is related to the refracting angle by the expression

$$\varepsilon^\circ = \beta^\circ(n - 1)$$

with the result that for a medium having an index of refraction of 1.5, in air,

$$\varepsilon^\circ = \frac{\beta^\circ}{2},$$

and for ophthalmic crown glass, having an index of refraction of 1.523,

$$\varepsilon^\circ = \beta(0.523).$$

If $\beta = 1°$,

$$\varepsilon^\circ = 1(0.523)$$
$$= 0.523$$

since

$$\varepsilon^\Delta = 100 \tan \beta(n - 1)$$

when $\beta = 1°$

$$\tan \beta = 0.0175$$

and

$$\varepsilon^\Delta = 100(0.0175)(0.523)$$
$$\varepsilon^\Delta = 0.915.$$

As a rule of thumb, for crown glass having an index of refraction of 1.523, a prism having a refracting angle of 1 degree produces *approximately 1 prism diopter* of deviation.

## 4.5. Effects of Prisms on Movements of the Eyes

The effects of prisms on movements of the eyes may be considered in terms of either *monocular* or *binocular* effects, and binocular effects may be considered in terms of resultant *horizontal* prismatic effects or in terms of resultant *vertical* prismatic effects.

**Monocular Prismatic Effects.** If one looks at a distant object with one eye occluded while a prism is introduced before the eye, the image of the object will be displaced toward the apex of the prism. If the person wishes to look at the image of the object, the eye must move through an angle equal to the angle of deviation of the prism (Figure 4-9). Base-in prism

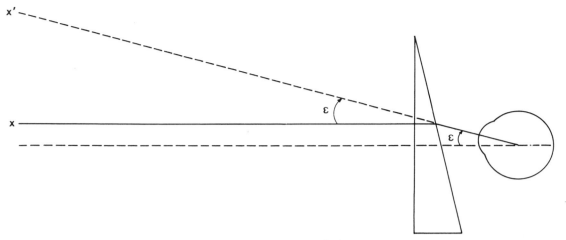

FIGURE 4-9. **The image of an object seen through a prism will be displaced toward the apex, at an angle equal to the angle of deviation of the prism.**

would cause the eye to move outward (abduction), and base-out prism would cause the eye to move inward (adduction).

**Binocular Prismatic Effects.** If, with both eyes open, one looks at a distant object while a base-in prism is placed before the right eye and a base-out prism of equal amount is placed before the left eye, in order to look at the object both eyes will move to the *right* by an equal amount, and the lines of sight of the two eyes will remain parallel. The movements of the eyes in which both eyes move in the same direction and of an equal amount are called *conjugate* movements, or *version* movements.

However, if a small amount of base-out prism is placed before each eye, in order to see a single image both eyes will move *inward* by the same amount. Similarly, base-in prism placed before each eye will cause the eyes to move *outward*. Eye movements in which the eyes move toward one another (convergence) or away from one another (divergence) are called *disjunctive* movements, or *vergence* movements.

If prisms bring about only conjugate movements, the *resultant prismatic effect* is considered to be equal to zero. However, if prisms cause a disjunctive or vergence movement, the *resultant prismatic effect* is the extent of the disjunctive movement.

**Resultant Horizontal Prismatic Effects.** When prisms are placed before each eye with their bases in the same direction for each eye (that is, both base in or both base out), the resultant prismatic effect can be found by adding the powers of the prisms (base in added to base in, or base out added to base out). If the bases are in opposite directions but of different powers, the power of the weaker prism is subtracted from that of the stronger prism, and the remainder

is the resultant prismatic effect. These principles will be illustrated by examples.

***EXAMPLE 1***
Given the prisms placed before the eyes,

> OD  3 prism diopters, base out
> OS  3 prism diopters, base out.

The resultant prismatic effect is found by adding the powers of the two prisms, and is therefore 6 prism diopters, base out.

***EXAMPLE 2***
Given the prisms placed before the eyes,

> OD  4 prism diopters, base in
> OS  2 prism diopters, base out

The resultant prismatic effect is found by subtracting the power of the weaker prism from that of the smaller prism, and is therefore 2 prism diopters, base in, OD.

**Resultant Vertical Prismatic Effects.** When vertical prisms are placed before each eye with their bases in the *same direction* for each eye but are of different powers, the resultant vertical prismatic effect is found by subtracting the power of the weaker prism from that of the stronger prism. The base of the resultant vertical prismatic effect is expressed as applying to the eye with the stronger prism. If the bases are in *opposite directions*, the resultant vertical prismatic effect is found by numerically adding the powers of the two prisms. Since a base-up effect for one eye is the same as a base-down effect for the other eye, the total prismatic effect may be expressed with respect to either eye (base-up for one eye or base-down for the other eye). These principles will be illustrated by examples.

### EXAMPLE 1

Given the prisms placed before the eyes,

> OD  3 prism diopters, base up
> OS  1 prism diopter,  base up.

The resultant prismatic effect is found by subtracting the power of the weaker prism from that of the stronger prism, and is therefore 2 prism diopters, base up, OD.

### EXAMPLE 2

Given the prisms placed before the eyes,

> OD  2 prism diopters,  base down
> OS  2 prism diopters,  base up.

The resultant prismatic effect is found by adding the powers of the two prisms and can therefore be expressed as 4 prism diopters, base down, OD, or as 4 prism diopters, base up, OS.

## 4.6. Prentice's Rule

A spherical lens can be considered as being made up of an infinite number of prisms stacked base to base (for a plus lens) or apex to apex (for a minus lens), as shown in Figure 4-10. For both plus and minus lenses, the power of a prism increases from the pole of the lens toward the periphery. For a given lens, the prismatic effect is a function of both the

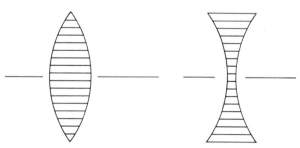

FIGURE 4-10. **A spherical lens considered as an infinite number of prisms stacked base to base (plus lens) or apex to apex (minus lens).**

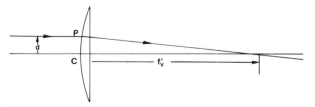

FIGURE 4-11. **Diagram used to derive Prentice's rule.**

distance of the point from the pole of the lens and the power of the lens.

This relationship, based upon the definition of the prism diopter and illustrated in Figure 4-11, is called *Prentice's rule*. As shown in this illustration, an incident ray from a distant object passing through a lens at a distance, $d$, from the lens pole, C, will cross the optic axis at the secondary focal point of the lens. For small angles,

$$\tan \varepsilon = \varepsilon_{rads} = \frac{d_m}{f_{V}'} = d_m F_V$$

and

$$\varepsilon_{rads} = d_m F_V$$

$$\varepsilon_{centrads} = 100 \, d_m F_V$$

$$= d_{cm} F_V$$

or since for thin prisms, prism diopters are equal to centrads,

$$\varepsilon_{prism \, diopters} = d_{cm} F_V. \qquad (4.7)$$

Prentice's rule simply states that the prismatic effect at any point on a spherical lens is equal to the distance of the point from the pole of the lens, in centimeters, multiplied by the power of the lens, or

$$P = dF, \qquad (4.8)$$

where $P$ = the prismatic power, in prism diopters; $d$ = the distance from the lens pole, in centimeters; and $F$ = the refracting power of the lens.

### EXAMPLE

A pair of spectacles having the prescription +3.00 DS, each eye, fits on the wearer's face in such a way that when used for distance vision the foveal line of sight for each eye passes through a point 5 mm nasal to the optical center, or pole, of the lens. What is the prismatic effect encountered by each eye?

$$P = dF$$

$$= 0.5 \, (+3.00)$$

$$= 1.5 \text{ prism diopters.}$$

In addition to being able to calculate the amount of prismatic power induced by an ophthalmic lens, it is important to know the direction of the *base* of the prism.

Specification of the direction of the base of a prism differs from specification of the axis of a cylinder, in the following manner. The position of the axis of a cylinder can be properly specified by using only the *upper half* of a circle (from 0° to 180°), with the zero position to the examiner's right, looking toward the patient, and with degree values increasing in a counterclockwise direction. Since the cylinder axis extends across the entire lens, the meridians 90° and 270° represent the same cylinder axis position. However, the base of a prism can assume *any position* on a circle and hence can vary in notation from 0° to 360°, as shown in Figure 4-12.

Although for a cylinder 0° and 180° identify the same axis position, by convention 180° is preferred to 0°. By contrast, a prism base will have *opposite* effects if specified at 0° (base in for the right eye or base out for the left eye) as opposed to 180° (base out for the right eye and base in for the left eye).

Oblique directions of prisms are most often described clinically in terms of horizontal and vertical components, using the cardinal directions, up, down, in, and out. *Base in* means the base of the prism is toward the nose (nasally), while *base out* means the base is away from the nose (temporally). For convenience, a circle for specifying the base of a prism is often divided into four quadrants as shown in Figure 4-13. It should be noted that for the *right eye*, in the 0° to 90° quadrant the base is *up and in*; for the 90° to

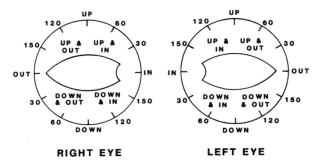

RIGHT EYE          LEFT EYE

FIGURE 4-14. **The English system of prism base notation, using two hemicircles.**

180° quadrant it is *up and out*; for the 180° to 270° quadrant it is *down and out*; and for the 270° to 360° quadrant it is *down and in*. For the *left eye* the quadrants are reversed, 0° to 90° being *up and out*; 90° to 180° being *up and in*; 180° to 270° being *down and in*; and 270° to 360° being *down and out*.

Another system, often used in England, is illustrated in Figure 4-14. In this system two 180° scales are used, one corresponding to the upper hemicircle and the other to the lower hemicircle. The zero position for the upper hemicircle starts on the right-hand side, while the zero position for the lower hemicircle starts on the left-hand side. Using the system illustrated in Figure 4-12, a prism for the right eye with a base located at 210° would be expressed, in the English system, as "30° down and out." Throughout this book the system of notation using the *entire 360°* will be used.

At any point other than the optical center, or pole, of a spherical lens, prism is induced according to Prentice's rule. The prismatic effect is found, as shown in the above example, simply by multiplying the distance (in centimeters) separating the point in question from the optical center by the power of the lens. The direction of the base−apex line of the prism lies along the line connecting the point in question to the optical center. The direction of the base is found by examining the *change in thickness* immediately surrounding the point along the line connecting the point to the optical center. The prism base direction is always the direction of *increasing thickness*. It is important to examine only the thickness changes on the line segment immediately surrounding the point in question. The direction of the base is then *projected upon a circle*, ignoring the remainder of the thickness changes of the lens.

One situation in which induced prismatic power is of importance occurs when a spectacle wearer *converges* to read while wearing lenses that are "cen-

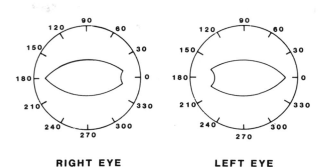

RIGHT EYE          LEFT EYE

FIGURE 4-12. **The base of a prism can assume any position on a circle, varying in notation from 0° to 360°.**

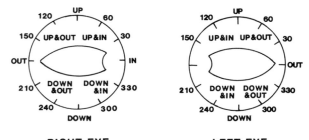

RIGHT EYE          LEFT EYE

FIGURE 4-13. **A circle for specifying the base of a prism may be divided into four quadrants.**

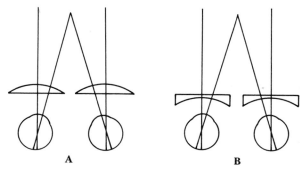

FIGURE 4-15. **Prismatic effects due to convergence of the eyes (when the lenses are centered for distance vision): A, base-out prismatic effect due to plus lenses; B, base-in prismatic effect due to minus lenses.**

tered" for the distance PD (interpupillary distance). As shown in Figure 4-15A, a wearer of plus lenses experiences a *base-out* (temporalward) prismatic effect when converging to read and, as shown in Figure 4-15B, a wearer of minus lenses experiences a *base-in* (nasalward) prismatic effect when converging to read (assuming in both cases that the lenses are centered for the distance PD). The amount and direction of this effect will be demonstrated by the use of examples, using Prentice's rule.

### EXAMPLE 1

Given a patient whose prescription is

$$OD \quad +3.00 \text{ DS}$$
$$OS \quad +3.50 \text{ DS}$$

having a distance PD of 64 mm and a near PD of 60 mm. If the lenses are centered for the distance PD, what is the prismatic effect experienced for reading, for each eye separately and for both eyes?

OD    $P = 0.2(3) = 0.6$ prism diopter, base out.
OS    $P = 0.2(3.5) = 0.7$ prism diopter, base out.

Both eyes: $P = 0.6 + 0.7 = 1.3$ prism diopters, base out.

### EXAMPLE 2

Given a patient whose prescription is

$$OD \quad -5.00 \text{ DS}$$
$$OS \quad -6.00 \text{ DS}$$

having a distance PD of 70 mm and a near PD of 65 mm. If the lenses are centered for the distance PD, what is the total prismatic effect for near work?

OD    $P = 0.25(5) = 1.25$ prism diopters, base in.
OS    $P = 0.25(6) = 1.50$ prism diopters, base in.

Both eyes: $P = 1.25 + 1.50 = 2.75$ prism diopters, base in.

When a patient looks upward or downward (through a point in each lens above or below the lens pole), a vertical prismatic effect is induced, as shown in the next example.

### EXAMPLE 3

Given a patient whose prescription for reading is

$$OD \quad +2.00 \text{ DS}$$
$$OS \quad +2.00 \text{ DS}$$

Since the lenses are worn only for reading, they are centered for the near PD. However, when the patient reads, each line of sight passes through a point in the lens 5 mm below the pole of the lens. What is the vertical prismatic effect for each eye and for both eyes (see Figure 4-16)?

OD $P = 0.5(2) = 1$ prism diopter, base up.
OS $P = 0.5(2) = 1$ prism diopter, base up.

In considering the vertical prismatic effect for both eyes, what we are concerned with is not the total prismatic effect (as in the case when adding base-in or base-out effects for the two eyes) but the *resultant* prismatic effect. We do this by asking the question "Is there a greater vertical prismatic effect for one eye than for the other eye?" In this example, the vertical prismatic effect is the same for each eye, so the resultant vertical prismatic effect is zero.

If, on the other hand, the patient's reading prescription is $+2.00$ DS for the right eye and $+3.00$ DS for the left eye, the vertical prismatic effect experienced by the left eye is

$$P = 0.5(3) = 1.5 \text{ prism diopters, base up,}$$

and the resultant prismatic effect is

$$1.5 \text{ base up} - 1.0 \text{ base up,}$$

or

$$0.5 \text{ prism diopter, base up, left eye.}$$

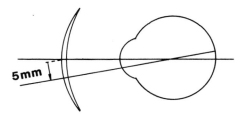

FIGURE 4-16. **A base-up prismatic effect is induced when a wearer of plus lenses reads through a point below the pole of the lens.**

## 4.7. Oblique Prismatic Effects

Our discussion of prismatic effects, so far, has been limited to effects in which the base-apex line is located in either the horizontal or the vertical meridian. However, oblique prismatic effects occur whenever the line of sight passes through a portion of the lens not horizontally or vertically in line with the pole of the lens. Such an effect is illustrated in the two diagrams in Figure 4-17, in which the line of sight passes, respectively, through points A, B, C, and D, each being equidistant from the center, O, for a plus lens, P (Figure 4-17A), and for a minus lens, M (Figure 4-17B).

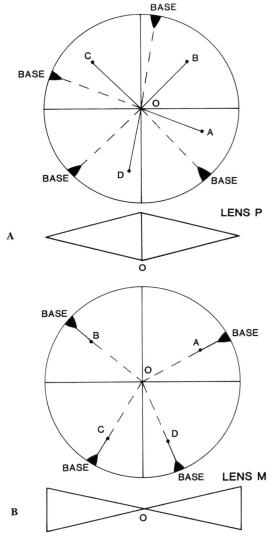

FIGURE 4-17. **Oblique prismatic effects: A, for a plus lens; B, for a minus lens.**

If we first consider the plus lens, it is clear that the prismatic effect for each point is equal to the distance (AO, BO, CO, or DO) multiplied by the power of the lens. Since the lens is a plus lens, the thickness decreases from point O outward. As we examine the thickness changes immediately surrounding points A, B, C, and D, it is obvious that the region central to each of these points is thicker than the region peripheral to each of the points. Hence the base direction is projected toward the center, O, and *through* the point O onto a circle surrounding the lens. It should be remembered that for a plus lens the projection must be made through the center, toward the perimeter of the circle.

If we now consider the minus lens, it is apparent that the prismatic effect for each point is equal to the distance (AO, BO, CO, or DO) multiplied by the power of the lens. Since the lens is a minus lens, the thickness increases from the point O outward. As we examine the thickness immediately surrounding the points A, B, C, and D, it is apparent that the region immediately peripheral to each of these points is thicker than the region immediately central to each of the points. Hence, the base direction is projected on a circle surrounding the lens.

The following example illustrates the procedures used in locating the base direction of a prismatic effect for plus and minus lens.

### EXAMPLE

Given a patient wearing a plus lens before the right eye and a minus lens before the left eye, as shown in Figure 4-18. Assume the patient moves his eyes to the point P on each lens, by looking 2 cm to the left and 1 cm upward from the pole, C. In each diagram, the direction of the base of the prism is indicated by an arrow. Inspection of the diagram for the right lens (Figure 4-18A) indicates that the tangent of angle $\theta$ is 0.5, with the result that angle $\theta = 180°$ plus 26.6°, or 206.6°,

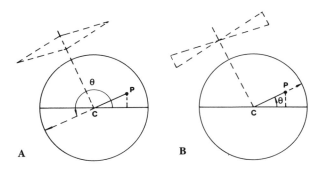

FIGURE 4-18. **Oblique prismatic effects for a patient wearing a plus lens before the right eye (A) and a minus lens before the left eye (B).**

and hence the base of the prism is located at 206.6°. For the left lens (Figure 4-18B), angle θ is equal to 26.6°, with the result that the base of the prismatic effect is located at 26.6°.

*Horizontal and Vertical Components of Oblique Prisms*

Another method of specifying the power of an oblique prism is to resolve the prism into horizontal and vertical components. Referring to Figure 4-19, ε represents an oblique prism with its base at θ, and $\varepsilon_h$ and $\varepsilon_v$ are its horizontal and vertical components. Expressions for these components are

$$\text{horizontal component, } \varepsilon_h = \varepsilon \cos \theta \qquad (4.9)$$

$$\text{vertical component, } \quad \varepsilon_v = \varepsilon \sin \theta. \qquad (4.10)$$

In a similar fashion two prismatic effects or components, one horizontal and one vertical, can be reduced to a single prismatic effect by the use of vector analysis. Since

$$\varepsilon_h = \varepsilon \cos \theta,$$

$$\varepsilon = \frac{\varepsilon_h}{\cos \theta}, \qquad (4.11)$$

and since

$$\varepsilon_v = \varepsilon \sin \theta,$$

$$\varepsilon = \frac{\varepsilon_v}{\sin \theta}. \qquad (4.12)$$

If both $\varepsilon_h$ and $\varepsilon_v$ are known, the prismatic effect, ε, can be found by making use of the Pythagorean theorem,

$$\varepsilon = \sqrt{\varepsilon_h{}^2 + \varepsilon_v{}^2}. \qquad (4.13)$$

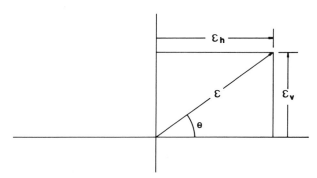

FIGURE 4-19. **Prismatic effects considered in terms of horizontal and vertical components.**

### EXAMPLE 1

What are the horizontal and vertical prismatic components of a prism having a power of 3 prism diopters base at 120°, placed before the right eye?

Since the base falls in the second quadrant, the bases of the horizontal and vertical components will be out and up, respectively. Since the angle of the prism base is 60° less than the horizontal meridian (that is, 120 = 180 − 60), the value of θ will be 60°.

$$\sin \theta = \frac{\varepsilon_v}{\varepsilon} \text{ and } \cos \theta = \frac{\varepsilon_h}{\varepsilon}.$$

$$\sin 60 = \frac{\varepsilon_v}{3}$$

$$0.866 = \frac{\varepsilon_v}{3}$$

$$\varepsilon_v = 2.6 \text{ prism diopters, base up,}$$

and

$$\cos 60 = \frac{\varepsilon_h}{3}$$

$$0.5 = \frac{\varepsilon_h}{3}$$

$$\varepsilon_h = 1.5 \text{ prism diopters, base out.}$$

### EXAMPLE 2

What single prism can replace the combination of a prism having a power of 4 prism diopters base in and a prism having a power of 2 prism diopters base down, for the right eye?

It is apparent that the base of the single prism will fall in the fourth quadrant.

$$\varepsilon = \sqrt{\varepsilon_h{}^2 + \varepsilon_v{}^2}$$

$$= \sqrt{16 + 4}$$

$$= \sqrt{20}$$

$$= \quad 4.47 \text{ prism diopters.}$$

Let angle θ be the angle from the prism base to the 360° meridian. Consequently,

$$\sin \theta = \frac{\varepsilon_v}{\varepsilon}$$

$$= \frac{2}{4.47}$$

$$= 0.447$$

and

$$\theta = 26.58°,$$

with the result that the direction of the prism base is

$$360 - 26.58 = 333.42°.$$

Hence, the single prism can be specified as follows:

4.47 prism diopters, base at 333.42°.

## 4.8. Obliquely Crossed Prisms

If a lens contains two prisms crossed at oblique angles, each prism is treated separately, being resolved into horizontal and vertical components, and these are then added (or subtracted as the case may be).

### EXAMPLE

Given a right lens having obliquely crossed prisms (Figure 4-20), one having a power of 2 prism diopters, base at 30°, and the other having a power of 3 prism diopters, base at 240°. Resolve the total prismatic effect into horizontal and vertical components.

For the 2 prism diopter prism having its base at 30°,

$$\varepsilon_h = \varepsilon \cos \theta = \varepsilon \cos 30$$

$$= 2(0.866)$$

$$= 1.732 \text{ prism diopters, base in,}$$

and

$$\varepsilon_v = \varepsilon \sin \theta = \varepsilon \sin 30$$

$$= 2(0.5) = 1 \text{ prism diopter, base up.}$$

For the 3 prism diopter prism having its base at 240°,

$$\varepsilon_h = \varepsilon \cos \theta = \varepsilon \cos 60$$

$$= 3(0.5) = 1.5 \text{ prism diopters, base out,}$$

and

$$\varepsilon_v = \varepsilon \sin \theta = \varepsilon \sin 60$$

$$= 3(0.866)$$

$$= 2.598 \text{ prism diopters, base down.}$$

The total prismatic effect is therefore

1.732 base in − 1.5 base out =

0.232 prism diopter, base in,

combined with

2.598 base down − 1 base up =

1.598 prism diopters, base down.

An alternative method of resolving obliquely crossed prisms involves the construction of a vector diagram. Figure 4-21 shows such a diagram, using the above example. First, a vector is constructed for the first prism, 2 units long and at an angle of 30°. A vector is then constructed for the second prism, 3 units long and at an angle of 240°. A parallelogram is then constructed, and the diagonal of the parallelogram represents the power and the base direction of the resultant prism.

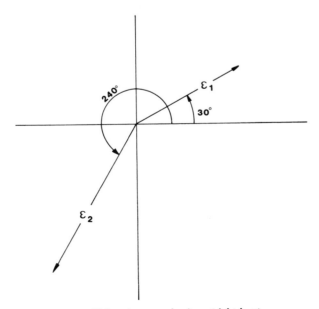

FIGURE 4-20. **Obliquely crossed prisms (right lens).**

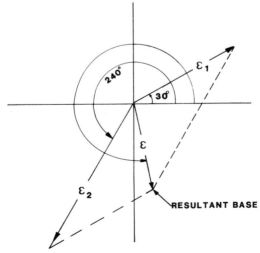

FIGURE 4-21. **Construction of a vector diagram for resolving obliquely crossed prisms.**

## 4.9. Effective Power of a Prism in an Oblique Meridian

When we talk about the amount of deviation of light caused by a prism (or the displacement of an object seen through the prism), we normally refer to the deviation or displacement in the base–apex meridian. However, there will be occasions when we want to know the effect (effective power) of a prism in a meridian other than the base–apex meridian, in particular the horizontal and vertical meridians.

Referring to Figure 4-22, if one looks through a base-down prism at a cross target, the horizontal line of the cross will appear to be displaced upward. If the observer stands at a distance of 1 m from the target, the amount of displacement in centimeters is equal to the power of the prism. If we now rotate the prism so that the base is in the position shown in

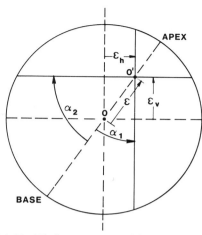

FIGURE 4-23. **Displacement caused by a prism in an oblique meridian.**

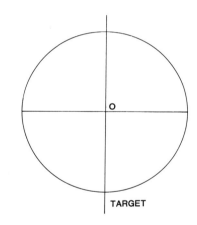

TARGET

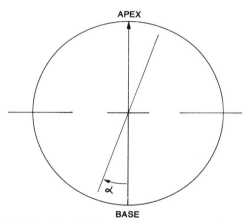

FIGURE 4-24. **The effective power of a prism in a meridian other than the base-apex meridian.**

Figure 4-23, the horizontal and vertical lines of the cross will be displaced. Referring to the diagram,

$$\cos \alpha_1 = \frac{\varepsilon_v}{\varepsilon}$$

and

$$\cos \alpha_2 = \frac{\varepsilon_h}{\varepsilon},$$

and by solving for $\varepsilon_v$ and $\varepsilon_h$, we can find the amount of vertical and horizontal displacement caused by the prism:

$$\varepsilon_v = \varepsilon \cos \alpha_1$$

and

$$\varepsilon_h = \varepsilon \cos \alpha_2.$$

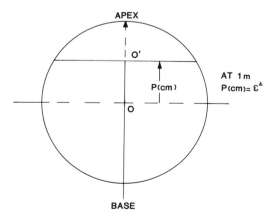

FIGURE 4-22. **Displacement caused by a prism in a principal section.**

As a general expression, therefore, the effective power of a prism in a meridian other than the base−apex meridian (Figure 4-24) is equal to the power of the prism times the cosine of the angle between the specified meridian and the base−apex meridian,

$$\varepsilon_\alpha = \varepsilon \cos \alpha .\qquad (4.14)$$

The use of equation (4.14) will be illustrated by the following examples:

### EXAMPLE 1

What is the prismatic effect along a direction that is 30° from the base−apex direction of a prism whose power is 3 prism diopters?

$$\varepsilon_\alpha = \varepsilon \cos \alpha$$
$$= 3 \cos 30$$
$$= 3(0.866)$$
$$= 2.598 \text{ prism diopters.}$$

### EXAMPLE 2

For a prism of any power, what is the prismatic effect along a line perpendicular to the base−apex direction?

$$\varepsilon_\alpha = \varepsilon \cos \alpha$$
$$= \varepsilon \cos 90$$
$$= \varepsilon(0)$$
$$= 0.$$

### EXAMPLE 3

For a prism of any power, what is the prismatic effect along a line 180° from the base−apex direction?

$$\varepsilon_\alpha = \varepsilon \cos \alpha$$
$$= \varepsilon \cos 180$$
$$= \varepsilon(-1)$$
$$= -\varepsilon.$$

## 4.10. Specification of Prismatic Effects: The Major Reference Point

When a prescription is written without any specification for prism, no resultant prismatic effects are expected when the eyes are in the *primary position*, that is, with the eyes looking straight ahead with the head erect. To accomplish this, the optical centers (poles) of the lenses should be located directly in front of the centers of the pupils. Prisms are often incorporated into spectacle prescriptions for the management of ocular motility problems. The amount of prism desired can often be obtained by mounting the spectacle lenses in such a way that the visual axes no longer pass through the poles of the lenses.

Fry[1] has defined the *major reference point* as the point on the lens that has the prismatic effect that is called for in the prescription. When no prism is prescribed, the major reference point is the optical center of the lens; but if prism is prescribed, the major reference point is the point on the lens manifesting the amount and direction of the prism called for in the prescription. In either case, the major reference point should fall directly in front of the center of the pupil. Implicit in the definition of the major reference point is the fact that the point not only has the prism called for in the prescription, but has the correct *dioptric power* as well. The location of the major reference point is usually specified in relation to the *geometrical center* of the lens.

## 4.11. Specification of Lens and Frame Sizes

Since January 1962, all frames made by members of the Optical Manufacturers' Association have been measured and designated by what is known as the *boxing system*. According to this system, the size of the frame (as well as the size of the lens within the frame) is specified by giving the horizontal and vertical dimensions of a rectangle that circumscribes the lens, as shown in Figure 4-25. The A measurement is the horizontal distance between the two vertical tangents at the lens bevel, and the B measurement is the vertical distance between the two horizontal tangents at the lens bevel. The horizontal A measurement is usually greater than the B measurement, and the

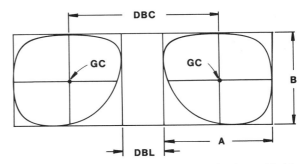

FIGURE 4-25. **The boxing system: The eye size is specified by giving the horizontal and vertical dimensions of a rectangle that circumscribes the lens. GC = geometrical center.**

difference between them is termed the *difference*. The horizontal *A* measurement is usually given as the "*eye size*" of the frame. The horizontal distance between the tangents at the lens bevels of the nasal side of each lens is called the *distance between lenses* (DBL): This measurement is the shortest distance between the two lenses.

The *geometrical center* of the frame aperture (or of the lens that fills it) is defined as the geometrical center of the circumscribing rectangle. This is easily found by constructing the diagonals of the "box": The intersection of the diagonals is the geometrical center. The intersection of the perpendicular bisectors of the horizontal and vertical (A and B) dimensions also determines the geometrical center.

**Distance between Centers.** The distance between centers (DBC) is the distance between the geometrical centers of the two apertures of the frame. This distance is often called the *frame PD*. The "A" measurement added to the DBL will always give the frame PD.

**Effective Diameter.** An additional specification provided by the frame manufacturer is the effective diameter, *ED* (Figure 4-26). The effective diameter of a lens is twice the distance from the geometrical center to the peak of the lens bevel that is farthest from the geometrical center. This measurement is useful in determining the *minimum blank size*, which is the smallest possible finished uncut lens size from which a finished lens having a particular prescription can be cut. The minimum blank size is determined by doubling the amount of decentration and adding this product to the effective diameter: The minimum blank size is therefore given by the expression

minimum blank size = ED + 2 (decentration).

If the minimum blank size exceeds the finished uncut lens size currently available for a particular prescription, oversize blanks must be ordered or the decentration must be provided by grinding prism onto the lenses.

***EXAMPLE***
Given the prescription OD +2.00 DS, OS +2.00 DS, PD = 66 mm, finished lens size = 54 by 45 mm, ED = 60 mm, and DBL = 20 mm. What is the minimum blank size?

Since the frame PD is 74 mm (54 + 20) and the patient's PD is 66 mm, the decentration per lens is 4 mm inward,

minimum blank size = 60 + 2(4) = 68 mm.

Realistically, it is best to make an allowance beyond the minimum blank size because the peripheral portions of the blank may not be suitable due to defects at the edge of the blank.

As lens and frame sizes have increased from year to year, the sizes of single-vision blanks have not kept up with the increase in the sizes of frames being used. Frequently, additional charges are made for providing oversize blanks or for grinding in the decentration.

Every optometrist should keep a current list of the readily available finished lenses including the prescription ranges and sizes. If a prescription can be filled from finished lenses kept in stock by the laboratory, extensive delays in obtaining the finished lenses as well as unexpected additional charges will be avoided.

**Lens Shape.** The shape of a lens is the contour made by the perimeter of the lens when the front surface of the lens is observed perpendicularly. The trade name of a frame implies a specific shape that is unique. There is currently no standardization for defining lens shape. The manufacturer of a given frame must make available *formers* (also called *patterns*) for use by laboratories in edging lenses to fit the frame. Under the boxing system, the center chuckhole of the former falls at the geometrical center of

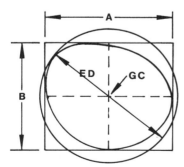

FIGURE 4-26. **The effective diameter (ED) of a lens is twice the distance from the geometrical center (GC) to the peak of the lens bevel that is farthest from the geometrical center.**

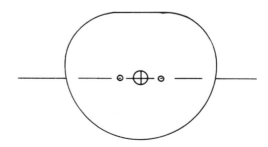

FIGURE 4-27. **The 0 to 180° line of a former.**

the rectangle enclosing the former. Thus the geometric center of the former and that of the lens coincide. A line running through the centers of the three chuckholes on the former gives the orientation of the 0 to 180° line (see Figure 4-27). Manufacturers usually identify formers by trade name or by a number.

## 4.12. Prismatic Power and Thickness

An expression for the change in thickness of a plano-prism between two points on the prism, P and Q, can be derived with the aid of Figure 4-28. In this diagram, a line parallel to the base has been constructed, extending from Q, as well as an additional line parallel to the flat face of the prism. Using Eq. (4.4), relating the refracting angle of the prism to its deviation in prism diopters,

$$\tan \beta° = \frac{\varepsilon}{100(n' - n)}.$$

Referring to the upper triangle in Figure 4-28,

$$\tan \beta° = \frac{t_P - t_Q}{\overline{QP}},$$

and setting the two right-hand terms equal to each other,

$$\frac{t_P - t_Q}{\overline{QP}} = \frac{\varepsilon}{100(n' - n)},$$

and

$$t_P - t_Q = \frac{\overline{QP}(\varepsilon)}{100(n' - n)}. \qquad (4.15)$$

This formula tells us that, for a plano-prism, the thickness difference between two points, P and Q (P being thicker than Q) is equal to the distance between the two points times the power of the prism in prism diopters, divided by 100 times the index difference.

**EXAMPLE**

If a 1 prism diopter prism is ground on a 50-mm lens blank having an index of refraction of 1.523, what is the thickness difference from one edge to the other?

$$t_P - t_Q = \frac{50(1)}{100(1.523 - 1)}$$

$$= 0.956 \text{ mm}.$$

As a rule of thumb, for a prism ground on a 50-mm blank, with $n = 1.523$, we can state that the thickness change across the prism is approximately *one millimeter per prism diopter of power.*

## 4.13. Thickness of a Lens/Prism

If a spherical ophthalmic lens contains prism power, calculations concerning thickness must include the thickness difference (between two points) due to the spherical lens power in addition to that due to the prismatic power. One method of creating prism power is to *decenter* the lens. As shown in Figure 4-29A, this is done by mounting an ophthalmic lens in a frame such that the pole, C, of the lens does not coincide with the geometrical center (that is, the lens is decentered). A decentered lens can be considered as a composite lens made up of a centered lens and an attached prism "on its back." By drawing a dashed line through the lens (as shown in Figure 4-29B) we are able to distinguish between thickness changes due to the *lens* and those due to the *prism.*

The point C', shown in Figure 4-29B, is used as a reference point, and is equivalent to the optical cen-

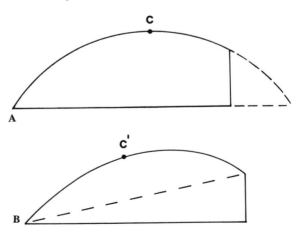

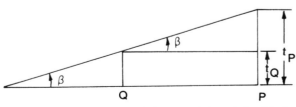

FIGURE 4-28. **Diagram for derivation of the relationship between prism power and change in thickness across the prism.**

FIGURE 4-29. **When a lens is decentered, it becomes a lens/prism. The diagonal line makes it possible to differentiate the thickness difference due to the lens from that due to the prism.**

ter (or pole) of the lens itself, without the prism behind it. The stated amount of prismatic effect (that is, the amount you would find by verifying the lens with a lensometer) is the prismatic effect at the point C′, which is solely due to the *prism* component. At any other point on the lens, the prismatic effect will be different than at point C′, because the *lens* component will produce an additional prismatic effect at all points other than the point C′.

We may use the following expression to find the thickness difference due to the lens only, between the point C at the geometrical center and a peripheral point P, by using Eq. (3.20),

$$t_C - t_P = \frac{F_A h^2}{2(n-1)}.$$

The thickness difference between points C and P due to the prism only, applied to Eq. (4.15), would be

$$t_C - t_P = \frac{\overline{CP}\varepsilon}{100(n-1)},$$

with the result that the equation for the thickness difference for the combination of the lens and the prism becomes

$$t_C - t_P = \frac{\overline{CP}^2 F_A}{2(n-1)} - \frac{\overline{CP}\varepsilon}{100(n-1)}. \quad (4.16)$$

As shown in Figure 4-30, we can establish a *sign convention* for use with this formula: If the point in question (P) is *thicker* than the prism at point C, the distance CP has a *plus* sign, but if the point is on a part of the prism that is *thinner* than the prism at point C, the distance CP has a *minus* sign. In other words, if the peripheral point, P, is located toward the *base* of the prism, CP has a plus sign, but if P is located toward the *apex* of the prism, CP has a minus sign.

### EXAMPLE 1
Given a left lens with the prescription +5.00 DS, decentered to obtain 2 prism diopters, base out, at the geometrical center of the lens. If the lens has a horizontal diameter of 40 mm, a nasal edge thickness of 1 mm, and an index of refraction of 1.50,

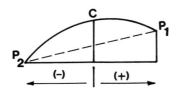

FIGURE 4-30. **Sign convention for changes in thickness of a decentered lens (a lens/prism).**

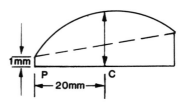

FIGURE 4-31. **Calculation of the thickness at the geometrical center of +5.00 DS decentered to obtain 2 prism diopters of base-out prism.**

what is the thickness at the geometrical center of the lens?

As shown in Figure 4-31, the lens (for the left eye) is decentered *out* in order to obtain base-out prism at the geometrical center.

$$t_C - t_P = \frac{(-0.02)^2(+5)}{2(1.5-1)} - \frac{(-0.02)(2)}{100(1.5-1)}$$

$$t_C - 0.001 = 0.002 - (-0.0008)$$

$$t_C = 0.0038 = 3.8 \text{ mm}.$$

It may be less confusing if Eq. (3.20) and Eq. (4.15) are solved separately and then the thickness differences are combined, always keeping in mind for each equation whether C or P is thicker.

**An Important Point.** For any decentered lens, the difference in edge thickness along the base-apex direction can be attributed solely to the prism. It will be recalled that a decentered lens is a centered lens with a prism on its back side: By using Eq. (4.15) when the edge thicknesses along the base-apex meridian are known, the prismatic effect at the geometrical center of the lens can be calculated.

### EXAMPLE 2
A −5.00D lens, 50 mm in diameter, has a nasal edge thickness of 4 mm and a temporal edge thickness of 1 mm. If the index of refraction is 1.50, what is the prismatic effect at the geometrical center?

If we consider this lens/prism as a centered −5.00 D lens with a prism on its back, then all of the prismatic effect at the geometrical center of the lens is due solely to the prism component, and

$$t_P - t_Q = \frac{\overline{QP}\varepsilon}{100(n'-n)}$$

$$0.004 - 0.001 = \frac{0.05\varepsilon}{100(0.5)}$$

$$0.05\varepsilon = 0.15$$

$$\varepsilon = 3 \text{ prism diopters}.$$

Therefore, the −5.00 D lens has 3 prism diopters of base-in prism at its geometrical center.

### 4.14. Prismatic Effects of Cylindrical Lenses

Prentice's rule can be applied to a cylindrical lens whose axis is either horizontal or vertical. As shown in Figure 4-32, in order to find the prismatic effect at a point, P, on a cylindrical lens, a line is drawn from P, perpendicular to the cylinder axis, and the prismatic effect is found by multiplying the distance, $d$ (in centimeters), by the power of the cylinder. For a plus cylinder the base is toward the cylinder axis, while for a minus cylinder the base is away from the cylinder axis.

The same principle applies for a cylindrical lens whose axis is *not* horizontal or vertical (Figure 4-33):

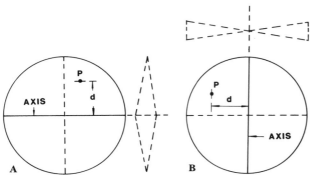

FIGURE 4-32. **Determining the prismatic effect at a point, P, on a cylindrical lens whose axes are horizontal and vertical.**

A line is drawn from the point, perpendicular to the cylinder axis. (Note that the sine square rule for calculating the power in an oblique meridian of a cylindrical lens *cannot* be used for calculating prismatic effects! This is because the deviation occurs only in the plane of a *principal section*.)

The only problem here is that we must have a method of determining the perpendicular distance, $d$, from the point in question to the cylinder axis. This may be done either by construction or by calculation.

Using the construction method, we draw a diagram to scale (Figure 4-33), indicating the horizontal and vertical meridians of the lens, the cylinder axis, and the point, P. We can then construct a triangle with $x'$ and $y'$ as the horizontal and vertical legs, and $d$ as the hypotenuse. The prismatic effect is then equal to the distance, $d$, multiplied by the power, $F_C$, of the cylinder. Since the prismatic effect along the hypotenuse of the triangle is equal to $dF_C$, the horizontal and vertical components of the prismatic effect may be found by the use of the expressions

$$\text{horizontal component} = x'(F_C) ,$$

$$\text{vertical component} = y'(F_C) .$$

The diagram shown in Figure 4-34 may be used to derive a formula for calculating the prismatic effect

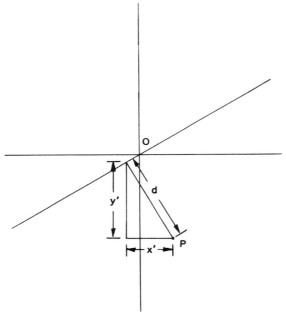

FIGURE 4-33. **Construction method for determining the prismatic effect at a point, P, on a cylindrical lens.**

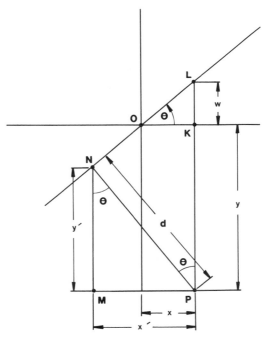

FIGURE 4-34. **Diagram for deriving the formula for calculating the prismatic effect at a point, P, on a cylindrical lens.**

at a point, P, on a cylindrical lens. The sign convention requires that the angle θ always be measured from the *nasal* (toward the temporal) side of the lens. This means that for a *right* lens, the angle θ is the same as the cylinder axis, but for a *left* lens the angle θ is equal to 180° minus the cylinder axis. The derivation is done on the basis of the similar triangles PNL and OKL.

In triangle PNL,

$$d = (y + w) \cos \theta,$$

and in triangle OKL,

$$w = x \tan \theta.$$

Substituting for $w$,

$$d = (y + x \tan \theta) \cos \theta \, ,$$

or

$$d = y \cos \theta + x \tan \theta \cos \theta$$

and since $\tan \theta \cos \theta = \sin \theta$,

$$d = y \cos \theta + x \sin \theta \, . \tag{4.17}$$

In this formula, the following sign convention must be used: If P is located nasally from the origin, O, $x$ has a *plus* sign; if P is located temporally from O, $x$ has a *minus* sign; $y$ has a *plus* sign if P is located below O, and has a *minus* sign if P is located above O. For the angle θ, when θ is acute, both sine and cosine are plus, but when θ is obtuse, sine is plus and cosine is minus.

We may also derive formulas for the horizontal and vertical components of the prismatic effect. For the horizontal component,

$$\varepsilon_h = F_C(x') \, ,$$

but in triangle MNP, $\sin \theta = x'/d$, and $\therefore x' = d \sin \theta$. Therefore,

$$\varepsilon_h = F_C(d \sin \theta) \, . \tag{4.18}$$

For the vertical component,

$$\varepsilon_v = F_C(y'),$$

but in triangle NMP, $\cos \theta = y'/d$, and $\therefore y' = d \cos \theta$. Therefore,

$$\varepsilon_v = F_C(d \cos \theta) \, . \tag{4.19}$$

The following sign convention applies to the horizontal and vertical components: $\varepsilon_h$ is positive for base out and negative for base in; $\varepsilon_v$ is positive for base up and negative for base down.

For a single *sphero-cylindrical* lens, it is necessary to find the horizontal and vertical components for the

sphere, as well as for the cylinder, and to add (or subtract) them to find the total prismatic effect in each meridian. Thus,

$$\text{total horizontal prism} = \varepsilon_{hC} + \varepsilon_{hS} \tag{4.20}$$

and

$$\text{total vertical prism} = \varepsilon_{vC} + \varepsilon_{vS}. \tag{4.21}$$

The two components are combined algebraically, adding base in to base in, base out to base out, base up to base up, and base down to base down. Base in is subtracted from base out (and vice versa), and base up is subtracted from base down (and vice versa). The procedure is then repeated for the second lens.

### Resultant Prismatic Effect (Prismatic Imbalance) between Right and Left Eyes

As noted in Section 4.5, when we consider prismatic effects for the two eyes we are concerned with *resultant* prismatic effects, or with prismatic *imbalance* between the two eyes. For horizontal prismatic effects, when the base of the prism is either *in* or *out* for *both* eyes, the resultant horizontal prismatic effect is found by adding the two amounts of the prism. However, for an equal amount of base in for one eye and base out for the other eye, the eyes simply make a *conjugate* (sideways) shift, to one side or the other, with no change in convergence. For vertical prismatic effects, an equal amount of base-up or base-down prism for each eye simply requires the eyes to make a conjugate movement (downward for base up or upward for base down), with the result that the prismatic effects "cancel" each other. However, when the bases of vertical prism have opposite directions, such as base up on the right eye and base down on the left eye, the resultant vertical prismatic imbalance is found by adding the amounts of the prism. Base-up prism in front of the right eye causes the right eye to turn downward, and base-down prism in front of the left eye causes the left eye to turn upward, causing what is known as *vertical divergence* to take place. These effects may be illustrated by the following examples:

OD 2 base out, OS 2 base out,
total effect = 4 base out.

OD 3 base out, OS 2 base in,
total effect = 1 base out.

OD 1 base down, OS 1 base down,
total effect = zero.

OD 1 base down, OS 2 base up,
total effect = 3 base down OD
or 3 base up OS.

The calculation of prismatic effects experienced by a wearer of sphero-cylindrical lenses will be illustrated by means of an example.

### EXAMPLE

Assume the prescription

OD +6.00 DS − 3.00 DC x 140
OS +2.00 DS − 3.00 DC x 120 .

When reading, the wearer looks through a point in each lens 11 mm down and 2.5 mm in (Figure 4-35). What are the prismatic effects for each eye, in the horizontal and vertical meridians, and what are the combined effects for both eyes?

For the *right eye* (see Figure 4-35A), using the sign convention, $x$, and $y$ will both have *positive* signs. For the *sphere*,

$$\varepsilon_h = F_S(x)$$

$$= +6(0.25) = +1.50$$

$$= 1.50 \text{ prism diopters, base out.}$$

$$\varepsilon_v = F_S(y)$$

$$= +6(1.1) = +6.60$$

$$= 6.60 \text{ prism diopters, base up.}$$

For the *cylinder*,

$$d = y \cos \theta + x \sin \theta$$

$$= 1.1 \cos 140 + 0.25 \sin 140$$

$$= -0.8426 + 0.1607 = -0.6819.$$

$$\varepsilon_h = F_C(d \sin \theta)$$

$$= -3(-0.6819)(0.6428)$$

$$= 1.315$$

$$= 1.315 \text{ base out.}$$

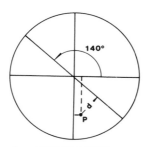

**A    RIGHT LENS**        **B    LEFT LENS**

FIGURE 4-35. **Prismatic effect occurring when wearer looks downward 11 mm and inward 2.5 mm (each eye) to read. See example in text.**

$$\varepsilon_v = F_C(d \cos \theta)$$

$$= -3(-0.6819)(-0.766)$$

$$= -1.567$$

$$= 1.567 \text{ base down.}$$

The total horizontal prismatic effect is given by

$\varepsilon_{hS} + \varepsilon_{hC}$: = 1.5 base out + 1.315 base out

$$= 2.815 \text{ base out.}$$

The total vertical prismatic effect is given by

$\varepsilon_{vS} + \varepsilon_{vC}$: = 6.60 base up + 1.567 base down

$$= 5.033 \text{ base up.}$$

For the *left eye* (Figure 4-35B), using the sign convention, both $x$ and $y$ are again positive. For the *sphere*,

$$\varepsilon_h = F_S$$

$$= 2(0.25) = +0.50$$

$$= 0.50 \text{ prism diopter, base out.}$$

$$\varepsilon_v = F_S(y)$$

$$= 2(1.1) = 2.2$$

$$= 2.2 \text{ prism diopters, base up.}$$

For the *cylinder*,

$$d = y \cos \theta + x \sin \theta$$

$$= 1.1 \cos 60 + 0.25 \sin 60$$

$$= 1.1(0.5) + 0.25(0.866)$$

$$= +0.7665 \text{ cm.}$$

$$\varepsilon_h = F_C(d \sin \theta)$$

$$= -3(0.7665)(0.866) = -1.99$$

$$= 1.99 \text{ prism diopters, base in.}$$

$$\varepsilon_v = F_C(d \cos \theta)$$

$$= -3(0.7665)(0.5) = -1.15$$

$$= 1.15 \text{ prism diopters, base down.}$$

The total horizontal prismatic effect is

$\varepsilon_{hS} + \varepsilon_{hC}$: = 0.5 base out + 1.99 base in

$$= 1.49 \text{ prism diopters, base in.}$$

The total vertical prismatic effect is

$\varepsilon_{vS} + \varepsilon_{vC}$: = 2.2 base up + 1.15 base down

$$= 1.05 \text{ base up.}$$

To find the resultant prismatic effect for both eyes:

resultant horizontal effect

$$= 2.185 \text{ base out} + 1.49 \text{ base in}$$

$$= 1.325 \text{ prism diopters, base out, right eye.}$$

resultant vertical effect

= 5.033 base up + 1.05 base up

= 3.983 prism diopters, base up, right eye.

## 4.15. Decentration

As described in Section 4.11, when using the boxing system the geometrical center of the lens coincides with the geometrical center of the rectangle enclosing the lens. In order to control prismatic effects the pole of the lens is usually displaced from the geometrical center. The displacement of the pole of the lens from the geometrical center is called *decentration*. A "centered" lens is one in which the pole falls at the geometrical center, while a "decentered" lens is one in which the pole is not located at the geometrical center.

Lenses are decentered to control prismatic effects: Decentration is used both to *create* prismatic effects and to *avoid* prismatic effects. When prismatic power is required in a prescription, it often can be obtained by decentration. When decentration *cannot* provide the desired prismatic effect, prism power may be *ground* on the lens by surfacing procedures. By grinding one surface at an appropriate angle to the other surface, differences in thickness occur across the opposing edges of a lens, creating the prismatic effect. While grinding the prism into a lens, the surface powers and thickness must also provide the correct back vertex power. Whether a finished lens has been decentered to provide prism or prism has been provided by grinding, if properly done the prismatic effect will be the same. Even if grinding is used to produce a prismatic effect, the finished lens will behave as if it were a decentered lens. However, the optical center of the lens may not even be located on the lens.

Decentering a finished uncut lens is usually more economical than grinding prism onto a lens. Decentration is done when the prismatic effect required is not large, or when the power of the lens is sufficiently great so that only a small amount of decentration is needed. If a large amount of prism is required, the prism must be produced by grinding. When prismatic power is produced by grinding, the surface powers may be chosen to optimize selected performance parameters of the lens/prism combination.

The following examples will illustrate the use of decentration in relation to the major reference point.

## Decentration of Spherical Lenses

### EXAMPLE 1

Given the following prescription,

OD −4.00 DS
OS −4.00 DS

PD = 64 mm, eye size = 48 mm, DBL = 20 mm, find (*a*) the locations of the major reference points of the lenses as related to the geometrical centers; (*b*) the amount and direction of the decentration.

(*a*) Since the frame PD is 68 mm (48 + 20) and the patient's PD is 64 mm, the centers of the pupils each fall 2 mm in from the geometrical center. Hence, the major reference point for each lens should be located 2 mm inward from the geometrical center of the lens.

(*b*) The major reference point is placed before the center of the pupil by the process of decentration. Since no prism is prescribed, the only point on the lens having zero prismatic effect is the pole of the lens. Consequently, the pole must be placed, for each lens, in front of the center of the pupil. In the finished lenses, the poles would be located 2 mm inward, for each lens, from the geometrical center. Hence, decentration is 2 mm in for each lens.

### EXAMPLE 2

Given the prescription,

OD −4.00 DS with 1 prism diopter, base in
OS −4.00 DS with 1 prism diopter, base in

PD = 64 mm, eye size = 50 mm, DBL = 20 mm, find (*a*) the location of the major reference points of the lenses as related to the geometrical centers; (*b*) the amount and direction of the decentration.

(*a*) Since the frame PD is 70 mm (50 + 20) and the patient's PD is 64 mm, the center of each pupil is located 3 mm inward from the geometrical center of the lens. Therefore the major reference points should fall 3 mm in, from the geometrical center, for each lens.

(*b*) For this problem it is easier to calculate the decentration in two distinct steps. The first step is to move the pole of the lens to the center of the pupil, which initially gives a *zero* amount of prism. In this problem the pole of each lens would first be moved 3 mm inward. For the second step, the pole of each lens is moved to provide the appropriate amount of prism before the center of the pupil. Since one prism diopter, base in, is required for each eye, by the use of Prentice's rule the pole of a −4.00 D lens must be displaced 2.5 mm to provide 1 prism diopter of prism. The direction of the displacement must be *out*, in order to achieve a base-in effect for each eye, wearing a minus lens.

In the first step the pole is moved *3 mm in* from the geometrical center and in the second step the pole is moved *2.5 mm out* from the center of each pupil. The net position of these two movements leaves the pole of each lens resting 0.5 mm in from the geometrical center. Hence, the decentration is *0.5 mm in* for each lens.

### EXAMPLE 3

Given the prescription,

OD −4.00 DS with 1 prism diopter, base out
OS +2.00 DS with 1 prism diopter, base out

PD = 64 mm, eye size = 48 mm, DBL = 20 mm,

find (*a*) the location of the major reference point for each lens relative to the geometrical center; (*b*) the amount and direction of decentration for each lens.

(*a*) Since the frame PD is 68 mm (48 + 20) and the patient's PD is 64 mm, the center of each pupil will be located 2 mm inward from the geometrical center of the lens. Therefore, the major reference points should be positioned *2 mm in* from the geometrical center, for each lens.

(*b*) Using the two-step method described in Example 2, the first step is to position the pole of each lens before the center of the pupil, which will provide zero prismatic effect initially. In this first step the pole would be moved *2mm in* from the geometrical center, for each lens. In the second step the pole is then moved from the position in front of the center of the pupil, for each eye, to provide the appropriate amount of prism. For the right eye, since 1 prism diopter, base out, is required, the pole of the −4.00 D lens must be displaced 2.5 mm, and the direction of the displacement must be *in* to achieve a base-out effect. In the first step the pole was moved *2 mm in* from the geometrical center and in the second step the pole was moved an additional *2.5 mm in*. The final position for these two movements is therefore *4.5 mm in* from the geometrical center of the lens. Hence, the *decentration* for the right lens is *4.5 mm in*.

Since 1 prism diopter, base out, is also needed for the left eye, in the second step the pole of the +2.00 D lens must be moved 5 mm from the center of the pupil. The direction of the displacement must be *out*, to provide a base-out effect before the left eye. For the left lens, in the first step the pole was moved *2 mm in* from the geometrical center, and in the second step the pole was moved *5 mm out* from the center of the pupil. The net position of the pole for these two movements is *3 mm out* from the geometrical center. Therefore, for the left lens the decentration is *3 mm out*.

### EXAMPLE 4

A +6.00 D lens, intended for the right eye, is to have a prismatic effect of 3 prism diopters base down and 2 prism diopters base in. What is the required decentration to bring about this prismatic effect?

The direction of the decentration can be found by visualizing where the major reference point must be placed in relation to the optical center of the lens. It should be remembered that we are concerned only with the *change in thickness* about the major reference point. In this example, the base-*down* and base-*in* prism are required for the right lens. Inspection of Figure 4-36 shows that for these prismatic effects to occur, the major reference point must fall in the *upper left* quadrant of the lens, and hence the optical center will have to be moved *downward* and *inward* from the major reference point.

The following rule-of-thumb is helpful in determining the direction of decentration in order to induce a given prismatic effect: If a lens (or a meridian of a lens) has *plus* power, the required decentration is in the *same* direction as that of the base of the prism; but if a lens (or meridian of a lens) has *minus* power, the required decentration is in the direction *opposite* to the base of the prism.

In the above example, the *amount* of decentration can be determined by the use of Prentice's rule,

$$\varepsilon = dF,$$

or

$$d = \frac{\varepsilon}{F}.$$

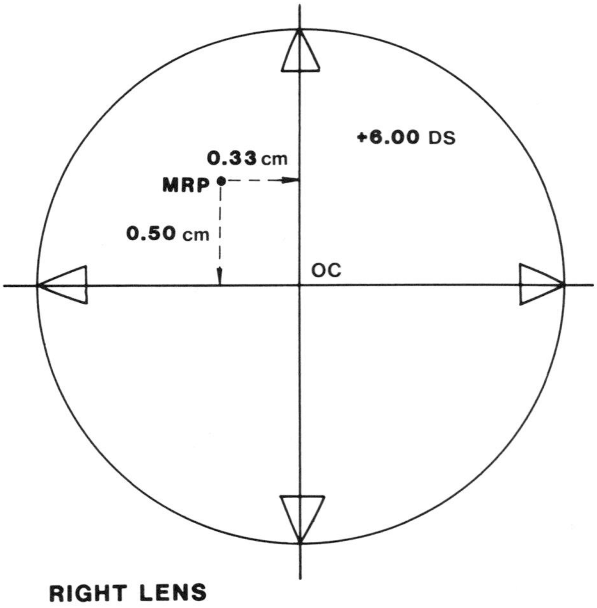

**RIGHT LENS**

FIGURE 4-36. A **+6.00** sphere, for the right eye, is to have a prismatic effect of 3 prism diopters base down and 2 prism diopters base in. For these prismatic effects to occur, the major reference point must fall in the upper left quadrant of the lens.

Therefore,

$$d_v = \frac{3}{6} = 0.50 \text{ cm downward,}$$

and

$$d_h = \frac{2}{6} = 0.33 \text{ cm inward.}$$

Hence, the required decentration is 0.50 cm down and 0.33 cm in.

### Decentration of Plano-Cylindrical Lenses

#### EXAMPLE 1

Given a right lens with the formula +3.00 DC axis 180. In what direction, and how much, must the lens be decentered to produce, at the major reference point, (a) 1 prism diopter, base down, or (b) 2 prism diopters, base up?

Since the prismatic effect of a cylindrical lens is always perpendicular to the cylinder axis (see Section 4.14), a cylindrical lens may be decentered to create a prismatic effect *only* when the prescribed base direction coincides with the direction of the power meridian. Therefore, a cylinder whose axis is horizontal, as the one in this example, can be decentered to produce only a *vertical* prismatic effect (base up or base down), whereas a cylinder whose axis is vertical (90°) can be decentered to produce only a *horizontal* prismatic effect (base in or base out). The solutions to this example, therefore, are as follows:

(a) In order to induce 1 prism diopter, base down, the lens must be decentered (using Prentice's rule)

$$d = \frac{\varepsilon}{F} = \frac{1}{3} = 0.33 \text{ cm downward.}$$

(b) In order to induce 2 prism diopters, base up, the lens must be decentered

$$d = \frac{\varepsilon}{F} = \frac{2}{3} = 0.67 \text{ cm upward.}$$

#### EXAMPLE 2

Given a right lens with the formula −2.00 DC axis 30. In what direction, and how much, must the lens be decentered to produce, at the major reference point, (a) 1 prism diopter, base at 120°, or (b) 2 prism diopters, base at 300°?

Since the prismatic effect of a cylindrical lens is perpendicular to the cylinder axis, a cylinder whose axis is 30° can be decentered to produce a prismatic effect whose base is *only* 120° or 300° (see Figure 4−37).

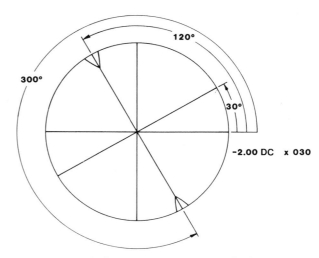

FIGURE 4-37.   **The lens, −2.00 DC axis 30, can be decentered to produce a prismatic effect whose base is either at 120 degrees or at 300 degrees.**

(a) In order to induce 1 prism diopter, base at 120°, the lens must be decentered

$$d = \frac{\varepsilon}{F} = \frac{1}{-2} = 0.5 \text{ cm in the 300° direction.}$$

(b) In order to induce 2 prism diopters, base at 300°, the lens must be decentered

$$d = \frac{\varepsilon}{F} = \frac{2}{-2} = 1.0 \text{ cm in the 120° direction.}$$

It is rare that the power meridian of a plano-cylindrical lens coincides exactly with the base-apex meridian of the prescribed prism. Therefore, if prismatic power is required in a plano-cylinder, it must usually be obtained by a *surfacing* operation; that is, by grinding the refractive surfaces of the lens at the inclination to one another that is required to produce the desired prismatic effect.

### Decentration of Sphero-Cylindrical Lenses Whose Principal Meridians Are Horizontal and Vertical

#### EXAMPLE

Given a right lens with the formula +2.50 DS −4.50 DC axis 90. In what direction, and how much, must the lens be decentered to produce, at the major reference point, 1 prism diopter base up and 1 prism diopter base in?

The first step is to determine the power in each of the two principal meridians, so that we may apply Prentice's rule to each meridian, using the rule-of-thumb for the direction of decentration.

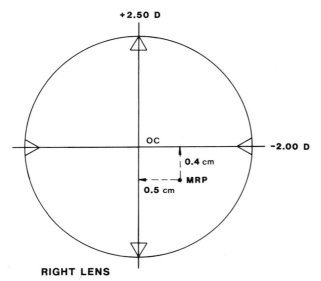

**RIGHT LENS**

FIGURE 4-38. **Decentration of the lens, +2.50 DS −4.50 DC axis 90, in order to produce a prismatic effect of 1 prism diopter base up and 1 prism diopter base in.**

As shown in Figure 4-38, the power in the vertical meridian is +2.50 D and that in the horizontal meridian is −2.00 D.

$$d_v = \frac{\varepsilon}{F_v} = \frac{1}{2.5} = 0.4 \text{ cm upward}$$

and

$$d_h = \frac{\varepsilon}{F_h} = \frac{1}{-2} = 0.5 \text{ cm outward.}$$

The decentration is therefore 0.4 cm upward and 0.5 cm outward.

## Decentration of Sphero-Cylindrical Lenses Whose Principal Meridians Are Oblique

When the principal meridians are oblique, a *graphical* method provides a much simpler method of arriving at a solution than a purely mathematical approach. Using the graphical method, the lens is considered in the crossed cylinder form. The desired prism is first resolved into components lying along the two principal meridians. The decentration is calculated separately for each principal meridian, and is then expressed in terms of vertical and horizontal components. In order to demonstrate both methods, the following example will be solved first by the graphical method and then by the mathematical method.

*EXAMPLE*
Given a left lens with the formula +3.00 DS −1.00 DC axis 30. In what direction, and how much, must the lens be decentered to produce, at the major reference point, 1 prism diopter base up and 3 prism diopters base in?

**Graphical Method.** With the aid of Figures 4-39 and 4-40, the graphical method is used in the following manner:

1. Construct vertical and horizontal lines which intersect the two principal meridians (30 and 120 degrees) at point R, as shown in Figure 4-39.

2. Using a convenient scale, mark off line segments RV to correspond to the direction and amount of the specified vertical prismatic effect and RH to correspond to the direction and amount of the specified horizontal prismatic effect:

   RV = 1 prism diopter base up.
   RH = 3 prism diopters base in.

3. Complete the rectangle RVMH.

4. Draw the diagonal RM. This represents the single resultant prismatic effect.

5. From point M, drop perpendiculars to the two principal meridians, forming line segments MA and MB.

6. The single original prism can now be resolved into prism components RA with its base at 210° and RB with its base at 120°. The lengths of line

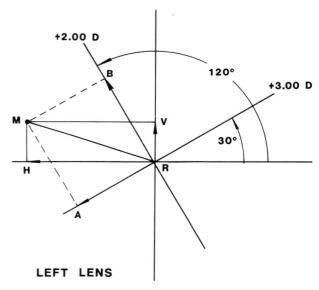

**LEFT LENS**

FIGURE 4-39. **Resolving the prismatic effect, 1 prism diopter base up and 3 prism diopters base in, for the lens +3.00 DS −1.00 DC axis 90, into components parallel and perpendicular to the cylinder axis. RV = 1.00 prism diopter; RH = 3.00 prism diopters; RA = 2.12 prism diopters; RB = 2.35 prism diopters.**

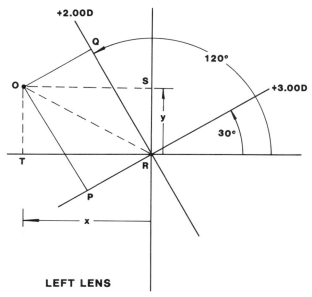

+2.00D

Q

120°

O

S

+3.00D

y

30°

T

R

P

x

**LEFT LENS**

FIGURE 4-40. **The graphical method of determining the direction and amount of decentration to produce a prismatic effect of 1 prism diopter base up and 3 prism diopters base in, for the lens +3.00DS −1.00 DC axis 90. RQ = 1.18 cm; RP = 0.706 cm; RS = 0.72 cm up; RT = 1.22 cm in.**

segments RA and RB represent the amount of prism along the two principal meridians. These are found to be:

$$\varepsilon_{210} = 2.12 \text{ prism diopters,}$$

and

$$\varepsilon_{120} = 2.35 \text{ prism diopters.}$$

7. The decentration required to produce the prismatic effect can now be determined by the use of Prentice's rule. Along the 210° meridian,

$$d = \frac{\varepsilon}{F} = \frac{2.12}{3} = 0.7067 \text{ cm,}$$

and along the 120° meridian,

$$d = \frac{\varepsilon}{F} = \frac{2.35}{2} = 1.18 \text{ cm.}$$

8. Draw a second diagram (Figure 4-40), again constructing vertical and horizontal lines which intersect the two principal meridians (30° and 120°) at point R.

9. Using a convenient scale, lay off the decentration along each principal meridian such that

RP = 0.7067 cm along the 210° meridian,

and

RQ = 1.18 cm along the 120° meridian.

10. Complete the rectangle QRPO, the corner O representing the position of the required optical center.

11. Drop perpendiculars from point O to the horizontal meridian (point T) and to the vertical meridian (point S).

12. The lengths of line segments RT and RS represent the amounts of horizontal and vertical decentration, respectively:

$$RS = y = 0.72 \text{ cm upward,}$$

and

$$RT = x = 1.22 \text{ cm inward.}$$

**Mathematical Method.** In Section 4.14, the total horizontal prismatic effect at a point in the periphery of a sphero-cylindrical lens was given by the expression

$$\text{total horizontal prism} = \varepsilon_{hS} + \varepsilon_{hC}$$

or

$$\varepsilon_h = xF_S + F_C (y \cos \theta + x \sin \theta) \sin \theta \quad (4.22)$$

and the total vertical prismatic effect at a point in the periphery was given by the expression

$$\text{total vertical prism} = \varepsilon_{vS} + \varepsilon_{vC}$$

or

$$\varepsilon_v = yF_S + F_C (y \cos \theta + x \sin \theta) \cos \theta. \quad (4.23)$$

The letters $y$ and $x$ are the vertical and horizontal decentrations of the optical center from the major reference point required to produce the specific prism at the major reference point, R.

Equations (4.22) and (4.23) can be expanded as follows:

$$\varepsilon_h = yF_C \sin \theta \cos \theta + x (F_S + F_C \sin^2 \theta). \quad (4.24)$$

and

$$\varepsilon_v = y(F_S + F_C \cos^2 \theta) + xF_C \sin \theta \cos \theta \quad (4.25)$$

These expressions form a pair of simultaneous equations. Solving these equations simultaneously or by transformation, we find that

$$x = \frac{\varepsilon_h F_S + \varepsilon_h F_C \cos^2 \theta - \varepsilon_v F_C \sin \theta \cos \theta}{F_S(F_S + F_C)}. \quad (4.26)$$

and

$$y = \frac{\varepsilon_v F_S + \varepsilon_v F_C \sin^2 \theta - \varepsilon_h F_C \sin \theta \cos \theta}{F_S(F_S + F_C)}. \quad (4.27)$$

Note that the sign convention developed in Section 4.14 is used in Eqs. (4.26) and (4.27): The angle $\theta$ is measured from the nasal side of the lens temporally to the axis; $\varepsilon_h$ is positive for base out and negative for base in; and $\varepsilon_v$ is positive for base up and negative for base down. When $y$ is positive, the decentration is up; when $y$ is negative, the decentration is down. When $x$ is positive, the decentration is out; when $x$ is negative, the decentration is in.

The mathematical method will now be used as an alternate method of solving the above example: Given a left lens with the formula $+3.00\,DS$ $-1.00\,DC$ axis 30. In what direction, and how much, must the lens be decentered to produce, at the major reference point, 1 prism diopter base up and 3 prism diopters base in?

$$y = \frac{\varepsilon_v F_S + \varepsilon_v F_C \sin^2 \theta - \varepsilon_h F_C \sin \theta \cos \theta}{F_S(F_S + F_C)}$$

$$= \frac{(1)(3) + (1)(-1)\sin^2 150 - (-3)(-1)\sin 150 \cos 150}{+3(3-1)}$$

$$= \frac{3 + (-0.25) + 1.299}{6} = \frac{4.049}{6} = 0.675 \text{ cm (upward)}.$$

$$x = \frac{\varepsilon_h F_S + \varepsilon_h F_C \cos^2 \theta - \varepsilon_v F_C \sin \theta \cos \theta}{F_S(F_S + F_C)}$$

$$= \frac{(-3)(3) + (-3)(-1)\cos^2 150 - (1)(-1)\sin 150 \cos 150}{+3(3-1)}$$

$$= \frac{-9 + 2.25 - 0.433}{6} = \frac{-7.183}{6} = -1.197 \text{ cm (inward)}.$$

## 4.16. Effects of Prisms on the Eyes

The effects that prisms have on the eyes, already referred to in Section 4.5, may be understood by performing a number of experiments with prisms. Either trial case prisms or square plastic prisms may be used.

### EXPERIMENT 1

Occlude one eye. While you fixate a distant object (such as a letter on a Snellen chart), place a 10 prism diopter prism, base down, in front of the open eye. This will bring about the situation shown in Figure 4-41: Light rays from the object will be deviated toward the base of the prism, with the result that the object will appear to be displaced toward the *apex* of the prism. If you choose to keep fixating the object, the eye *must* move upward. Indeed, it is difficult *not* to allow the eye to move upward. If a classmate looks behind the occluder, he or she will see that the occluded eye *also* moves upward. This is a *conjugate* movement, that is, a movement in which both eyes move in the same direction.

### EXPERIMENT 2

Repeat Experiment 1, but with both eyes open. Diplopia (double vision) will occur: You will be able to fixate either the upper or the lower image of the object, but you will be unable to fuse them into one.

### EXPERIMENT 3

Repeat this experiment, again with both eyes open, but with only a 2 prism diopter prism, base down. Although you may briefly see double, the images will quickly coalesce into one. This is because a *vertical fusional vergence* movement has been made: The eye seeing through the prism turns upward, in order to keep the images on the foveas of the two eyes. Since the visual system's ability to make vertical fusional vergence movements is not very great, any prism stronger than 3 or 4 prism diopters (base down or base up) usually results in diplopia, as in Experiment 2. A vergence movement, in which one eye turns in one direction and the other eye makes no movement or moves in the opposite direction, is called a *disjunctive* movement.

### EXPERIMENT 4

With both eyes open, while watching a distant object, place a 5 prism diopter prism, base out, in

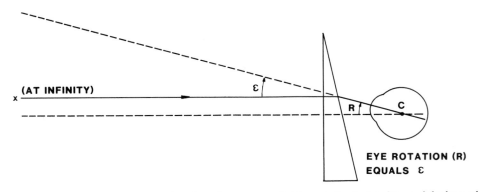

FIGURE 4-41. **A 10 prism diopter prism, base down, causes light rays to be deviated toward the base of the prism, with the result that the object will appear to be displaced toward the apex.**

front of each eye (see Figure 4-42A). Although the distant object may blur slightly, it will remain single due to the fact that the eyes will make a *positive fusional vergence* movement. This, again, is a disjunctive movement, since the eyes move in opposite directions (one to the right, the other to the left). If the object appears to blur, this is because the accommodative mechanism has been stimulated in order to maintain single binocular vision.

### EXPERIMENT 5

Repeat Experiment 4 but with the prism oriented in the base-in direction (see Figure 4-42B). Since 10 prism diopters is about the maximum amount of *negative fusional vergence* (fusional divergence) of which the eyes are capable, you will probably see double in this experiment. If you continue to see single, repeat the experiment with stronger prisms (e.g., 7 or 8 prism diopters for each eye). Diplopia will most likely occur.

### EXPERIMENT 6

Again while fixating a distant object, place a 5 prism diopter prism in front of each eye with the bases in the same direction—first with both bases to the right, then with both bases to the left, and then with both bases up and again with both bases

down. In each case, single vision will remain, due to the eyes making a *conjugate* movement (to the left with base-to-the-right prism, and to the right with base-to-the-left prism, downward with base-up, or upward with base down).

Summarizing, there are two kinds of eye movements that can occur when prisms are used: (1) conjugate movements (also called version movements), in which both eyes move in the same direction, and (2) disjunctive movements (also called vergence movements), in which the two eyes move in opposite directions.

## 4.17. Effectivity of a Prism

We can speak of the effectivity (or *effective power*) of a prism, just as we can talk about the effective power of a lens. It will be recalled (Chapter 3) that as a lens is moved away from the eye its effective power increases (for a plus lens) or decreases (for a minus lens). When we speak of the effectivity of a prism, we are concerned with the *angular rotation* of the eye (inward, outward, upward, or downward) in relation to the power of the prism.

The effectivity of a prism for a distant object is illustrated in Figure 4-43A. As shown in this diagram, the rotation of the eye (R) is equal to the angle

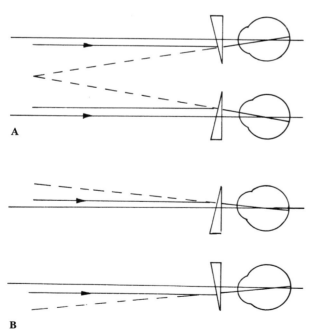

A

B

FIGURE 4-42. **(A) A 5 prism diopter base-out prism in front of each eye causes the eyes to make a positive fusional vergence movement. (B) a 5 prism diopter base-in prism in front of each eye causes the eyes to make a negative fusional vergence movement.**

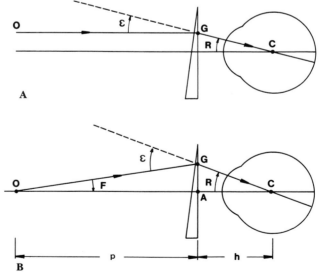

A

B

FIGURE 4-43. **Diagram for use in deriving the formula for the effectivity of a prism. Note that for a distant object (A) there is no loss of effectivity, whereas for a near object (B) a loss of effectivity occurs.**

of deviation of the prism (ε), with the result that, for distance fixation, there is no loss in effectivity.

The situation existing for near vision is shown in Figure 4-43B. As shown in this diagram, angle $R$ is smaller than angle ε, so the prism has lost effectivity. Note that, in this diagram, $h$ represents the distance between the prism and the center of rotation for the eye, and $p$ represents the distance from the prism to the object of regard. A formula for determining the effectivity of a prism, when used for near vision, may be derived as follows:

$$\text{angle } \varepsilon = \text{angle } R + \text{angle } F$$

$$\text{angle } F = \text{angle } \varepsilon - \text{angle } R$$

$$\tan R = \frac{\overline{AG}}{h}, \text{ and } \tan F = \frac{\overline{AG}}{p}$$

$$\overline{AG} = h \tan R = p \tan F$$

$$hR = pF \quad \text{(for small angles only)}$$

$$F = \frac{hR}{p}.$$

Substituting $\varepsilon - R$ for $F$,

$$\varepsilon = R + \frac{hR}{p}.$$

$$\varepsilon = R(1 + h/p)$$

$$= \frac{R(p + h)}{p}$$

$$R = \frac{\varepsilon p}{(p + h)}. \qquad (4.28)$$

**EXAMPLE**
What is the effectivity of a 15 prism diopter prism, located 25 mm in front of the center of rotation of the eye, when the wearer reads at a distance of 40 cm from the prism?

$$R = \frac{15(0.40)}{0.40 + 0.025}$$

$$= \frac{6}{0.425}$$

$$= 14.12 \text{ prism diopters.}$$

The prism, therefore, has lost approximately 1 prism diopter of effectivity. Since prisms are usually prescribed in powers of no more than a few prism diopters, this loss of effectivity, for near vision, usually presents no problem. It does, however, assume some importance for rotating prisms

used in refractors or phoroptors, since these prisms are used in powers exceeding 15 prism diopters. Also, the value of $h$ is greater for refractors and phoroptors than for a spectacle lens, since Risley prisms are located several millimeters in front of the lens apertures. If the object is located at infinity, $R$ is equal to ε. If $h$ approaches zero in value, the value of $R$ approaches that of ε.

## 4.18. Risley Prisms

Risley prisms, also called *rotary prisms*, are prisms used in refractors or phoroptors to measure phorias and fusional vergences. Each Risley prism is made up of two prisms, one behind the other, which are geared so that they rotate together when a knurled knob is turned. The carrier in which the two prisms are mounted may be positioned with the knurled knob oriented either vertically or horizontally. When the knob is oriented vertically (as shown in Figure 4-44) *base-in* or *base-out* prism power may be introduced, whereas when the knob is oriented horizontally *base-up* or *base-down* prism may be introduced.

When the knob is oriented vertically and is turned so that the scale reading is zero (as in Figure 4-45A), one of the prisms is oriented in the base-in position and the other is oriented in the base-out position, so that no horizontal prismatic effect is present. When the knob is turned in the *base-in* direction, the prisms rotate in opposite directions (one clockwise, the other counterclockwise) but both bases rotate in a *nasalward* direction, causing the desired base-in effect. Similarly, when the knob is turned in the *base-out* direction, the prisms again rotate in opposite directions, but in such a way that both bases move *temporally*, causing a base-out effect. As shown in Figures 4-45B and C, in the two extreme positions, after a rotation of 90 degrees for each prism, a maximum amount of base-in or base-out prismatic effect is ob-

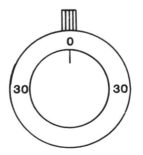 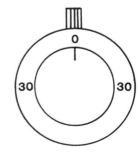

FIGURE 4-44. **Risley prisms oriented vertically so that base-in or base-out prism power can be introduced.**

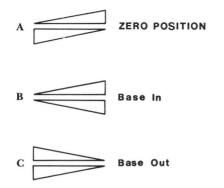

A    ZERO POSITION

B    Base In

C    Base Out

FIGURE 4-45. **Introduction of horizontal prism: A, no prismatic effect; B, maximum base-in effect; C, maximum base-out effect.**

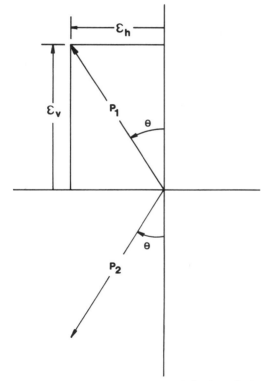

FIGURE 4-46. **Diagram for derivation of the formula for the deviation caused by a pair of Risley prisms.**

tained; if each prism has a power of 15 prism diopters, the maximum achievable effect is 30 prism diopters. The same applies, of course, for the introduction of base-up or base-down prism.

In Figure 4-46, $P_1$ and $P_2$ represent the positions of the base-apex lines of the two prisms as base-in or base-out prism power is introduced. If the length of the line $P_1$ (or $P_2$) is equal to the power of 1 prism unit, the length of the line $\varepsilon_h$ indicates the power of a prism unit for that position of the base-apex line: For example, when $P_1$ is oriented vertically, $\varepsilon_h = 0$, but when $P_1$ is oriented horizontally, $\varepsilon_h = P_1$. For any position of $P_1$ and $P_2$, *for each prism,*

$$\varepsilon_h = P \sin \theta, \qquad (4.29)$$

and the deviation for *both* prisms can be represented by

$$2\varepsilon_h = P_1 \sin \theta + P_2 \sin \theta$$
$$= (P_1 + P_2)(\sin \theta)$$

since $P_1 = P_2$

$$2\varepsilon_h = 2 P \sin \theta. \qquad (4.30)$$

As for the vertical prismatic effect, $\varepsilon_v$, note that for any position of the prisms the values of $\varepsilon_v$ are equal for the two prisms, but opposite in direction. These values, therefore, cancel each other out, so the only resulting prismatic effect is base in or base out.

The same principle applies, of course, to the use of Risley prisms in producing vertical prismatic effects. With the knob oriented horizontally, the amount of base-up or base-down prismatic effect that is introduced is given by the equation

$$2\varepsilon_v = 2 P \sin \theta, \qquad (4.31)$$

with the lateral components cancelling each other out.

## 4.19. Fresnel Press-on Prisms

In 1970, the Optical Sciences Group of San Rafael, California, developed a series of Fresnel Press-on prisms—thin plastic membrane prisms ranging in powers from 0.5 to 30 prism diopters. These prisms are an adaptation of the hand-ground lenses designed by the French engineer and physicist Augustine Fresnel in 1821. Fresnel's lenses were originally used for lighthouse beacons.

Since refraction occurs only at the surface of an optical element and depends primarily on the angle between the two surfaces, the intervening (nonrefracting) portion of a conventional optical element can be removed without appreciably affecting the refracting power. The angle between the two surfaces remains constant across the prism.

The Fresnel Press-on prism consists of a series of identical small plastic prisms lying parallel and adjacent to each other on a thin plastic base membrane (see Figure 4-47). Each of the small prisms shown in Figure 4-47B has the same deviation power as the conventional prism shown in Figure 4-47A. However, the Fresnel Press-on prism is only 1 mm thick, or about one-tenth the thickness of a solid conventional prism of the same power.

Fresnel Press-on prisms are made of a specially formulated plasticized polyvinyl chloride (PVC) material with a refractive index of 1.525, very close to that of ophthalmic crown glass. The prisms are manufactured by a process of either injection or compression molding, using molds cut usually from brass. Only one surface of the membrane is impressed with the prismatic grooves, the other surface being flat. The resulting flat surface of the thin flexible membrane can be applied without adhesive material to the back surface of a conventional glass lens. The prisms are designed for in-office application: The membrane is first cut to conform to the size and shape of the lens, and is then applied to the back surface of the lens while the lens and membrane are immersed in water. After drying, the membrane is highly adherent to the glass surface, and can be removed only by immersing the lens in water and peeling the membrane off.

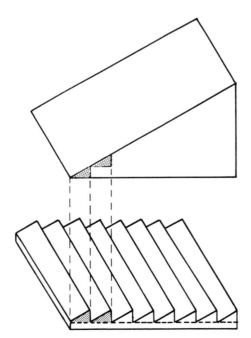

FIGURE 4-47. **A Fresnel Press-on prism.**

Press-on prisms can be used clinically as a permanent correction—for example, as a substitute for conventional prisms when the required prism power is not available or is subject to a long delay. However, these prisms are applied most commonly for temporary use, in diagnosing a binocular vision problem or as an aid in orthoptic training. They are also useful when an oculomotor condition varies with time—in a situation in which the amount of prism required may undergo frequent changes.

The greatest advantages of Press-on prisms are their thinness, their lightness of weight, and their in-office application. Their disadvantages include a slight loss of visual acuity and contrast for the wearer and the striated appearance (to observers) of the grooves.

There are a number of reasons for the visual acuity loss. As with conventional prisms, Fresnel prisms suffer from *distortion* and *chromatic aberration*. A comparison of distortion for conventional prisms and for Fresnel prisms has been made by Adams et al.[2] They broke the complex aberration of distortion into the following five simple distortions:

1. horizontal magnification
2. vertical magnification
3. curvature of vertical lines
4. asymmetric horizontal magnification
5. change in vertical magnification with horizontal angle

They found that the Fresnel membrane prisms produced substantially less horizontal and vertical magnification and slightly less curvature of vertical lines than the conventional prisms. Conventional prisms demonstrated slightly less asymmetric horizontal magnification and change of vertical magnification with horizontal angle than the Fresnel prisms.

A second source of visual acuity loss is *scattering of light*, caused by the grooves. This scattering of light causes both a loss of contrast and the presence of bothersome reflections. The most important factor determining resolution is the width of the grooves: Resolution improves with decreasing groove width, but if the grooves are *too* narrow, diffraction becomes a problem. As for the striated appearance, to an observer, this can be improved to some extent by using a lightly tinted or photochromic lens.

When a prismatic component is to be included in a patient's prescription, the advantages and disadvantages of Press-on prisms should be carefully weighed, to determine the suitability of using Press-on prisms.

# References

1. Fry, G. A. The Major Reference Point in Single Vision Lenses. *Amer. J. Optom. Arch. Amer. Acad. Optom.*, Vol. 24, pp. 1–7, 1947.
2. Adams, A. J., Kapash, R.J., and Barkan, E. Visual Performance and Optical Properties of Fresnel Membrane Prisms. *Amer. J. Optom. Arch. Amer. Acad. Optom.*, Vol. 48, pp. 289–297, 1971.

# Questions

1. What is the refracting angle of a prism of 1 prism diopter having the following indices of refraction?
   (a) 1.42
   (b) 1.53

2. Compute the prismatic power in the 60th meridian of a prism of 3 prism diopters base at 90.

3. A patient wears +2.00 DC x 180, with the cylinder axis passing through the geometrical center. During reading, the lines of sight pass through the lens 10 mm down and 2 mm in from the geometrical center of the lens. What is the prismatic effect?

4. Compute the equivalent prism for the following combinations of prisms:
   (a) OD 2 prism diopters base out/4 prism diopters base down
   (b) OS 2 prism diopters base up/1 prism diopters base in

5. Compute the horizontal and vertical components of the following prisms:
   (a) OD 3 prism diopters base at 30
   (b) OS 2 prism diopters base at 240

6. A Risley prism consisting of two prisms of 12 prism diopters is mounted before an eye. Each prism is rotated 30° from the zero position. What is the prismatic effect obtained before the eye?

7. The refracting angle of a prism is 8° and the index of refraction is 1.55. What deviation will it produce in degrees for a pencil of light from infinity?

8. The refracting angle of a prism is 10° and the index of refraction is 1.65. What is the deviation in prism diopters it will produce for a pencil of light from infinity?

9. Compute the horizontal and vertical prismatic effects for a point located 12 mm below and 2 mm in from the geometrical center of each of the following lenses:
   (a) OD +3.00 DS
   (b) OS −2.00 DC x 90
   (c) OD +2.00 DC x 180

10. Calculate the vertical and horizontal prismatic effects for the given points on the following lenses:
    (a) 10 down and 2 in: OD +5.00 DS/−2.00 DC x 45
    (b) 10 down and 5 out: lens same as in (a)
    (c) 5 up and 5 in: OS −5.00 DS/+2.00 DC x 120

11. A prescription is as follows:
    OD −2.50 DS
    OS +2.00 DS/−5.00 DC x 30
    If the reading points are 10 mm below and 3 mm in each from the distance optical centers, what are the vertical and horizontal prismatic imbalances for the two eyes at the reading level?

12. The pole of a finished lens is located 5 mm directly below point A and 10 mm nasalward of point B. Point A has a prismatic power of 2 prism diopters base up and point B has a prismatic power of 3 prism diopters base in. What is the refracting power ($R_x$) of this lens?

13. When a patient looks through a point 6 mm above the optical center of one of his spectacle lenses at an object 6 m away, he discovers that the image of the object is displaced 12 cm upward. On looking through a point 8 mm to his left of the optical center at the same object, the image is displaced 24 cm to his right. What is the refracting power ($R_x$) of the above lens?

14. If a 15 prism diopter lens is placed 25 mm in front of the center of rotation of the eye, what is the actual rotation made by the eye when viewing an object at 20 cm from the prism?

15. Given:
    OD +2.00 DS with 1 prism diopter base in
    OS +4.00 DS with 1 prism diopter base in
    frame size = 48 mm
    DBL = 20 mm
    PD = 62 mm
    After the lens is cut the decentration is found to be 8 mm in on the right lens. Where is the major reference point on the right lens located?

16. If a +6.00 D lens (index = 1.50) has been decentered in 4 mm and the horizontal box size of the finished lens is 56 mm, what is the difference in edge thickness along the horizontal meridian?

**17.** A prescription of +4.00 D is made up into a lens with an effective diameter of 50 mm. If the unfinished blank is 60 mm, what is the maximum prismatic effect obtainable by decentration at the geometrical center of the finished lens?

**18.** Given the following:
   a right lens = +6.00 D
   index of refraction = 1.5
   size of lens = 50.0 mm round
   nasal thickness = 1.0 mm
   temporal thickness = 2.8 mm
   Edge thickness at the 90° meridian = edge thickness at the 270° meridian
   In relation to the geometric center, where is the horizontal location of the pole?

**19.** A frame has the following boxed dimensions:
   horizontal (A) = 54 mm
   vertical (B) = 48 mm
   effective diameter (ED) = 59 mm
   DBL = 16 mm
   If the patient's PD is 64 mm, what is the minimum blank size required if no allowance is made for cutting and edging operations?

**20.** The $R_x$:
   OD −2.50 DS/1 prism diopter base out
   OS +5.00 DS/1 prism diopter base out
   The prism is obtained by decentration:
   PD = 60/56 mm
   DBL = 20 mm
   horizontal box size = 42 mm
   What is the decentration (amount and direction) of the right and left lenses, respectively?

**21.** A single-vision uncut +3.00 D sphere 60 mm round with knife edges is decentered to produce 3 prism diopters base out before the left eye. The index of refraction is 1.50. What is the approximate thickness at the major reference point?

**22.** Two finished lenses ($n$ = 1.50) have a power of +5.00 D, a round shape with a diameter of 60 mm, and a 1-mm minimum edge thickness. During surfacing and finishing, lens A was not decentered, while lens B was decentered 4 mm in. Assuming spherical surfaces, what is the thickness at the geometrical center of lens A in comparison to that of lens B?

# CHAPTER FIVE

# The Correction of Ametropia

In the first four chapters of this text, ophthalmic lenses have been considered with little or no reference to the eyes for which the lenses were intended; and although a detailed presentation of the optical system of the human eye is more suitable for a textbook on physiological optics than for this textbook, it is appropriate at this time to introduce the concepts and principles involved in the use of lenses for the correction of ametropia.

It should be understood that the term "correction" is misleading, when used in conjunction with ametropia. Ophthalmic lenses (or contact lenses) do not "correct" the ametropia: The term "compensate" would be more appropriate.

As we shall see, the power of a lens for the correction of ametropia depends not only on the optical properties of the eye (specifically, the location of the "far point" of the eye), but also on the distance between the correcting lens and the eye. The required lens power, therefore, may change significantly when an ametrope is corrected with contact lenses rather than glasses.

No discussion of the use of lenses for the correction (compensation) of ametropia can progress very far without a discussion of *accommodation*. The statement "with accommodation at rest" is an integral part of any definition of emmetropia or ametropia. Thus, we find that emmetropia, myopia, and hyperopia are defined in terms of the positions of the *far point of accommodation* and the *near point of accommodation*. We find also that the accommodative state of the eye varies with the mode of correction (glasses vs. contact lenses), varies from one eye to the other in anisometropia, and even varies in the two principal meridians of the eye in astigmatism.

An additional concept that must be dealt with, in a discussion of the correction of ametropia with lenses, is the change in retinal image size brought about by the correcting lens. This change in retinal image size is of little consequence for spherical ametropia of about the same amount for both eyes, but it can often become a significant problem when astigmatism or anisometropia is present.

## 5.1. The Schematic Eye

The refractive state of a given eye is determined by the values of the individual refractive components of the eye and their relationships with one another. These refractive components are corneal refracting power, anterior chamber depth, lens refracting power, and the axial length of the eye. The values of these refractive components are determined both by their radii of curvature (in the case of the cornea and the lens) and by their indices of refraction. For the healthy eye, the indices are thought to show little variation from one eye to another; but the radii of curvature of the cornea and the lens, as well as the anterior chamber depth and the axial length of the eye, are known to vary widely from one eye to another.

It is convenient to characterize the refractive components of the eye, together with its focal points, principal points, and nodal points, in terms of what is known as a *schematic eye*. Such "paper and pencil" eyes are useful in performing a number of calculations relative to the optics of the eye, and are particularly useful for calculations involving *magnification*.

Although a large number of schematic eyes have been designed, the best known, and most detailed, is the "exact" schematic eye (shown in Figure 5-1) designed by Gullstrand.[1] This schematic eye differs from others by stipulating indices of refraction and radii of curvature for both the nucleus and the cortex of the lens, and by representing the cornea as having both front and back surfaces. Gullstrand supplied values for all of the "constants" of this eye, in both the unaccommodated and the accommodated state. Representative values, in the unaccommodated state, are given in Table 5-1. As shown in Figure 5-1, the principal planes of this eye are located just behind the cornea, while the nodal planes straddle the posterior surface of the lens. The total refracting power of the eye is 58.64 D, and the refractive state is 1.00 D of hyperopia.

Gullstrand's "simplified" schematic eye, shown in Figure 5-2, represents the lens as having just one pair

TABLE 5-1
**Schematic Eye Data (Unaccommodated)**

|  |  | Gullstrand's "Exact" Schematic Eye | Gullstrand's "Simplified" Schematic Eye |
|---|---|---|---|
| Indices | Cornea | 1.376 | — |
|  | Aqueous | 1.336 | 1.336 |
|  | Lens cortex | 1.386 | — |
|  | Lens nucleus | 1.406 | 1.413 |
|  | Vitreous | 1.336 | 1.336 |
| Positions | Cornea, anterior | 0 | 0 |
|  | Cornea, posterior | 0.5 mm | — |
|  | Lens, anterior | 3.6 mm | 3.6 mm |
|  | Lens, posterior | 7.2 mm | 7.2 mm |
| Radii | Cornea, anterior | 7.7 mm | 7.8 mm |
|  | Cornea, posterior | 6.8 mm | — |
|  | Lens, anterior | 10.0 mm | 10.0 mm |
|  | Lens, posterior | −6.0 mm | −6.0 mm |
| Powers | Cornea | 43.05 D | 42.74 D |
|  | Lens | 19.11 D | 21.76 D |
|  | Complete eye | 58.64 D | 60.48 D |
| Focal points | Anterior ($A_1F$) | −15.70 mm | −14.99 mm |
|  | Posterior ($A_1F'$) | 24.38 mm | 23.90 mm |
| Axial length |  | 24.00 mm | 23.90 mm |

SOURCE: H. H. Emsley, *Visual Optics*, Vol. 1., Hatton Press, London, 1963, pp. 343–348.

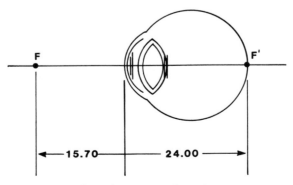

FIGURE 5-1. **Gullstrand's "exact" schematic eye.**

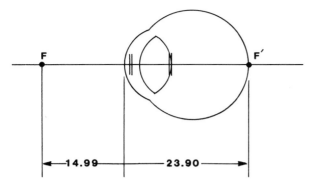

FIGURE 5-2. **Gullstrand's "simplified" schematic eye.**

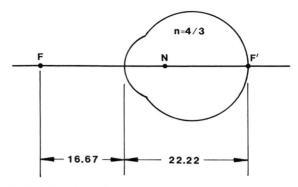

FIGURE 5-3. **The reduced eye.**

of refracting surfaces and a single index of refraction. In addition it represents the cornea as having a single refracting surface, the radius of the anterior surface having been increased slightly to compensate for the small amount of refraction that would occur at its posterior surface. The "constants" of this eye are shown in Table 5-2. As shown in this table, the index of refraction of the aqueous and vitreous are the same as those given for the exact schematic eye (1.336), whereas the index of refraction of the lens (1.413) is greater than that for either the nucleus or the cortex of the lens for the "exact" schematic eye. One advantage of the simplified eye is that the values of many of its components are easy to remember: The radii of curvature of the cornea and the two lens surfaces are, respectively, 7.8, 10.0, and 8.0 mm, and the anterior chamber depth and the lens thickness are both 3.6 mm. Convenient numbers to keep in mind for the anterior focal length of the eye (measured from the cornea) and the axial length of the eye are, respectively, 15 and 24 mm, and the total refracting power of the eye can be considered as 60 D (even though the value given for the exact schematic eye is somewhat less than this amount and that given for the simplified schematic eye is somewhat more).

TABLE 5-2
**Principal Point and Nodal Point Data for Gullstrand's Simplified Schematic Eye**

|                  |        | Unaccommodated Eye | Accommodated Eye |
| ---------------- | ------ | ------------------ | ---------------- |
| Principal points | $A_1P$  | 1.6 mm             | 1.8 mm           |
|                  | $A_1P'$ | 1.9 mm             | 2.1 mm           |
|                  | $F_1P$  | 16.6 mm            | 14.4 mm          |
| Nodal points     | $A_1N$  | 7.1 mm             | 6.6 mm           |
|                  | $A_1N'$ | 7.4 mm             | 7.0 mm           |

Source: H. H. Emsley, *Visual Optics*, Vol. 1, Hatton Press, London, 1963, p. 436 (rounded to first decimal place).

The "reduced eye," shown in Figure 5-3, is further simplified. The total refraction of the reduced eye is considered to take place at the front surface of the cornea, located about 2 mm behind the "real" cornea (actually the location of a single principal point), and the eye is considered to have a single nodal point located about 7 mm behind the "real" cornea.

## 5.2. Emmetropia and Ametropia

When ametropia is discussed, consideration is given to a single eye and accommodation is assumed to be at rest, with the result that the refractive power (or vergence) of the eye is at its minimum value.

An *emmetropic* eye is one in which, with accommodation at rest, infinity and the retina are conjugate points or, in other words, one in which parallel incident rays of light come to a focus on the retina. A standard emmetropic eye (such as Gullstrand's schematic eye) has a "standard" refractive power and a "standard" axial length. As shown in Figure 5-4, emmetropia may result when an eye has

(A) a standard refractive power and a standard axial length
(B) an excessive refractive power and a shorter than normal axial length
(C) a deficient refractive power and an excessive axial length

An *ametropic* eye is defined as an eye in which, with accommodation at rest, infinity and the retina are *not* conjugate, or as one in which parallel rays of light fail to come to a focus on the retina. If parallel rays of light focus in front of the retina the eye is *myopic*, whereas if they focus behind the retina the eye is *hyperopic*.

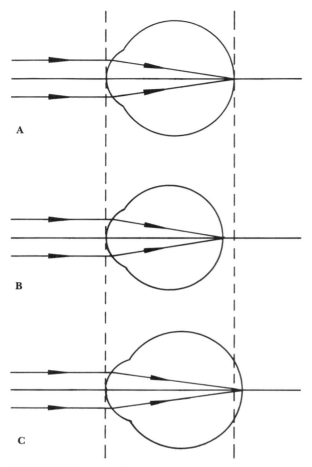

FIGURE 5-4. **Emmetropic eyes.**

An eye having *axial ametropia* is one that has the "standard" refractive power but an axial length that is either too long, having axial myopia (Figure 5-5A), or too short, having axial hyperopia (Figure 5-5B). An eye having *refractive ametropia* is one that has the "standard" axial length but one of the refractive components (the cornea or the crystalline lens) provides either excessive refractive power, resulting in refractive myopia (Figure 5-6A), or too little refractive power, resulting in refractive hyperopia (Figure 5-6B).

The distinction between axial and refractive ametropia may best be understood by considering the eye as having two "lengths": the *axial* length (the physical length of the eyeball) and the *focal length* of the eye's optical system, that is, the distance from the secondary principal plane to the secondary focal plane of the eye. These two "lengths" are shown in Figure 5-7.

If the axial length of the eye and the secondary focal length of the eye's optical system both correspond to those of the schematic eye, the eye will be emmetropic (although Gullstrand's "exact" schematic eye has 1.00 D of hyperopia, we will consider for the present that the schematic eye is emmetropic). If the axial length of the eye is shorter or longer than that of the schematic eye, while the secondary focal length corresponds to that of the schematic eye, the eye will have *axial* hyperopia or myopia, respectively. If the axial length of the eye in question corresponds to that of the schematic eye but the focal length of the optical system is either longer or shorter than that of the schematic eye, the eye will have *refractive* hyperopia or myopia, respectively.

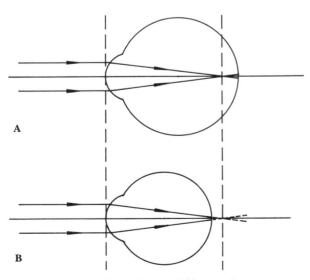

FIGURE 5-5. **A, axial myopia; B, axial hyperopia.**

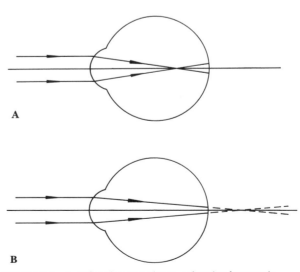

FIGURE 5-6. **A, refractive myopia; B, refractive hyperopia.**

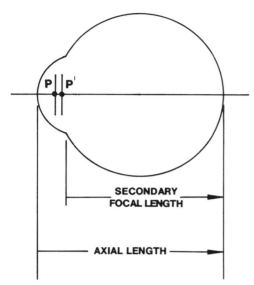

FIGURE 5-7. **The axial length (physical length) of the eyeball as compared to the secondary focal length of the eye's optical system.**

correction a +3.00 D lens (and is therefore hyperopic). Such ambiguity can be avoided by stating that an eye has "3.00 D of myopia" or "3.00 D of hyperopia." In any event, the context in which the terminology is used will usually make the meaning clear.

## 5.3. The Far and Near Points of Accommodation

Accommodation may be defined as the adjustment of the optical system of the eye for vision at various distances, resulting in the retina being conjugate to the object of regard.

The *far point of accommodation* (punctum remotum, or $M_r$) is defined as the object point that is conjugate with the retina when accommodation is fully relaxed, in which case an object placed at the far point will be imaged on the retina. As shown in Figure 5-8, the far point of accommodation for an emmetropic eye is at infinity (Figure 5-8A); the far point of accommodation for an uncorrected myopic eye is a *real* object point in front of the eye (Figure 5-8B); and the far point of accommodation for an uncorrected hyperopic eye is a *virtual* object point located behind the eye (Figure 5-8C).

The *near point of accommodation* (punctum proximum, or $M_p$) is defined as the object point that is

It should be understood that the terms axial and refractive ametropia are, to a great extent, only abstractions. It has been found by Sorsby, Benjamin, Davey, Sheridan, and Tanner[2] that emmetropic eyes have a wide range of axial lengths, corneal refracting powers, and lens refracting powers. They also found that while eyes having ametropia up to about ±4.00 D have axial lengths within the same range as those of emmetropic eyes, eyes having ametropia in excess of ±4.00 D almost always have axial lengths outside the emmetropic range. Therefore, eyes having low refractive errors can be considered as having a *combination* of axial and refractive ametropia, whereas high refractive errors (particularly those well above ±4.00 D) can be considered as having relatively pure *axial* ametropia.

A point of confusion sometimes arises in designating the refractive error of a given eye. Since a myopic eye has an excessive amount of refractive power in relation to its axial length, while a hyperopic eye is deficient in refractive power in relation to its axial length, a myopic eye is sometimes considered to be a *plus* eye (requiring a minus lens) while a hyperopic eye is considered as a *minus* eye (requiring a plus lens). Although most writers make it clear whether they are referring to a myopic eye or a hyperopic eye, a statement that an eye "has a refractive error of +3.00 D" is ambiguous, since it could mean either that the eye has an excess of refractive power of 3.00 D (and is therefore myopic) or that it requires for its

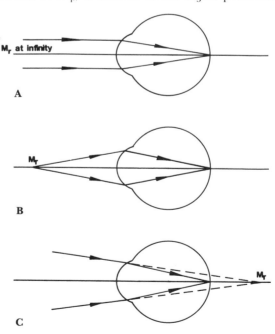

FIGURE 5-8. **Far point of accommodation: A, for an emmetropic eye; B, for an uncorrected myopic eye; C, for an uncorrected hyperopic eye.**

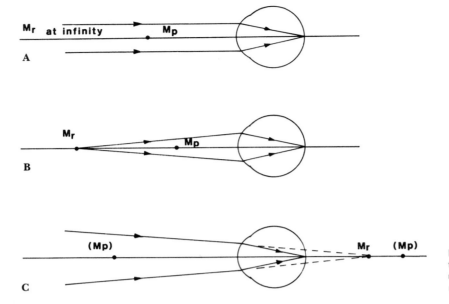

FIGURE 5-9. **Near point of accommodation: A, for an emmetropic eye; B, for an uncorrected myopic eye; C, for an uncorrected hyperopic eye.**

conjugate to the retina with accommodation fully in play. As shown in Figure 5-9, for both an emmetropic eye (Figure 5-9A) and an uncorrected myopic eye (Figure 5-9B), the near point of accommodation is a real object point located in front of the eye, while for an uncorrected hyperopic eye (Figure 5-9C) the near point of accommodation may be either a real object point located in front of the eye or a virtual object point located behind the eye (depending upon both the amount of hyperopia and the amplitude of accommodation).

The amount of hyperopia that falls within the range of the patient's accommodation is known as *facultative* hyperopia; the amount of hyperopia that cannot be overcome by accommodation is called *absolute* hyperopia.

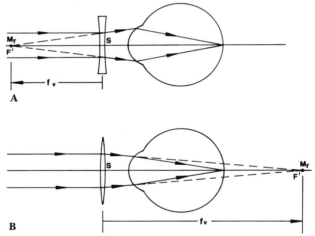

FIGURE 5-10. **The correction of ametropia: A, myopia; B, hyperopia.**

## 5.4. The Correction of Spherical Ametropia

The function of a correcting lens, whether for myopia or hyperopia, is to artifically place the far point of the eye at infinity with the result that infinity (rather than a real object point in front of the eye or a virtual object point behind the eye) will be conjugate to the retina. In order for this to occur, the secondary focal point of the correcting lens must coincide with the far point of the eye. As shown in Figure 5-10A, for a myopic eye both the far point of the eye and the secondary focal point of the correcting lens are located *in front of* the eye. Figure 5-10B shows that for a hyperopic eye both the far point of the eye and the secondary focal point of the correcting lens are located *behind* the eye. Although the word "correct" is used to describe the function of a spectacle lens, as was mentioned before, the lens does not "correct" the refractive error: The term "compensate" would be more appropriate.

It should be noted, as shown in Figure 5-10, that the correcting spectacle lens is considered to be placed in a plane known as the *spectacle plane*, considered to be located at or near the primary focal plane of the eye (which, for Gullstrand's exact schematic eye, is 15 mm in front of the corneal apex). However, the correcting lens is more often placed *closer* to the eye than the primary focal plane, in which case the focal length for a minus lens must be longer (a lens of weaker power), and the focal length for a plus lens must be shorter (a lens of stronger power). Thus we have a change in the *effective power* of the lens. As discussed in Section 3.5, the change in effective power of a lens, as it is moved closer to the eye, can be determined either by comparing the focal lengths without the use of a formula or by the use of the effective power formula given in equation (3.7).

The correction of myopia and hyperopia by the use of spherical lenses will be illustrated by the following examples:

### EXAMPLE 1

An uncorrected myope has a far point of accommodation located 26.5 cm in front of the corneal plane. What would be the required power of the correcting lens if the lens is to be located (*a*) 15 mm in front of the corneal apex; (*b*) 10 mm in front of the corneal apex? See Figure 5-11A.

(*a*)  The power of the correcting lens placed 15 mm from the corneal apex would be the reciprocal of the secondary focal length of the lens.

$$f' = 0.265\,\text{m} - 0.015\,\text{m} = -0.25\,\text{m}.$$

$$F = \frac{1}{-0.25} = -4.00\ \text{D}.$$

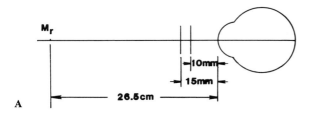

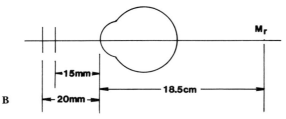

FIGURE 5-11. **The effective power of a lens: the correction of a myopic eye (A) and a hyperopic eye (B) with lenses located at two different distances from the corneal apex.**

(*b*) The focal length of the lens, if located 10 mm in front of the corneal apex, would be

$$f' = -0.265\,\text{m} - 0.010\,\text{m} = -0.255\,\text{m}$$

and

$$F = \frac{1}{-0.255} = -3.92\ \text{D}.$$

Alternatively, we may work Example 1 as an effectivity problem, using Eq (3.7), with a change in *d* of 5 mm.

$$F_B = \frac{F_A}{1 - dF_B} = \frac{-4.00}{1 - 0.005(-4.00)} = -3.92\,\text{D}.$$

### EXAMPLE 2

An uncorrected hyperopic eye has a far point of accommodation located 18.5 cm behind the corneal apex. What would be the required power of a correcting lens located (*a*) 15 mm in front of the corneal apex; (*b*) 20 mm in front of the corneal apex? See Figure 5-11B.

(*a*) The power of the correcting lens, placed 15 mm from the corneal apex, would be the reciprocal of the secondary focal length of the lens.

$$f' = 0.185\,\text{m} + 0.015\,\text{m} = +0.20\,\text{m}.$$

$$F = \frac{1}{+0.20} = +5.00\,\text{D}.$$

(*b*) The focal length of the correcting lens, for a 20-mm fitting distance, would be

$$f' = 0.185\,\text{m} + 0.020\,\text{m} = +0.205\,\text{m}$$

and

$$F = \frac{1}{+0.205} = +4.88\ \text{D}.$$

As an effectivity problem, assigning to *d* a value of −5mm,

$$F_B = \frac{+5.00}{1 - (-0.005)(+5.00)} = +4.88\,\text{D}.$$

## 5.5. Range and Amplitude of Accommodation

Accommodation, like refraction, is normally specified as if it occurred in the *spectacle plane*. Accommodation may be specified in terms of (*a*) range or (*b*) amplitude.

*Range of accommodation* is defined as the linear distance traversed by the point of conjugacy in moving from the far point to the near point of accommodation. It is the distance over which accommodation is available, and hence an object can be seen clearly only if it is located at the far point or near point of accommodation or at any point between the far and near points. Referring to Figure 5-12, the range of accommodation (in either the uncorrected or the corrected state) is given by the statement:

The range of accommodation is equal to the distance from the far point of accommodation to the near point of accommodation.

An *emmetropic* eye that has a finite (real) near point of accommodation (Figure 5-12A) has an infinite range of accommodation (extending from infinity to the near point of accommodation). For a *myopic* eye, the far point and the near point will both always be *real* object points, located at finite distances in front of the eye (Figure 5-12B) with the result that the range of accommodation will always be less than infinity.

If a *hyperopic* eye (having the far point of accommodation behind the eye) has the near point of accommodation located at a finite distance in front of the eye (Figure 5-12C), the range of accommodation is infinite, as it is for the emmetropic eye: The point of conjugacy moves (to the right) from the far point to infinity, as a *virtual* object point, and then from infinity to the near point of accommodation (still moving to the right) as a *real* object point. (Since a real object can be placed only in "real" space, it is obvious that an object cannot be placed at the far point of accommodation of a hyperopic eye or at the near point of accommodation of a hyperopic eye if the near point is located behind the eye.)

*Amplitude of accommodation* is defined as the dioptric difference between the far and near points of accommodation. For an eye in either the uncorrected or the corrected state, the amplitude of accommodation is given by the expression,

$$\text{amplitude of accommodation} = \frac{1}{M_rS} - \frac{1}{M_pS}, \quad (5.1)$$

where $M_rS$ is the linear distance from the far point of

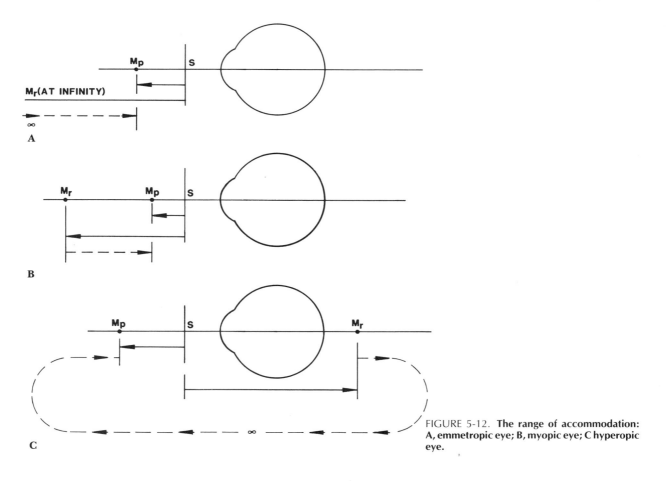

FIGURE 5-12. **The range of accommodation:** A, emmetropic eye; B, myopic eye; C hyperopic eye.

accommodation to the spectacle plane, and $M_pS$ is the linear distance from the near point of accommodation to the spectacle plane.

Typical values of the range and amplitude of accommodation will be illustrated by the following examples:

### EXAMPLE 1

Given an emmetropic eye (having its far point at infinity) as shown in Figure 5-13A, having a near point of accommodation located at a distance of 10 cm in front of the spectacle plane. What are (*a*) the range and (*b*) the amplitude of accommodation?

(*a*) The range of accommodation extends from infinity to a distance of 10 cm from the spectacle plane; therefore it is equal to infinity.

(*b*) The amplitude of accommodation

$$= \frac{1}{M_rS} - \frac{1}{M_pS}$$

$$= \frac{1}{\text{infinity}} - \frac{1}{-0.010}$$

$$= 0 + 10.00 = +10.00 \text{ D.}$$

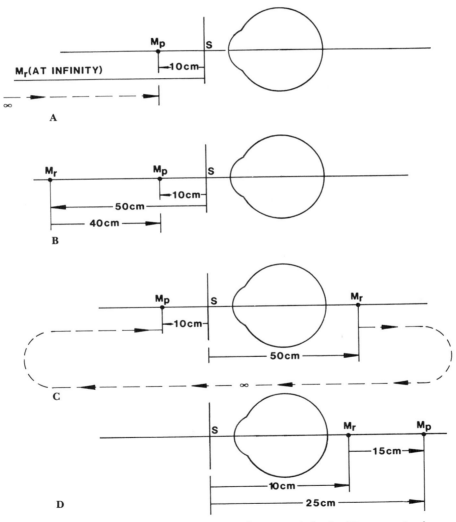

FIGURE 5-13. Determination of range and amplitude of accommodation for (A) an emmetropic eye, (B) an uncorrected myopic eye, and (C,D) uncorrected hyperopic eyes (see Examples 1, 2, 3, and 4 in text).

This example shows that for an emmetropic eye, having its far point of accommodation located at infinity, the *range* of accommodation is equal to infinity minus a finite number, which is, of course, still infinity. The amplitude of accommodation, on the other hand, is equal simply to the reciprocal of the near point of accommodation.

### EXAMPLE 2

Given an uncorrected myopic eye (Figure 5-13B) having a far point of accommodation located 50 cm in front of the spectacle plane and a near point of accommodation 10 cm in front of the spectacle plane. What are (a) the range and (b) the amplitude of accommodation?

(a) The range of accommodation extends from 50 to 10 cm from the spectacle plane, and is therefore equal to 40 cm.

(b) The amplitude of accommodation

$$= \frac{1}{-0.50} - \frac{1}{-0.10}$$

$$= -2.00 + 10.00 = 8.00 \, \text{D}.$$

### EXAMPLE 3

Given an uncorrected hyperopic eye (Figure 5-13C) having a far point of accommodation located 50 cm behind the spectacle plane and a near point of accommodation located 10 cm in front of the spectacle plane. What are (a) the range and (b) the amplitude of accommodation?

(a) The range of accommodation extends from 50 cm behind the spectacle plane to 10 cm in front of the spectacle plane, and is therefore equal to infinity.

(b) The amplitude of accommodation

$$= \frac{1}{0.50} - \frac{1}{-0.10}$$

$$= 2 + 10 = 12.00 \, \text{D}.$$

This example shows that, if an uncorrected hyperopic eye has a finite, or *real*, near point of accommodation, the range of accommodation (like that of the emmetropic eye) will be "infinite."

### EXAMPLE 4

Given an uncorrected hyperopic eye (Figure 5-13D) having a far point of accommodation located 10 cm behind the spectacle plane and a near point of accommodation located 25 cm behind the spectacle plane. What are (a) the range and (b) the amplitude of accommodation?

(a) The range of accommodation extends from 10 cm behind the spectacle plane to 25 cm behind the spectacle plane, and is therefore equal to 15 cm.

(b) The amplitude of accommodation

$$= \frac{1}{0.10} - \frac{1}{0.25}$$

$$= 10.00 - 4.00 = 6.00 \, \text{D}.$$

It should be understood that, since both the far and near points of the eye in this example are located behind the spectacle plane, it is impossible to place a real object at either of these points.

## 5.6. Spectacle Refraction versus Ocular Refraction

As discussed in Section 5.4, a lens that corrects an ametropic eye (with accommodation relaxed) places the image of an infinitely distant object at the far point of the eye. The image formed by the correcting lens becomes an object for the optical system of the eye; and, by definition, if an object lies at the far point of the eye, the image of that object formed by the eye will be focused on the retina.

The amount of ametropia is indicated by the dioptric value of the distance from the far point of accommodation to the spectacle plane. As already discussed, the focal length of a correcting lens depends upon the position of the lens in front of the eye. In practice, spectacle lenses are located as close to the eyes as possible, the limiting factors being anatomical features such as eyelashes, eyebrows, nose, cheeks, and the mechanical fitting aspects of the frame. The typical vertex distance is approximately 13 mm, with the spectacle plane therefore lying closer to the eye than the eye's anterior focal plane.

The power of the correcting lens, specified at the spectacle plane, is termed *spectacle refraction*. On occasion it may be useful to calculate the ametropia in relation to the first principal plane of the eye, in which case it is referred to either as *ocular refraction* or as *principal plane refraction*.

The objective of a clinical refraction is to determine the power of the correcting lens in the spectacle plane. Spectacle refraction is therefore of greater interest to clinicians than ocular refraction, and when reference is made by clinicians to the refractive state of the eye (without any qualification), spectacle refraction is assumed.

When a contact lens is placed on the cornea, the difference between ocular refraction and the refraction brought about by the contact lens is very small, since the distance between the corneal apex and the first principal plane of the eye is only slightly more than 1 mm (1.35 mm, for Gullstrand's "exact" schematic eye). A logical (but not often used) term for the power of a contact lens, specified as the power required for a lens placed in contact with the corneal apex, is *contact lens refraction.*[3]

Although the difference involved in measuring the refractive state at the spectacle plane rather than at the first principal plane of the eye is negligible for small amounts of ametropia, it assumes large proportions for large amounts of ametropia. This can be demonstrated by calculating the ocular refraction for eyes having spectacle refractions of +10.00 D and −10.00 D, assuming in each case that the refractive examination was performed at a distance of 15 mm from the primary principal plane of the eye.

Referring to Figure 5-14, for an eye whose spectacle refraction is +10.00 D, the secondary focal length of a correcting lens located at the primary principal plane of the eye would be

$$10 \text{ cm} - 1.5 \text{ cm} = +8.5 \text{ cm},$$

and the refracting power of the correcting lens, or the ocular refraction, would be

$$\frac{1}{+0.085} = +11.76 \text{ D}.$$

Referring to Figure 5-15, for an eye whose spectacle refraction is −10.00 D, the secondary focal length of a correcting lens located at the primary principal plane of the eye would be

$$-10 \text{ cm} - 1.5 \text{ cm} = -11.5 \text{ cm},$$

and the ocular refraction would be

$$\frac{1}{-0.115} = -8.70 \text{ D}.$$

For these two examples, ocular refraction differs from spectacle refraction by more than 1.75 D for the 10.00 D hyperope and by more than 1.25 D for the 10.00 D myope. For each of these examples, *contact lens refraction*, if it were to be calculated, would differ little from ocular refraction.

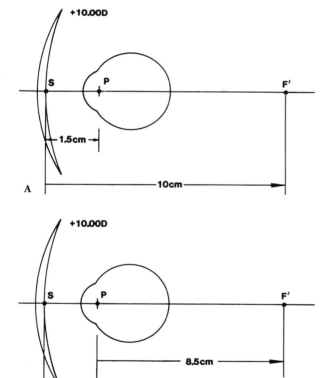

FIGURE 5-14. **For a 10.00 D hyperope, secondary focal length of the correcting lens, (A) measured from the spectacle plane, is 10 cm, B, measured from the primary principle plane, is 8.5 cm.**

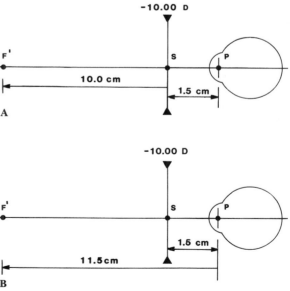

FIGURE 5-15. **For a 10.00 D myope, secondary focal length of the correcting lens, (A) measured from the spectacle plane, is 10 cm, (B) measured from the secondary principle plane, is 11.5 cm.**

## 5.7. Spectacle Accommodation versus Ocular Accommodation

For the sake of convention, accommodation is usually considered to take place at the spectacle plane. However, since it represents a *change in ocular refraction*, accommodation actually takes place at the first principal plane of the eye. The fact that accommodation takes place at the first principal plane must be taken into consideration when determining the amount of accommodation necessary for a corrected ametrope.

The amount of ocular accommodation required of an eye (see Figure 5-16) can be determined by the use of the formula

$$\text{ocular accommodation} = V_d - V_n, \qquad (5.2)$$

where $V_d$ is the vergence of light in the principal plane for an object at a distance of 6 m (assumed to be infinity), and $V_n$ is the vergence of light in the principal plane for an object at near distance (usually assumed to be at 40 cm). Since, in any near-point situation, $V_n$ has a less plus or greater minus value than $V_d$, the eye must supply the amount of accommoda-

tion equal to the difference between the two vergence values in order for the near-point object to be clearly focused. For example, for an emmetropic eye (Figure 5-17A) viewing a distant object the vergence of light at the principal plane is equal to

$$V_d = \frac{1}{\text{infinity}} = 0,$$

and the vergence for an object at a distance of 40 cm from the spectacle plane (considering a distance of 15 mm from the spectacle plane to the first principal plane of the accommodated eye), as shown in Figure 5-17B, is equal to

$$V_n = \frac{1}{-0.40 - 0.015} = -2.41 \text{ D.}$$

This is only 0.09 D less than the value of 2.50 D required for the spectacle plane.

Let us now consider the amount of ocular accommodation required for the corrected 10.00 D hyperope and the corrected 10.00 D myope, for the 40-cm distance.

For the 10.00 D hyperope (Figure 5-18A), it will be recalled from Section 5.6 that the ocular refraction for this eye (the vergence of light at the first principal plane for an object at infinity) is

$$V_d = +11.76 \text{ D.}$$

The vergence of light at the spectacle plane for an object at 40 cm (Figure 5-18B) can be found by first adding the vergence of the light rays entering the lens ($-2.50$ D) to the vergence impressed by the lens

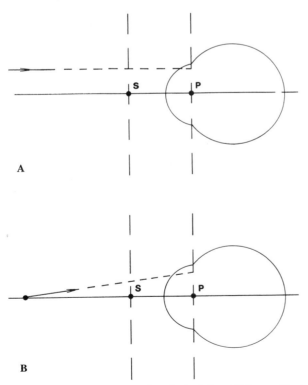

A

B

FIGURE 5-16. A, vergence at the principal plane (P) for an object at infinity ($V_d$); B, vergence at the principal plane (P) for a near object ($V_n$).

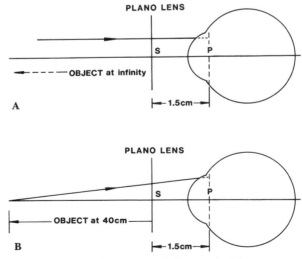

PLANO LENS

OBJECT at infinity

A

1.5cm

PLANO LENS

OBJECT at 40cm

B

1.5cm

FIGURE 5-17. Ocular accommodation required for an emmetropic eye.

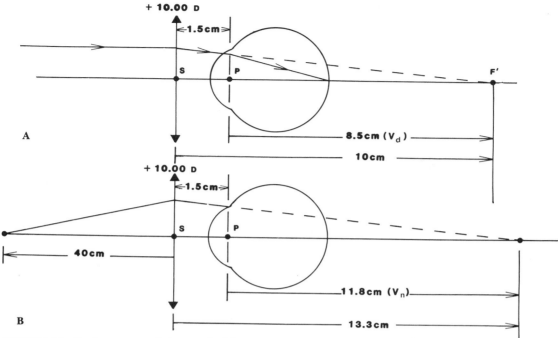

FIGURE 5-18. **Ocular accommodation required for a corrected 10.00 D hyperopic eye.**

(+10.00 D), resulting in a vergence upon leaving the lens of +7.50 D. What we have done here is to apply the formula

$$L' = L + F$$
$$= -2.50 + 10.00$$
$$= +7.50 \text{ D}.$$

The distance from the spectacle plane to the image is found by taking the reciprocal of +7.50 D, which is +0.133 m. Subtracting from this the 15-mm fitting distance, we have

$$+0.133 - 0.015 = +0.118 \text{ m},$$

which is the distance from the principal plane to the image. Therefore,

$$V_n = \frac{1}{+0.118} = +8.47 \text{ D},$$

and

$$\text{ocular accommodation} = V_d - V_n$$
$$= +11.76 - 8.47$$
$$= +3.29 \text{ D}.$$

For the 10.00 D myope (Figure 5-19A), recall from Section 5.6 that the ocular refraction for this

eye (i.e., the vergence of light at the principal plane for a distant object) was

$$V_d = -8.70 \text{ D}.$$

The vergence of light at the spectacle plane for an object at 40 cm (Figure 5-19B) can be found by adding the vergence of the light rays entering the lens (−2.50 D) to the vergence impressed by the lens (−10.00 D), resulting in a vergence upon leaving the lens of −12.50 D, that is,

$$L' = L + F$$
$$= -2.50 + (-10.00)$$
$$= -12.50 \text{ D}.$$

Taking the reciprocal of −12.50 D, we have an image distance from the spectacle plane of −0.08 m. Adding to this the 15-mm fitting distance, we have

$$-0.08 - 0.015 = -0.095 \text{ m}$$

for the distance from the principal plane to the image. Therefore,

$$V_n = \frac{1}{-0.095} = -10.53 \text{ D},$$

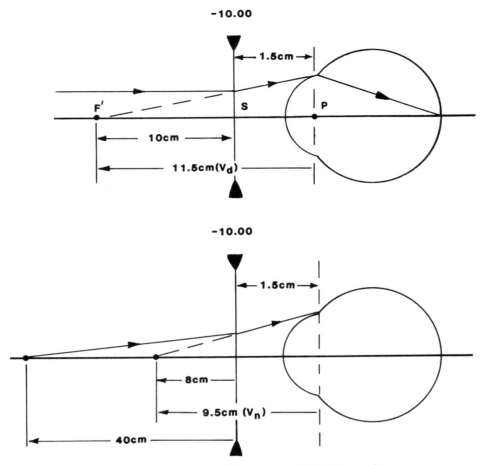

FIGURE 5-19. **Ocular accommodation required for a corrected 10.00 D myopic eye.**

and

$$\text{ocular accommodation} = V_d - V_n$$
$$= -8.70 - (-10.53)$$
$$= +1.83 \text{ D.}$$

Summarizing, we find that the ocular accommodation required for the 10 D hyperope is 3.29 D, while that required for the 10 D myope is only 1.83 D, as compared to the 2.41 D of accommodation required of an emmetrope. The 10 D hyperope must therefore accommodate about 1.00 D *more* than the emmetrope, while the 10 D myope needs to accommodate about 0.50 D *less* than the emmetrope. We usually think of 0.25 D as being the minimum change in stimulus to accommodation that would have an effect on the individual's visual functioning (a "just-noticeable-difference" step), so, as a rule of thumb, for a reading distance of 40 cm the amount of in-

creased accommodation for a hyperope or decreased accommodation for a myope reaches 0.25 D for about 3.00 D of hyperopia or myopia.

The increase in the amount of ocular accommodation required by the spectacle-wearing hyperope is of little consequence for the young hyperope with an adequate amplitude of accommodation. However, when the amplitude of accommodation decreases as the onset of presbyopia approaches, a 4.00 D hyperope would be expected to require a bifocal addition at an *earlier age* than a 4.00 D myope because of the greater accommodative demand.

Since a contact lens fits on the cornea rather than about 13 mm in front of it, and therefore is less than 2 mm in front of the first principal plane of the eye, the newly presbyopic hyperope will have to accommodate *less* while wearing contact lenses than while wearing spectacles, and the newly presbyopic myope will have to accommodate *more* while wearing contact lenses than while wearing spectacles. Consequently, fitting contact lenses to a myope who is an incipient presbyope may appear to hasten the onset of the presbyopia.

## 5.8. Accommodation and Effectivity in Anisometropia

The principle of effectivity applies also to anisometropia. We assume (according to Hering's law of equal innervation to the ocular muscles) that the two eyes accommodate equally. However, if one eye has 10 D of hyperopia and the other has 10 D of myopia, and if each eye is corrected by means of a spectacle lens for distance vision, one eye would be out of focus by more than 1.25 D (as compared to the other) when accommodating at a 40-cm distance. What actually happens in such a case is that the accommodative mechanism responds with no more than the amount of accommodation required for the myopic eye (in fact, it is well established that most people routinely exhibit a *lag of accommodation* for near work, accommodating about 0.50 or 0.75 D less than required by the stimulus).

### EXAMPLE

As a more probable example, consider a patient whose distance prescription is

OD +2.00 DS
OS −4.00 DS

(*a*) What is the induced anisometropia at 40 cm for a young patient having a high amplitude of accommodation? (*b*) For a presbyopic patient, what is the power of the add, for a distance of 40 cm, that would require 1.00 D of ocular accommodation for each eye? (*c*) What is the induced anisometropia for an absolute presbyope who requires a +2.50 D add for a distance of 40 cm?

(*a*) For the young patient, we must calculate the values $V_d$ and $V_n$ for each eye. To calculate $V_d$, we first use the formula $L' = L + F$.

*For the right eye*, the vergence at distance is given by

$$L' = L + F$$
$$= 0 + 2.00 = +2.00 \text{ D},$$
$$l' = \frac{1}{+2.00} = +0.50 \text{ cm}$$

at the spectacle plane.

The image formed by the lens is the object for the eye; so the object distance, $l$ (for the first principal plane of the eye), is given by

$$l = +0.50 - 0.015$$
$$= +0.485 \text{ m}.$$

Therefore,

$$V_d = \frac{1}{0.485} = +2.062 \text{ D}.$$

To find the value of $V_n$, we use the relationship

$$L' = L + F$$
$$= -2.50 + 2.00 = -0.50 \text{ D},$$
$$l' = \frac{1}{-0.50} = -2.00 \text{ m}$$

at the spectacle plane.

The near-object distance, $l$, for the principal plane of the eye is given by

$$l = -2.00 - (0.015) = -2.015 \text{ m}.$$

$$V_n = \frac{1}{-2.015} = -0.496 \text{ D}.$$
$$\text{ocular accommodation} = V_d - V_n$$
$$= +2.062 - (-0.496)$$
$$= 2.56 \text{ D}.$$

*For the left eye*, the vergence at distance is given by

$$L' = L + F$$
$$= 0 + (-4.00) = -4.00 \text{ D},$$
$$l' = \frac{1}{-4.00} = -0.25 \text{ m}$$

at the spectacle plane.

The object distance, $l$ (for the first principal plane of the eye), is given by

$$l = -0.25 - 0.015$$
$$= -0.265 \text{ m}.$$

Therefore,

$$V_d = \frac{1}{-0.265} = -3.77 \text{ D}.$$

To find the value of $V_n$,

$$L' = L + F$$
$$= -2.50 + (-4.00) = -6.50 \text{ D},$$
$$l' = \frac{1}{-6.50} = -0.1538 \text{ m}$$

measured from the spectacle plane.

The near-object distance, $l$, measured from the principal plane of the eye, is given by

$$l = -0.1538 - (0.015) = -0.1688 \text{ m}.$$

$$V_n = \frac{1}{-0.1688} = -5.92 \text{ D.}$$

ocular accommodation $= V_d - V_n$

$$= -3.77 - (-5.92)$$

$$= 2.15 \text{ D.}$$

The induced anisometropia at 40 cm is equal to the accommodation required by the right eye less the accommodation required by the left eye, or,

$$2.56 - 2.15 = 0.41 \text{ D.}$$

We would expect the patient to accommodate no more than the amount required for the myopic eye, or 2.15 D. In any event, if either the right eye or the left eye is clearly focused, the opposite eye will be out of focus by 0.41 D. If a patient wore this prescription and did a significant amount of near work, for maximum binocular efficiency single-vision lenses for near incorporating an add of approximately +0.37 D for the hyperopic eye should be prescribed. The prescription for near work would be

OD +2.37 D<br>OS  −4.00 D.

(b) For the presbyopic patient who is required to use 1.00 D of ocular accommodation for each eye, at a distance of 40 cm, we may use the relationship

required ocular accommodation $= V_d - V_n$.

For an add that would require 1.00 D of ocular accommodation for each eye, $V_d - V_n$ must be equal to 1.00 D for each eye.

For the right eye, as in Example (a), $V_d = +2.062$; and $V_d - V_n$ must equal +1.00 D. Therefore,

$$V_n = V_d - 1.00 = 2.062 - 1.00$$

$$= +1.062 \text{ D,}$$

$$l_n = \frac{1}{+1.062} = +0.942 \text{ m,}$$

from the principal plane.

This distance is the reciprocal of the vergence of light, for the 40-cm distance, at the principal plane of the eye, and therefore is the distance from the principal plane to the image formed by the spectacle lens. In order to find the distance from the spectacle plane to the image, we must add 0.015 m to this value,

$$+0.942 + 0.015 = +0.957 \text{ m,}$$

this being the image distance from the spectacle plane, and by taking the reciprocal of this dis-

tance, we can find the image vergence at the spectacle plane,

$$\frac{1}{+0.957} = +1.045 \text{ D.}$$

In order to determine the power of the lens needed to require 1.00 D of accommodation for the right eye, we use the formula

$$L' = L + F$$

$$+1.045 = -2.50 + F$$

$$F = +3.545 \text{ D.}$$

To find the *add* needed for the right lens, we subtract from this the power of the distance prescription, +2.00 D,

$$\text{add} = +3.545 \text{ D} - 2.00 \text{ D} = +1.545 \text{ D.}$$

For the left eye, as in Example (a),

$$V_d = -3.77 \text{ D.}$$

$$V_d - V_n = +1.00 \text{ D.}$$

$$V_n = -3.77 - 1.00 = -4.77 \text{ D.}$$

$$l_n = \frac{1}{-4.77} = -0.2096 \text{ m.}$$

This is the distance from the principal plane of the eye to the image formed by the spectacle lens. In order to find the distance from the spectacle plane to the image, we must subtract 0.015 m from this value,

$$-0.2096 - (-0.015) = -0.1946 \text{ m,}$$

this being the image distance from the spectacle plane, and we can find the image vergence at the spectacle plane by taking the reciprocal of this distance,

$$L_{S}' = \frac{1}{-0.1946} = -5.14 \text{ D.}$$

To determine the lens needed to permit 1.00 D of accommodation for the left eye, we use the formula

$$L' = L + F$$

$$-5.14 = -2.50 + F$$

$$F = -2.64 \text{ D.}$$

To find the add for the left lens, we subtract from this the power of the distance prescription, −4.00 D,

$$\text{add} = -2.64 \text{ D} - (-4.00 \text{ D}) = +1.36 \text{ D.}$$

In summary, the distance prescription would be

OD +2.00 DS<br>OS −4.00 DS

and the near prescription that would require 1.00 D of ocular accommodation for each eye would be

$$OD +3.55\ DS$$

$$OS -2.64\ DS$$

The difference in the powers of the lenses for the two eyes, for the distance prescription, is 6.00 D, and the difference in the powers for the near prescription to require 1.00 D of ocular accommodation at 40 cm is 6.19 D. The induced anisometropia while looking at 40 cm without an add is 0.41 D. This reduces to approximately 0.19 D (the difference between the powers of the adds) when each eye is permitted to accommodate 1.00 D.

(c) For the absolute presbyope wearing a +2.50 D add for 40 cm, the patient would exert no accommodation with the result that there would be *no* induced anisometropia. The bifocal portion of the lens can be considered as a +2.50 D lens in front of the distance lens. The +2.50 D addition lens renders the rays from an object at 40 cm parallel upon the distance lens (Figure 5-20). The parallel rays impinging upon the distance lens are then focused at the far point, and finally on the retina.

These three examples demonstrate that the induced anisometropia is at a maximum when no add is worn; that the wearing of a bifocal add reduces the amount of induced anisometropia by reducing the amount of accommodation required; and that a maximum add (e.g., a +2.50 D add for a 40-cm reading distance) reduces the amount of induced anisometropia to zero.

## 5.9. Accommodation and Effectivity in Astigmatism

The concepts of accommodation and effectivity may be applied to astigmatism in the same manner in which they apply to anisometropia, if astigmatism is considered to be a "special case" of anisometropia, confined to one eye, with the refracting power differing in the two principal meridians. We would expect, then, that the effectivity changes demonstrated for anisometropia can also be demonstrated for astigmatism.

Hofstetter[1] has pointed out that the cylindrical correction necessary for distance vision may not be adequate for near work. O'Brien and Bannon[5] determined the correction for astigmatism for each of fifty eyes at both distance and near by each of three techniques, and found a constant increase in the power of the cylinder, of from 8% to 10%, for near vision by all three methods. The increase was always in the direction (and in the approximate amount) to be expected as a result of loss of effectivity of the lens power prescribed for distance when used for an accommodating eye for near vision.

In practice, the astigmatism correction determined for distance vision is considered sufficient for correction of astigmatism at the near point. The fact is, however, that in large astigmatic errors for young persons, the astigmatic correction for distance vision represents a substantial undercorrection for near vision. Keeping in mind the idea that 0.25 D represents a "just-noticeable-difference" step, a patient who has from 3.00 to 4.00 D of astigmatism would be out of focus by about 0.25 D in one meridian as compared to the other, when accommodating for a distance of

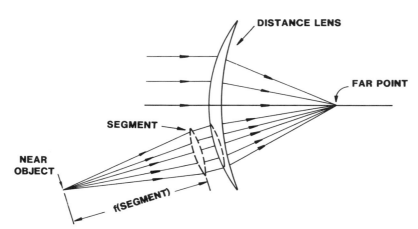

FIGURE 5-20. **With a +2.50 D addition, the rays of light from an object at 40 cm are parallel when striking the distance portion of the lens.**

40 cm. If we take an extreme case (which is very unlikely to occur) in which the refraction of the eye is +5.00 D in one principal meridian and -5.00 D in the other, one meridian would be out of focus by more than 0.75 D as compared to the other meridian.

In addition to changes in power of the astigmatic correction based on effectivity, changes in cylinder *axis* may occur, due to the depression and convergence of the visual axes for near work (Hughes[6]). In cases of high astigmatism, changes in the amount and/or axis of the astigmatic correction should be investigated. If a patient does extensive near work, a separate pair of spectacles should be prescribed, based upon an examination for astigmatism (both axis and power) at the near point.

## 5.10. Retinal Image Size in Uncorrected Ametropia

In determining the size of the retinal image, one is concerned primarily with finding the *linear distance* between two image points that correspond to two known object points.

Several methods are available for determining retinal image size. However, when dealing with an ametropic eye in the uncorrected state, the *chief ray* method has the advantage that the size of the retinal image is unaffected by *image blur*: This is because a chief ray (as shown in Figure 5-21) passes through the center of the entrance pupil of the eye (E) with the result that it passes through the center of the retinal blur circle formed by the object point. Using the chief ray method, the retinal image size is found by determining the subtense of the retina of the angle formed at the center of the exit pupil (E′) by a pair of chief rays. The chief rays originate from a pair of object points, whose separation represents the *size of the object* under consideration.

For simplicity, one of the object points (point P in Figure 5-21) is placed on the optic axis of the eye, whereas the second object point, Q, is an off-axis

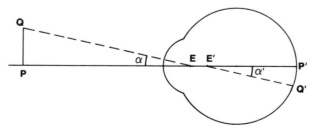

FIGURE 5-21. **Chief ray method of determining retinal image size.**

point. The size of the retinal image, Q′P′, is determined by tracing the chief ray from the object point Q through the eye to the image point Q′, on the retina. This ray is directed toward the center of the entrance pupil, E, and emerges into the vitreous as if it came from the center of the exit pupil, E′.

### Retinal Image Size in Uncorrected Axial Ametropia

Referring to Figure 5-22, the angle α′ made between the optic axis and the chief ray as it emerges from the exit pupil forms a *constant ratio* with the angle α between the optic axis and the incident chief ray. The retinal image size is therefore directly proportional both to the angle that the incident chief ray makes with the optic axis, α, and to the axial length of the eye. Consequently, for a given object size, the size of the retinal image of an uncorrected axially ametropic eye varies *only* with the axial length of the eye.

As shown by the three retinal positions in Figure 5-22, the retinal image is of "normal" size in the emmetropic eye, smaller than "normal" in an axially hyperopic eye, and larger than "normal" in the axially myopic eye. It follows that if the optical system of the eye (made up of the cornea, the anterior chamber, and the lens) is similar to that of the emmetropic schematic eye, the greater the amount of axial hyperopia the smaller the retinal image, while the greater the amount of axial myopia the larger the retinal image.

### Retinal Image Size in Uncorrected Refractive Ametropia

When the ametropia is refractive rather than axial, the refractive power of the eye's optical system varies from one eye to another, but the axial length is equal to that of the standard (schematic) emmetropic eye. As shown in Figure 5-23, the power of the eye's refractive system has no effect on the angle of incidence of the chief ray as it passes through the entrance pupil, so the retinal image size is essentially the same for uncorrected refractive myopia, emmetropia, and uncorrected refractive hyperopia.

## 5.11. Retinal Image Size in Corrected Ametropia

### Axial Ametropia

It will be recalled that a ray passing through the primary focal point of any lens or optical system will

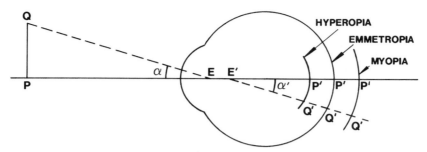

FIGURE 5-22. **In uncorrected axial ametropia, the size of the retinal image varies only with the axial length of the eye.**

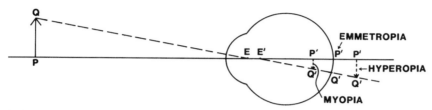

FIGURE 5-23. **In uncorrected refractive ametropia, the size of the retinal image, in both myopia and hyperopia, is essentially the same as in the emmetropic eye. The size of the retinal image is given by the distance between the centers of the blur circles formed at points P′ and Q′ on the retina.**

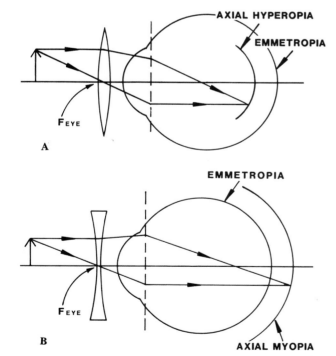

FIGURE 5-24. **If an eye has purely axial ametropia, a correcting lens placed in the primary focal plane of the eye will provide the eye with the same retinal image size as the (theoretical) emmetropic eye. A, axial hyperopia; B, axial myopia.**

be refracted in such a way as to emerge from the optical system *parallel to the optic axis.* Therefore, as shown in Figure 5-24, if the correcting lens for an *axially ametropic eye* is placed in the primary focal plane of the eye, the size of the retinal image is determined *only* by the distance from the optic axis to the ray passing through the primary focal point of the eye. Therefore, in all cases of purely axial ametropia (whether myopia or hyperopia), the size of the retinal image formed by the combination of the eye and the correcting lens is the same as that of the emmetropic schematic eye. Consequently, if it is known that an eye has *purely axial* myopia or hyperopia, a correcting lens placed at the primary focal plane of the eye will provide that eye with the same retinal image size as that of the (theoretical) emmetropic eye.

A comparison of Figures 5-22 and 5-24 shows, however, that for the eyes in question the correcting lens *enlarges* the size of the retinal image for an axially hyperopic eye, and *reduces* the size of the retinal image for an axially myopic eye.

*Refractive Ametropia*

When the correcting lens for an eye having refractive ametropia is placed at the primary focal point of the eye, as shown in Figure 5-25, the situation occurring for the axially ametropic eye does not prevail. Rather

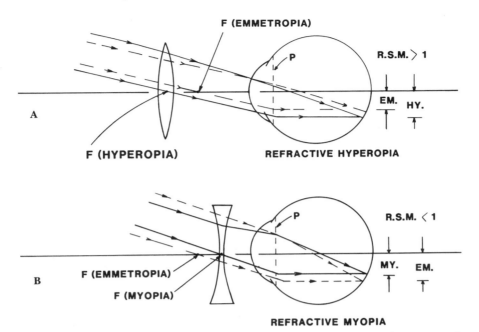

FIGURE 5-25. **For an eye having purely refractive ametropia, the retinal image for a corrected myopic eye will be smaller than that for an emmetropic eye, while the retinal image for a corrected hyperopic eye will be larger than for an emmetropic eye.**

than the retinal image size being basically similar for myopia, emmetropia, and hyperopia, Figure 5-25A shows that the retinal image for the corrected refractively hyperopic eye will be *larger* than that for the emmetropic eye, while Figure 5-25B shows that the retinal image for the corrected refractively myopic eye will be *smaller* than that for the emmetropic eye.

However, a comparison of Figures 5-23 and 5-25 shows that for the eye in question the correcting lens *enlarges* the size of the retinal image for a refractively hyperopic eye, and *reduces* the size of the retinal image for a refractively myopic eye—exactly the same situation occurring with an *axially* myopic or hyperopic eye.

### Magnification by a Spectacle Lens

The magnification brought about by a spectacle lens will be discussed in more detail in Chapter 10. For the present, two commonly used terms will be defined:

**Spectacle Magnification.** This is the magnification brought about by a correcting lens; it compares the retinal image size with the correcting lens to that without the lens. It is due to both the *power* and the *shape* (or *form*) of the lens. As shown in the above examples, for both axial and refractive ametropia a plus lens enlarges the retinal image size of the eye in question, whereas a minus lens reduces the retinal image size.

**Relative Spectacle Magnification.** This is the comparison of the size of the retinal image of the eye, with the correcting lens, to the size of the retinal image of the theoretical (schematic) emmetropic eye. As discussed above and shown in Figures 5-24 and 5-25, relative spectacle magnification depends upon whether the ametropia is axial or refractive, the image size being essentially the same as that for the emmetropic eye in *axial* ametropia, but magnified in hyperopia and minified in myopia, as compared to the emmetropic eye, in *refractive* ametropia.

### References

1. Emsley, H.H. *Visual Optics*, Vol. 1, pp. 323–348. The Hatton Press, London, 1963.
2. Sorsby, A., Benjamin, B., Davey, J. B., Sheridan, M., and Tanner, J. M. *Emmetropia and Its Aberrations: A Study of the Correlation of the Optical Components of the Eye*. Medical Research Council Special Report Series, No. 203, H. M. Stationery Office, London, 1957.

3. Grosvenor, T. *Contact Lens Theory and Practice*, Ch. 5. Professional Press, Chicago, 1963.
4. Hofstetter, H. W. Correction of Astigmatism for Near Work. *Amer. J. Optom. Arch. Amer. Acad. Optom.*, Vol. 23, pp. 121–131, 1945.
5. O'Brien, J. M., and Bannon, R. E. Accommodation and Astigmatism. *Amer. J. Ophthal.*, Vol. 30, pp. 289–296, 1947.
6. Hughes, W. Changes in Axis of Astigmatism with Accommodation. *Arch. Ophthal.*, Vol. 26, pp. 742–749, 1941.

## Questions

1. Given a pair of lenses:
   OD +6.00 D
   OS plano

   The lenses are 15 mm from the primary principal planes of the eyes. Assume the left eye is in focus for an object at 25 cm and that differential accommodation is impossible. What is the uncorrected ocular anisometropia for an object *at 25 cm* from the spectacle plane?

2. Given a pair of lenses:
   OD +2.00 D
   OS −4.00 D

   Assume that the spectacle plane is 15 mm from the primary principal plane of the eyes and that differential accommodation is impossible. The target is located at 40 cm from the spectacle plane. What are the adds for the right eye and left eye, respectively, that would require 1.00 D of ocular accommodation for each eye?

3. (*a*) This distance correction is
   OD plano
   OS −5.00 D

   The lenses are 20 mm from the primary principal planes of the eyes. The reading distance is 25 cm from the spectacle plane. The left eye is given a +2.00 D add. What must the power of the add of the right lens be in order that both eyes will be in focus for the near distance simultaneously? Assume that the eyes cannot accommodate differentially.
   (*b*) In the above problem what is the anisometropia for an object at 25 cm without the use of the bifocal? Assume the eyes cannot accommodate differentially.

4. What is the stimulus to accommodation for a person wearing a far-point correction under the following conditions? Lens power = +10.00 D; distance from the lens to the primary principal plane = 15 mm; distance from the spectacle plane to the object = 40 cm.

5. Rework No. 4 assuming a +1.00 D bifocal add.

6. The following lens corrects an eye for infinity: +2.00 DS −7.00 DC x 90. If the horizontal meridian of the eye is in focus for an object 25 cm from the spectacle plane, what $R_x$ is needed to bring the vertical meridian into focus? Assume the spectacle is 20 mm from the primary principal plane of the eye and that astigmatic accommodation is impossible.

7. The following lens corrects the eye for infinity: +3.00 DS +5.00 DC x 180. Assume the spectacle plane is 15 mm from the primary principal plane of the eye and that astigmatic accommodation is impossible. If the vertical meridian is in focus for an object 30 cm from the spectacle plane, what lens power is needed to bring the horizontal meridian into focus?

8. A 3 D hyperope has an amplitude of accommodation of 6 D. When uncorrected, where is the near point of accommodation located?

9. If the accommodative amplitude is 5 D, and the far point of accommodation is 20 cm, what is the range of accommodation?

10. A patient has a punctum remotum of 100 cm when looking through a +4.00 D lens. When looking through a −2.00 D lens the punctum proximum is 20 cm. What is the amplitude of accommodation?

11. A 50-year-old patient with findings indicating an amplitude of accommodation of 3.00 D is given a bifocal prescription providing a +2.00 D add. Through the reading portion of the lenses, where is the near point of accommodation?

12. Light from a patient's retina emerges from his unaccommodated eye, passes through a +6.00 D lens, and then comes to a focus 25 cm in front of his spectacle plane. What is the patient's refractive error?

13. A patient's distance subjective is +1.50 DS OU (both eyes). With trial lenses correcting the distance ametropia and an add of +2.50 D, her range of clear vision at near is from 40 to 25 cm. What is this patient's amplitude of accommodation?

**14.** An eye, with no lens in place, has a near point of accommodation of 25 cm. The conjugate focus of the retina is located 20 cm in front of the eye when using a +2.50 D lens and half of the amplitude of accommodation. What is the refractive error of the eye?

**15.** An eye, with no lens in place, has a near point of accommodation of $33\frac{1}{3}$ cm. The conjugate focus of the retina is located 50 cm in front of the eye when using a +2.00 D lens and half of the amplitude of accommodation. What is the amplitude of accommodation of the eye?

# CHAPTER SIX

# Aberrations and Ophthalmic Lens Design

Ophthalmic lenses are subject to a number of aberrations, which degrade their performance in varying degrees. These include chromatic aberration, which occurs as a result of the heterochromatic nature of ambient light (whether sunlight or artificial light), and the monochromatic aberrations, which occur whenever incident light is not confined to a narrow bundle of paraxial rays. It is therefore desirable that the design of ophthalmic lenses take these aberrations into consideration, minimizing their effects to the extent that seems to be practical.

Chromatic aberration is due, for the most part, to the material from which the lens (or prism) is made, and therefore can be controlled to some extent by the judicious selection of materials. The monochromatic aberrations, on the other hand, are due less to the properties of the material than to factors such as the size of the lens aperture, the angle the incident rays make with the optic axis of the lens, and the position of the lens in relation to the eye. They are controlled by varying the front and back surface powers and the thickness of the lens, as well as by varying vertex distance.

Although an understanding of the principles of lens design was evident early in the nineteenth century, the development of commercially available lenses designed to minimize the monochromatic aberrations began to take place almost a century later. During the early decades of the present century, each major lens manufacturer developed its own series of "corrected curve" lenses, and the responsibility of the practitioner was simply that of selecting the "trade name" lens that he or she judged was the best for the patient at hand. However, in recent years, recommendations made by a committee of the American National Standards Institute, together with a decrease in the influence of the major lens manufacturers, have combined to place a greater responsibility for ophthalmic lens design on the shoulders of the individual practitioner.

# INTRODUCTION

## 6.1. Laws of Geometrical Optics

In geometrical optics we study a number of laws which explain the way light behaves. According to the law of rectilinear propagation of light, a *ray* of light is the direction along which light travels; in a homogeneous, isotropic medium, the rays travel in straight lines. The rays which constitute a *beam* of light behave independently: When two rays of light intersect, they continue along their course without interaction. According to the laws of reflection, the incident ray, the reflected ray, and the line normal to the reflecting surface at the point of incidence all lie in the same plane, and the angle of reflection is equal to the angle of incidence, each lying on opposite sides of the normal. According to the laws of refraction, the incident ray, the refracted ray, and the line normal to the surface at the point of incidence all lie in the same plane and the sine of the angle of incidence forms a constant ratio with the sine of the angle of refraction, the relationship known as Snell's law.

In geometrical optics we accept these laws as established properties of light. These laws hold true, but they involve a certain amount of oversimplification. For instance, a ray of light, although a convenient device for representing the direction of propagation of a wavefront, is an abstraction and in fact does not really exist. In geometrical optics and in ophthalmic lens theory we have treated optical systems as if they were perfect imaging systems. In doing so, we have assumed (1) that all rays originating from one object point reconverge to form one conjugate image point (real or virtual) after treatment by the optical system; (2) that a plane object situated at right angles to the optical axis is imaged point for point in a similar plane in image space; and (3) that the image is an exact geometrical reproduction of the object.

The above assumptions are valid only if we restrict ourselves to the following practices: (1) using *monochromatic* light, for which the optical system would have a single index of refraction and would have no chromatic aberration; (2) limiting the rays involved in image formation entirely to rays whose paths throughout the system are confined to a small cylindrical region immediately surrounding the optical axis of the system, known as the *paraxial* region. A paraxial ray lies near the optical axis and hence the angles and the ray heights are very small.

The sine of an angle may be expressed in terms of the angle, in radians, in the following expression:

$$\sin \alpha = \alpha - \frac{\alpha^3}{3!} + \frac{\alpha^5}{5!} - \frac{\alpha^7}{7!} \cdots$$

If the angles are small, the first term can be used without serious error and this assumption forms the basis for paraxial theory, also known as *first-order theory*. The error involved in equating the angle, expressed in radians, to its sine reaches approximately 1% for an angle of 15°. First-order theory is useful, since any well-designed optical system will nearly conform to first-order calculations, and hence first-order optical imagery provides a reference against which to measure the extent of the aberrations.

In actual practice, it is often found necessary to extend the limits of effective rays beyond the paraxial region. The failure of rays to arrive at unique positions based on object point–image point correspondence of first-order theory creates the various aberrations which either impair the quality of the image or serve to deform the image.

An approximate conception of the deviation (aberration) from perfect image formation can be obtained by using the first and second numbers of the expression for sin α, that is,

$$\sin \alpha = \alpha - \frac{\alpha^3}{3!}.$$

This approximation is called *third-order theory*, and provides accuracy to four decimal places for values up to 23°, or about 0.4 radian.

Seidel's[1] mathematically complicated calculations of third-order monochromatic theory addressed five aberrations: spherical aberration, coma, oblique astigmatism, curvature of image, and distortion. In Seidel's calculations the aberrations, that is, the departure from the imagery produced by paraxial (Gaussian) theory, are delineated by five different sums, $S_1$, $S_2$, $S_3$, $S_4$, and $S_5$, which are the coefficients of the terms of the equations. When the more oblique rays perform as those in the paraxial region, the value of each of the five terms is zero. If Seidel's first condition, $S_1$, is satisfied (that is, if $S_1 = 0$) no spherical aberration is present. If his second condition is satisfied ($S_2 = 0$) then no coma exists; if the third condition is satisfied ($S_3 - S_4 = 0$) the oblique astigmatism is eliminated; if the fourth condition is satisfied ($S_3 = S_4 = 0$) curvature of image is eliminated; and if $S_5 = 0$, distortion is eliminated. It should be understood that each succeeding term is meaningless unless the preceding terms have been satisfied, that is, reduced to zero.

Seidel's mathematical expressions provide the lens designer with approximate criteria for initial construction of the image. However, an accurate representation of the image is completed by trigonometrically tracing pertinent rays through the system.

Computers have reduced the time and effort necessary for such a process.

Although detailed ophthalmic lens design is beyond the scope of this textbook, a basic understanding of aberrations is required to evaluate the performance of ophthalmic lenses and to provide some insight into the problems confronting the lens designer. For this purpose, each aberration will be discussed as if it were the only aberration present, and as if no interaction occurs among aberrations.

# CHROMATIC ABERRATION AND ACHROMATIC LENSES

## 6.2. Chromatic Aberration

Chromatic aberration occurs solely due to the nature of the *material* from which a lens or prism is made, and is due to the fact that the material has differing indices of refraction for light of differing wavelengths. Chromatic aberration can be considered either as *longitudinal* (axial) or as *transverse* (lateral) aberration.

*Longitudinal (axial) chromatic aberration* refers to the fact that the secondary focal length of a lens will be different for each of the monochromatic constituents of white light. If white light is incident upon a lens (as shown in Figure 6-1) and if filters are successively used to isolate various wavelengths, it will be found that the shortest wavelength (violet) undergoes the greatest refraction and the longest wavelength (red) undergoes the least refraction. Since longitudinal chromatic aberration requires an optical system that *changes the vergence* of incident light, it is not produced by a prism (unless the prism also has refracting power).

The *bichrome* test, a test used to determine the endpoint in subjective refraction, takes advantage of

the longitudinal chromatic aberration of the eye: One side of the chart is projected through a red filter and the other side through a green filter, with the result that if the eye is in focus for the yellow portion of the spectrum (to which the eye is most sensitive) the red and green charts will appear to be equally distinct. However, if the red side of the chart is more sharply focused than the green side, this indicates that the eye is relatively *myopic* for yellow light; if the green side of the chart is more sharply focused, the eye is relatively *hyperopic* for yellow light.

*Transverse (lateral) chromatic aberration*, in the case of a lens, can be expressed (1) as the differences in the image size (linear magnification) of chromatically formed images or (2) as angular dispersion (differences in prismatic effects). Whether expressed in linear or angular terms, transverse chromatic aberration occurs because the equivalent focal length of a lens is different for components of light having different refractive indices.

1. *Linear Magnification Differences.* Since the secondary focal length of a lens will be different for each of the monochromatic constituents of white light, the axial image positions of these constituents will differ and hence, for an extended object, their image sizes will differ (Figure 6-2). When expressed in this manner, transverse chromatic aberration is often referred to as the *chromatic difference in magnification.* If the longitudinal chromatic aberration of a lens has been corrected by the use of an achromatic doublet (see Section 6.7), transverse chromatic aberration may still be present. Even though the blue-colored image (for example) occupies the same position on the axis

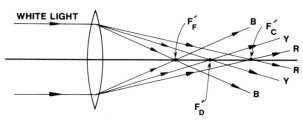

FIGURE 6-1. **Longitudinal (axial) chromatic aberration (C.A.).**

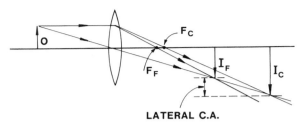

FIGURE 6-2. **Transverse (lateral) chromatic aberration (C.A.); also referred to as the chromatic difference in magnification.**

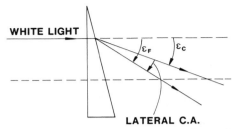

FIGURE 6-4. **A prism can manifest transverse chromatic aberration (C.A.) but cannot manifest longitudinal chromatic aberration.**

of the lens as the red-colored image, the principal planes for the blue and red light are not located in the same positions, with the result that the two images will not be the same size.

2. *Differences in Prismatic Effects.*

a. *Lenses.* The transverse chromatic aberration of a lens can be expressed in terms of the differences in prismatic effects created by (for example) blue and red light. Referring to Figure 6-3, the prismatic effect for blue light is equal to $dF_F$, and the prismatic effect for red light is equal to $dF_C$, with $d$ being the distance from the optic axis of the lens. Expressed in prism diopters, the lateral chromatic aberration is therefore equal to

$$dF_F - dF_C = d(F_F - F_C).$$

b. *Prisms.* Since a prism does not change the vergence of light, it does not manifest axial chromatic aberration. It does, however, manifest angular dispersion (lateral chromatic aberration), as shown in Figure 6-4. The angular dispersion is defined as the difference in prismatic effects for blue and red light, that is,

$$\varepsilon_F - \varepsilon_C.$$

Chromatic aberration will be discussed further in Section 6.4 (for prisms) and in Section 6.6 (for lenses).

Due to the presence of chromatic aberration, a wearer of high-powered lenses may occasionally see *color fringes* around objects. The amount of chromatic aberration is dependent upon the material of

which the lens is made: For example, it is relatively high for *flint* glass, used for the manufacture of some bifocal segments.

Since chromatic aberration requires light containing a range of wavelengths it does not occur with *monochromatic* light. However, light falling upon an ophthalmic lens is very seldom monochromatic, with the result that all substances from which lenses and prisms are made are subject to chromatic aberration. Since it is due to the material itself, chromatic aberration cannot be reduced or eliminated by changing the form (the amount of *bend*) of the lens. However, it can be controlled by the use of a *doublet*, which is a lens made of two kinds of glass. Chromatic aberration does not occur in reflection, because the angle of reflection is the same for all wavelengths of light; nor does it occur when light is refracted by parallel plates of glass.

## 6.3. Chromatic Dispersion

When we specify the index of refraction of a substance, we normally specify the index for the *sodium D* line. This is the Fraunhofer line having a wavelength of 589 nm. For example, ophthalmic crown glass is specified as having an index of refraction of 1.523: This is the index of refraction for the sodium D line. The index of refraction of the D line, $n_D$, is defined as follows:

$$n_D = \frac{\text{velocity of sodium light in a vacuum}}{\text{velocity of sodium light in the medium}}.$$

In order to discuss chromatic dispersion, we must consider the index of refraction of two additional spectral lines: $n_C$ refers to the index of refraction of the Fraunhofer C line, at the red end of the spectrum and having a wavelength of 656 nm; and $n_F$ refers to the index of refraction of the Fraunhofer F line, having a wavelength of 486 nm.

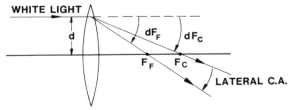

FIGURE 6-3. **Transverse chromatic aberration (C.A.) expressed as the difference in prismatic effects created by red and blue light.**

*Mean refractivity* is given by the expression

$$n_D - 1,$$

while *mean dispersion* is given by the expression

$$n_F - n_C.$$

It is usually (although not always) true that the higher the mean refractive index $n_D$ the greater the mean dispersion ($n_F - n_C$). The *dispersive power* of a medium is given by the ratio of mean dispersion to mean refractivity,

$$\frac{n_F - n_C}{n_D - 1}.$$

The reciprocal of dispersive power is known as *refractive efficiency*, or *constringence*, and is indicated by the Greek letter $\nu$ (and is therefore called the *nu value*):

$$\nu = \frac{n_D - 1}{n_F - n_C}. \tag{6.1}$$

In comparing the nu values of two media, if the media have the same mean dispersion the one having the higher mean refractivity will have the higher nu value, but if the two media have the same refractivity, the medium having the lower mean dispersion will have the higher nu value (and the lower dispersive power).

The advantage of using nu value (rather than dispersive power) to specify the chromatic dispersion of a lens is that it is expressed as a *whole number*, and since nu value is the reciprocal of dispersive power, the most desirable lens materials are generally those with the higher nu values. For example, the nu value of ophthalmic crown glass is 59, whereas that of flint glass is 30.

## 6.4. Chromatic Aberration in Prisms

As shown in Section 6.2, the transverse chromatic aberration of a prism can be expressed as the difference in the deviating power of the prism for the Fraunhofer F and C lines, or

transverse chromatic aberration $= \varepsilon_F - \varepsilon_C$.

Since the refracting angle and the deviating power of a prism are related by the expression

$$\varepsilon^\circ = \beta^\circ(n - 1),$$

it follows that

$$\varepsilon_D = \beta(n_D - 1)$$
$$\varepsilon_F = \beta(n_F - 1)$$
$$\varepsilon_C = \beta(n_C - 1)$$

and

$$\varepsilon_F - \varepsilon_C = \beta(n_F - 1) - \beta(n_C - 1)$$

Since

$$\beta = \frac{\varepsilon_D}{n_D - 1}$$

$$\varepsilon_F - \varepsilon_C = \frac{\varepsilon_D(n_F - 1)}{n_D - 1} - \frac{\varepsilon_D(n_C - 1)}{n_D - 1}.$$

$$\varepsilon_F - \varepsilon_C = \frac{\varepsilon_D(n_F - n_C)}{n_D - 1}.$$

Since

$$\frac{1}{\nu} = \frac{n_F - n_C}{n_D - 1},$$

$$\varepsilon_F - \varepsilon_C = \frac{\varepsilon_D}{\nu}. \tag{6.2}$$

This equation tells us that the chromatic dispersion of a prism is equal to the deviation of the D line divided by the nu value of the material from which the prism is made. For example, for a prism made of crown glass having a nu value of 60, and having a deviating power of 5 prism diopters,

$$\frac{\varepsilon_D}{\nu} = \frac{5}{60} = 0.08 \text{ prism diopter.}$$

For a prism made of a given material, the chromatic dispersion increases in direct proportion to the power of the prism.

In view of the fact that the transverse chromatic aberration of a prism is equal to

$$\varepsilon_F - \varepsilon_C,$$

and since equation (6.2) shows that

$$\varepsilon_F - \varepsilon_C = \frac{\varepsilon_D}{\nu},$$

it follows that the transverse chromatic aberration of a prism is equal to

$$\frac{\varepsilon_D}{\nu}.$$

## 6.5. Achromatic Prisms

An achromatic prism is a prism that produces deviation without dispersion, and is constructed by placing two prisms made of different materials in direct contact with the bases oriented in opposite directions. Such a prism can be made achromatic for two wavelengths, usually for the F and C lines. The rays of light for the two wavelengths for which the prism is achromatic will emerge parallel, as shown in Figure 6-5. In order to be achromatic, the chromatic dispersion of the two prisms must be equal, or

$$\frac{\varepsilon_1}{v_1} = \frac{\varepsilon_2}{v_2}.$$

Since the prisms have their bases in opposite directions, the final deviation, $\varepsilon$, is given by

$$\varepsilon = \varepsilon_1 - \varepsilon_2$$

and

$$\varepsilon_1 = \frac{v_1}{v_2}(\varepsilon_2) = \frac{v_1}{v_2}(\varepsilon_1 - \varepsilon)$$

and also

$$\varepsilon_2 = \frac{v_2}{v_1}(\varepsilon_1) = \frac{v_2}{v_1}(\varepsilon_2 + \varepsilon).$$

The deviating powers of the two prisms are given by the derived expressions

$$\varepsilon_1 = \frac{v_1 \varepsilon}{v_1 - v_2} \tag{6.3}$$

and

$$\varepsilon_2 = \frac{v_2 \varepsilon}{v_1 - v_2}, \tag{6.4}$$

which can be used to design an achromatic prism.

**EXAMPLE**
Given two prisms, one having an index of refraction of 1.50 and a nu value of 60, and the other having an index of refraction of 1.60 and a nu

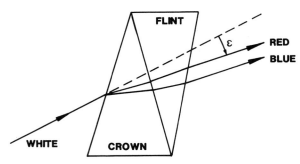

FIGURE 6-5. **An achromatic prism.**

value of 40. Design an achromatic prism having a power of 2 prism diopters.

$$\varepsilon_1 = \frac{60(2)}{60 - 40} = 6 \text{ prism diopters}$$

and

$$\varepsilon_2 = \frac{40(2)}{60 - 40} = 4 \text{ prism diopters.}$$

Chromatic aberration is responsible for the color fringes sometimes seen with high-power prisms during phoria and fusional vergence testing. However, for the low-power prism components that are occasionally used in ophthalmic lenses (usually no stronger than 2 or 3 prism diopters), chromatic aberration seldom presents a problem. In any event, chromatic doublets are not even seriously considered as ophthalmic appliances, since both thickness and weight would be unacceptable.

## 6.6. Chromatic Aberration in Lenses

*Longitudinal Chromatic Aberration*

The formula for the longitudinal chromatic aberration of a lens can be derived by using the lens maker's formula for the Fraunhofer D, F, and C lines:

$$F_D = (n_D - 1)(R_1 - R_2),$$
$$F_C = (n_C - 1)(R_1 - R_2),$$
$$F_F = (n_F - 1)(R_1 - R_2).$$

Therefore,

$$F_F - F_C = (n_F - 1)(R_1 - R_2) - (n_C - 1)(R_1 - R_2),$$
$$F_F - F_C = (n_F - n_C)(R_1 - R_2).$$

Since

$$R_1 - R_2 = \frac{F_D}{n_D - 1}$$

$$F_F - F_C = \frac{n_F - n_C}{n_D - 1}(F_D)$$

Since

$$\frac{n_F - n_C}{n_D - 1} = \frac{1}{v},$$

$$F_F - F_C = \frac{F_D}{v}. \tag{6.5}$$

This expression shows that, as with a prism, the longitudinal chromatic aberration of a lens varies directly with power. For example, for a $-6.00$ D lens having a nu value of 60,

$$\text{longitudinal chromatic aberration} = \frac{-6}{60} = -0.10 \text{ D.}$$

### Transverse Chromatic Aberration

In Section 6.2, the angular expression for transverse chromatic aberration (for a lens) was found to be

$$d(F_F - F_C).$$

Since in formula (6.5)

$$F_F - F_C = \frac{F_D}{\nu},$$

the transverse chromatic aberration of a lens may therefore be expressed as follows:

$$\text{transverse chromatic aberration} = d\,\frac{F_D}{\nu}. \qquad (6.6)$$

## 6.7. Achromatic Lenses

In order for a lens to be achromatic, two lenses of different materials, one convex and one concave, having different nu values, must be used (see Figure 6-6). The power of each lens must be such that

$$F_T = F_1 + F_2$$

and

$$\frac{F_1}{\nu_1} = \frac{-F_2}{\nu_2},$$

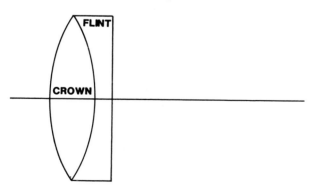

FIGURE 6-6. **An achromatic lens.**

or

$$F_1 = \frac{\nu_1}{\nu_2}(-F_2)$$

and

$$-F_2 = \frac{\nu_2}{\nu_1}(F_1).$$

Substituting,

$$F_1 = F_T - F_2 = F_T + \frac{\nu_2}{\nu_1}(F_1)$$

and

$$F_2 = F_T - F_1 = F_T + \frac{\nu_1}{\nu_2}(F_2).$$

The refracting powers are given by the derived expressions. For the first lens,

$$F_1\left(\frac{\nu_1 - \nu_2}{\nu_1}\right) = F_T$$

and

$$F_1 = \left(\frac{\nu_1}{\nu_1 - \nu_2}\right)F_T. \qquad (6.7)$$

For the second lens,

$$F_2\left(\frac{\nu_2 - \nu_1}{\nu_2}\right) = F_T$$

and

$$F_2 = -\left(\frac{\nu_2}{\nu_1 - \nu_2}\right)F_T. \qquad (6.8)$$

### EXAMPLE
Given two lenses, one having an index of refraction of 1.50 and a nu value of 60, and the other having an index of refraction of 1.60 and a nu value of 40. Design an achromatic lens having a refracting power of $-2.00$ D.

$$F_1 = \left(\frac{60}{60 - 40}\right)(-2)$$

$$= \left(\frac{60}{20}\right)(-2) = -6.00 \text{ D.}$$

$$F_2 = -\left(\frac{40}{60 - 40}\right)(-2)$$

$$= \left(\frac{-40}{20}\right)(-2) = +4.00 \text{ D.}$$

The only use for achromatic doublets in ophthalmic lens design is for some high-powered lenses such as those used as aids for low vision. For ordinary ophthalmic lenses, doublets would be excessively thick and heavy. Achromatic lenses correct only for longitudinal chromatic aberration, leaving some "lateral color" uncorrected.

One possible reason that the eye tolerates longitudinal chromatic aberration is that the *eye itself* has about 1 D of longitudinal chromatic aberration. In addition, the eye is most sensitive (in photopic vision) to a wavelength of about 555 nm, with a decrease in sensitivity toward the blue and red regions. Finally, it has been hypothesized that the accommodative responses made by the eye are at least to some extent based upon the eye's chromatic aberration interval. In any event, the visual system seems to be well adapted to its own chromatic aberration and it apparently adapts easily to the additional chromatic aberration caused by an ophthalmic lens.

There is, however, one situation in which lateral color may sometimes be a problem: Wearers of fused bifocal lenses sometimes complain of seeing color fringes, particularly when flint segments are used. This problem will be discussed in Chapter 8.

## THE MONOCHROMATIC ABERRATIONS

The five monochromatic aberrations, as originally defined by Seidel, are spherical aberration, coma, oblique astigmatism, curvature of image, and distortion.

*Spherical aberration* occurs when a pencil of light is refracted by a large-aperture optical system, occurring as a result of the fact that different "zones" of the aperture have different focal lengths. It affects the sharpness of image points.

*Coma* occurs when oblique rays are refracted by a large-aperture optical system. It also affects the sharpness of image points.

*Oblique astigmatism* occurs when oblique rays are refracted by a small-aperture system, and affects both sharpness of image points and image position. When oblique astigmatism has been corrected, curvature of image is still present.

*Curvature of image* manifests itself as a curved image surface for a flat object surface, and primarily affects image position. However, if the curvature of the image surface does not match that of the screen on which the image is formed, peripheral portions of the image will be out of focus and therefore blurred.

*Distortion* occurs when the magnification of an extended object varies with its distance from the optical axis. It affects image shape and lateral position, but not image clarity.

The center of rotation of the moving eye, together with the pupil, serves to limit the bundle of rays received by the spectacle-lens/eye system, with the result that spherical aberration and coma are usually of little concern in lens design. As for the aberrations affecting narrow bundles of rays, only *oblique astigmatism* and *curvature of image* need to be taken into consideration in routine ophthalmic lens design: Distortion is a significant aberration only for high-powered lenses and for unusual cases of anisometropia.

### 6.8. Spherical Aberration

Spherical aberration occurs as a result of the fact that each zone of a lens has its own focal length for rays originating on the optic axis, varying from the optic axis out to the edge of the lens, as shown in Figure 6-7. Normally, the deviating power of a lens increases toward the periphery. This is an inherent property of any lens (given by Prentice's rule) but in the case of spherical aberration the deviating power increases too rapidly, or not rapidly enough, to maintain a point focus for a point object. Spherical aberration may be either *positive* or *negative*. In positive spherical aberration (shown in Figure 6-7), peripheral rays have a shorter focal length than central rays; in negative spherical aberration, the central rays have the shorter focal length. Of the two, positive spherical aberration is the most common. The three-dimensional surface which encloses the totality of the refracted rays is called a *caustic surface*. The caustic surface is a symmetrical funnel-shaped figure with the small end of the funnel (called the *cusp*) located at the paraxial image.

*Longitudinal* (or *axial*) spherical aberration refers to the separation of the paraxial focus and the pe-

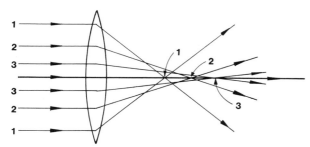

FIGURE 6-7. **Spherical aberration.**

ripheral focus, in diopters, along the optic axis. *Lateral* (or *transverse*) spherical aberration is specified in terms of the diameter of the *confusion disc* at the paraxial focus, the confusion disc being defined as the image formed on a screen placed at any point along the caustic.

Spherical aberration is a problem only for *large-aperture* optical systems: Longitudinal spherical aberration increases with the square of the aperture, while lateral spherical aberration increases with the cube of the aperture. In addition to controlling the size of the aperture, spherical aberration may be controlled by varying the "bend" of the lens, that is, by manipulating the curvature of each of the two surfaces without changing the refracting power of the lens. As a general rule, spherical aberration is kept to a minimum when rays traverse the lens symmetrically, in which case the refraction is equally divided between the two surfaces.

Spherical aberration is not of great importance for ophthalmic lenses of moderate power, since the eye uses only a small portion of the lens at any one time due to the limiting aperture of the pupil. However, for lenses of high power (+10.00 D or more, as required by aphakics), it can be serious. In these lenses spherical aberration can be controlled by the use of *aspheric* surfaces. An aspheric surface is an ellipsoid or parabolic surface, which gradually flattens in curvature from the center to the edge. Aspheric lenses may be made of glass, but these lenses are very difficult to grind and are therefore very expensive. An easier way to manufacture an aspheric surface is to *mold* it, using a plastic material for the lens and a glass mold.

Without the use of an aspheric surface, a minimum of spherical aberration occurs when a *crossed* lens is used. This is a lens in which the front surface power is greater than the back surface power by a factor of 6, or

$$\frac{F_1}{F_2} = 6.$$

When made in the form of a plus lens, this would be a bi-convex lens. Although minimizing spherical aberration, such a lens would allow considerable amounts of oblique astigmatism and distortion to occur.

As a practical matter, spherical aberration is taken into consideration in ophthalmic lens design only for high-powered convex lenses, for which plastic aspheric lenses are routinely used.

## 6.9. Coma

Coma, like spherical aberration, occurs as a result of the fact that each zone of a lens has its own focal length. However, whereas spherical aberration occurs for beams of light that are parallel to the optic axis of a lens, coma occurs for *oblique* beams (see Figure 6-8). The result is that the image of a point object resembles a *comet*, or a teardrop. Another difference between spherical aberration and coma is that, while the imagery in spherical aberration has an axis of symmetry, which coincides with the optic axis or with the *chief ray* (which is defined as the ray that passes through the center of the entrance pupil), the imagery in coma is not symmetrical with respect to the optic axis or the chief ray.

Like spherical aberration, coma is a problem mainly for large-aperture optical systems, and can be largely ignored in the design of spectacle lenses because of the limiting effect of the pupil. Factors that can be used to control coma, in addition to aperture size, are the form of the lens and the angle of obliquity.

In an optical system for which spherical aberration is corrected, coma will be corrected when, for all rays proceeding from an object point, the ratio of the sines of the slope angles ($\alpha$ and $\alpha'$), of each pair of

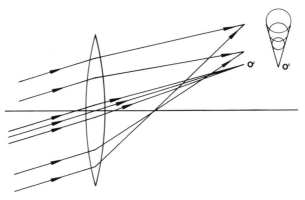

FIGURE 6-8. **Coma.**

corresponding incident and emergent rays, shall be constant. This is shown by the expression

$$\frac{\sin \alpha}{\sin \alpha'} = \frac{y}{y'} = \text{constant}.$$

This relationship is known as *Abbe's sine condition*.

If an optical system is corrected for spherical aberration for a pair of conjugate points, M and M', and the sine condition is satisfied for those points, an extra-axial object point, Q, lying close to the axis and located in a plane normal to the axis at M, will produce an image, Q', that is free from coma. The points M and M' are called *aplanatic points*. Only one pair of aplanatic points exists for any optical system.

## 6.10. Oblique Astigmatism

Oblique astigmatism (also called *radial* astigmatism or *marginal* astigmatism) occurs when a narrow pencil of light from an object passes obliquely through a spherical surface: Rays of light do not form a point focus, but form the characteristic *interval of Sturm*, containing two line foci and the circle of least confusion, as if the surface were toric rather than spherical. It is called oblique astigmatism to distinguish it from the astigmatism produced by a cylindrical or toric lens for an object situated on the optic axis. The amount of astigmatism between the two focal lines is the dioptric separation of the two line foci. The size of the circle of least confusion and the corresponding lengths of the focal lines are determined by the amount of astigmatism and the size of the aperture.

The terminology necessary for describing this aberration may be understood by referring to Figure 6-9. The *tangential plane* (or median plane) is the plane containing the chief ray from the object point and the optic axis; the *tangential focus*, T (primary focus or meridional focus), is the focus formed by the rays in the tangential plane. The *sagittal plane* (or equatorial plane) is the plane containing the chief ray from the object point that is also perpendicular to the tangential plane; the *sagittal focus*, S (equatorial or secondary focus), is the focus formed by the rays in the sagittal plane.

It should be understood that the tangential focus lies in the sagittal plane, and that the sagittal focus lies in the tangential plane. The amount of oblique astigmatism is equal to the dioptric separation between the tangential focus, T, and the sagittal focus, S.

As described in the following section (6.11), when a lens forms an image of a plane object, the image lies along a curved surface, known as the *Petzval* surface, which is located at the secondary focal point of the lens. When oblique astigmatism is present, a *true* Petzval surface does not exist: It can exist *only* if oblique astigmatism has been eliminated.

When oblique astigmatism exists, both the tangential (T) and the sagittal (S) foci fall on the *same side* of the theoretical Petzval surface: They both may be either in front of or behind the Petzval surface, but they can never straddle it.

The relationship between the tangential and sagittal surfaces in regard to the Petzval surface, P, is such that

$$\frac{\overline{PT}}{\overline{PS}} = \frac{3}{1}, \qquad (6.9)$$

or, T is three times as far from P as is S (see Figure 6-10), and

$$\overline{PS} = \frac{\overline{ST}}{2}.$$

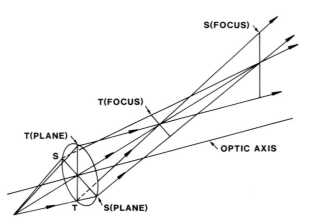

FIGURE 6-9. **Oblique astigmatism.**

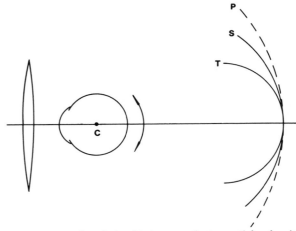

FIGURE 6-10. **The relationship between the tangential and sagittal image surfaces and the Petzval surface.**

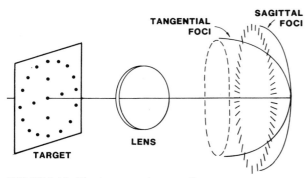

FIGURE 6-11. **The teacup and saucer diagram.**

Oblique astigmatism exists only for *point objects*, not for extended objects. However, an extended object may be considered to be made up of an infinite number of points, each of which is subject to oblique astigmatism. As is shown in Figure 6-11, for an object made up of an infinite number of points, each point gives rise to both a tangential and a sagittal focus. If we take all of the tangential foci together, they will form a paraboloidal surface; in a like manner, all of the sagittal foci, taken together, form a second paraboloidal surface. The curvature of the tangential surface is always steeper than that of the sagittal surface (Figure 6-11); the surface formed by the tangential foci has the appearance of a *teacup*, while the surface formed by the sagittal foci has the appearance of a *saucer*, with the result that the diagram shown in Figure 6-11 is known as the "teacup and saucer" diagram. The teacup and saucer surfaces represent the loci of all tangential and sagittal images.

The two surfaces touch each other, on the optic axis, at the point conjugate to a point object on the optic axis. Between the two surfaces one may construct a third surface, which represents the position of the circle of least confusion for each point object. For a spherical lens, the entire teacup and saucer configuration is symmetrical about the optic axis, with the result that these surfaces can be represented by sections in a plane including the optic axis (as shown in Figure 6-10).

Fortunately, it is possible to produce spherical lenses that are free from oblique astigmatism for a given position of the center of rotation of the eye. Jalie[2] has derived the following expression for the back surface power, $F_2$, of a lens of power $F$ for which there is no oblique astigmatism:

$$F_2 = \frac{F}{2} - 13.85 \pm \sqrt{30 - 2.87F - 0.182F^2} \,.$$

$$(6.10)$$

He assumed that the viewing distance is infinity, that the center of rotation of the eye is located 27 mm behind the lens, and that the index of refraction of the lens is 1.523. The range of lens powers over which this equation is effective is approximately from $-23.00$ to $+7.25$ D. Beyond these limits spherical spectacle lenses cannot be made entirely free of oblique astigmatism, but there is a form of lens which produces a *minimum* amount of oblique astigmatism. By differentiating Eq. (6.10) and setting the differential coefficient equal to zero, we obtain

$$F_2 = \frac{F}{2} - \frac{L_2'(n^2 - 1)}{n + 2} \,.$$

If $n = 1.523$ and if $L_2'$ (the dioptric value of the distance from the lens to the center of rotation of the eye) = 37 D, then the lens form required to minimize astigmatism is given by

$$F_2 = \frac{F}{2} - 13.85. \qquad (6.11)$$

It is apparent that there are *two* solutions for Eq. (6.10): the positive root and the negative root. If the negative root is used, that is,

$$F_2 = \frac{F}{2} - 13.85 - \sqrt{30 - 2.87F - 0.182F^2} \,,$$

$F_2$ will have a *shorter* (or steeper) radius of curvature than if the plus root is used, that is,

$$F_2 = \frac{F}{2} - 13.85 + \sqrt{30 - 2.87F - 0.182F^2} \,.$$

The steeper lens form obtained by using the negative root is called the *Wollaston* form, named after William Wollaston, who in 1804 studied image formation experimentally by using different forms of lenses. He concluded that the best image for oblique rays was produced by the use of very steep surface powers. The shallow lens form obtained by using the positive root is called the *Ostwalt* form, named after F. Ostwalt, of France, who in 1898 presented calculations for lenses which were free from oblique astigmatism, and concluded that there were two lens forms which gave zero astigmatism—a shallow form that was ultimately named after him, and the deeper form similar to that of Wollaston.

If, within the range of lens powers over which oblique astigmatism is zero, both the Ostwalt and Wollaston forms are plotted, an ellipse is generated, which is known as *Tscherning's ellipse* (shown in Figure 6-12). In 1904 M. Tscherning presented equations providing the theoretical basis for the treatment of oblique astigmatism.

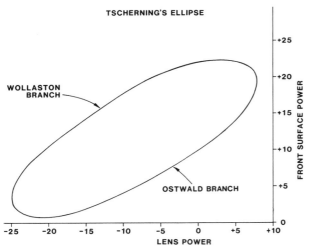

FIGURE 6-12. **Tscherning's ellipse, plotting both the Ostwalt and Wollaston forms of "zero oblique astigmatism" lenses.**

Theoretically, for a lens of power $F$, there are two forms of the lens, the shallow Ostwalt form and the steeper Wollaston form, that would produce zero oblique astigmatism. The Wollaston form not only eliminates oblique astigmatism but would minimize *distortion* as well (see Section 6.12). However, the Wollaston form is more expensive, is difficult to manufacture, is less desirable cosmetically because of the bulbous appearance, and is more difficult to mount, support, and retain in a frame. The Ostwalt form, on the other hand, is used as a basis for the design of modern ophthalmic lenses.

**The Tscherning Ellipse.** By way of review, Tscherning's ellipse is the locus of points showing the relationship between the power of a surface (either front or back) and the power of the lens for which oblique astigmatism is eliminated. A unique Tscherning's ellipse is based upon specific assumptions of the lens designer, such as (*a*) the fixation distance; (*b*) the angle of view; (*c*) the stop distance (the distance from the back pole of the lens to the center of rotation of the eye); (*d*) the index of refraction of the lens; and (*e*) the thickness of the lens. As assumptions are altered, the ellipse expands and shrinks, and moves to a slightly different position. For each set of assumptions the ellipse changes in location and form, and hence it is possible to construct a *family of ellipses*.

For ophthalmic lenses two assumptions alter the Tscherning ellipse in a significant way. If the stop distance is increased, the form of the lens becomes *shallower*; if the fixation distance is reduced (as in near vision) the lens form again becomes *shallower*. If other assumptions remain constant, a change in fixa-

tion from far to near requires that a lens must be made flatter by about 2.00 D, if zero oblique astigmatism is desired.

## 6.11. Curvature of Image

An ophthalmic lens does not form a plane image for a plane object: It forms a *curved* image, as shown in Figure 6-13. The curvature of the image is obviously a function of the correcting lens, and the image surface is known as *Petzval's surface*. The radius of curvature of the image surface was shown by Petzval and Coddington[3] to be equal to the index of refraction of the lens times the secondary focal length of the lens,

$$r_{PS} = -nf'. \qquad (6.12)$$

The surface of the image formed by the eye, like the surface of the image formed by the lens, is also a *curved* surface. It will be recalled that the function of an ophthalmic lens is to form an image at the *far point* of the eye: As the eye rotates, a spherical surface, known as the *far-point sphere*, is traced out (see Figure 6-14). The center of curvature of the far-point sphere is assumed to be a point, the *center of rotation* of the eye, located approximately 27 mm behind the spectacle plane.

Ideally, the image formed by an ophthalmic lens, for a moving eye, will be formed on the far-point sphere of the eye. However, the aberration of curvature of image will be absent only when the *image surface* (Petzval's surface) corresponds to that of the *far-point sphere*. As shown in Figure 6-14, the radius of curvature of the far-point sphere (for either a myopic eye or a hyperopic eye) is given by

$$r_{FPS} = 0.027 - f'. \qquad (6.13)$$

Therefore, the image surface will correspond to the

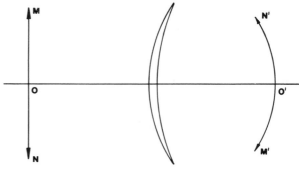

FIGURE 6-13. **Curvature of image.**

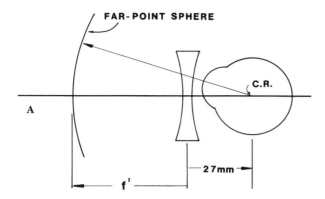

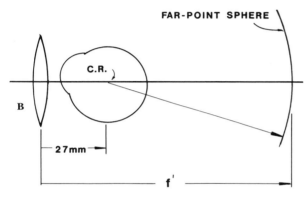

FIGURE 6-14. **The far-point sphere: A, for a myopic eye; B, for a hyperopic eye. C.R. = center of rotation.**

far-point sphere only when

$$r_{FPS} = r_{PS},$$

or when

$$0.027 - f' = -nf'. \qquad (6.14)$$

By rearranging terms, we can solve for $f'$:

$$f' - nf' = 0.027$$

$$f' = \frac{0.027}{1 - n}.$$

For a lens having an index of refraction of 1.523, the image surface and the far-point sphere will coincide when

$$f' = \frac{0.027}{-0.523} = -0.516 \text{ m},$$

$$F = -19.37 \text{ D}.$$

For any lens whose power is *other than* $-19.37$ D (assuming a center of rotation distance of 27 mm), in the

absence of oblique astigmatism little can be done to cause the Petzval surface of the lens to correspond to the far-point sphere of the eye (although it will be noted that, for a given power of lens, using a lens having a lower index of refraction will reduce the radius of curvature of the Petzval surface, making it correspond more closely to the far-point sphere). The following examples show the relationship between the far-point sphere and the Petzval surface for commonly used lens powers:

### EXAMPLE 1

Given a $+2.00$ D sphere, located at a distance of 27 mm in front of the center of rotation of the eye. How does the radius of curvature of the image surface compare to that of the far-point sphere of the eye?

The far-point sphere of the eye and the secondary focal plane of the $+2.00$ D sphere are located 500 mm behind the spectacle plane, as shown in Figure 6-15A. The radius of curvature of the far-point sphere is therefore

$$r_{FPS} = 27 - 500 = -473 \text{ mm},$$

and the radius of curvature of the image surface is

$$r_{PS} = -1.523(500) = -761.5 \text{ mm}.$$

The image surface is therefore found to be *flatter* than the far-point sphere of the eye that the lens is designed to correct.

### EXAMPLE 2

Given a $-2.00$ D sphere, located at a distance of 27 mm from the center of rotation of the eye. How does the radius of curvature of the image surface compare to that of the far-point sphere of the eye?

The far-point sphere and the secondary focal plane of the $-2.00$ D sphere are located 500 mm in front of the spectacle plane, as shown in Figure 6-15B. The radius of curvature of the far-point sphere is therefore

$$r_{FPS} = 27 - (-500) = 527 \text{ mm},$$

and the radius of curvature of the image surface is

$$r_{PS} = -1.523(-500) = 761.5 \text{ mm}.$$

Again, the image surface is flatter than the far-point sphere.

We may now ask the question: In which of the cases does the *greater problem* occur? Inspection of Figures 6-15A and B shows that, when the hyperopic eye rotates into a secondary position, the lens now *undercorrects* the hyperopia (the focal length being too long); when the myopic eye rotates into a second-

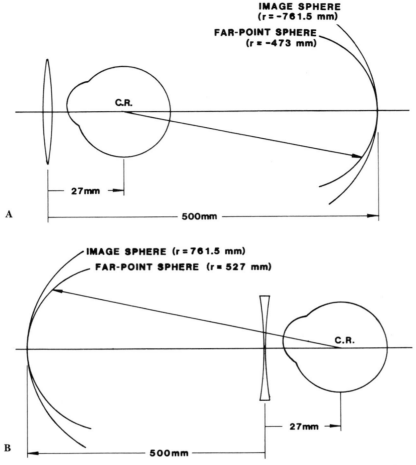

FIGURE 6-15. **Relationship between the image surface and the far point sphere: A, for a 2.00 D hyperope; B, for a 2.00 D myope. C.R. = center of rotation.**

ary position, the lens likewise *undercorrects* the myopia. Although the error is slightly greater for the hyperopic eye than for the myopic eye, the hyperope has the advantage that vision can be cleared up by accommodating slightly, whereas the myope has the disadvantage that accommodating only makes matters worse. This problem can be solved by slightly *overcorrecting* the myopic eye, so that a small amount of accommodation must be used for straight-ahead vision; but when the eye rotates to a secondary or tertiary position, vision will be clear without the need for accommodation.

Other than to slightly overcorrect the myopic eye, we have no method of controlling curvature of image when oblique astigmatism is absent. The far-point sphere, being a property of the eye, can't be changed; the image surface, being a property of the correcting lens, can be altered only slightly by changing the index of refraction of the lens material, but cannot be altered by changing the form of (or "bending") the lens. Only if oblique astigmatism is permitted to exist in the spectacle lens can curvature of image be controlled. This will be discussed further in Section 6.14.

If we wish to steepen the image surface of a single lens, the radius of curvature of the Petzval surface must be as short as possible, which for a given lens requires a low refractive index. If two lenses can be used, one plus lens and one minus lens (similar to an achromatic doublet), curvature of image will be eliminated when

$$n_1 f_1' + n_2 f_2' = 0.$$

However, from a strictly practical point of view, changing the index of refraction or using a doublet would seldom, if ever, be seriously considered for an ophthalmic lens.

## 6.12. Distortion

Distortion is concerned not with sharpness of the image but with the ability of an optical system to reproduce faithfully an image of the geometrical shape of the object. Distortion occurs when the magnification of an extended image varies with the distance of the corresponding object from the optical axis. When the ratio of the image size (y') to the object size (y) has a constant value for all object sizes, then no distortion exists and the condition of *orthoscopy* is achieved. If, however, the ratio of image size to object size varies for different object sizes, distortion (rather than orthoscopy) is present.

Distortion occurs as either *pincushion* distortion or *barrel* distortion. Pincushion distortion occurs if the image size to object size ratio increases with an increase in object size (Figure l6-16A), whereas barrel distortion occurs if the image size to object size ratio decreases with an increase in object size (Figure 6-16B).

Spectacle lenses, especially those of high power, produce distortion. For plus spectacle lenses the virtual image suffers *pincushion* distortion, whereas for minus spectacle lenses the virtual image suffers *barrel* distortion.

When formed by a centered, or *coaxial*, optical system, distortion has radial symmetry about the optical axis. Pincushion and barrel distortion are examples of symmetrical distortion formed when an object is *square* in shape. However, a *circular* object, having its center on the optical axis, will be imaged as a circle: In this special case, distortion does not alter the shape of the image, but it does alter the *proportions* of the image. For example, if an object consisting of two concentric circles is affected by pincushion distortion, the ratio of the respective diameters of objects would be less than the corresponding diameters of the image. Hence, the image is not a distortion-free reproduction of the object, even though the respective images (of concentric circles) are true circles (see Figure 6-17).

If an optical system is not centered, distortion may not be symmetrical. A *prism* produces asymmetrical

distortion, as shown in Figure 6-18. Since the optical elements of the human eye are not centered, the distortion of the retinal image may be asymmetrical.

A *chief ray* (or *principal ray*) of a pencil of light is one that passes through the center of an aperture. The chief ray from each point on an extended object forms an angle of incidence ($\alpha$) and an angle of refraction ($\alpha'$) with the optical axis. When tan $\alpha'$/tan $\alpha$ is a constant for every chief ray, the image will be

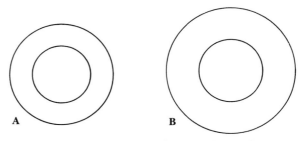

FIGURE 6-17. **Distortion of an object consisting of two concentric circles: A, object: B, pincushion distortion. Distortion does not alter the shape of the image, but alters its proportions.**

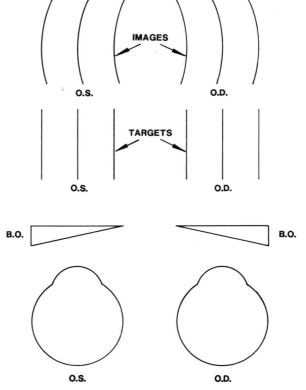

FIGURE 6-18. **Asymmetrical distortion, produced by base-out prisms.**

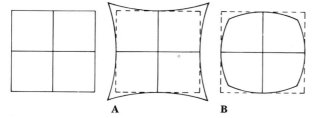

FIGURE 6-16. **Distortion of a square object: A, pincushion distortion; B, barrel distortion.**

uniformly magnified throughout the field of view, and the image will be free of distortion. This is known as fulfilling the *tangent condition* for orthoscopy. The constancy of tan α'/tan α for all corresponding points of the image and the object is the criterion for geometrical congruence of the figures.

The positions of the *stops* in an optical system greatly influence the amount and type of distortion. If an aperture stop is placed with its center coincident with the optical center of a thin convex lens, shown in Figure 6-19, no distortion occurs. However, if a stop is placed behind a thin convex lens, the real image will manifest pincushion distortion (Figure 6-20A). If the stop is placed in front of the lens, barrel distortion will be produced (Figure 6-20B). Distortion is often reduced in complex optical systems by separation of the optical components of the system and by proper placement of the stops.

As with spherical aberration and coma, distortion is a problem mainly for lenses of high power. Distortion tends to falsify the positions of objects, causing vertical lines to wave as the line of sight intersects different zones of the lens. The situation in which distortion most often causes a problem occurs with the newly aphakic patient—an individual who has just begun to wear strong plus lenses following cataract surgery. The patient experiences severe pincushion distortion, and eventually learns that the effects of distortion (such as doorways appearing to be convex toward the center) can be minimized by looking straight ahead through the centers of the lenses.

Distortion is largely ignored in the design of ophthalmic lenses. Although distortion can be minimized or even eliminated by using very steep back curves (the Wollaston form, as discussed in Section 6.10), such lenses are difficult to manufacture and are cosmetically unacceptable. However, even moderately steep back surfaces of the Ostwalt form (−3.00 or −4.00 D) are of some help in controlling distortion in high plus lenses, and the practice of using aspheric surfaces (to eliminate spherical aberration) in these lenses is also of some help.

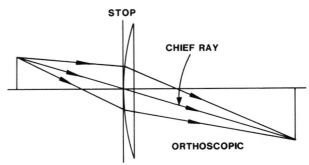

FIGURE 6-19. **No distortion occurs when the center of an aperture stop coincides with the optical center of a thin convex lens; the condition is known as orthoscopy.**

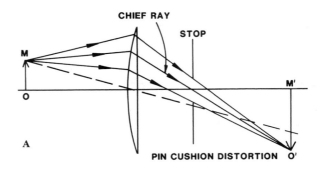

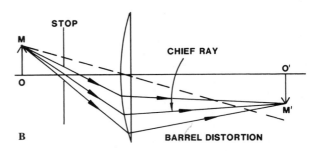

FIGURE 6-20. **A, pincushion distortion caused by a stop placed behind a plus lens. B, barrel distortion caused by a stop placed in front of a plus lens.**

## PRINCIPLES OF LENS DESIGN

### 6.13. Introduction

The *far point* of the eye has been defined (Chapter 5) as the point conjugate to the fovea with accommodation relaxed. As the eye moves in various directions of gaze, it rotates about a fixed point, the *center of rotation*. The fovea and the far point also rotate with the eye, the far point generating a surface concentric with the center of rotation, called the *far-point sphere* (described in Section 6.11 and shown in Figure 6-14). In order for a sharp image of a distant object point to be formed on the fovea for all directions of gaze, the image formed by the spectacle lens must be a point image falling on the far-point sphere. To be seen clearly, all rays entering the eye from a point object must be focused on the fovea. Hence, the chief ray must also pass through the center of rotation of the eye.

Although the pupil of the eye actually moves around the center of rotation for all directions of gaze, the light behaves as if the eye had a *fixed stop*, the size of the pupil, situated at the center of rotation of the eye. With this concept the problem of lens design is greatly simplified, since the optical system of the eye can be ignored: The lens designer simply *replaces* the optical system of the eye with the far-point sphere and a stop aperture located at the center of rotation.

If a patient is refracted at a known vertex distance and if a lens of the correct back vertex power is fitted at that vertex distance, the prescription will be correct when the patient looks along the optic axis of the lens. As the eye rotates behind the lens, so that the line of sight passes through different portions of the lens, it is, of course, desirable that the prescription should be correct for oblique angles of gaze. However, the radius of curvature of the back surface of an ophthalmic lens is usually longer than the radius of the arc generated by the motion of the cornea (Figure 6-21). This results in lenses with different base curves having different vertex distances for identical viewing angles, making it difficult to compare the performance of one lens to that of another. For example, if a patient requires a +3.00 DS prescription at a vertex distance of 13 mm measured along the optic axis of a lens, he will require a correction of +3.00 D at 13 mm in front of the cornea regardless of

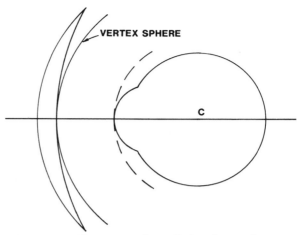

FIGURE 6-22. **The vertex sphere: Each point on the vertex sphere is equidistant from the cornea, as the eye rotates.**

head or eye position (that is, regardless of where the visual axis passes through the lens).

To avoid the problem of differing lens powers for differing positions of the visual axis, calculations are referred to a theoretical spherical surface, known as the *vertex sphere* (Figure 6-22). Each point on the vertex sphere is equidistant from the cornea, as the eye turns to look in various directions of gaze. The vertex sphere is tangent to the back pole of the lens, and its center of curvature is located at the center of rotation of the eye. One of the objects of lens design is to provide an accurate correction over this theoretical surface. An important property of the vertex sphere is that it remains in the same position regardless of changes in *lens form*, as long as the lens is always placed at the same vertex distance. The vertex sphere provides a reference point for a given angle of view, from which the tangential and sagittal powers can be compared for different lens forms.

The task of the lens designer, therefore, is to design a lens, placed at a specified distance in front of the center of rotation of the eye, with its optic axis passing through the center of rotation, such that narrow pencils of light (passing through peripheral as well as central portions of the lens) are refracted in such a manner that they pass through the stop position at the center of rotation and converge upon (for a positive lens) or diverge from (for a negative lens) the far-point sphere of the eye.

## 6.14. Lens Design Variables

The ophthalmic lens designer appears to have at his disposal the following variables:

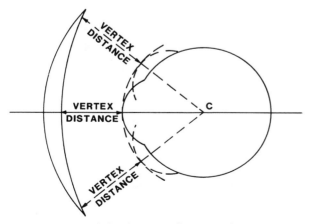

FIGURE 6-21. **Variation in vertex distance with movement of the eye.**

1. The number of lenses per aperture
2. The index of refraction
3. Lens thickness
4. Vertex distance
5. Lens form

However, if we examine each of these apparent variables, it becomes obvious that some of them are *fixed*, in ophthalmic lens design, or are not easily subject to variation in any practical manner.

1. *Number of Lenses.* Because of weight and appearance, a single lens is routinely used. Doublets are almost never utilized.

2. *Index of Refraction.* Index of refraction is not considered as a design variable, because of cost and because of a need to standardize.

3. *Lens Thickness.* Because of factors such as workability, stability, safety, appearance, and weight, thickness is not usually considered to be a variable. For a given power, thickness may, however, be varied within rather narrow limits.

4. *Vertex Distance.* This variable is also subject only to minor variations. Too great a vertex distance severely limits the field of view of a lens, in addition to being unsightly, and extremely short vertex distances are impractical because of the wearer's anatomical features.

5. *Lens Form.* The radius of curvature of the front surface, $r_1$, or that of the back surface, $r_2$, may be varied. Once a value has been set for either the front or the back surface radius of curvature (for a lens of a given back vertex power), the value for the other radius of curvature is fixed. Radius of curvature (which we think of in terms of the surface powers, $F_1$ and $F_2$) can be varied within rather wide limits. Lens form, therefore, is a variable that lens designers may make use of in minimizing aberrations. Fortunately, the aberrations of oblique astigmatism and curvature of image can be controlled by proper choice of lens form.

The lens designer, therefore, essentially has only one variable—lens form, or "bending"—to control lens aberrations. With only one degree of freedom, it is impossible to control *all* aberrations. Spectacle lenses suffer from all of the aberrations (both chromatic and monochromatic), but these aberrations are not all equally detrimental. *Chromatic aberration* is usually of little importance and is largely ignored: The eye itself suffers from chromatic aberration, and adaptation to this aberration very likely occurs. In addition, the eye has its greatest sensitivity in the central portion of the visible spectrum, with the extremes of the visible spectrum being relatively ineffective. Correction by a doublet results in both a heavy and an unattractive lens. *Spherical aberration* and *coma* are aperture-dependent aberrations; because of the small effective aperture at the center of rotation of the eye, these aberrations have relatively small effects in comparison to those of the other monochromatic aberrations. *Distortion* is a problem only for high-powered lenses and for large amounts of anisometropia. Since distortion can be controlled only by severe bending of the lens, it is ignored except in unusual circumstances.

The aberrations of greatest importance in the design of ophthalmic lenses are *oblique astigmatism* and *curvature of image*. Of the two, oblique astigmatism is usually of greater concern. It is largely controlled by varying the form of the lens, and in the process curvature of image can also be modified.

With only one degree of freedom, the lens designer may address the correction of either oblique astigmatism or curvature of image, or may make a compromise between the two. *Corrected curve*, or *best-form*, are adjectives used to describe a lens that is designed to control oblique astigmatism, curvature of image, or both, for oblique directions of view through the lens. A "corrected-curve" lens should have a rationale underlying its design. For lenses of identical power, different lens forms arise from different philosophical concepts regarding the intent of the design. Each manufacturer should clearly state the lens design philosophy so that the optometrist may intelligently select the lens form best suited to a particular patient.

Two schools of thought exist in the design of ophthalmic lenses. One approach is to concentrate solely on the reduction of oblique astigmatism and to ignore the residual curvature of image error. Lenses made on the basis of this approach are known as *point-focal* lenses. On the other hand, if primary consideration is given to reduction of curvature of image while allowing a residual amount of oblique astigmatism to exist, the lens design will conform to Percival's design philosophy. Percival, an English ophthalmologist, recommended this approach, believing that the correction of curvature of image was more important than the elimination of oblique astigmatism.

While lens designers hold different views as to which aberration should be maximally corrected for optimum performance, the majority seem to favor the correction of oblique astigmatism with only a secondary interest in curvature of image error.

As stated in Section 6.11, it is possible to eliminate oblique astigmatism with the point image of a point object falling on a curved surface, known as the

Petzval surface. The ideal lens design would produce a point image of a distant object on the far-point sphere. The discrepancy between the Petzval surface and the far-point sphere determines the amount of *curvature of image error*, often called the *power error*. Basically, oblique astigmatism is an astigmatic error, while curvature of image is a spherical error.

As discussed in Section 6.11, the radius of curvature of the Petzval surface is equal to $-nf'$, and the radius of curvature of the far-point sphere is equal to $0.027 - f'$ (assuming a distance of 27 mm from the back vertex of the lens to the center of rotation of the eye), so the Petzval surface and the far-point sphere will coincide only when

$$0.027 - f' = -nf',$$

or when

$$f' = \frac{0.027}{1 - n},$$

or when

$F = -19.37$ D (assuming an index of refraction of 1.523).

For other lens powers for which oblique astigmatism is eliminated, images of object points will fall on the Petzval surface but the Petzval surface will not coincide with the far-point sphere.

For corrected hyperopia, the Petzval surface is always *flatter* (having a longer radius of curvature) than the far-point sphere, so *undercorrection* of the hyperopia occurs. If accommodation is utilized, the Petzval surface may be made to coincide with the far-point sphere for various angles of gaze. For corrected myopia, the Petzval surface is again *flatter* than the far-point sphere, and, again, *undercorrection* occurs. If the myopia is properly corrected, relaxation of additional accommodation is not possible, with the result that objects seen in oblique directions of gaze (through peripheral portions of the lens) will be blurred. A slight overcorrection of the patient's myopia, however, will require a small amount of accommodation for axial objects, but will allow objects in oblique directions to be seen clearly.

In the absence of oblique astigmatism, curvature of image error is the dioptric difference separating the Petzval surface and the far-point sphere. On the other hand, if oblique astigmatism is present, the *amount* of oblique astigmatism is indicated by the dioptric separation between the tangential and sagittal foci. The algebraic mean of the tangential and sagittal powers, that is,

$$\frac{F_\mathrm{T} + F_2}{2}$$

is called the *mean oblique power*, and determines the dioptric position of the circle of least confusion which lies between the tangential and sagittal foci for an oblique pencil of rays. The difference between the mean oblique power and the back vertex power is known as the *mean oblique error*.

If the circle of least confusion (for the tangential and sagittal images) falls on the far-point sphere, the mean oblique error, or *power error*, is zero: This satisfies the Percival design philosophy.

The lens designer is not restricted to a pure choice between the elimination of oblique astigmatism (point-focal lenses) and creating a zero mean oblique error (Percival form). A lens designer may decide upon a *compromise* design, that is, limiting the amount of oblique astigmatism and at the same time reducing the mean oblique error by placing the circle of least confusion near the far-point sphere.

The lens form needed to eliminate oblique astigmatism for distance vision is not the same as that needed for near vision. Most lens designers who follow the point-focal design method tend to select the form needed for *distance* vision. The form necessary for the elimination of oblique astigmatism (point-focal form) for *near* vision would be 2.00 D flatter than the form needed for distance vision. The Percival form needed to bring the mean oblique error to zero, for distance vision, is somewhat flatter than the point-focal form for distance vision: In fact, the Percival form for *distance* vision would be essentially free of oblique astigmatism for *near* vision.

## 6.15. Design Assumptions

When the eye turns behind a stationary spectacle lens, the fovea turns with it, but the pencil of light forming the foveal image traverses different zones of the spectacle lens. The designer's role is to design a lens so that the performance of the lens in these different zones is as satisfactory as possible.

The designer must first formulate a design philosophy. Since there is no single *best-form* for a lens of a given power which eliminates both oblique astigmatism and curvature of image error simultaneously, the principle of design must be articulated.

The designer must make assumptions about how the lens will be used and worn. If the conditions under which the lens is used and worn match the assumptions made by the designer, the lens will perform as intended. Most lens designers make the following classic assumptions:

1. It is assumed that the center of rotation of the eye is a fixed point with respect to the cornea.
2. It is assumed that the optic axis of the correcting lens passes through the center of rotation of the eye.
3. It is assumed that the foveal chief ray passes through the center of rotation of the eye.

On the basis of these assumptions, the optics of the eye can be ignored (as already pointed out), with the exception of the eye's center of rotation and the far-point sphere. The center of rotation is considered to be the stop in the system, through which all rays must pass.

Assumptions for which there is less agreement include:

1. *The Viewing Distance.* It is well known that different base curves are required for distance and near vision. This principle applies to both point-focal designs and zero mean error (Percival) designs. As a general rule, the closer the viewing distance, the shallower the lens form.

2. *Angle of Obliquity.* Oblique power errors increase with increasing obliquity. For example, oblique power errors at 15 degrees of ocular rotation are only 25% of the oblique power errors for 30 degrees of ocular rotation. The most common choice for the angle of obliquity is *30 degrees*. The greater the angle of obliquity, the flatter the base curve must be.

3. *The Stop Distance.* The stop distance is the distance from the back pole of the spectacle lens to the center of rotation of the eye. It is made up of the vertex distance plus the distance from the cornea to the center of rotation. The magnitude of the stop distance has become increasingly controversial. Early lens designers used 25 mm as the stop distance, but 27 mm has now become a typical value. Generally, myopes are expected to have larger stop distances than hyperopes. As lenses have increased in size, the vertex distance component of the stop distance has increased. As the stop distance increases, the lens form must be shallower.

4. *Lens Thickness.* Lens thickness is usually ignored in approximate expressions for calculating the extent of an aberration. When it is taken into account, none of the aberrations is significantly affected.

## 6.16. The Base Curve of a Lens

Lens form is usually discussed in terms of the base curve of the lens in question. The term *base curve* has been given many meanings and is often loosely and incorrectly used. When used in connection with ophthalmic lens design, the term base curve refers to a standard curvature ground onto a lens for a grouping of powers.

This definition has two meanings: (1) A lens designer may specify a +6.00 D base curve for every lens in a series of lenses. Using this system, there would be only one base curve for the whole series. (2) A designer may specify a different base curve for each lens in the series, having, in effect, an almost infinite number of base curves. In practice, the concept of base curve amounts to a compromise between the use of a single base curve for all lenses and a different base curve for each lens. For example, a −5.50 D base curve may be used for lenses between powers of +2.00 and +3.00 D, and so forth.

A corrected-curve lens *series* is made up of a group of base curves, each used over a range of prescription powers. The base curve can be designated in terms of the refracting power of either the *front* or the *back* surface of the lens. A small number of front surface base curves results in a large number of back surface powers, whereas a small number of back surface base curves results in a large number of front surface powers. The number of base curves available for a lens series serves as an index of the precision of the design. However, when the number of base curves of two or more corrected-curve lens series are compared, care should be taken that the base curves refer to the *same surface* (front or back) of the lens. It should be understood that the number of base curves, in itself, provides insight into neither the philosophy of the lens design nor the quality of the manufactured product.

Since several lenses are grouped together in a lens series, it is essential that there be available a means of indicating their relationships, the intended use of the semifinished blanks, and grinding instructions for local prescription laboratories. In order to meet this need, the manufacturer provides the laboratory with a "surfacing chart," as shown in Figure 6-23.

The surfacing chart serves two purposes: (1) It provides information regarding the *surface power* of the particular manufacturer's lenses. This allows the practitioner to make comparisons between lenses available from different manufacturers. (2) It furnishes surfacing instructions, so that local optical laboratories can select semifinished blanks and surface the lenses according to the designer's intent.

Lens manufacturers supply lenses to optical laboratories in two forms: (1) *Finished uncut lenses*, with both surfaces finished, ready to be cut, edged, and mounted in the frame. Unless the manufacturing

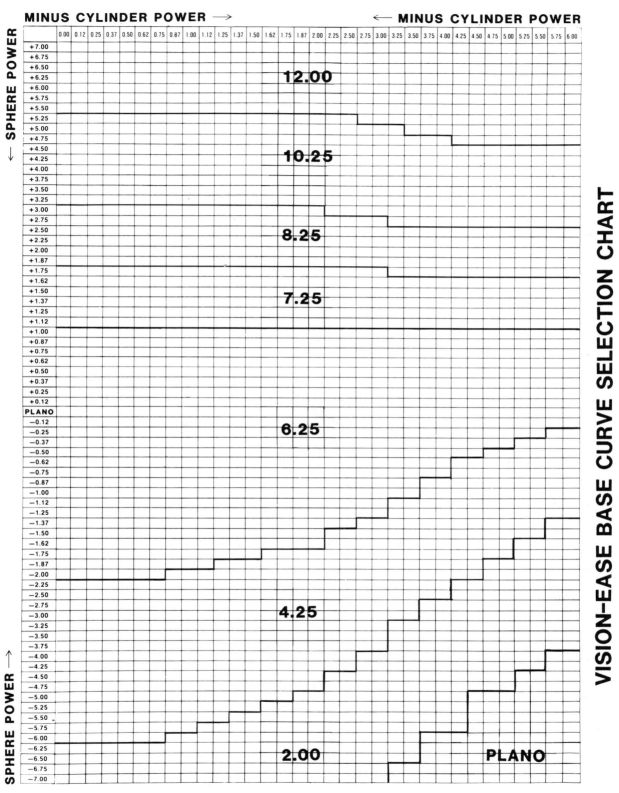

FIGURE 6-23. **A surfacing chart. This chart allows the practitioner to compare series of lenses available from different manufacturers.** (From Vision-Ease Corp., St. Cloud, MN.)

practice is known, a finished lens provides no clue as to which curve (the front or the back) is the base curve. (2) *Semifinished lenses*, having a finished side and an unfinished side, with the base curve on the finished side. The location of the base curve (whether the manufacturer finished the front or the back side of the lens) differs for different types of lenses. For *uncorrected spheres* (those for which there has been no attempt to control oblique astigmatism), the base curve for a plus lens is on the *back* side of the lens, while for a minus lens the base curve is on the *front* of the lens. Standard base curves for these lenses are −6.00 D for plus lenses and +6.00 D for minus lenses. *Uncorrected sphero-cylinders* are always made in front toric form, so the sphere is ground on the back side, and the base curve is the flattest of the two curves on the front side of the lens.

The base curve situation is somewhat different for *corrected* lenses—lenses for which oblique astigmatism or curvature of image, or both, have been con-

trolled. For corrected *spheres*, the base curve is usually on the front, for both plus and minus lenses. Exceptions to this rule occur for some higher powers of lenses. For corrected *sphero-cylinders*, the toric curve may be on the front surface (positive toric) or on the back surface (negative toric). If made in positive toric form the base curve is the flattest curvature on the front surface; but if made in the negative toric form the base curve is the spherical front surface. There are distinct advantages to *back-toric* lenses, which will be discussed later in this chapter.

All *multifocal* lenses have the base curve ground on the side of the lens that provides the additional power for near vision. This is true of fused, one-piece, and progressive addition multifocals. All toric multifocals are negative torics, with the exception of the back-surface Ultex multifocal, and therefore have both the spherical surface (the base curve) and the added power for near on the front side of the lens.

## HISTORY AND EVOLUTION OF LENS DESIGN

### 6.17. Spherical Lenses

The earliest ophthalmic lenses were bi-convex in form, bi-concave lenses being developed later. Both of these lenses were popular because they were easy to manufacture. The surface powers for bi-convex and bi-concave lenses are weaker than for any other form, and both sides of these lenses have the same curvature. As manufacturing skills increased, "flat" lenses were produced, a plus lens having a flat back surface and a minus lens having a flat front surface.

In 1804, William Wollaston, an English physician and chemist, proposed a convex-concave (meniscus) form of lens, and established its superiority over the plano-spherical and equi-sided bi-convex and bi-concave forms by investigating the image formation produced by different forms of lenses. Wollaston did not provide the theoretical proof of the superiority of this lens, but demonstrated its performance experimentally. His more deeply curved lens form gave a wider field of vision, and Wollaston called the lenses *Periscopic*, or "to-look-around" lenses. He concluded that the best form of spectacle lens was one designed so that a small oblique pencil of light would make equal angles with the two surfaces of the lens.

Theoretically the Wollaston design was a sound one, and a patent was issued for the Periscopic lens in 1804. However, due to the deep curvatures required, the lenses proved to be difficult to manufacture.

It was not until the year 1874 that Rodenstock, in Germany, introduced lenses that approached the periscopic concept of Wollaston. In 1867 the German firm of Nitsche and Gunther introduced the 1.25 D uniform-surface lens, commonly referred to as the "periscopic" lens. In this series, a plus lens had a −1.25 D back surface and a minus lens had a +1.25 D front surface. This series of periscopic lenses was a standard lens for many years. Nitsche and Gunther later offered another landmark lens, the 6.00 D base curve lens, and the term *meniscus*, meaning "moon-shaped," was adopted for these lenses. For plus spherical lenses, the back surface had a power of −6.00 D, while for minus spherical lenses the front surface had a power of +6.00 D. For lenses not designed to control aberrations (noncorrected lenses), the 6.00 D uniform lenses, still referred to as meniscus lenses, are still in use.

In 1898, F. Ostwalt, a French oculist, published the results of calculations for a series of spherical lenses which were designed to eliminate oblique astigmatism. His calculations produced lenses of much shallower design than those of Wollaston. In the period 1904–1908, Tscherning, who first realized the significance of the center of rotation of the eye as a reference point in lens design, demonstrated that for an extended range of lens powers (from approximately +7.00 to −22.00 D), oblique astigmatism could be eliminated by either of two forms of "bent" lenses, the shallower Ostwalt form and the deeper Wollaston form. Further, he found that if the surface powers of both forms were plotted graphically, they formed an *ellipse*. As described in Section 6.8 and shown in Figure 6-11, this is now known as *Tscherning's ellipse*. Tscherning's calculations applied only to infinitely thin lenses and to the elimination of oblique astigmatism for paraxial rays. Although lenses of the Wollaston form produce less distortion than those of the Ostwalt form, the Ostwalt form results in a shallower lens, and these lenses are both easier to produce and cosmetically superior. For these reasons, modern lenses conform to the Ostwalt branch of the Tscherning ellipse.

In 1908, Moritz von Rohr, extending the work of Allvar Gullstrand, began research in spectacle lens design with the intent of eliminating oblique astigmatism. He chose the rigorous, tedious method of trigonometric ray tracing to evaluate the performance of the lenses. Von Rohr made the following assumptions: (1) The primary thrust was to eliminate oblique astigmatism; (2) the distance from the back pole of the lens to the center of rotation of the eye (the stop distance) was 25 mm; (3) the viewing angle was 35° for plus lenses and 30° for minus lenses; (4) the viewing distance was infinity; (5) each lens had a specific thickness; and (6) sphero-cylindrical lenses would be manufactured in plus toric form.

Zeiss Optical Company patented von Rohr's lens design as the *Punktal* (point-forming) lens, in 1911, and started production in 1913. Each individual lens power was ground with an optimum curvature, which resulted in the need for several thousand separate surface curvatures. Since each lens of the Punktal series had its own surface powers, about 15,000 uncut lenses were necessary to service most prescriptions. The cost of such a lens inventory was prohibitive, and consequently many optical laboratories failed to carry the full range of uncut lenses. Unfortunately, special factory orders and ensuing delays were common. These lenses were not widely used in the United States, but were well received in Europe.

Spectacle lens correction as conceived and implemented by von Rohr provided a classic foundation for lens design for many years. Von Rohr was also credited with creating the *back vertex system* of specifying ophthalmic lens power, as well as the concept of *vertex refraction*.

In 1947 the Punktal lens series was recomputed by Wolfgang Roos. With emphasis still on the correction of oblique astigmatism, Roos decided to correct oblique astigmatism not only at infinity but also for near distances as well. Lens sizes had increased significantly since von Rohr's calculations. Roos extended the range of correction for oblique astigmatism beyond the viewing angle of 30°. Setting a tolerance of 0.20 D of oblique astigmatism, the new Punktal lenses provided correction of oblique astigmatism for distances from infinity to 25 cm and for lens diameters of up to 50 mm. However, he attached more importance to the correction for near objects than for distant objects. This adaptation of the design, for larger angles of obliquity and for viewing distances nearer than infinity, produced lenses that were shallower in form than the original Punktal lenses. However, by increasing the tolerance for oblique astigmatism, Roos greatly decreased the number of lens blanks required for the new lens series.

Currently, Punktal lenses by Zeiss are made in both positive toric form and negative toric form. Positive toric form lenses are used for most plus lenses and low-powered minus lenses; negative toric lenses are used for strong minus lenses. In the low-powered plus lenses and weak minus lenses in which the cylindrical component exceeds 4.00 D, the lenses are available in either positive toric form or negative toric form. The transition range is designed to provide uniform construction in cases of anisometropia, in which magnification properties and esthetic considerations become important.

In 1919, Edgar Tillyer patented a lens design in which both oblique astigmatism and curvature of image (power error) were considered. American Optical Company made these lenses available to the profession in 1926 under the trade name *Tillyer*. The calculations for the Tillyer lens series were based upon a viewing angle of 30°, a distance from the back surface of the lens to the center of rotation of 27 mm, and vision for objects at infinity.

Since oblique astigmatism and power error cannot be reduced to zero with the same lens design, Tillyer decided to let oblique astigmatism become as large as 0.12 D for weak prescriptions and 0.25 D for strong prescriptions as long as the power error could be reduced. He also decided to permit both aberrations to reach this tolerance level if the number of base curves could be reduced. He discovered that he could maintain these tolerance levels with a single base curve over a range of prescriptions, thus laying the groundwork for a "stepped," or "clustered," base curve series for corrected-curve lenses. Because of the consideration given to power error, Tillyer lenses were *flatter* than Punktal lenses. For spherical powers ranging from +7.00 to −20.00 D and cylinders up to 6.00 D, nineteen base curves were used. For powers of +7.00 to −5.00 D, the Tillyer lenses were made in positive toric form with the base curve being the flatter curve on the front surface, while from −5.00 to −20.00 D the toric surface was placed on the back and the base curve was the spherical curve on the front surface.

In 1928, Bausch and Lomb Optical Company introduced the *Orthogon* lens. The Orthogon lens was designed by Wilbur Rayton, who modified the Punktal calculations in order to group lenses together on a common base curve. The fundamental principle in the Orthogon lens design, like the Punktal design from which it was derived, was the correction of oblique astigmatism. No correction of curvature of image was incorporated into the design. Calculations were based on a distance from the back surface to the center of rotation of 25 mm, a viewing angle of 30°, and vision for objects at infinity. In most cases oblique astigmatism was around 0.05 D, but was permitted to increase to 0.25 D in rare instances. The Orthogon lenses were made in positive toric form with the base curve usually being the flatter curve on the front of the lens. However, for lenses from +6.00 to +8.00 D, the base curve was on the back of the lens. Orthogon single-vision lenses are supplied in uncut and semifinished lenses in prescriptions from +8.00 to −12.00 D, utilizing twelve base curves. The Orthogon lens is slightly *steeper* than the Tillyer lens.

The *Widesite* lens, a product of Shuron Optical Company, was designed on the basis of fourteen base curves, all lenses being made in positive toric form. For semifinished sphero-cylindrical lenses, the front surfaces were finished in five base curves: +5.50, +6.50, +7.00, +8.00, and +9.00 D. For semifinished spheres the back surfaces were ground on nine base curves: −5.00, −5.25, −5.75, −6.50, −7.00, −7.75, −8.75, −9.25, and −12.00 D. Finished lenses were supplied with the same base curves as semifinished lenses. While primarily correcting for oblique astigmatism, correction was limited to 5% of the axial power of the lens.

In 1920 the Kurova corrected-curve lens was introduced by Continental Optical Company. The lens was probably the earliest corrected-curve lens made in this country, although the first lenses were not based on a complete lens design analysis. In 1925, F. E. Duckwall redesigned the lenses. The lens calculations were based on a distance from the back surface

of the lens to the center of rotation of 25 mm, a viewing angle of 30°, and vision for objects at infinity. The goal in the Kurova lens design was to reduce oblique astigmatism to insignificant levels and, at the same time, to minimize power error (curvature of image error). In the order of importance, errors were listed as follows: marginal astigmatism first, plus power error next, and minus power error last. The Kurova lens series was made in positive toric form with the base curve being the flatter curve on the front surface. The lenses were made in both finished and semifinished form, with thirty-nine base curves ranging from +2.50 to +12.50 D.

In 1950 the *Normalsite* corrected-curve series was introduced by the Titmus Optical Company. The lens was designed by Foster Klingaman, who based his calculations on a stop distance of 27 mm, a viewing angle of 30°, and an object distance of 13 feet. The Normalsite lens was the first American lens to correct for an object distance closer than infinity. Primary consideration was given to the correction of oblique astigmatism at the intermediate distance of 13 feet, while secondary consideration was given to the curvature of image error, also at 13 feet. Since near object distances require flatter lens forms than infinite object distances, and since consideration of curvature of image error also requires flatter lens forms, the Normalsite lenses were flatter than most other lenses on the market. For lenses having powers from +7.00 to −8.00 D, the lens is produced with a positive toric surface with the base curve being the shallower front surface curve. For lenses from −8.25 to −13.00 D, the lenses are made in negative toric form, with the base curve being the spherical curve on the front surface.

In early 1964 Univis Lens Company, primarily known for their multifocal products, introduced the *Best-Form* lens, a negative toric lens. The lens was subsequently updated and renamed the *Uni-form* lens. The lens designer was E. W. Bechtold. His calculations were based on a stop distance of 27 mm, a viewing angle of 30°, and a viewing distance of infinity. The lens had negative toric construction, with the base curve being the spherical curve on the front surface. Primary attention was given to the correction of oblique astigmatism. The performance of Uni-Form lenses was determined by MTF (modulation transfer function) testing, which involves the ratio of the contrast of the image to the contrast of the object. The Uni-Form lenses were discontinued in 1971.

Closely following the introduction of the Best-Form lens in 1964, American Optical Company placed on the market a negative toric corrected-curve lens called the *Tillyer Masterpiece* lens. This lens was designed by J. K. Davis, H. H. Fernald, and A. W. Rayner. With the availability of high-speed computing equipment, calculations were made in terms of the following considerations: (1) stop distances of 27 to 33 mm (which include 90% of the population), (2) viewing angles of 20°, 30°, and 40° from the optical axis, (3) object distances of infinity and 30, 16, and 13 inches, and (4) ray plots for each principal meridian and for meridians 45° from the principal meridians. Aberrations considered were oblique astigmatism, curvature of image, and lateral chromatic aberra-

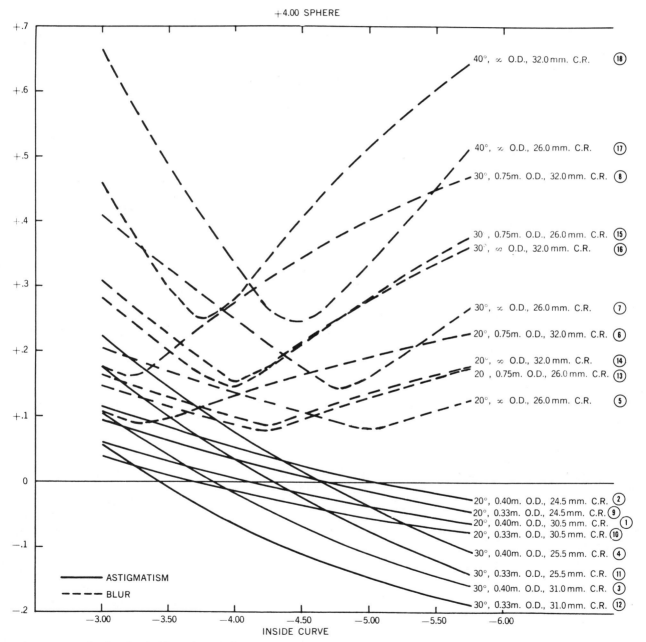

+4.00 SPHERE

| Label | Curve |
|---|---|
| 40°, ∞ O.D., 32.0 mm. C.R. | 18 |
| 40°, ∞ O.D., 26.0 mm. C.R. | 17 |
| 30°, 0.75m. O.D., 32.0 mm. C.R. | 8 |
| 30°, 0.75m. O.D., 26.0 mm. C.R. | 15 |
| 30°, ∞ O.D., 32.0 mm. C.R. | 16 |
| 30°, ∞ O.D., 26.0 mm. C.R. | 7 |
| 20°, 0.75m. O.D., 32.0 mm. C.R. | 6 |
| 20°, ∞ O.D., 32.0 mm. C.R. | 14 |
| 20°, 0.75m. O.D., 26.0 mm. C.R. | 13 |
| 20°, ∞ O.D., 26.0 mm. C.R. | 5 |
| 20°, 0.40m. O.D., 24.5 mm. C.R. | 2 |
| 20°, 0.33m. O.D., 24.5 mm. C.R. | 9 |
| 20°, 0.40m. O.D., 30.5 mm. C.R. | 1 |
| 20°, 0.33m. O.D., 30.5 mm. C.R. | 10 |
| 30°, 0.40m. O.D., 25.5 mm. C.R. | 4 |
| 30°, 0.33m. O.D., 25.5 mm. C.R. | 11 |
| 30°, 0.40m. O.D., 31.0 mm. C.R. | 3 |
| 30°, 0.33m. O.D., 31.0 mm. C.R. | 12 |

ASTIGMATISM
BLUR

INSIDE CURVE

FIGURE 6-24. **Plot showing the blur index and oblique astigmatism for a +4.00 D. lens. (From J. K. Davis, H. G. Fernald, and A. W. Rayner, The Tillyer Masterpiece Lens, a Technical Discussion, American Optical Corp., 1964.)**

tion. A single "blur index" was developed mathematically, in which the contribution of each aberration to the blur was summed, given an index of bl·r for each possible design of the lens. The lens design with the smallest blur index would, theoretically, provide the best image. Figure 6-24 shows a plot of a +4.00 D lens with eighteen different curves, displaying the blur index and the oblique astigmatism for different angles of view, object distances, and stop distances. The blur index is shown with broken lines, and astigmatism is shown with solid lines. It is obvious that the base curve selected must be a compromise. Visual inspection indicates that the range of −4.00 to −4.25 D is the best compromise. In practice, computer programs, with priorities set by the lens designer, would indicate the best compromise.

Consideration of longer stop distances, object distances closer than infinity, and power errors resulted in a lens noticeably shallower (1.50 to 2.00 D) than the original positive toric Tillyer lens. The lens was not an unqualified success, as some patients were found to be unhappy with it. Subsequently, the lens was redesigned, resulting in a *steeper* lens. The second generation of Masterpiece lenses was called the *Masterpiece II*.

Shuron Optical and Continental Optical merged, forming the Shuron-Continental Division of Textron, Inc. Shuron's positive toric Widesite lens was integrated with Continental's positive toric Kurova series, forming a single Kurova positive toric series. In 1966 Shuron-Continental introduced a negative toric series, the *Kurova Shursite*. The calculations for the Shursite lens were based upon stop distances of 26.5 mm (for a power of +8.00 D) to 31.00 mm (for a power of −8.00 D), depending upon lens power; viewing angles of 20°, 30°, and 40°; and object distances of infinity and 14 inches. Selection of base curves was made on the basis of compromising between the correction of oblique astigmatism, power error, and lateral chromatic aberration. The lens is manufactured in negative toric form, with the base curve being the spherical curve on the front surface. Twenty base curves range from +0.50 to +12.00 D. The Shursite lens "bendings" fall roughly midway between the steeper Uni-Form lens and the original, shallower, Tillyer Masterpiece lens. The bendings of the Shursite negative toric lenses do not depart significantly from those of Shuron-Continental's Kurova positive toric series.

## 6.18. Sphero-Cylindrical Lens Design

A high percentage of lens prescriptions incorporate a cylindrical component for the correction of astig-matism. Bannon and Walsh[4] found, in a study of 200 patients, that 83% had a sufficient amount of astigmatism to warrant its correction. The lenses needed for the correction of astigmatism possess two principal meridians which parallel the principal meridians of the eye. The power of the lens is a minimum in one principal meridian and a maximum in the other. It is customary to think of a lens which corrects astigmatism as being made up of a spherical component and a cylindrical component, in much the same way that the correction obtained by the use of a refractor consists of a spherical element and a cylindrical element.

In practice, the surfaces of lenses used for the correction of astigmatism are surfaces of revolution (see Chapter 2). Most lenses for the correction of astigmatism have a spherical surface on one side and a toric surface on the other, with the sum of the surface powers in each principal meridian fixed in order to render the image of the lens/eye system stigmatic for axial vision.

In a real sense, a sphero-cylindrical lens has *two* powers. Light from an object point on the optical axis incident upon a sphero-cylindrical lens forms an astigmatic pencil after refraction, and passes (in succession) through two focal lines. In order to correct the eye for axial gaze, the two astigmatic focal lines must fall on the respective far points for the two principal meridians of the eye. When the eye rotates to view objects not located on the optical axis, each far point generates a corresponding *far-point sphere*. In order to form on the retina a point image for a point object for all directions of gaze, the astigmatic focal lines formed by the lens should, after refraction, fall upon the respective far-point spheres.

The sphero-cylindrical lens is, of course, subject to the same aberrations as is the spherical lens. As the eye looks obliquely through peripheral portions of a lens, the aberrations of oblique astigmatism and curvature of image are superimposed upon the astigmatic pencil produced by the lens, resulting in removal of the line foci from their ideal positions on the respective far-point spheres. Hence, the astigmatic effect for rays passing through peripheral portions of the lens differs from that along the optical axis of the lens. The discrepancy between the image surfaces formed by the line foci and their respective far-point spheres will be found to vary with different directions of gaze. Thus, the teacup and saucer figure associated with oblique astigmatism generated by a spherical lens becomes a much more complex figure, defying simple conceptual description.

In an attempt to provide the same effective power of the astigmatic lens over the entire lens, we have available the same method as that used in dealing with a spherical lens, that is, "bending" the lens. Since the lens has a spherical surface on one side and a toric surface on the other, it is not possible to cause both image surfaces to coincide with their corresponding far-point spheres. While it is possible to equalize the amount of astigmatism in each of the principal meridians for a given angle of view, the amount of astigmatism may differ from the intended astigmatic correction (the correction that is present along the optical axis). When the cylinder component is small, the form suitable for the spherical component is often used. For larger cylindrical components, the powers of the two principal meridians are averaged, which results in a compromise in performance for each principal meridian.

## 6.19. Negative versus Positive Toric Lenses

Although all corrected-curve sphero-cylindrical lenses were originally designed as positive toric lenses, many have now been redesigned as *negative* torics. One advantage of a negative toric lens is that most multifocal lenses are negative torics (having the bifocal addition on the front) and, if during the pre-presbyopic years an individual wears negative toric lenses adaptation to a different lens form will not be necessary when multifocal lenses are first worn.

However, a more important advantage of negative toric lenses has to do with *spectacle magnification*. Spectacle magnification is given by the formula

$$SM = \left( \frac{1}{1 - F_1 \frac{t}{n}} \right) \left( \frac{1}{1 - hF_V} \right), \qquad (6.14)$$

where $F_1$ = front surface power, $t$ = thickness, $n$ = the index of refraction, $h$ = the distance from the back pole of the lens to the entrance pupil of the eye, and $F_V$ = back vertex power of the lens. The first member of the right-hand side of the equation relates to the form of the lens, and is called the *shape factor*, while the second member relates to the power of the lens and is called the *power factor*.

For a positive toric lens, there are two values of $F_1$ and two values of $F_V$. Therefore, the power of the front surface, $F_1$, contributes to the magnification difference between the two meridians. However, for a negative toric lens, $F_1$ has the same value for both meridians, with the result that the meridional magnification is less than that for a positive toric lens.

In addition to the optical advantages of negative toric lenses, they can be made more attractive than positive toric lenses. Negative toric lenses having a significant cylindrical component can be edged with

the bevel placed at a constant distance from the spherical front surface so that the thick edges of the most minus meridian are hidden either within the frame or behind the frame. Such a lens is also easier to mount. Since the bevel follows the spherical contour of the front surface, the entire extent of the bevel lies in a single plane, making the lens easier to mount (and retain) in the frame.

The first manufacturer to mass-produce a negative toric single-vision lens was American Optical, who introduced the Tillyer Masterpiece lens, already described. Shuron-Continental also makes a negative toric lens, known as the *Kurova Shursite* minus cylinder lens, introduced in 1966.

## 6.20. Design of High Plus Lenses

While it is possible to eliminate oblique astigmatism in ophthalmic lenses ranging from approximately $-23.00$ to $+7.50$ D by using spherical surfaces, beyond this range the elimination of oblique astigmatism is not possible (as long as spherical surfaces are used). Although the number of myopic patients having refractive errors greater than $-23.00$ D is extremely small, the number of aphakic patients requiring a refractive correction over $+8.00$ D has been increasing, along with life expectancy. Contact lenses and intraocular lenses are alternatives to conventional correction with spectacle lenses: However, many aphakic patients are elderly and may not be psychologically or physically able to cope with contact lenses, whereas intraocular lenses are still not accepted by all patients. For the time being, a percentage of aphakic patients will continue to select spectacle lenses.

Oblique astigmatism in lenses over $+8.00$ D may be controlled by using *aspheric* surfaces. These are surfaces in which the power of the lens is progressively reduced toward the periphery. In optical terminology, a surface is described as aspheric when it is axially symmetrical and is formed by the rotation of a portion of an ellipse, a parabola, or a hyperbola. In a strict sense, cylindrical and toric surfaces (having two planes of symmetry) are called *nonspherical* surfaces, to distinguish them from *aspherical* surfaces. Most aspherical ophthalmic lenses have ellipsoidal surfaces or surfaces that closely approximate an ellipsoidal surface. In the case of most aspheric surfaces commonly used for ophthalmic lenses, the curvature, extending radially in all directions from the center, becomes progressively *flatter*. The curvature in the opposite meridian (circumferentially) also becomes shallower, for succeeding zones, but the change is much less rapid than in the radial direction. Thus, an aspheric surface is astigmatic away from its center, and the astigmatism of the surface is used to balance the astigmatism originating from looking obliquely through a strong plus lens.

On the basis of Gullstrand's work, Zeiss introduced an aspherical back surface lens called the *Katral* lens, in 1909. The lenses proved to be difficult to manufacture and they were expensive. Consequently, they are no longer available.

In 1958 an American ophthalmologist, David Volk, was successful in producing excellent aspheric glass spectacle lenses, called *Conoid* lenses. With wide acceptance of CR-39 plastic lens material, the cost of production of aspheric lenses greatly decreased, as these surfaces could be replicated by a molding process rather than by grinding. Many optical manufacturers now produce plastic aspheric lenses. These lenses will be discussed further in Chapter 11.

# THE OPTOMETRIST'S ROLE IN LENS DESIGN

The optometrist who places an order for eyewear and dispenses the eyewear to the patient is ultimately responsible for the performance of the lenses. Prior to 1972, the majority of single-vision spectacle lenses were mass-produced as uncut lenses. The manufacturers of such lenses selected the objectives of lens design and developed a series of lenses to achieve these objectives (as described in Section 6.17). Manufacturers provided information to optometrists for the purpose of aquainting them with the advantages and uniqueness of their particular lens design. Optometrists were encouraged to order lenses by trade name. The major manufacturers of lenses were also manufacturers of ophthalmic equipment, and operated branch ophthalmic laboratories throughout the country. The manufacturers, therefore, designed the lenses and sold the optometrist his or her examination equipment, and the manufacturer's local laboratory supplied the finished glasses. The responsibility for the performance of the lenses, then, rested solely on the manufacturer, with the result that the optometrist learned, by experience, where quality and service could be obtained.

## 6.21. Changes in the Optical Industry and in ANSI Standards

In recent years the optical industry has undergone many changes. Because of financial reasons, the major lens manufacturers no longer maintain branch optical laboratories. Lens product identification and product loyalty have diminished greatly. Other factors, also, have affected the optometrist's relationship with optical laboratories. The emphasis upon negative toric lenses and the proliferation of larger and larger lens sizes have made it difficult and costly for a laboratory to maintain an adequate stock of uncut and semifinished lenses for a large variety of corrected-curve lenses.

A significant change affecting lens design occurred when the ANSI (American National Standards Institute) Z-80.1-1972 Standard for ophthalmic lenses was revised in 1979. Unless an optometrist chose a criterion other than the correction of oblique astigmatism and curvature of image, and so requested in the order to the optical laboratory, the (heretofore existing) 1972 ANSI Standard required the laboratory to supply a lens with a base curve that would limit the unwanted cylinder and sphere power at certain points along the meridians of maximum and minimum power. These points, designated by ANSI as points A and B, were selected so that they each subtended an angle of 30 degrees with the optic axis of the lens, as measured from the center of rotation of the eye. By selecting the base curve to meet the 1972 ANSI Standard, the optical laboratory determined the basis of lens design. The ANSI Z-80.1-1979 Standard has eliminated the requirement that unwanted cylinder and sphere in the periphery of an ophthalmic lens be held to specific tolerances. The ANSI Z-80.1-1979 Standard, therefore, has become a specification, or *laboratory* standard, rather than a design, or *manufacturing* standard.

The optical laboratory (rather than the manufacturer) now has the prerogative of selecting the base curve for a lens unless the base curve has been specified by the practitioner who ordered the lens. While most laboratories are likely to continue to supply lenses that minimize oblique astigmatism and curvature of image, it is possible that, in filling a given prescription, the laboratory may not have in stock either a corrected-curve uncut lens or a semifinished lens from which a proper corrected-curve lens can be made. Further, it is difficult to insert deeply curved plus lenses into large frames, so it is possible that the laboratory may resort to shallower lens forms when plus lenses are ordered.

## 6.22. Base Curve Specification

It is obvious that the practitioner should become more involved in lens design than has been necessary in the past: When ordering lenses for a patient, he or she should specify either the *base curve* for each lens or the *trade name* of the lenses to be supplied. The optometrist should be well informed concerning ophthalmic lens design, and should obtain from lens manufacturers accurate and specific information concerning the optical performance of their lenses. Fry,[5,6] in a series of papers, has summarized what an optometrist should know about the characteristics of lenses in order to keep them relatively free from oblique astigmatism and curvature of image. In another series of papers, Davis [7–9] has attempted to help the practitioner identify the properties of lenses designed to have superior performance.

Fry[10] has proposed that the spherical equivalent of each lens can be computed, and that either Table 6-1 or Figure 6-25 (both reproduced from Fry's pa-

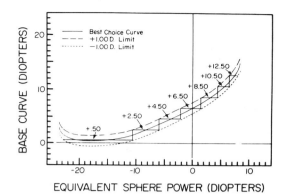

FIGURE 6-25. **Diagram for use in selecting the base curve (spherical front curve), proposed by G. A. Fry. (From G. A. Fry, *J. Amer. Optom. Assoc.*, Vol. 50, p. 562, 1979.)**

per) be used for selecting the base curve on the spherical front side of a given lens. The laboratory should be allowed a tolerance of ±0.75 D in base curved selections, as specified in the ANSI Z-80.1-1979 Standard.

On occasion, an optometrist may wish to order base curves other than those provided on uncut stock or semifinished lenses. Surfacing charts are available from manufacturers, listing the base curves for each power range. Specific information such as the ideal curves for certain stop and object distances, and tolerances for astigmatism and power errors, should be acquired to assist in evaluating various curvatures and in ordering special base curves when it is deemed desirable.

The optometrist can determine two principal variables influencing lens design calculations: lens *stop distance* and *object distance*. Normally, the lens designer makes assumptions concerning these variables: The more nearly these assumptions are met when the lenses are worn, the better the performance of the lenses.

The stop distance is the sum of the vertex distance and the distance from the cornea to the eye's center of rotation. The practitioner can approximate the stop distance by measuring the vertex distance and estimating the center of rotation distance. To obtain the vertex distance, measure the eyewire distance with the selected frame in place and add the sagittal distance of the lens to that sum. With a vertical lens dimension of 42 mm, the sagittal distance will be approximately 0.75 mm for every 2.00 D of minus back surface power. The center of rotation distance can be estimated, since it is related to the axial length of the eye. If we add the average dioptric value of the corneal curvature to the mean spherical refractive

TABLE 6-1
**Base Curves for Spheres and Back Toric Glass Lenses (Index = 1.523)**

| Base Curve | Ranges of Power for the Spherical Equivalent[a] | |
|---|---|---|
| | Lower Limit | Upper Limit |
| +12.50 | +6.399 | — |
| +10.50 | +4.387 | +6.399 |
| +8.50 | +1.446 | +4.387 |
| +6.50 | −2.106 | +1.446 |
| +4.50 | −5.880 | −2.106 |
| +2.50 | −10.503 | −5.880 |
| +0.50 | — | −10.503 |

Source: G.A. Fry, *J. Amer. Optom. Assoc.*, Vol. 50, p. 562, 1979. As described by Fry, the spherical equivalent is first computed, and this table (or Figure 6-25) can be used for selecting a lens with a given base curve on the front. The lab must be allowed a tolerance of ±1.00 D in selecting from stock of semifinished lenses the lenses to be used. In the case of anisometropia it is recommended that for the weaker lens the same base curve be used as for the lens with the stronger power.

error, and then subtract 42.50 D from the result, we obtain the value of the axial ametropia. For example, if a patient's keratometer findings are 43.00 D at 180 combined with 44.00 D at 90 and the spherical refractive error is −4.00, we would first add the mean keratometer finding (43.50 D) to −4.00 D, giving us 39.50, and, subtracting 42.50 from this result, would find the value of the patient's axial ametropia to be −3.00 D.

The center of rotation distance is about 15 mm for an average emmetropic eye. This value changes approximately 0.15 mm for each diopter of axial ametropia. Thus, the center of rotation distance increases with axial myopia and decreases with axial hyperopia. For the example given above, the center of rotation distance would be 15 mm + 3(0.15 mm) = 15.45 mm.

When the actual lens stop distance is greater than that assumed by the lens designer, a flatter base curve provides better lens performance. The base curve should be specified to be approximately 0.25 D flatter or steeper for each millimeter that the stop distance is longer or shorter than the value anticipated by the designer. The minimum significant change in base curve is approximately 0.50 D.

The patient's preferred viewing distance can also be used as a basis for changing the base curve. A long viewing distance generally requires a steeper base curve than a short viewing distance. If lenses are prescribed or used primarily for one distance, the practitioner should consider this fact when selecting a base curve.

Since oblique astigmatism cannot be completely eliminated in both principal meridians of a toric lens with a spherical surface in one side and a toric surface on the other, stock lens design tends to balance the residual astigmatic error for the two principal meridians; however a lens can be designed to improve the performance in one principal meridian at the expense of the other. If one principal meridian is used a great deal more than the other, the curves should be selected according to the prescription in the meridian of greatest interest.

All lenses function best when the optic axis passes through the eye's center of rotation (the sighting center). *Pantoscopic tilt* brings the bottom of the lens closer to the cheeks, and consequently tilts the optic axis of the lens. For the optic axis to pass through the center of rotation the pole of the lens must be dropped 1 mm below the center of the pupil for every 2° of pantoscopic tilt as discussed in Section 2:20.

By keeping these variables in mind, the practitioner can provide patients with well-designed lenses that will provide optimum performance.

## References

1. Southall, J.P.C. *Mirrors, Prisms and Lenses*, 3rd edition, pp. 545–550, The Macmillan Company, New York, 1933.
2. Jalie, M. *The Principles of Ophthalmic Lenses*, 3rd ed., pp. 410–411. Association of Dispensing Opticians, London, 1977.
3. Southall, J.P.C. *Mirrors, Prisms and Lenses*, 3rd edition, pp. 538–540, The Macmillan Company, New York, 1933.
4. Bannon, R.E., and Walsh, R. On Astigmatism, Part IV. *Amer. J. Optom. Arch. Amer. Acad. Optom.*, Vol.22, pp. 263–277, 1945.
5. Fry, G.A. Computation of Unwanted Sphere and Cylinder in the Periphery of a Spectacle Lens. *Amer. J. Optom. Physiol. Optics*, Vol. 54, pp. 606–616, 1977.
6. Fry, G.A. Choosing the Base Curve for an Ophthalmic Lens. *Amer. J. Optom. Physiol. Optics*, Vol. 55, pp. 238–248, 1978.
7. Davis, J.K. Corrected Curve Lenses and Lens Quality. *Amer. J. Optom. Arch. Amer. Acad. Optom.*, Vol. 39, pp. 135–148, 1962.
8. Davis, J.K., Fernald, H.G., and Rayner, A.W. An Analysis of Ophthalmic Lens Design. *Amer. J. Optom. Arch. Amer. Acad. Optom.*, Vol. 41, pp. 400–421, 1964.
9. Davis, J.K. Stock Lenses and Custom Design. *Amer. J. Optom. Arch. Amer. Acad. Optom.*, Vol. 44, pp. 776–800, 1967.
10. Fry, G.A. The Optometrist's New Role in the Use of Corrected Curve Lenses. *J. Amer. Optom. Assoc.*, Vol. 50, pp. 561–562, 1979.

## Questions

1. A thin prism of refracting angle 8° is made of glass for which the refractive indices are: D line = 1.50, C line = 1.48, F line = 1.53. What is the dispersive power of the glass?

2. A thin prism of refracting angle 4° is made of glass for which the refractive indices are: D line = 1.50, C line = 1.48, F line = 1.53. What is the nu value of the glass?

3. What will be the chromatic aberration of a +6.00 D crown glass lens with a nu value of 65?

4. A thin prism of refracting angle 8° is made of glass for which the refractive indices are: D line = 1.50, C line = 1.48, F line = 1.53. What is the chromatic aberration of the prism?

5. A +5.00 D lens has indices of 1.52, 1.53, and 1.54 for the C, D, and F Fraunhofer lines, respectively. What is the amount of the chromatic aberration?

6. It is required to make an achromatic lens of +5.00 D, using crown glass of index 1.50 with a nu value of 70, and flint glass for which the index is 1.60, with a nu value of 40. What are

the powers of the crown and flint lenses, respectively?

7. It is required to make an achromatic prism of 3 prism diopters, using a crown glass for which the index of refraction = 1.52, dispersive power = 1/60, and a flint glass for which the index of refraction = 1.61, dispersive power = 1/38. Find the powers of the prism components.

8. The equivalent focal length of a positive achromatic doublet (composed of two lenses with zero separation) is 24 cm. If the converging crown glass lens has a focal length of 18 cm, what is the focal length of the diverging flint glass lens?

9. A cemented achromatic lens of +2.00 D is made of crown glass (index of refraction = 1.5; nu = 60) and flint glass (index of refraction = 1.6; nu = 40). The exposed (uncemented) surface of the flint glass has a power of −1.60 D. The curvatures of the cemented surfaces are equal. What is the power of the exposed (uncemented) surface of the crown glass?

10. When looking 30° from the center of a +7.00 D lens the sagittal focal power is +6.88 D and the tangential focal power is +7.48 D. What is the curvature of field or power error?

11. When looking up 30° from the center of a −9.00 D lens at a cross formed of horizontal and vertical lines the sagittal focal power is −8.90 D and the amount of oblique astigmatism is −0.36 D. What is the amount of blur expressed in diopters of the horizontal lines on the cross?

12. A +3.00 D lens (index = 1.523) is prescribed. Assume the lens is located 27 mm from the center of rotation. What is the radius of curvature of the Petzval surface?

13. A −3.00 D lens (index = 1.523) is prescribed. Assume the lens is located 27 mm from the center of rotation. What is the radius of the far-point sphere?

14. A lens of index 1.50 is designed to eliminate marginal astigmatism. If the lens is worn 25 mm in front of the center of rotation of the eye, at what power will the Petzval surface and the far-point sphere coincide?

# CHAPTER SEVEN

# Absorptive Lenses and Lens Coatings

The human eye has evolved in the presence of solar radiation, which includes not only the visible spectrum, extending from approximately 400 to 700 nm, but also the neighboring ultraviolet and infrared regions of the spectrum. Radiation in both the ultraviolet and infrared regions has been found to have harmful effects on the ocular tissues; fortunately, these effects can be controlled, to a great extent, if absorptive lenses are worn.

Lenses made of clear ophthalmic crown glass or clear CR-39 plastic transmit only about 92% of incident visible light, as a result of the loss by reflection at the front and back surfaces; due to the absorption of some of the ultraviolet and infrared radiation, they transmit somewhat smaller amounts of radiation in these regions of the spectrum. Absorption in the ultraviolet and infrared regions (or even in the visible spectrum) can, however, be increased by the addition of certain "absorbers" to the batch of glass or plastic.

Absorptive lenses may be considered in terms of a number of categories. These include lenses designed for general wear, absorbing the spectrum evenly; lenses designed to selectively absorb either ultraviolet or infrared radiation, or both; and lenses designed to selectively absorb "bands" of radiation within the visible spectrum. In addition, lenses (called "photochromic" lenses) have been developed which have the property of varying the absorption of radiation on the basis of the ambient illumination level, increasing in absorption in high levels of illumination and decreasing in lower levels of illumination.

Lens reflections may present a problem, both for the wearer of the lenses and for an observer. From the wearer's point of view, reflections from the lens surfaces may be annoying, and can be minimized by the use of an antireflective coating. Reflections annoying to the observer include the "rings" seen near the edges of minus lenses. A number of methods for minimizing these reflections will be described.

# _____ EFFECTS OF RADIATION ON THE EYE _____

## 7.1. The Nature of Light

*Radiation* is a physical term defining the transfer of energy through space, from an "emitter" or "radiator" to a receiver. When light is emitted by a source and is subsequently absorbed by a receptor, a net transfer of energy occurs. The sun is a radiator, producing energy which radiates through space in all directions. The sun's radiation is called *electromagnetic* radiation, because it consists of an oscillatory electric field and an oscillatory magnetic field which are perpendicular to one another and to the direction of propagation of the radiation. At any point along the path of propagation the oscillations of the electric and magnetic fields are synchronized with each other: Their maxima and minima occur simultaneously. These forces oscillate about their positions in a transverse direction, akin to the movement of a cork bobbing from ripples on the surface of a body of water. The electromagnetic fields do not move in the direction of propagation. On the contrary, it is the *wave* that advances.

In a vacuum, all electromagnetic energy travels at a velocity of $10^{10}$ meters per second (or 186,000 miles per second). However, it is slowed down in any medium other than a vacuum. The velocity of light in any medium, X, is an indication of the *optical denseness*, or *refractive index*, of the medium, as expressed in the following formula:

$$\text{refractive index of medium X} = \frac{\text{velocity of light in a vacuum}}{\text{velocity of light in medium X}} \quad (7.1)$$

or

$$\text{velocity of light in medium X} = \frac{\text{velocity of light in a vacuum}}{\text{refractive index of medium X}}$$

Thus, when light travels from a vacuum to another medium, the index of refraction of that medium is the factor by which the velocity of light is *reduced* in that medium. Not only is the velocity reduced in various media, light is also absorbed differentially in various media.

Historically, the nature of light has been closely studied and debated. While the nature of light is still not clearly understood, it has been shown that electromagnetic radiation possesses both *particle-like* and *wave-like* characteristics. According to the particle theory, radiant energy consists of discrete units of energy, called *quanta*, or *photons*. The particle theory accounts for the interchange of energy between radiation and matter, for emission and absorption of radiation, and for the photoelectric effect. Conversely, the wave theory is used to explain the phenomena of refraction, reflection, interference, and polarization. These theories are used interchangeably, one theory explaining some observations or phenomena, and the other theory explaining others. The two theories, therefore, complement rather than compete with each other.

Light traveling through a transparent medium is more easily represented as *waves*, whereas interactions occurring between light and matter can be more readily explained by considering light as if it were made up of *particles*.

The wavelike property of radiant energy allows light to be described in terms of standard wave motion parameters (Figure 7-1). If two points on adjacent waves have the same displacement and position, the points are said to be *in phase*. Examples of pairs of points that are in phase are adjacent maxima (crests), adjacent minima (troughs), or alternate zeros. The distance between any two points in phase of adjacent waves, in a vacuum, is referred to as the *wavelength* (λ). The wavelength is generally specified in *nanometers* (nm) (formerly called millimicrons), a nanometer being equal to $10^{-9}$m. If a point on a wave moves in such a manner that it retraces its path at regular intervals of time, its motion is described as *periodic*.

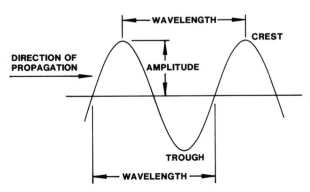

FIGURE 7-1. **The parameters of wave motion.**

Each vibration is called a *cycle*, and the number of cycles of a particular wave train emitted from, passing through, or arriving at a given point is termed the *frequency*. Hence,

$$\text{frequency} = \frac{\text{velocity in a vacuum}}{\text{wavelength in a vacuum}}. \quad (7.2)$$

The frequency of a wavefront is a more fundamental attribute than are its velocity or its wavelength: Frequency is a property of the emitting source, and remains constant regardless of the medium through which the radiation may travel. It is apparent from formula (7.2) that the wavelength is always directly proportional to the velocity; when one is reduced by

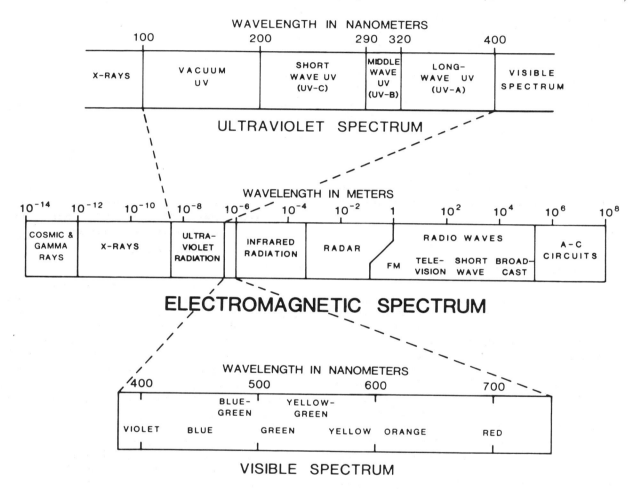

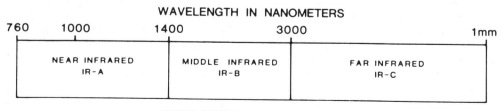

FIGURE 7-2. **The electromagnetic spectrum. (From D. Pitts, *J. Amer. Optom. Assoc.*, Vol. 52, p. 12, 1981.)**

a factor of *n* (refractive index), the other is reduced by the same factor.

In quantum theory, each photon has associated with it a *quantum* of energy, whose magnitude is directly proportional to the frequency with which the emission occurs. Therefore, since the energy (momentum) of a photon is directly proportional to its frequency, and since the frequency is inversely related to the wavelength, the shorter wavelengths of electromagnetic radiation possess higher energy, or momentum, per photon than the longer wavelengths of radiation.

The components of the electromagnetic radiations form a continuous range of radiations differing in frequency, and therefore also differing in wavelength, called the *electromagnetic spectrum*. The electromagnetic spectrum has been divided arbitrarily into regions depending on wavelength. The enormous range of electromagnetic radiation (Figure 7-2) extends from the very short cosmic rays (as short as $10^{-13}$) through gamma rays, X-rays, ultraviolet radiation, visible light, infrared radiation, and radio rays (as long as $10^{6}$ m).

## 7.2. The Visible Spectrum

*Light* may be defined as a form of electromagnetic radiant energy that is capable of stimulating the retinal photoreceptors, giving rise to the sensation of vision. The visible spectrum occupies a narrow band of the total electromagnetic spectrum, extending from approximately 380 to 760 nm. There are no sharp cutoffs at the upper and lower limits, so it is difficult to determine the range of wavelengths included in the visible spectrum with any degree of certainty. It varies with the level of illumination, the clarity of the crystalline lens of the eye, and other factors relative to the observer. Within the specified boundaries, radiation reaching the retina acts as a physical stimulus to produce electrical impulses which are conducted via the optic nerve to the occipital cortex of the brain, providing the sensation of vision, even in color! Color vision is a mental process: the rays of light are not colored. The response of the eye is not uniform within the visible spectrum, the *luminous efficiency*, or *luminosity*, of light varying with wavelength. If we plot the average luminosity as a function of the wavelength, we generate the photopic *luminosity curve*, shown in Figure 7-3.

If we consider an *octave* to be a doubling of the frequency (as in music), the visible spectrum consists of only 1 octave of a total of 60 octaves in the electromagnetic spectrum.

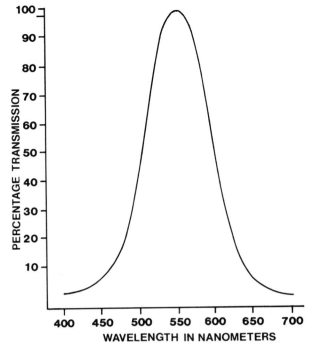

FIGURE 7-3. **The photopic luminosity curve.**

Although the sun is the source of most of the electromagnetic radiation occurring naturally in the atmosphere, on the surface of the earth we are subjected to only a small portion of the sun's radiation because much of it is filtered out by water particles, ozone, oxygen, and carbon dioxide. Oxygen and the ozone layer remove the deadly short-wavelength ultraviolet (up to 250 nm), while water and carbon dioxide decrease the amount of infrared radiation at the earth's surface (most infrared radiation above 3000 nm is filtered out by the earth's atmosphere). Thus we are regularly exposed to some ultraviolet radiation, the visible spectrum, and the infrared portion of the electromagnetic spectrum. While exposure to radiation bordering on the visible spectrum does not give rise to the sensation of vision, these bands of radiation are worthy of our attention because they can exert harmful effects on the eyes.

Ultraviolet radiation extends approximately from 10 to 380 nm. Because certain bands of ultraviolet radiation are associated with particular biological effects, the ultraviolet (UV) spectrum is arbitrarily subdivided into three bands: UV-A extends from 380 to 320 nm, UV-B from 320 to 290 nm, and UV-C from 290 to 200 nm. Sometimes UV-A is called *near UV*, UV-B is called *middle UV*, and UV-C is called *far UV*, because of their positions relative to the visible spectrum.

The infrared spectrum extends from 760 to $10^6$ nm, or 1 mm. The infrared (IR) spectrum is also divided into three portions: IR-A extends from 760 to 1400 nm, IR-B extends from 1400 to 3000 nm, and IR-C extends from 3000 nm to 1 mm. It is of interest that although the visible spectrum occupies only 1 octave, ultraviolet radiation occupies almost 4 octaves and infrared radiation occupies slightly more than 7 octaves. The divisions of UV and IR spectra are summarized in Table 7-1.

In addition to the electromagnetic radiation produced by the sun, manufactured radiation sources emit ultraviolet, visible, and infrared radiation. While many of the manmade sources have beneficial uses, they may also be sources of potential hazard if not used properly.

TABLE 7-1
**Divisions of the Ultraviolet and Infrared Spectra**

|  | *Ultraviolet* | *Infrared* |
| --- | --- | --- |
| A (near) | 380–320 nm | 760–1,400 nm |
| B (middle) | 320–290 nm | 1,400–3,000 nm |
| C (far) | 290–200 nm | 3,000–1,000,000 nm |

## 7.3. Classification of Radiation Effects

*Draper's law* states that for radiation to have an effect on a substance through which it travels, it must be *absorbed* by the substance. Radiation has *no* effect (beneficial or deleterious) on a substance through which it is completely transmitted or by which it is completely reflected. For example, radiation in the region of the visible spectrum causes the sensation of vision because of the fact that it is absorbed by the photo-pigments of the retina.

When radiation is absorbed by an ocular tissue, various effects are produced by the transfer of radiant energy to the molecules and atoms of the absorbing tissue. The absorbed energy can effect the visual apparatus in the following ways:

1. *The Thermal Effect.* This is a heating effect, brought about by the absorption of radiant energy by the molecule, raising the molecule from the resting state to an excited state. A return to the resting state can occur by the dissipation of heat arising from molecular collision. Solar retinopathy, caused by looking directly at a solar eclipse, is an example of a thermal lesion.

2. *The Photochemical Effect.* When radiant energy is absorbed, the molecule that absorbs it may decompose or chemically react to produce a unique chemical product. This is the effect that, in the visible spectrum, produces a chemical reaction in the retina initiating the sensation of vision. However, harmful photochemical effects can occur with other ocular tissues, as with photokeratitis produced by excessive absorption of ultraviolet radiation by the cornea.

3. *Photoluminescence (Fluorescence).* Fluorescence is the phenomenon in which radiant energy of one wavelength is absorbed and radiant energy of a different wavelength (usually longer) is emitted, as described by *Stoke's law.* Thus, the emitted radiation has a lower energy level than the absorbed radiation. The crystalline lens is capable of visible fluorescence, when illuminated by ultraviolet light.

## 7.4. Concentration of Radiant Energy by the Eye

Absorption of radiant energy by a specific tissue depends on its molecular structure and its chemical composition. However, the amount of damage arising from radiant energy absorption is proportional to the radiant energy absorbed per unit of mass or volume.

As radiant energy passes through the eye, it is attenuated in a number of ways. These include (*a*) absorption by the (apparently transparent) ocular media; (*b*) scattering within the eye; (*c*) reflection by the various optical interfaces; and (*d*) loss due to the aberrations of the eye's optical system. Most absorptive losses occur in the anterior structures of the eye. Offsetting this attenuation is the concentration of energy brought about by the focusing effects of the cornea and the lens.

The concentration of radiant energy within the eye is also dependent upon the size of the pupil and the angular extent of the source. With increasing pupil size, the illumination on the retina increases by the square of the pupillary radius. As shown in Figure 7-4A, for a point source of high intensity, refraction by the eye's optical system will concentrate the energy on the retina, causing tissue damage, while having little effect on the cornea and the lens. An example of this type of damage is *solar retinopathy*, following exposure to a solar eclipse. For extended sources, the radiant energy is sharply concentrated in the lens, as shown in Figure 7-4B, but widely distributed on the retina. It should be understood that the possibility of a source of radiation causing damage to ocular tissues depends both on the *intensity* and on the *angular size* of the source of the radiation; while a small source of low intensity is usually harmless to the retina, an extended source (such as a desert

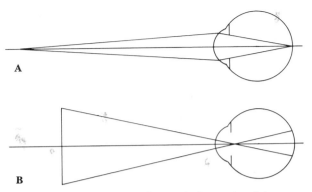

FIGURE 7-4. **Concentration of energy in the eye: A, point source; B, extended source.**

or a body of water) of the same intensity may provide a dangerous concentration of radiant energy in the lens.

## 7.5. Absorption of Radiation by the Ocular Tissues

The *tear layer* absorbs only a small amount of radiation, absorbing ultraviolet radiation below about 290 nm and infrared radiation above about 3000 nm, and therefore transmitting radiation from approximately 290 to 3000 nm. The *cornea*, like the tear layer, absorbs ultraviolet radiation below about 290 nm and infrared radiation above 3000 nm. However, the cornea has *partial* transmission for ultraviolet in the 290- to 315-nm range and for infrared in the 1000- to 3000-nm range. The cornea therefore has relatively high transmission in the range extending from about 315 to 1000 nm, which of course includes the longer ultraviolet wavelengths, all of the visible spectrum, and the shorter infrared wavelengths. The transmission of the cornea (particularly for the shorter wavelengths) decreases markedly with age.

The *aqueous humor* absorbs very little radiation, with the result that any radiation that is transmitted by the cornea is transmitted by the aqueous humor also, and passes to the iris and the lens. In the *iris*, the uveal pigment absorbs radiation and converts it into heat. This conversion can be accompanied by a marked contraction of the pupil, probably due to the release of histamine. The *lens*, like the cornea, has variable absorption properties, depending upon age. The lens of a child absorbs ultraviolet radiation below about 310 nm and infrared radiation beyond 2500 nm, thus transmitting ultraviolet radiation between about 310 and 380 nm, all of the visible spectrum, and infrared radiation up to 2500 nm. The lens of an older adult, on the other hand, absorbs almost all radiation below about 375 nm, therefore transmitting very little ultraviolet radiation. However, there is no change in the absorption of infrared radiation with increasing age.

The *vitreous* largely absorbs radiation below 290 nm and above 1600 nm, therefore transmitting to the retina radiation in the range from 290 to 1600 nm. However, as the crystalline lens absorbs more and more ultraviolet radiation with increasing age, the amount of ultraviolet radiation available to the vitreous gradually decreases.

The radiation received by the *retina* is obviously that radiation transmitted by the vitreous. Whereas the ultraviolet radiation received by the retina decreases in amount with age, the infrared radiation does not decrease in amount: 94% of the infrared radiation of 770 nm reaches the retina, falling off to 90% at 900 nm and to a very low level beyond 1500 nm.

The *eye as a whole* absorbs (and transmits) about the same amount of visible radiation as it would if it were a layer of water approximately 1 inch thick. The transmission of radiation by the ocular media is summarized in Table 7-2.

TABLE 7-2
**Transmission of Radiation by the Ocular Media**

|  | Ultraviolet | Visible | Infrared |
|---|---|---|---|
| Tear layer | 290–380 | 380–760 | 760–3,000 |
| Cornea | 290–380[a] | 380–760 | 760–3,000[a] |
| Aqueous | 290–380[a] | 380–760 | 760–3,000[a] |
| Lens (child) | 310–380[a] | 380–760 | 760–2,500[a] |
| Lens (older adult) | 375–380[a] | 380–760[a] | 760–2,500[a] |
| Vitreous | 290–380[a] | 380–760 | 760–1,600[a] |

[a]There is *partial* transmission for some of the wavelengths in the indicated range.

## 7.6. Effects of Ultraviolet Radiation

The primary ocular effect resulting from absorption of ultraviolet radiation of 300 nm and below is *photochemical* damage to the corneal epithelium. This is known as *photo-ophthalmia*, or as *photokeratitis* or *photoconjunctivitis*. Since the corneal epithelium absorbs most of the ultraviolet radiation, the corneal damage is confined to this layer.

These effects occur after a latent period which may vary from 30 minutes to 24 hours, the length of time depending upon the intensity of exposure. In addition to being a latent effect, the photochemical effect tends to be a *cumulative* effect, repeated exposures with intermissions being equivalent to a single long exposure as long as the intermissions are sufficiently short (24 hours or less) to keep physiological healing from occurring.

In acute photokeratitis the patient experiences a foreign body sensation, photophobia, lacrimation, blepharospasm, redness, and edema. The clinical picture is also seen in "snowblindness," which occurs with long exposure to ultraviolet radiation reflected from large areas of snow. "Welder's flash," or *welder's keratitis*, is a similar condition experienced by arc welders who strike an arc before lowering the protective helmet.

Photokeratitis is self-limiting, the acute symptoms disappearing within 24 to 48 hours: The corneal rehabilitation is remarkably rapid. Permanent damage is rare, and occurs only with extremely high intensity exposure.

It is widely thought that repeated, long-continued exposure to ultraviolet radiation is a causative factor in the development of *pterygia*, growths of vascular and connective tissue into the epithelium of the bulbar conjunctiva and the cornea. Ultraviolet radiation has also been implicated in the development of *pingueculae*, which are small, yellowish elevations of the bulbar conjunctiva.

The crystalline lens is continuously exposed to near ultraviolet radiation (transmitted by the tear layer, cornea, and aqueous) in the 290- to 380-nm range. One of the cumulative effects of this radiation is the formation of lens pigments which cause an increasing yellow coloration of the lens nucleus. This deepening coloration further increases the absorption of ultraviolet radiation in the 290- to 380-nm range, as well as the absorption of visible light. The pigments are largely produced in the nucleus, and lead to a decrease in the light transmission of the lens as one grows older.

Even though only small amounts of ultraviolet radiation reach the lens, the cumulative effects of ultraviolet exposure, over a period of many years, may be responsible for producing lens opacities: in particular, the brown or *brunescent* cataract of the nucleus. The avascular lens, with its inefficient metabolic system, is vulnerable, apparently because the repair mechanisms available are not as well developed as those of the cornea or the retina.

As already described, the phenomenon of photoluminescence (fluorescence) occurs when a substance absorbs radiation of one wavelength and emits radiation of a different (usually longer) wavelength. Visible fluorescence of the lens was noted more than 100 years ago. When radiant energy is absorbed, the absorbing molecules are raised to more active, more excited states: Fluorescence is a mechanism of deactivation through the emission of photons.

The fluorescence of the ocular lens is a well-known phenomenon: If an ultraviolet light source (a "black light," such as that used to evaluate the fit of a contact lens) is held in front of the eye, the lens fluoresces with an eerie greenish-yellow color. A source of 370 to 390 nm is most effective for its production. The amount of fluorescence of the lens increases slowly with increasing age. The absorption of ultraviolet light by the ocular lens has been linked to the formation of fluorescent material within the lens. As the amount of this material, called *fluorogens*, increases with time, there is a parallel increase in the yellowish color of the lens. This may be one of the factors associated with the aging of the lens.

In the normal eye, the retina is shielded from much of the ultraviolet radiation (as described in the previous section) by the filtering action of the cornea and the lens: Under ambient solar radiation, the small amount of ultraviolet radiation reaching the retina is not likely to cause any serious retinal damage. Nevertheless, it is possible that repeated exposure over a period of many years may lead to some degree of damage, due to a slow cumulative effect. When the lens has been removed, because of a cataract, the aphakic eye is subjected to ultraviolet radiation in the 300- to 380-nm range which had previously been filtered out by the lens; in addition, the amount of *visible* radiation increases in aphakia. Thus the increased concentration of ultraviolet and visible radiation on the retina increases the potential for photochemical or thermal damage.

## 7.7. Effects of Infrared Radiation

Wavelengths longer than 3000 nm do not reach the earth's surface, due to absorption by water and carbon dioxide in the atmosphere. Even for wavelengths shorter than 3000 nm, photon energy decreases with increasing wavelength with the result that the region of infrared radiation involved in ocular damage extends only from 780 to 2000 nm. Most of the radiation over 1400 nm is absorbed by the tears and the cornea. At the present time there is little evidence that absorption of ambient solar infrared radiation is damaging to the ocular tissues. However, severe retinal damage from solar infrared radiation can occur by looking at or near the sun, as when observing an eclipse of the sun or when driving an automobile toward the rising or setting sun. Spotters of aircraft, who search the skies for military aircraft, also can encounter ocular damage due to infrared radiation. *Manmade* sources of high-intensity infrared radiation, such as carbon, tungsten, and xenon arc lamps, photoflood lamps, and some laser sources produce levels of infrared radiation significantly above corresponding radiation obtained from the sun.

If the eye is subjected to high-intensity sources of infrared radiation, absorption of this radiation can cause rotational and vibrational changes in the molecules, producing a thermal lesion. Whereas the photochemical damage due to ultraviolet radiation occurs only after a latent period, thermal damage due to infrared occurs immediately. Thermal lesions can be manifested in the *cornea*, as coagulation, leading to opacification; in the *iris*, as congestion, depigmentation, and atrophy due to the absorption of infrared radiation by the iris pigment; in the *lens* as exfoliation of the lens capsule, coagulation of protein, and the production of cataract; and in the *retina* as a necrotic burn.

The potential for the development of cataract from infrared radiation has long been known in workers exposed to high levels of heat radiation such as glassblowers and blast furnace operators. As mentioned previously, the radiation from an extended source, upon undergoing refraction by the cornea and the anterior lens surface, will be concentrated in the posterior region of the lens and then distributed over a wide area of the retina. This concentration of radiation from an extended source at the posterior zone of the lens produces thermal effects that, over a period of years, may result in a posterior cortical cataract.

Anterior lenticular lesions are often found, adjacent to thermal lesions of the iris, following exposure to infrared radiation. The infrared radiation is absorbed by the pigmented epithelium of the iris, is converted to heat, and is then transmitted to adjacent tissues.

A retinal burn occurs when the eye is exposed to a point-like source of infrared radiation of high intensity, such as the sun. Eclipse blindness (sun blindness) arises as a result of observing an eclipse of the sun without adequate ocular protection. The refracting power of the cornea and lens will greatly increase the total amount of radiant energy striking the retina per unit area, compared to that incident per unit area upon the cornea. The infrared radiation is absorbed by the pigment epithelium of the retina and by the choroid, resulting in a thermal lesion involving both of these tissues. The lesion often occurs in the *macular* area, producing a small, dense, central scotoma (a blind area within the field of vision), loss of visual acuity, and metamorphopsia (distortion of perceived images). Large numbers of cases of solar retinopathy tend to occur following an eclipse of the sun, despite widely disseminated warnings against direct observation of a solar eclipse.

## 7.8. The Effects of Visible Radiation

Being transparent, the cornea, aqueous, lens, and vitreous all transmit the great majority of radiation in the visible spectrum (380 to 760 nm). The absorption of these wavelengths by the photoreceptors of the retina initiates the photochemical and neural processes giving rise to the sensation of *vision*. Since human eyes have become adapted to the visual environment under which they have evolved, normal levels of visible light resulting from ambient solar radiation are not normally considered to be hazardous. However, unusually high levels of radiation within the visible spectrum (whether solar or manmade radiation) can cause both photochemical and thermal injury to the retina. The visible wavelengths are not only transmitted by the cornea and lens but are focused on the retina so that the radiation per unit area reaching the retina is markedly higher than that incident upon the cornea. If the offending source emits both visible and near infrared radiation, much of the visible radiation and a smaller but significant amount of the infrared radiation will be transmitted to the retina and absorbed by the photoreceptors, retinal pigment epithelium, and choroid. The short-wavelength end of the visible spectrum (where the energy level is high) tends to produce photochemical damage; the longer wavelengths of the visible spectrum tend to produce both photochemical and thermal damage; and infrared, of course, produces only thermal damage.

Even at levels that are not capable of producing retinopathy, prolonged absorption of visible radiation may have undesirable functional effects. While the normal *dark adaptation* process requires from 30 to 40 minutes, exposure to strong sunlight for three or four hours or more so thoroughly bleaches the retinal photopigments that a significantly longer dark-adaptation time is required for a period of 24 hours or more following exposure.[1] In a situation in which viewing requirements are critical (as when returning home at dusk, after a day at the beach) filtering lenses should be worn during the day, in order to preserve the retina's ability to adapt to low illumination levels at night.

## 7.9. Other Forms of Radiation

The discussion, up to this point, has been concerned only with the effects of ultraviolet, visible, and infrared radiation. However, it should not be inferred that other regions of the electromagnetic radiation spectrum are harmless. Long-wavelength microwaves and diathermy, and ionizing radiation such as X rays and gamma rays, are known to produce adverse effects on the eyes and on other living tissues. However, these sources are usually present only in special installations, and systems of protection are normally present for such installations. The basic protection principle involves *prevention* of propagation of the radiation, by some form of shielding.

## 7.10. Recommended Levels of Retinal Illumination

For a given level of ambient illumination, the amount of illumination reaching the retina is controlled by the pupil. If we consider the pupil size to vary from about 1 mm in very bright illumination to about 7 mm in low illumination, the pupil's control of retinal illumination extends over a range of $7^2$, or about 50 to 1.

However, the wide range of illumination under which the retina is capable of functioning is due more to *retinal adaptation* than to variation in pupil size. The retina is capable of adapting its sensitivity by a factor of 20,000, from the lowest to the highest illumination levels.

Although the eye is capable of functioning in such an incredibly wide range of illumination levels, high levels of illumination *can* cause discomfort. According to Richards[2], an ideal level of illumination is 400 ft-lamberts. This level of illumination would occur on a typical summer day if one were reading under the shade of a big tree. However, on a bright summer day (without the shade tree), the level of illumination

may be much higher than 400 ft-lamberts. For example, on a bright, sunny Texas day in the middle of July, the illumination level can reach 10,000 ft-lamberts. This exceeds the "comfortable" 400 ft-lambert level by a factor of 10,000/400, or 25/1, and would require a filter having a transmittance of 1/25, or 4%. Even with a more typical illumination level of 5000 ft-lamberts, the "comfort" level is exceeded by a factor of 12.5/1, requiring a filter having a transmittance of 8%.

In summary, undesirable effects result from excessive levels of illumination, not only in the ultraviolet and infrared regions but also in the visible region of the spectrum. Although most patients who request "sunglasses" or other tinted lenses do so because of visual discomfort, it is important for these lenses to be designed to control ultraviolet and infrared radiation as well as controlling visible light.

**Glare.** Many people think they need sunglasses in order to protect their eyes from discomfort caused by "glare." The term *glare* refers to the presence of one or more areas in the field of vision that are of sufficient brightness to cause an unpleasant sensation, a temporary blurring of vision, or a feeling of ocular fatigue. Glare therefore occurs as a result of the fact that the illumination in a part of the visual field is much greater than the level of illumination for which the retina is adapted. An extreme case of glare often occurs during night driving: While the retina is adapted to a very low level of illumination, the headlights on an oncoming car can emit a level of illumination that is sufficiently greater than the level of adaptation to cause extreme discomfort.

As a practical matter, glare can be said to occur when the ratio between the highest level of illumination in the visual field and the background illumination exceeds a ratio of 3 to 1. The importance of background illumination, in order to avoid glare, was emphasized in the early days of television: In viewing television, discomfort is likely to occur if the level of illumination of the television screen exceeds three times that of the background. In this situation glare can be avoided by having a low level of illumination (rather than complete darkness) in the room.

Tinted lenses or other filters are, unfortunately, of no help in controlling glare. If a "hot spot" in the visual field is ten times as bright as the background illumination, the use of a filter having 50% transmittance (or any other percentage transmittance) will be of no help: *Everything* in the visual field, both the hot spot and the background, will be reduced by the same percentage. However, *Polaroid* lenses (to be discussed in Section 7.14) can be useful in controlling glare, since they have the property of filtering out reflected light.

# ABSORPTIVE LENSES

## 7.11. Reflection, Absorption, and Transmission

When light is incident upon a lens, some of it is reflected by each of the lens surfaces, some of it is absorbed by the lens, and the remainder is transmitted. As shown in Figure 7-5, the percentage of the incident light that leaves the lens after passing through it (the transmitted light) is found by calculating the percentage light lost by reflection at the front surface, the percentage lost by absorption, and the percentage lost by reflection at the back surface.

The amount of light *reflected* by a lens surface, in air, is given by the Fresnel equation

$$I_R = \frac{(n' - n)^2}{(n' + n)^2} (I) . \qquad (7.3)$$

For clear ophthalmic crown glass, having an index of refraction of 1.523, the light lost by reflection at the front surface is

$$\frac{(0.523)^2}{(2.523)^2} = \frac{0.2735}{6.3655}$$

$$= 0.043 = 4.3\%.$$

Since clear ophthalmic crown glass absorbs no light, the percentage of incident light remaining when the light reaches the back surface of the lens is

$$100 - 4.3 = 95.7\%,$$

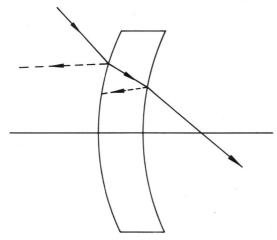

FIGURE 7-5. **Reflection, absorption, and transmission by a lens.**

and the light lost by reflection at the back surface of the lens is

$$(0.957) \frac{(0.523)^2}{(2.523)^2}$$

$$= 0.041 = 4.1\%,$$

and the light transmitted by the lens is

$$100 - 4.3 - 4.1 = 91.6\%.$$

For a plastic lens (CR-39 material) having an index of refraction of 1.498, the light loss at the front surface is

$$\frac{(0.498)^2}{(2.498)^2} = \frac{0.248}{6.240}$$

$$= 0.040 = 4.0\%,$$

and the light loss at the back surface is

$$(0.96)(0.40)$$

$$= 0.038 = 3.8\%.$$

The light transmitted by the lens, assuming that no light is absorbed, is

$$100 - 4.0 - 3.8 = 92.2\%,$$

or about 0.6% higher than transmitted by an ophthalmic crown glass lens.

The amount of light *absorbed* by a lens is given by Lambert's law of absorption, which states that for an absorptive material such as a tinted lens layers of equal thickness absorb equal quantities (or percentages) of light regardless of the intensity of the light.

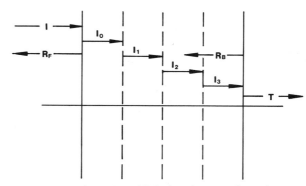

FIGURE 7-6. **Absorption of light by a lens (Lambert's law).**

Figure 7-6 shows a lens divided into several layers of equal thickness. The intensity of light incident upon the front surface (the first "layer" or "unit") of the lens is designated as $I$; the intensity continuing past the front surface is designated as $I_0$ (which is equal to $I-R$). While passing through the first layer, the light is transmitted by the amount indicated by a factor, $q$, the *transmission factor*. For example, if $q = 0.5$ unit, the intensity of light leaving the first unit will be 0.5 of the intensity at the front surface of the lens (after reflection). If we call the second layer of the lens $I_1$, the third $I_2$, and so forth,

$$I_1 = I_0(q)$$

$$I_2 = I_1(q) = I_0(q)^2$$

$$I_3 = I_2(q) = I_0(q^3)$$

and

$$I_x = I_0(q^x). \tag{7.4}$$

The total transmission of the lens, $T$, will then be

$$T = T_x - R_B,$$

$R_B$ designating the light lost by reflection at the back surface of the lens.

### EXAMPLE

Given an ophthalmic lens, 6 mm thick, having a transmission factor, $q$, of 0.8 per 2 mm of thickness, and an index of refraction of 1.523. What is the transmission, $T$, of this lens?

The light lost by reflection at the front surface is

$$R_F = \frac{(0.523)^2}{(2.523)^2} = 0.043.$$

The intensity of light incident upon the first layer of the lens, after reflection is

$$I_0 = I - R_F$$
$$= 1 - 0.043$$
$$= 0.957.$$

The total amount of light lost by absorption will be

$$I_3 = I_0(q^3)$$
$$= 0.957(0.8^3)$$
$$= 0.957(0.512)$$
$$= 0.49.$$

The light reflected by the back surface, $R_B$, is

$$R_B = (0.49)\frac{(0.523)^2}{(2.253)^2}$$
$$= (0.49)(0.043)$$
$$= 0.021$$

and the total transmission, $T$, of the lens is equal to

$$T = I_3 - R_B$$
$$= 0.49 - 0.021$$
$$= 0.469 = 46.9\%.$$

## 7.12. Opacity

Additional terms that are used in connection with absorptive lenses are *opacity* and *density*. Opacity is defined as the reciprocal of transmission,

$$O = \frac{1}{T}. \tag{7.5}$$

For the lens in the above example, having a transmission of 46.9%,

$$O = \frac{1}{0.469} = 2.13.$$

If light passes through a number of lenses, one after another, the *ultimate transmission* is found by multiplying the separate transmissions of each of the lenses,

$$T_U = (T_1)(T_2)(T_3) \cdots,$$

and the *ultimate opacity* is found by multiplying the separate opacities of the lenses,

$$O_U = (O_1)(O_2)(O_3) \cdots.$$

### EXAMPLE

An automobile driver wears a tinted lens having a transmission of 80%, while looking through a windshield having a transmission of 70%. What are (a) the ultimate transmission and (b) the ultimate opacity of the windshield-lens combination?

(a) The ultimate transmission is equal to

$$T_U = (T_1)(T_2)$$
$$= (0.80)(0.70)$$
$$= 0.56 = 56\%.$$

(b) The ultimate opacity is equal to

$$O = (O_1)(O_2)$$
$$= \left(\frac{1}{0.80}\right)\left(\frac{1}{0.70}\right)$$
$$= (1.25)(1.43) = 1.79.$$

## 7.13. Density

In solving problems concerning the transmission of light by a lens it is convenient to translate the transmission into the optical *density* of the material. The use of the concept of density is convenient because, if the density is stated for a given thickness of material, it may be found for any other thickness by the use of simple proportion. Once the transmission of a material is known, the density may be found by use of the relationship

$$\text{density} = \log_{10} \text{opacity}$$

$$= \log \frac{1}{T}$$

$$= \log 1 - \log T$$

$$= 0.0 - \log T.$$

$$\text{Density} = -\log T. \qquad (7.6)$$

The calculation of the transmission of a given lens, when losses by reflection are considered, is mathematically simplified when transmission is stated in terms of density. Since the transmission of a medium is the ratio of transmitted light to incident light, the transmission of each surface can be expressed as a density. Hence, for a simple lens the total density of the lens may be considered to consist of the two *surface densities* and a *media density*. The density of the front surface of the lens is designated by

$$D_S = \log \frac{1}{I_O}, \qquad (7.7)$$

where $I_O = I - R_1$.

The media density is the density per unit thickness, $(\log 1/q)$, multiplied by the number of units, $(x)$, with the result that

$$\text{media density} = (x)(D/\text{unit}).$$

Since the total density of a lens is the sum of the densities of the parts, then the total density is given by the expression

$$D = 2 \text{ surface densities} + (\text{number of units}) \\ (\text{density per unit})$$

$$D = 2D_S + (x)(D/\text{unit}). \qquad (7.8)$$

### EXAMPLE 1

A lens 2 mm thick has a refractive index of 1.523 and a total transmission of 30%. If a transmission of 50% is desired, what must be the thickness of the lens?

The density of the 2-mm-thick lens is found by the formula

$$D = -\log T$$

$$= -\log 0.3$$

$$= -(-0.523) = 0.523.$$

The reflection from the front surface is found by

$$R_1 = \left(\frac{n' - n}{n' + n}\right)^2 = \left(\frac{0.523}{2.523}\right)^2$$

$$= 0.043 = 4.3\%.$$

The transmission, $T$, of the front surface is given by

$$T = I - R_1 = 100 - 4.3 = 95.7\%.$$

The front surface density is equal to

$$D_S = -\log T = -\log 0.957$$

$$= -(-0.0191) = 0.0191.$$

The two surface densities (front and back) are equal to

$$2(0.0191) = 0.0382.$$

The total density of the lens is given by the relationship

$$0.523 = 2D_S + D_{\text{med}};$$

therefore,

$$D_{\text{med}} = 0.523 - 0.0382 = 0.4848.$$

Since the lens is 2 mm thick, the density per millimeter of thickness is

$$\frac{0.4848}{2} = 0.2424.$$

The density of a lens with a transmission of 50% is

$$D = -\log T = -\log 0.5 = -(-0.301) = 0.301.$$

$$D = 2D_S + (\text{number of units})(D/\text{unit})$$

$$0.301 = 0.0382 + (\text{number of units})(0.2424),$$

or

$$(0.2424)(\text{number of units}) = 0.301 - 0.0382$$

$$= 0.2628.$$

and,

$$(\text{number of units}) = \frac{0.2628}{0.2424} = 1.08.$$

The thickness of the lens is therefore equal to *1.08 mm*.

### EXAMPLE 2

If the above lens is made 3.5 mm thick, what would be its transmission?

Total density = $2D_S$ + (number of units)($D$/unit)

$$= 0.382 + (3.5)(0.2424)$$
$$= 0.0382 + 0.8484$$
$$= 0.8866.$$

Therefore, the density of this lens, if made in a thickness of 3.5 mm, would be 0.8866.

$$\log T = -D = -0.8866.$$

The antilog of $-0.8866 = 0.1298$,

and the transmission of the lens is therefore

12.98%.

## 7.14. Methods of Manufacturing Absorptive Lenses

An *absorptive lens* is one which is used for the specific purpose of reducing the amount of transmitted light or radiant energy, thus acting as a *filter*. Absorptive lenses are sometimes referred to as *tinted*, or *colored*, lenses, since they are not usually clear and colorless, as are lenses made of white ophthalmic crown glass. The absorption may be *uniform* (or neutral), absorbing visible light of all wavelengths, or *selective*, absorbing some wavelengths more than others.

Presently, the major forms of absorptive lenses produced by lens manufacturers are: (1) tinted solid glass lenses; (2) glass lenses with surface coatings; (3) tinted plastic lenses; (4) photochromic lenses; and (5) Polaroid lenses.

### 1. Tinted Solid Glass Lenses

The principal ingredients of white ophthalmic crown glass are silica, soda, and lime, to which small amounts of potassium, aluminum, and barium oxides are added in order to provide the desired physical and chemical properties. To produce a tinted lens, one or more *metals* or *metallic oxides* are introduced into the basic batch at the start of the process. The spectral transmission characteristics of the finished lenses are controlled by the quantities of these metals and metallic oxides present at the initial batch stage. The concentration of metals and metallic oxides incorporated to produce colors (even dense colors) in lenses is less than 1%. The elements most commonly used and the colors they produce are:

| | |
|---|---|
| iron | green |
| manganese | pink |
| cobalt | blue |
| cerium | pinkish brown |
| nickel | brown |
| uranium | yellow |
| chromium | green |
| gold | red |
| silver | yellow |
| didymium | pink |
| vanadium | pale green |

The color imparted by the addition of an absorptive substance is of no particular significance, but an incidental by-product. The color of the lens does not indicate the specific absorptive characteristics of the lens for the ultraviolet or infrared regions of the spectrum. The apparent color of a lens depends upon those portions of the visible spectrum that are transmitted: If a lens transmits more of the green portion of the visible spectrum than it does of the blue, the lens will appear to be green by transmitted light. When we speak of the color of a filter, we mean the apparent color of a "standard white" object seen through it. This apparent color is useful for determining the general effect upon the visible portion of the spectrum. If the transmission is *uniform* across the visible spectrum, the lens will appear to be a *neutral gray*, and any colors viewed through it will not be appreciably changed. However, the overall *brightness* of the scene will appear to decrease. If the transmission over the visible spectrum is not uniform but is *selective*, the lens will appear to have a distinct color, and the color of any object viewed through the lens may be altered.

Different absorptive substances may produce lenses of similar color, but the spectral transmission curves for the lenses will differ. When one is provided with a spectral transmission curve for a lens, it is not possible to determine what the color of the lens will be. Although the spectral transmission curve of a lens is the major consideration, the color of the lens is useful for identification purposes, and may have important psychological and cosmetic ramifications. In addition, the *depth* of the color provides a general indication of the level of transmission of the visible spectrum.

Solid glass tinted lenses were the earliest absorptive lenses to be produced. These lenses have many advantages, compared to other methods of producing absorptive lenses: They may be produced in large quantities at low cost; the transmission is affected very little by surface scratching; there is an absence of the reflections that are associated with surface coatings; and no special equipment is needed for surfacing and finishing the lenses. Disadvantages include a variation in transmission from center to edge, for lenses of high power; a variation in transmission from one eye to the other for patients with large amounts of anisometropia; the permanence of the tint (it cannot be removed); and the large inventory of semifinished lenses required to service a wide range of prescriptions.

## 2. Glass Lenses with Surface Coatings

A lens may be tinted by depositing a thin metallic oxide on the surface of the lens. The coating is deposited on the lens by an evaporation process conducted under a vacuum at high temperatures. Due to the high temperature required, the vacuum coating process cannot be used with plastic lenses.

The density of the coated lens depends upon the thickness of the metallic oxide coating. Since the index of refraction of the metallic oxides is higher than the index of refraction of the underlying glass, the amount of light reflected from the absorptive coating is greater than the amount reflected from the uncoated front surface of the glass (note that the coating is deposited on the *back* surface of the glass). To reduce the amount of light reflected by the coating, an antireflective coating of magnesium fluoride is placed on the metallic oxide coating. When the lens is thoroughly cleaned and the coating properly applied, the resulting hardness is equal to that of ophthalmic crown glass. The coating has high resistance to wear and to chemical interaction.

If the coating is applied uniformly across the surface of the lens, the transmission will be uniform from edge to edge. Metallic coatings may also be applied to solid glass tinted lenses in the following circumstances: (1) to darken a lens that the patient believes is too light; (2) if a solid glass tinted lens of high power is found to provide unsatisfactory performance or appearance due to the variation in transmission from the center to the edge, the variation can be minimized by the application of a metallic oxide coating on either the front or back surface of the lens.

## 3. Tinted Plastic Lenses

As already pointed out, plastic lenses cannot be surface coated by evaporation because they would be deformed by the high temperature required. Consequently, plastic lenses are tinted by dipping them in a solution containing the appropriate organic dye. The resulting density is dependent upon the nature of the dye and the time the lens is immersed in the solution. In order to achieve a particular tint and transmission, the lens may be dipped into several different tinted solutions. Since the dye penetrates the surface layer of the lens to a uniform depth, the lenses are of uniform density regardless of the variation in thickness from the center to the edge. If the tint is found to be too dark, or for any reason needs to be changed, some of the tint can be removed by dipping the lens in a bleaching solution.

## 4. Photochromic Lenses

In 1964, Corning Glass Works introduced glass lenses that were photochromic, that is, lenses that darken when exposed to long-wavelength ultraviolet radiation. Photochromic glass, having an index of refraction of 1.523, contains microscopic crystals of silver halide. Upon absorption of the long ultraviolet radiation, the silver halide crystals decompose into silver and halogen atoms and the lens becomes darker. The rigid matrix of the glass holds the silver and halogen in close proximity, and upon removal of the activating ultraviolet radiation the silver and halogen atoms recombine into silver halide crystals with the result that the lens becomes lighter. The darkening rate is temperature dependent; the lower the temperature the faster and deeper the amount of darkening. Hence, the degree of darkening is dependent on a number of factors, including the intensity of the radiation, the length of exposure, and the ambient temperature. The rate of *fading* is dependent upon the composition of the glass, the temperature (thermal bleaching—the higher the temperature, the faster the fading), and exposure to wavelengths longer than those used for darkening (optical bleaching). In general, lenses darken faster than they fade. The cycle of darkening and fading is apparently inexhaustible. Photochromic lenses will be discussed further in Section 7.21.

## 5. Polaroid Lenses

As discussed earlier, electromagnetic radiation is transmitted by a process exhibiting both particle-like and wave-like behavior. The wave-like behavior is useful for explaining the phenomenon of polarization.

Ordinarily, a light beam is composed of many *wave trains*, each with its plane of vibration oriented in a random direction, but always at right angles to the direction of propogation. An end-on view of the wave can be represented diagrammatically, as shown in Figure 7-7A. The beam of light is circularly symmetrical, and is said to be *unpolarized*. Some types of

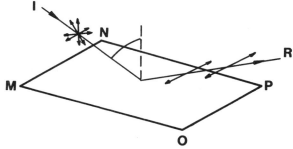

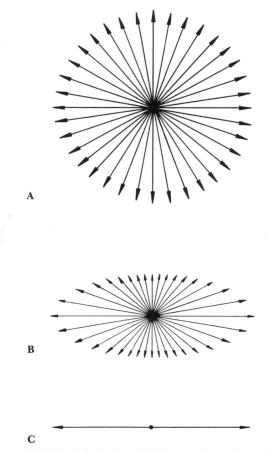

**A**

**B**

**C**

FIGURE 7-7. **Polarization of light: A, end-on view of an un-polarized beam; B, a partially polarized beam; C, a plane-polarized, or linearly polarized, beam.**

crystals, such as quartz, tourmaline, and calcite, have the property of either partially or totally suppressing the vibrations in a particular direction. When the vibration is partially suppressed in a particular direction, the beam is not circularly symmetrical and the wave train is said to be *partially polarized* (Figure 7-7B). When the vibration is restricted to a single plane, the transmitted light is said to be *plane polarized*, or *linearly polarized* (Figure 7-7C).

When direct sunlight (which is unpolarized) is specularly reflected by particles or surface materials, it becomes partially or completely polarized, depending on the angle of incidence and the nature of the reflecting material. Materials which best polarize light by reflectance are generally nonconductors, called *dielectrics*, such as glass, pavement, sand, and snow. The light which is reflected from a dielectric surface will be completely polarized at a specific an-

gle of incidence (called Brewster's angle), at which the angle between the associated refracted and reflected rays at the surface is 90°. This occurs when the tangent of the angle of incidence is equal to the index of refraction of the medium upon which the light is incident, that is,

$$\tan i = n.$$

For specularly reflected light from ophthalmic crown glass of refractive index 1.523, Brewster's angle is about 57°; for water, having an index of 1.333, Brewster's angle is about 53°. The plane of polarization (vibration) of the reflected light is parallel to the reflecting surface, or perpendicular to the plane of the incident light, as shown in Figure 7-8. Hence, for a horizontal reflecting surface the plane of polarization is horizontal. Specularly reflected light is highly localized, is intense, and appears to be *white* in color when reflected from a mirrorlike surface. Because of the veiling glare, it provides no useful information about the color, texture, or other properties of the underlying surface. In effect it *obscures* the surface, and if intense may be visually uncomfortable. When the angle of incidence is near Brewster's angle, the reflected light is strongly polarized, but at adjacent angles the reflected light will consist of both polarized and unpolarized light (partial polarization).

If a polarizing lens (called an *analyzer*) with its plane of polarization in the vertical meridian is placed in the path of a beam of light reflected from a horizontal surface, it will absorb the reflected polarized light. The elimination of specularly reflected horizontally polarized light improves visual acuity, by reestablishing the natural balance of light intensities, and restores fidelity to the surface color.

Polarizing lenses offer distinct advantages to fishermen who experience glare from the surface of the water, to motorists who find the reflected glare from highway surfaces annoying, and to skiers faced with large expanses of snow.

FIGURE 7-8. **The plane of polarization of reflected light is parallel to the reflecting surface.**

Polaroid filters are made by heating and stretching a thin sheet of polyvinyl alcohol to about four times its original length. The stretching aligns the molecular structure into long chains parallel to the direction of stretch. The sheet is then passed through a weak iodine solution, and the iodine molecules diffuse into the polyvinyl layer and attach themselves to the chains of long molecules, thereby creating a polarizing filter. The thin polarizing sheet is then laminated between two layers of coated cellulose acetate butyrate and is pressed to the desired curvature. For glass lenses, the polarizing material is laminated between two layers of glass which may be tinted and surfaced to any desired power.

The transmission of a polarized lens is dependent not only upon the degree of plane polarization but also upon the absorptive nature of the laminated material. Standard polarized sunglasses are uniform-density lenses from the center to the edge, because the tinted layer has a uniform thickness. Special additives in the tint coating increase the absorption of ultraviolet.

The unaided eye does not usually distinguish between polarized and unpolarized light. However, if a diffuse blue background is viewed through a rotating plane polarizer, a yellowish double-ended brushlike figure is seen, originating from the point of fixation, upon the blue background. The brushes rotate with the plane of vibration of the polarizer. This phenomenon, known as *Haidinger's brushes*, is thought to be due to the double refraction effect of the radial nerve fibers of Henle, surrounding the fovea. It is used, in the diagnosis of amblyopia and strabismus, to detect eccentric fixation.

The permanent physiological effects of polarized light on the eye (if any) are unknown.

## 7.15. Categories of Absorptive Lenses and Specification of Transmission

Absorptive lenses may be arbitrarily considered in terms of the following categories:

1. Lenses, designed for general wear, that absorb the spectrum evenly, absorbing little more ultraviolet, visible light, or infrared radiation than is absorbed by clear ophthalmic crown glass.
2. Lenses that selectively absorb ultraviolet radiation and transmit the visible spectrum more or less evenly.
3. Lenses that selectively absorb both ultraviolet and infrared radiation, while having little effect on the visible spectrum.

4. Lenses that selectively absorb portions of the visible spectrum in a nonuniform manner.
5. Lenses designed for occupational use, selectively absorbing particular wavelength bands or having extremely high absorption.
6. Lenses whose absorption characteristics vary with illumination (photochromic lenses).
7. Miscellaneous absorptive lenses.
8. Cosmetic lenses.

**Specification of Transmission.** The transmission properties of a lens are illustrated by means of a *spectral transmission curve*, such as those shown for clear glass and plastic lenses in Figures 7-9A and B. A spectral transmission curve shows the percentage transmission for all of the visible spectrum and for portions of the ultraviolet and infrared wavelengths. These curves depict the amount of radiant energy transmitted as a percentage of the radiant energy incident upon the lens, as a function of wavelength. The single percentage number given for a particular lens refers to the area under the curve within the visible spectrum.

It is important to understand that manufacturers' transmission data (such as those shown in Figure 7-9) are based on a lens thickness of 2 mm. For thicker (or thinner) lenses, the transmission will be lower (or higher). This should be kept in mind by the practitioner when he prescribes or recommends absorptive lenses: For example, a hyperope whose lenses may have an average thickness of 4 mm will suffer almost twice the loss of light transmission for a given tint as will a myope or low hyperope whose lenses will be only about 2 mm thick.

## 7.16. General-Wear Lenses Absorbing the Spectrum Evenly

Lenses in this category are very lightly tinted, and absorb little or no more ultraviolet, visible, or infrared radiation than clear ophthalmic glass. Usually having flesh-colored tints, these lenses are known as "indoor" tints or as "cosmetic" tints. Since transmission across the spectrum is fairly uniform for these lenses, color values of objects are changed very little. Representative transmission curves of lenses in this category, shown in Figures 7-10A and B, are remarkably similar, falling very close to one another. Some of these lenses contain ingredients such as cerium oxide, rendering them slightly more absorptive for the shorter wavelengths than light crown glass.

The lenses in this category are so similar in appearance that it is often difficult to distinguish one

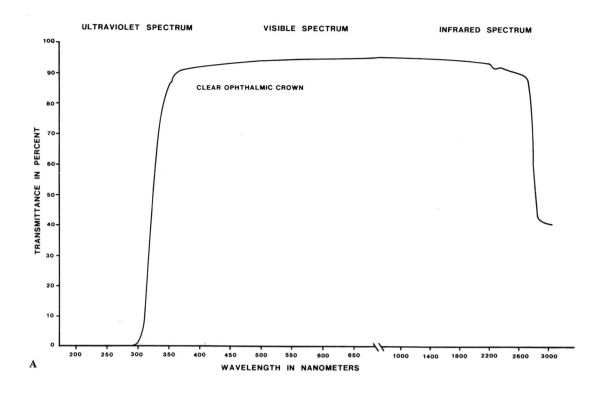

A

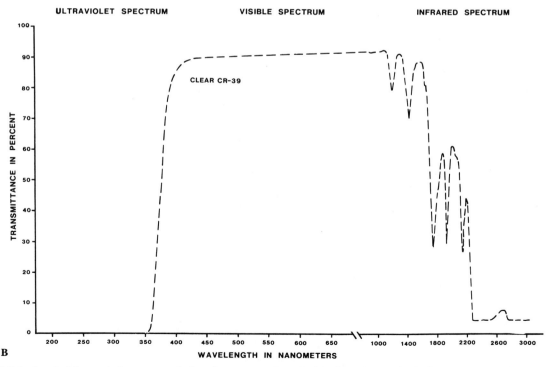

B

FIGURE 7-9. **Spectral transmission curves: A, for clear crown glass; B, for clear CR-39 plastic. (From D. Pitts and B. Maslovitz, unpublished data.)**

195

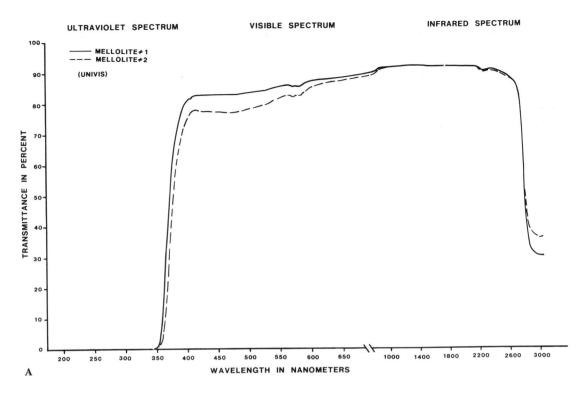

A

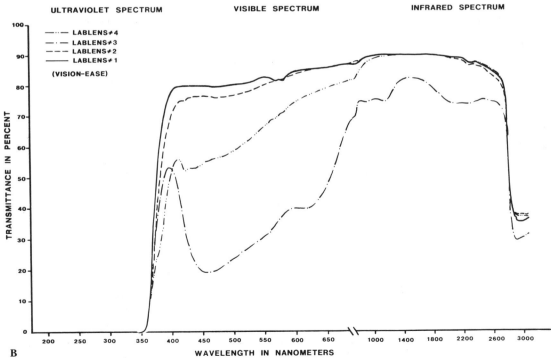

B

FIGURE 7-10. **Spectral transmission curves for general-wear lenses: A, Univis Mellolite; B, Vision-Ease Lablens. (From D. Pitts and B. Maslovitz, unpublished data.)**

manufacturer's lens from another, especially under the low illumination levels found in an examination room, or even to determine whether the lens is tinted at all! One method of detecting a very light tint is to place the pair of glasses on a white sheet of paper; another is to remove one lens from the frame—if the lens is viewed through its edge, a distinct color may be seen.

When such a lens is being replaced, it may be wise to replace *both* lenses, since the color differences are so subtle. If only one lens is to be ordered, it is a good idea to send the glasses to the laboratory, so that color and density of tint can be matched.

The value of these lenses has long been debated, some optometrists prescribing them on a fairly regular basis and others prescribing them only with great reluctance. Even though such a lens transmits almost as much light as clear crown glass, many wearers are convinced that the lenses make their vision more comfortable and that the tint is beneficial. The origin of the reported vision comfort is not clearly understood. Perhaps the effects measurable at any one moment are small, but the comfortable feeling is the sum of these effects during continuous wear. Knoll[3] has suggested that one of the factors could be the attenuation of the intensity of reflected images, especially those reflected images formed after multiple traverses through the lens.

**Clinical Note:** If a patient has been wearing a light tint and likes it, it is usually best *not* to arbitrarily remove the tint. When ordering a new prescription, the tint should be duplicated as nearly as possible.

Patients who frequently ask for light tints for general wear include the following: (1) lightly pigmented individuals, including albinos; (2) patients with high refractive errors, especially aphakics and myopes; (3) patients who work under poorly designed artificial illumination conditions; (4) patients who are not in good general health; (5) patients who have neurasthenic or neurotic tendencies.

Patients who work under poorly designed illumination conditions usually report a "glare" which interferes with vision and produces discomfort. When the visual environment includes extended light sources such as large windows or brightly illuminated walls, the reflected images will have a large angular subtense. This can produce veiling glare which will lead to a reduction of contrast in the visual field. The patient may have concluded that *fluorescent lighting* is the cause of the problem, and some practitioners routinely prescribe a pinkish-brown tint for patients who work under fluorescent lighting.

Because the color of the lenses tends to match skin tones, the lenses are often considered to have a defi-

nite *cosmetic* value, removing some of the "glassiness" of the lenses. When used with fused bifocal lenses, the tint tends to camouflage the bifocal segment.

Lenses available in this category include the American Optical *Cruxite* lens, the Univis *Mellolite* (Figure 7-10A), the Titmus *Velvetlite*, and the Vision-Ease *Lablens* (Figure 7-10B). The lightest of these lenses have an average transmission (throughout the visible spectrum) of approximately 88%, or only 4% less than that of clear crown glass. Although American Optical has discontinued their *darker* flesh-colored tints, other companies continue to make such lenses. These include the Mellolite 3, the Tonetex 3, and the Lablens 3 and 4 (Figure 7-10B).

For general-purpose wear, and especially when the lenses will be used for *night driving*, only the lightest of tints should be used. It appears that for some patients these lenses may be "habit forming": Once having worn such a tint, the patient may request a *darker* tint. The patient may wish to use the darker tint for both general wear and as a substitute for sunglass lenses. However, any fixed tinted lens (i.e., not photochromic) giving significant reduction of illumination outdoors is *too dark* for inside use and for nighttime wear.

Most roadways are poorly illuminated at night, and the introduction of a tint can further reduce the illumination. The natural illumination of the stars and the moon provide light for determining the outline of the roadbed and the contours of the surrounding landscape so essential in country driving. The wearing of darkly tinted lenses (often combined with a tinted windshield) can create an illuminated field that is greatly restricted, being confined to the paths of the headlight beams. Even a small reduction in the illumination available for night driving can decrease the "seeing distance" (the distance at which a nonilluminated object may be seen) in front of the vehicle. If the seeing distance is less than the braking or stopping distance for a particular speed, one can drive into a stalled or slowly moving vehicle instead of avoiding it.

One method of maximizing the transmission of an absorptive lens that will be used for night driving is to place an antireflective coating on a lens having one of the lightest general-wear tints: The increase in transmission due to the coating will be sufficient to offset the loss due to absorption by the tint, with the result that the lens will have a transmission equal to that of clear crown glass. The coating has, of course, the additional advantage of appreciably reducing the intensity of light reflected by the lens surfaces.

If a patient is uncomfortable in outdoor illumination, the lightest of the general-wear tints are not likely to be of much help. Since the darker general-wear tints are responsible for the problems just discussed (inadequate illumination for indoor wear and particularly for night driving) a separate pair of glasses may be prescribed. An alternative would be a pair of glasses having photochromic lenses.

The darker shades of the neutral density tints (#2, #3, and #4 tints) may be useful for photophobia associated with albinism or for pathological conditions affecting the cornea, conjunctiva, or iris. The transmission curves of these darker tints often differ significantly from those of the lighter tints: They absorb disproportionately more radiation in the short-wavelength visible region and therefore tend to take on a reddish-brown color.

## 7.17. Lenses That Selectively Absorb Ultraviolet Radiation While Transmitting the Visible Spectrum in a Uniform Manner

As noted in Section 7.6, exposure of the cornea to ultraviolet radiation can result in photokeratitis. Wavelengths shorter than 290 nm primarily affect the corneal epithelium; wavelengths in the 290- to 315-nm range cause damage to the corneal stroma, Descemet's membrane, and the endothelium, and can also produce a secondary uveitis.[4]

It was also noted in Section 7.6 that ultraviolet radiation is thought to be responsible for the production of cataracts. Recent epidemiological studies of the incidence of cataracts in various geographic and climatic regions support the concept of a relationship between sunlight and nuclear senile cataracts.[5–8] Specifically, UV-A (280–320 nm) has been implicated as the causative factor, on the basis of biochemical, photochemical, and histological studies.[9–12] Both the ultraviolet-absorbing pigments in the lens and the fluorescence of the lens have been found to increase with age, and it has been postulated that fluorescent substances in the lens are responsible for other changes including the darkening of the lens which leads to the brunescent (brown) form of senile cataract.[13,14] Whereas environmental, nutritional, and genetic factors are also known to play a role in the etiology of cataracts, epidemiological and experimental data suggest that ultraviolet radiation is an important factor.

Protection from solar and artificially produced near ultraviolet radiation is recommended for individuals who are repeatedly subjected to ultraviolet radiation during occupational, medical, or recreational activities. Pitts[15] has pointed out that exposure to ultraviolet radiation under normal living conditions can be considered safe, due to the fact that the eyes are protected by the eyebrows (being set below and behind the eyebrows and recessed in the orbits); furthermore, since the line of sight is ordinarily not directed toward the sun, the energy of the ultraviolet radiation reaching the eyes is attenuated by atmospheric absorption and scattering. However, in environments where the nominal intensities of near ultraviolet radiation are exceeded, protection should be provided against the potential hazards of ultraviolet exposure. The amount of ultraviolet radiation increases with increasing altitude, with the result that outdoor activities such as skiing and mountain climbing can greatly increase the level of ultraviolet exposure. Solar reflections from large expanses of snow, water, or sand contain significant amounts of ultraviolet radiation. While the most common source of ultraviolet radiation is sunlight, a host of artificial light sources used in manufacturing and in health-related activities emit potentially hazardous levels of ultraviolet radiation. These sources include fluorescent lamps, mercury discharge lamps, lasers, and welding arcs.

Whereas the crystalline lens absorbs most of the ultraviolet radiation in the 320- to 380-nm range,[16] the *aphakic* eye is more vulnerable to ultraviolet radiation damage. The absorption of ultraviolet radiation by the pigment epithelium of the retina and by the choroid enhances the potential for photochemical and thermal damage. For example, Ham et al.,[17] working with aphakic monkeys, found that in the absence of the lens there was sufficient UV-A radiation in the environment to damage the retina.

As described in Section 7.6, the aging of the eye is accompanied by a decrease in the transmission of visible light, due to accumulation of fluorescent pigment in the lens. Therefore, removal of the lens increases not only the amount of ultraviolet radiation reaching the retina but also the amount of *visible* radiation. Increased retinal sensitivity to damage from the short-wavelength visible radiation, in the aphakic eye, has been demonstrated.[18] Cystoid macular edema is a well-known complication following cataract surgery, and it is possible that it may be caused by the increased amount of UV-A and visible radiation reaching the retina of the aphakic eye.

Many chemicals sensitize the skin and the eye to radiation in the ultraviolet and other regions of the spectrum. The list of photosensitizing chemicals includes not only pharmaceutical compounds used in routine health-care procedures, but also chemicals used for industrial and agricultural purposes. Common chemical photosensitizers include Tetracycline, sulfonamide, Gresofulvin, and Phenothiazine.[19] Patients taking such medications should avoid long exposure to sunlight. Ocular protection should consist of wearing ultraviolet-absorbing lenses.

In order to prevent possible ocular damage and to minimize discomfort and the loss of visual function, eye protection from ultraviolet radiation should be provided for certain individuals. Ultraviolet-absorbing lenses should be prescribed for any patient who encounters excessive exposure to ultraviolet radiation in industrial, health-care, or recreational activities. Protection from ultraviolet radiation should also be provided for patients taking photosensitizing pharmaceuticals (as described above). Furthermore, since the retina of the aphakic eye is subjected to increased levels of ultraviolet and visible radiation, the aphakic patient should wear lenses which absorb ultraviolet radiation and significantly reduce the visible radiation, especially in the blue end of the visible spectrum.

Ocular protection from ultraviolet radiation can be provided by means of either absorptive or reflective filters. Absorptive filters are manufactured by adding substances to the lens material, whereas reflective filters are designed by coating the lens with the appropriate reflecting material.

Pitts[15] has recommended that an ultraviolet filter should absorb all radiation up to 380 nm. The absorption characteristics of clear spectacle lenses differ slightly from one manufacturer to another, and differ for different materials (i.e., glass vs. plastic). As shown in Figure 7-9A, clear ophthalmic crown glass transmits ultraviolet radiation beginning at a wavelength of approximately 300 nm, and the transmission rises rapidly for the longer ultraviolet wavelengths. However, the plastic material CR-39 (Figure 7-9B) transmits ultraviolet radiation only from about 350 nm upward, the transmission also rising rapidly for the longer ultraviolet wavelengths.

Polycarbonate (discussed in Chapter 1) is a plastic material with very high impact resistance, available in normal prescription powers. Hard coatings are necessary to protect the relatively soft surface. Recently, the coating has been formulated to incorporate special ultraviolet absorbers that screen out all ultraviolet radiation below 380 nm (see Figure 7-11). Thus,

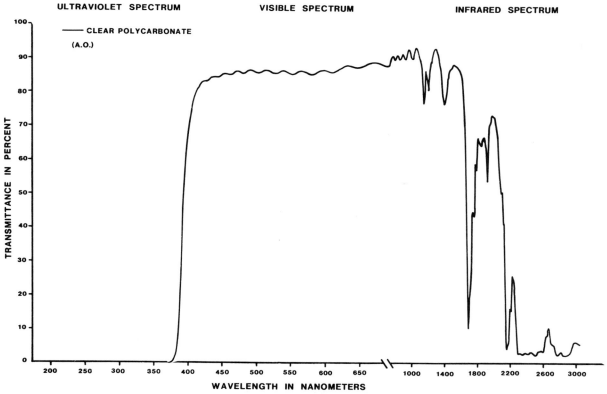

FIGURE 7-11. Spectral transmission curve for clear polycarbonate lens with coating which incorporates ultraviolet absorbers. (From D. Pitts and B. Maslovitz, unpublished data.)

the coated polycarbonate lens is not only virtually indestructible and resistant to surface abrasion, but also provides adequate protection from ultraviolet radiation.

It is obvious that neither glass nor the more commonly used CR-39 plastic provides adequate absorption of all ultraviolet radiation (i.e., up to 380 nm). However, many products have been introduced for this purpose. It is important that a lens manufacturer supply a spectral transmission curve for a lens allegedly designed to eliminate ultraviolet radiation; it should be understood that the specific absorption and transmission properties of a lens *cannot* be determined by the *color* of the lens. Tints that fail to adequately eliminate ultraviolet radiation can cause more harm than good. If the lowered transmission in the visible spectrum causes the pupil to dilate behind a tinted lens, higher levels of ultraviolet radiation will reach the crystalline lens than would have been the case if the patient had worn an *untinted* lens and the pupil had been in its normal, undilated state.

**CR-39 Plastic Lenses.** Several CR-39 plastic lenses containing ultraviolet absorbers are currently available. These include the Optical Radiation *UV-400*

(Figure 7.12), which eliminates all radiation shorter than 400 nm; the Vision-Ease *UV-Lite* (Figure 7-13), which absorbs 95% of all radiation in the 250- to 400-nm range; the *UVS* lens by Silor Optical (Figure 7-14), which absorbs 94−96% of the radiation at the 400-nm level; the Varilux 2 *UVX-Light* lens by Multi-Optics, which absorbs all radiation below 400 nm; and the Recreation Innovations *NoIR* lenses (#101, #102, #111, and #112, shown in Figure 7-15), which also absorb all radiation below 400 nm.

**Glass Lenses.** Glass lenses which transmit less than 5% of ultraviolet radiation below 400 nm include the American Optical *Hazemaster* (Figure 7-16), the Vision-Ease *Striking Yellow* (Figure 7-17), and the Bausch and Lomb *Kalichrome*. These are all *yellow* lenses, and have been promoted both as shooting lenses and as night-driving lenses. They will be discussed further in Section 7.19.

The *Spectra-Shield Human II* lens is a unique glass lens, coated with twenty-eight layers of a dielectric material on the concave surface of the lens. It is designed to reflect all ultraviolet radiation below 400 nm, and all infrared radiation above 700 nm.

*Photochromic* lenses from Corning Glass Works (in-

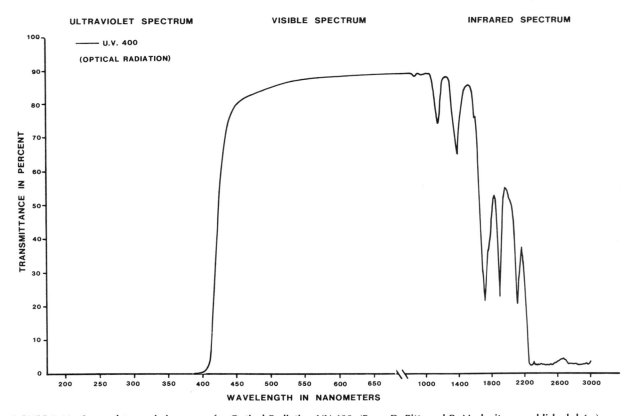

FIGURE 7-12. **Spectral transmission curve for Optical Radiation UV-400. (From D. Pitts and B. Maslovitz, unpublished data.)**

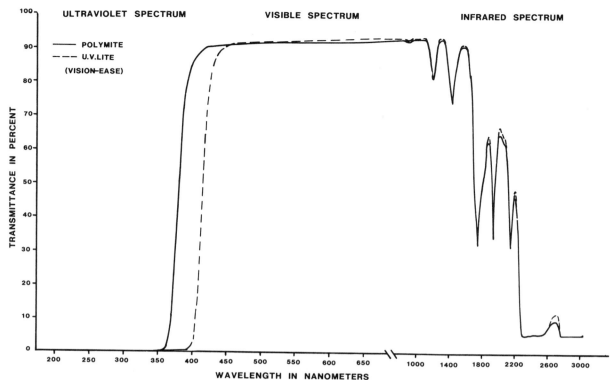

FIGURE 7-13. Spectral transmission curves for Vision-Ease Polymite (clear CR-39) and UV-Lite lenses. (From D. Pitts and B. Maslovitz, unpublished data.)

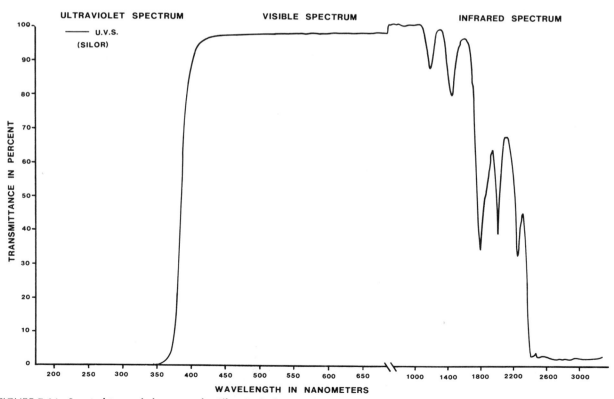

FIGURE 7-14. Spectral transmission curve for Silor Optical UVS. (From D. Pitts and B. Maslovitz, unpublisned data.)

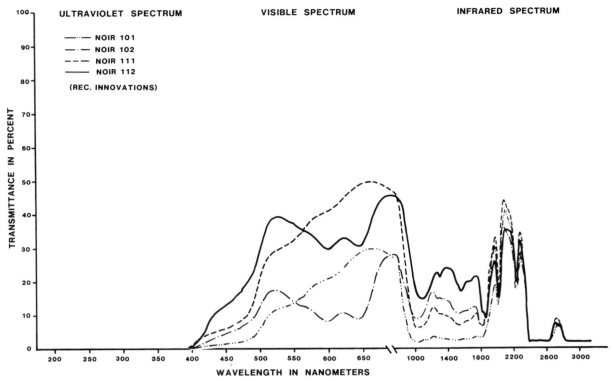

FIGURE 7-15. **Spectral transmission curves for Recreational Innovations NoIR #101, #102, #111, and #112. (From D. Pitts and B. Maslovitz, unpublished data.)**

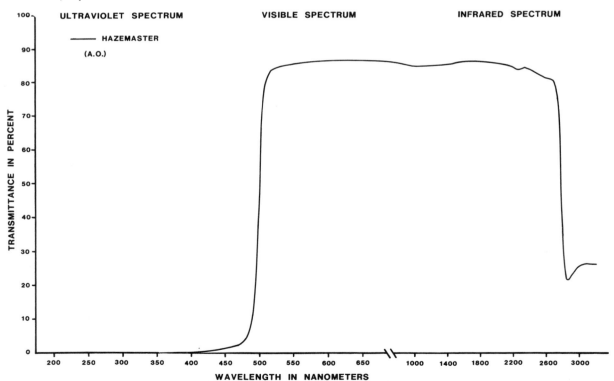

FIGURE 7-16. **Spectral transmission curve for American Optical Hazemaster. (From D. Pitts and B. Maslovitz, unpublished data.)**

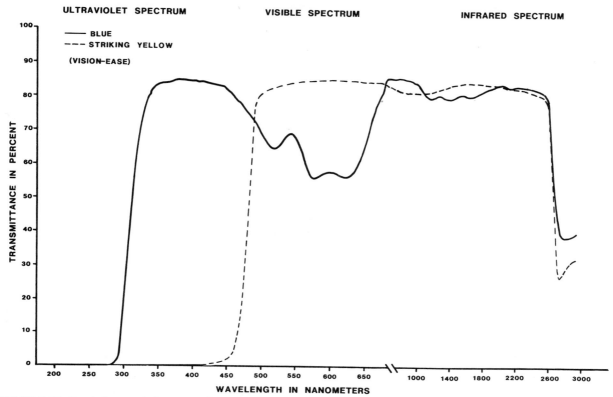

FIGURE 7-17. **Spectral transmission curves for Vision-Ease Blue and Striking Yellow. (From D. Pitts and D. Maslovitz, unpublished data.)**

cluding *Photogray II, Photogray Extra, Photobrown Extra,* and *Photosun II*) all transmit *no* ultraviolet radiation in the 290- to 315-nm range, in the fully darkened state, and transmit less than 3% in the 315- to 380-nm range. Photochromic lenses will be discussed in Section 7.21.

### 7.18. Lenses That Selectively Absorb Both Ultraviolet and Infrared Radiation While Absorbing a Substantial Amount of Visible Radiation in a Relatively Uniform Manner

Most of the lenses in this category absorb sufficient radiation in the visible spectrum to be considered as "sunglass" lenses. The American Bureau of Standards has defined a sunglass lens as one that transmits less than 67% of incident light. This is an arbitrary definition: In order to afford sufficient protection for bright daylight conditions, transmission should be on the order of 10 to 20%.

Some of the lenses in this category are listed in Table 7-3, and their transmission curves are shown in the accompanying diagrams, as indicated in the table. The majority of these lenses are not sufficiently dark to wear in extremely high illumination levels: As noted previously, exposure to 10,000 ft-lamberts (equal to 34,260 cd/m$^2$) of illumination would require a lens whose absorption is in the neighborhood of 4%. The only lenses that are close to this range are the darker lenses such as the American Optical *True-Color* and *True-Tone D* (Figure 7-18) and *Cosmetan* (Figure 7-19), Univis *Sun-Gray #3* (Figure 7-20), Vision-Ease *Gray #3* (Figure 7-21), Shuron-Continental *Neutrex #3* (Figure 7-22), and Recreational Innovations *NoIR #107, #108,* and *#109* (Figure 7-23).

Luria[20] conducted a study to determine which neutral filter was most preferred in bright sunlight in summer and winter. In summer, when the light level was about 14,000 cd/m$^2$, the preferred optical density of the filter was 1.0 (transmission of 10%). In the winter, against a bright snowy background, the light

TABLE 7-3
**Ultraviolet- and Infrared-Absorbing Lenses**

| Color | Trade Name | Material | Transmission Curve |
|-------|-----------|----------|--------------------|
| Green | American Optical True-Tone | Glass | Fig. 7-18 |
|  | Recreational Innovations NoIR #107, |  |  |
|  | #108, and #109 | Plastic | Fig. 7-23 |
|  | American Optical Calobar D | Glass | Fig. 7-25 |
|  | Vision-Ease Green #3 | Glass | Fig. 7-26 |
| Gray | American Optical True-Color D | Glass | Fig. 7-18 |
|  | Univis Sun-Gray #2 | Glass | Fig. 7-20 |
|  | Univis Sun-Gray #3 | Glass | Fig. 7-20 |
|  | Vision-Ease Gray #3 | Glass | Fig. 7-21 |
|  | Shuron-Continental Neutrex #3 | Glass | Fig. 7-22 |
| Brown | Optical Radiation UV-400 | Plastic | Fig. 7-12 |
| or | Vision-Ease UV-Lite | Plastic | Fig. 7-13 |
| tan | Silor Optical UVS | Plastic | Fig. 7-14 |
|  | Recreational Innovations NoIR #101, |  |  |
|  | #102, #111, and #112 | Plastic | Fig. 7-15 |
|  | American Optical Cosmetan | Glass | Fig. 7-19 |
|  | American Optical Cosmalite | Plastic | Fig. 7-19 |
|  | Vision-Ease Tan #3 | Glass | Fig. 7-24 |
| Yellow | American Optical Hazemaster | Glass | Fig. 7-16 |
|  | Vision-Ease Striking Yellow | Glass | Fig. 7-17 |
| Blue-green | Therminon | Glass | Fig. 7-27 |

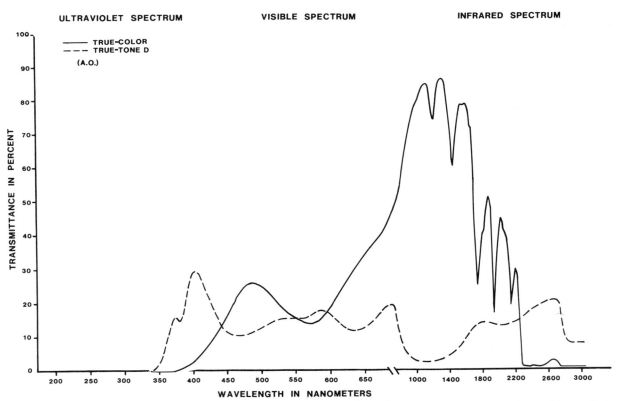

FIGURE 7-18. **Spectral transmission curves for American Optical True-Color and True-Tone D. (From D. Pitts and B. Maslovitz, unpublished data.)**

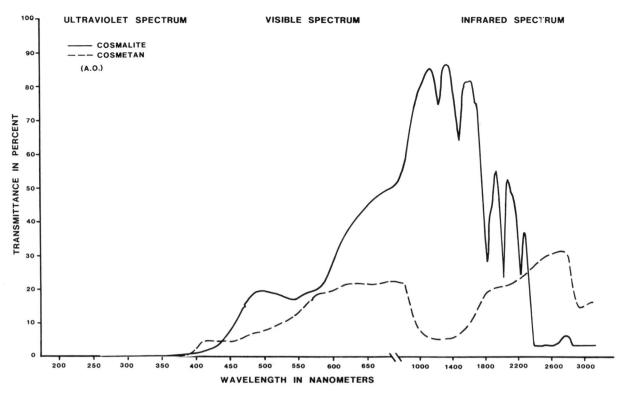

FIGURE 7-19. Spectral transmission curves for American Optical Cosmalite and Cosmetan. (From D. Pitts and B. Maslovitz, unpublished data.)

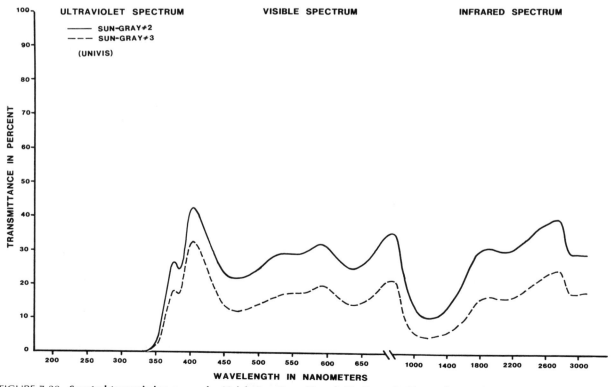

FIGURE 7-20. Spectral transmission curves for Univis Sun-Gray #2 and #3. (From D. Pitts and B. Maslovitz, unpublished data.)

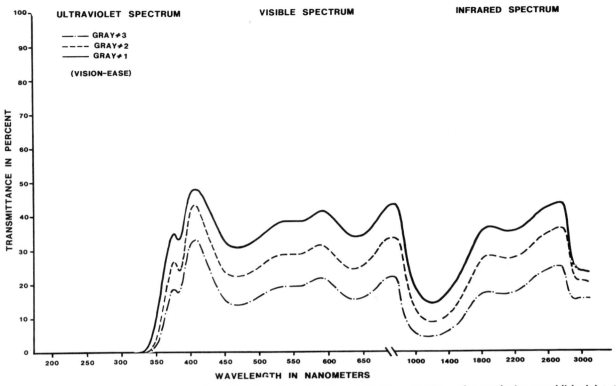

FIGURE 7-21.  **Spectral transmission curves for Vision-Ease Gray #1, #2, and #3. (From D. Pitts and B. Maslovitz, unpublished data.)**

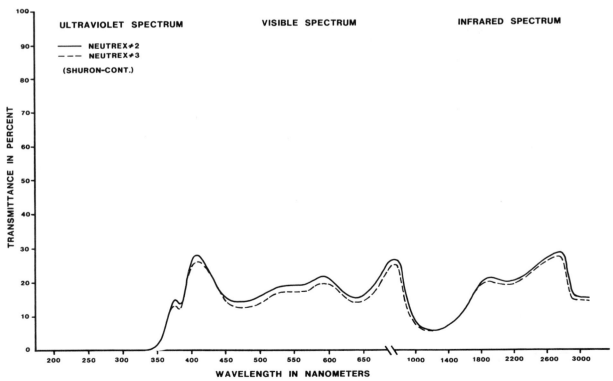

FIGURE 7-22.  **Spectral transmission curves for Shuron-Continental Neutrex #2 and #3. (From D. Pitts and B. Maslovitz, unpublished data.)**

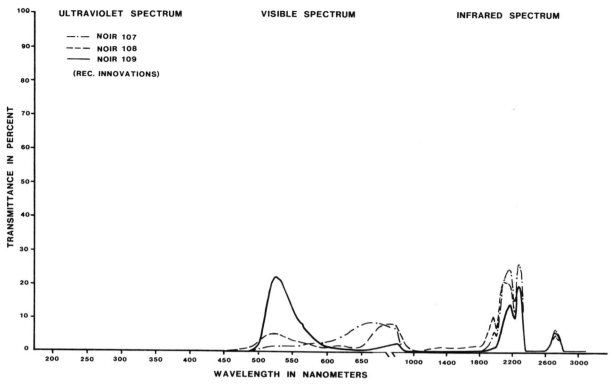

FIGURE 7-23.  Spectral transmission curves for Recreational Innovations NoIR #107, #108, and #109. (From D. Pitts and B. Maslovitz, unpublished data.)

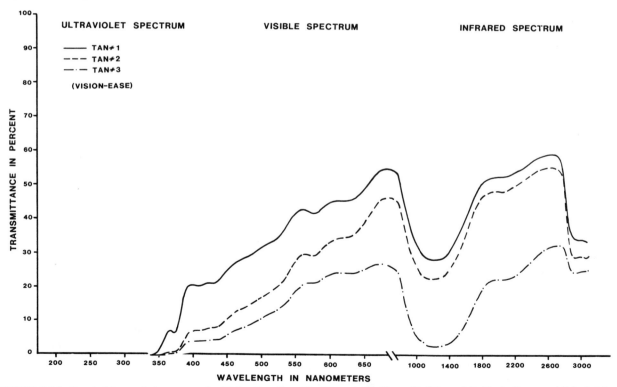

FIGURE 7-24.  Spectral transmission curves for Vision-Ease Tan #1, #2, and #3. (From D. Pitts and B. Maslovitz, unpublished data.)

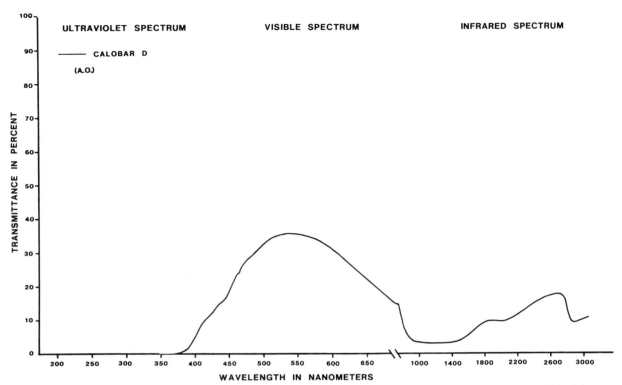

FIGURE 7-25. **Spectral transmission curve for American Optical Calobar D. (From D. Pitts and B. Maslovitz, unpublished data.)**

levels were 17,000 to 27,500 cd/m², and the preferred optical density was 1.3 (transmission of 4.8%). In both cases the light level was reduced to about 1000 to 1400 cd/m² at the eye, with the preferred filter. The optical density most preferred ranged from 1.05 to 1.60 (8.9 to 2.5% transmission). The preferred optical densities reported by Luria are considerably darker than the density of most sunglasses that are currently available commercially. This study confirms the observation of Richards[2] that 400 ft-lamberts (1370 cd/m²) is the optimum level for comfortable vision.

It should be understood that none of the lenses in this category should be used as *general-wear* lenses. Their transmission is *much too low* for safe use in night driving, and if darkly tinted lenses are used for *indoor* wear, the eyes become adapted to such a low level of illumination that the lenses tend to lose their effectiveness for outdoor wear.

**Ultraviolet Absorption.** Whereas all of the lenses in this category have excellent absorption in the 200- to 300-nm portion of the ultraviolet region, it should be noted that some of them have inadequate absorption in the longer, 300- to 400-nm range. The only lenses listed in Table 7-3 having less than 5% transmission in this region are the American Optical *Cos-*

*metan* and *Cosmalite* (Figure 7-19), the Recreational Innovations *NoIR #107*, *#108*, and *#109* (Figure 7-23), and the Vision-Ease *Tan #3* (Figure 7-24). Some of the gray lenses such as the American Optical *True-Color D* (Figure 7-18) and the Univis *Sun-Gray #2* and *#3* (Figure 7-20) transmit more than 25% in this region.

**Infrared Absorption.** Many of the lenses listed in Table 7-3 have very good absorption in the infrared region of the spectrum. For example, lenses having less than 15% transmission in the 700–1500 range are green *glass* lenses including the American Optical *Calobar D* (Figure 7-25) and the Vision-Ease *Green #3* (Figure 7-26), and some *plastic* lenses including the *NoIR #107*, *#108*, and *#109* (Figure 7-23). However, some of the lenses listed in the table have very *poor* absorption in the infrared region: The American Optical *True-Tone* (Figure 7-18) and *Cosmalite* (Figure 7-19) plastic lenses have transmissions exceeding 75% in this range.

Pitts, Cullen, and Dayhaw-Barker[21] have made the point that the level of exposure to infrared radiation required to cause ocular damage is quite high compared to the required level of exposure to ultraviolet radiation. On the other hand, Clark[22] has taken the point of view that lenses having high infra-

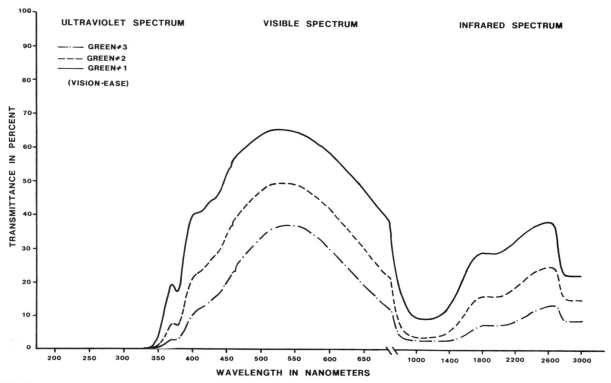

FIGURE 7-26.  Spectral transmission curve for Vision-Ease Green #1, #2, and #3. (From D. Pitts and B. Maslovitz, unpublished data.)

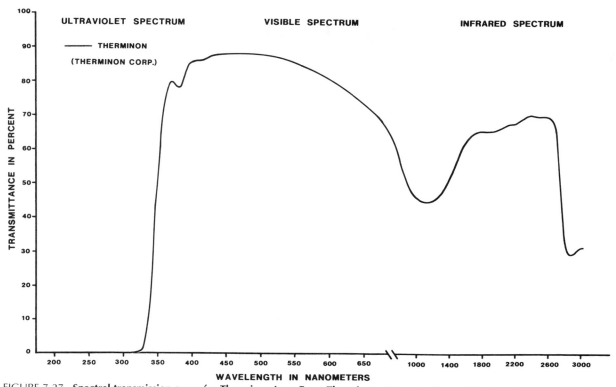

FIGURE 7-27.  Spectral transmission curve for Therminon Lens Corp. Therminon. (From D. Pitts and B. Maslovitz, unpublished data.)

209

red transmission can be hazardous: Since the reduced visible light transmission causes the pupil to
dilate (as noted also in the discussion of ultraviolet
radiation) the retina may actually receive *more* infrared radiation than if an absorptive lens were not
worn. In another paper[23] Clark concluded that there
is no advantage in absorptive lenses being *colored*,
and recommended that absorptive lenses should be a
neutral gray rather than colored. However, it should
be noted that, of the presently available lenses, only
the *dark green* lenses consistently reduce transmission
in the infrared region of the spectrum.

## 7.19. Lenses That Selectively Absorb Portions of the Visible Spectrum

Although not commonly used, lenses that selectively
absorb portions of the visible spectrum are available.
These include both yellow and bluish-green lenses.
The yellow lenses, previously described in Section
7.16, include the American Optical *Hazemaster* (Figure 7-16), the Vision-Ease *Striking Yellow* (Figure
7-17), and the Bausch and Lomb *Kalichrome*. The
bluish-green lenses include the Therminon Corp.
*Therminon* (Figure 7-27) and the Titmus Optical
*Infra-Bar*.

The Hazemaster, Striking Yellow, and Kalichrome absorb almost *all* illumination below 500 nm,
and therefore greatly reduce the *scattering* of light
(since scattering is due mainly to the wavelengths in
the blue end of the spectrum). Since the retina is most
sensitive to radiation of 555 nm, it has been suggested that lenses having high relative transmission
at or near this wavelength will aid visual acuity at
night. Consequently, yellow lenses have been promoted as "night-driving lenses."

Colors are known to have different subjective effects upon different observers. For some people, yellow glasses create a brighter environment and a sense
of seeing better, even though the lenses obviously
absorb *some* light and therefore reduce the brightness
of the environment. Wright[24] attributes this observation to the association of yellow color with sunlight and hence with a high level of illumination.
Other observers find yellow glasses disquieting and
uncomfortable, and report that they see better without them.

Since short-wavelength radiation is scattered more
by fog and haze than is long-wavelength radiation,
yellow light should reduce the scattering effects of
fog and haze. Theoretically, yellow light should also
reduce the chromatic aberration of the eye. However, since the retina has a decreased sensitivity to
short-wavelength radiation, the yellow lenses are not
as helpful in reducing the effects of chromatic aberration of the eye as they are in applications in photography. In fact, at low levels of illumination when
maximal sensitivity of the eye shifts from 555 to 510
nm, yellow glasses can have a negative effect on vision due to the decreased sensitivity of the eye for the
longer wavelengths.

A number of researchers have studied the effect
of yellow lenses on visual performance, and the results of these studies have been negative. Luckiesh
and Holladay[25] found that yellow light does not
provide an advantage in foggy or misty weather.
Richards[26] found a small but statistically significant
loss of vision occurring with yellow lenses. According
to Blackwell,[27] yellow driving glasses reduce visual
detection distances by 33% as compared with wearing no glasses. Davey[28] found that glare-recovery
time was 29% longer with yellow glasses than without. Often, there is barely sufficient illumination for
night driving, and the loss of illumination when yellow lenses are worn is potentially dangerous: Expert
opinion advises that *yellow lenses should not be worn for
night driving*.

Yellow lenses have also been recommended for
hunting and skeet shooting, since scattered light may
create a problem in these activities. Bierman[29] tested
the shooting ability of fifty men, with and without
yellow lenses. The majority were not better marks-
men with the yellow lenses, while many were worse.
It is of interest that, for the nineteen men who liked
the yellow lenses, scores were higher with the lenses
than without them; while for the twenty-six men who
disliked the lenses, scores were higher without the
lenses. These lenses should therefore be prescribed
for shooting only when, for a given patient, they have
proven to be beneficial.

The pale bluish-green lenses, Therminon (Figure
7-27) and Infra-Bar, have about the same transmission as crown glass in the visible spectrum, but have a
gradual reduction of transmission in the infrared
region. These lenses are claimed to be *cool* lenses, and
their use has been proposed for patients whose eyes
are constantly exposed to highly reflective white surfaces, and for cooks and others who must work in
environments in which there is an excessive amount
of infrared radiation. Some wearers of these lenses
tend to become "hooked" on the lenses, and are
unhappy if the practitioner unknowingly switches
them to another pale tint or to clear lenses.

One of the earliest absorptive lenses to be introduced was the *Crookes* series of glass lenses, developed by Sir William Crookes. These are smoke-

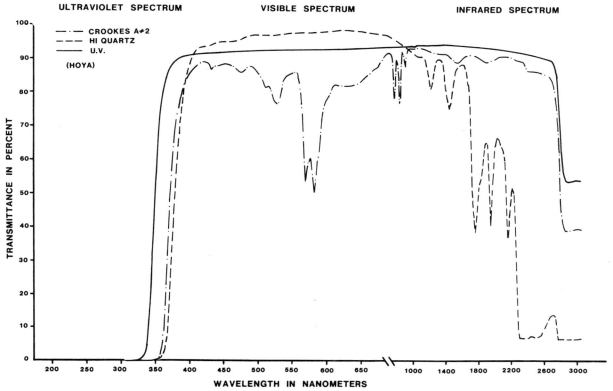

**FIGURE 7-28.** Spectral transmission curve for Hoya Crookes A #2. Note the absorption bands in the yellow region of the spectrum. (From D. Pitts and B. Maslovitz, unpublished data.)

colored lenses that absorb all ultraviolet radiation below 350 nm, and are unique in that they have additional absorption bands in the yellow region of the visible spectrum. These absorption bands result from the incorporation of didymium (a mixture of the elements neodymium and praseodymium) as well as cerium oxide. The spectral transmission curve of the Hoya *Crookes A #2* lens is shown in Figure 7-28.

## 7.20. Absorptive Lenses Designed for Occupational Use

When protective glasses are worn in industry, the glasses must not only protect the wearer from potential mechanical and chemical injury, but also from possible radiant energy damage. Specially designed absorptive lenses are available for a number of occupations including blast furnace operation, glass blowing, and welding.

Current standards[30] require that an absorptive lens for industrial workers must almost completely absorb both the ultraviolet and infrared radiation from harmful sources. Consequently, absorption is high for both the blue and red regions of the visible spectrum, with maximum transmittance in the bluish-green or greenish-yellow regions of the spectrum (with the result that these lenses are bluish-green or greenish-yellow in color).

The proper function of an absorptive lens designed for industrial use involves not only the protection of the wearer from ultraviolet and infrared radiation but also the reduction of the radiation in the visible spectrum to a comfortable level so that the industrial operation can be performed as efficiently as possible. The luminous transmittance of a filter determines its appropriateness for a particular industrial task.

Welding lenses are available in two general categories: *Class I* welding lenses are general-purpose lenses, while *Class II* welding lenses are used in operations producing high radiant flux at the wavelengths of the sodium lines (589.3 nm). Class II lenses are similar to Class I lenses in regard to the reduction of ultraviolet and infrared radiation, but differ from

Class I lenses in that they contain *didymium*, which reduces the transmittance in the 589.3-nm region to a relatively low value.

The high absorption of welding glasses for the ultraviolet and infrared regions of the spectrum provides protection from any source of radiant energy that may be encountered in an industrial environment, as long as the proper density is used. Since individual needs may differ, the appropriate density for a particular operation may be personally selected on a trial basis for optimum results. Within reasonable limits, the industrial worker should choose the density which seems suitable for his eyes and for the nature of the welding process.

The luminous transmission of a welding lens is specified in terms of a *shade number*. Shade numbers increase with increasing absorption (decreasing transmission). The relation between optical density, luminous transmittance, and shade number is given by the formula

$$\text{optical density} = \log_{10} \frac{1}{T},$$

where $T$ (transmittance) is expressed as a decimal.

$$\text{optical density} = \frac{3}{7} (\text{shade number} - 1),$$

or

$$\text{shade number} = \frac{7}{3} (\text{optical density}) + 1.$$

Shade numbers run in discrete steps from 1.5 to 14. Tolerances in transmittances permit each transmittance value to fall within a particular shade number. The shade numbers and percentage transmittances are as follows:

| Shade Number | Percentage Transmission |
|---|---|
| 1.5 | 61.1 |
| 1.7 | 50.1 |
| 2.0 | 37.3 |
| 2.5 | 22.8 |
| 3.0 | 13.9 |
| 4.0 | 5.18 |
| 5.0 | 1.93 |
| 6.0 | 0.72 |
| 7.0 | 0.27 |
| 8.0 | 0.10 |
| 9.0 | 0.037 |
| 10.0 | 0.0139 |
| 11.0 | 0.0052 |
| 12.0 | 0.0019 |
| 13.0 | 0.00072 |
| 14.0 | 0.00027 |

Lenses having shade numbers from 1.5 to 3.0 are useful for people who are exposed to strong light from cutting and welding operations but don't perform these operations themselves. Shade number 4 is intended for the same use as shade numbers 1.5 to 3.0, but for conditions involving greater luminous flux. Shade number 5 is intended for light gas cutting and welding, and for light electronic spot welding.

Shade numbers 6 and 7 are intended for gas cutting, medium gas welding, and for arc welding up to 30 amperes. Shade number 8 is intended for heavy gas welding and for arc welding and cutting, when using over 30 amperes but not exceeding 75 amperes. Shade numbers 9 through 14 are quite dark, and are used for arc welding and cutting over 75 amperes.

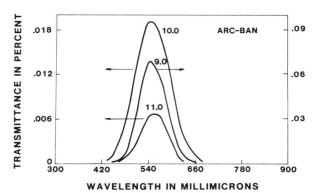

FIGURE 7-29. **Spectral transmission curves for Bausch and Lomb Arc-Ban welding lenses. In this diagram and in Figures 7-30, 7-31, and 7-32, the arrow on each transmission curve points to the appropriate scale. (From R. Stair, Spectral Transmission Properties and Use of Eye Protective Glass, National Bureau of Standards Circular No. 471, 1948.)**

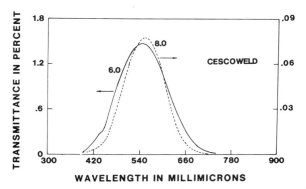

FIGURE 7-30. **Spectral transmission curves for Chicago Eye Shield Co. Cescoweld welding lenses. (From R. Stair, Spectral Transmission Properties and Use of Eye Protective Glass, National Bureau of Standards Circular No. 471, 1948.)**

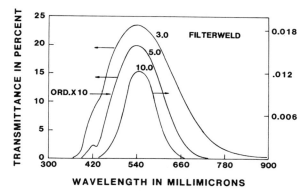

FIGURE 7-31. **Spectral transmission curves for American Optical Filterweld welding lenses. (From R. Stair, Spectral Transmission Properties and Use of Eye Protective Glass, National Bureau of Standards Circular No. 471, 1948.)**

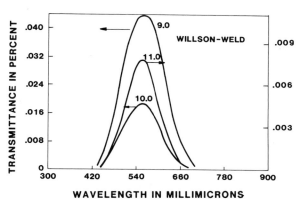

FIGURE 7-32. **Spectral transmission curves for Willson Products Willson-Weld welding lenses. (From R. Stair, Spectral Transmission Properties and Use of Eye Protective Glass, National Bureau of Standards Circular No. 471, 1948.)**

Trade names for welding lenses include *Arc-Ban* (Figure 7-29), *Cescoweld* (Figure 7-30), *Filterweld* (Figure 7-31), and *Willson-Weld* (Figure 7-32). For shade numbers 1.5 to 3.0, the ANSI Z87.1-1979 Standard[30] specifies that the luminous transmittances of a pair of lenses shall be within 10% of each other. For shade numbers 4.0 through 14.0, luminous transmittances of a pair of lenses shall be within 20% of each other.

## 7.21. Photochromic Lenses

Lenses made of photochromic glass have the remarkable property of darkening upon exposure to sun-

light and reverting to the original state in the absence of sunlight. The photochromic properties of these lenses are due to very minute *silver halide* crystals that are precipitated from the homogeneous melt during initial cooling or during subsequent reheating at temperatures between the annealing and softening points, for a long enough period of time for the crystals to grow to optimum size.

The photochromic process is basically the same as the photolytic dissociation of silver halide into silver and halogen that occurs in silver halide photographic material, the difference being that the photographic process is *irreversible* whereas the photochromic process occurring in glass is reversible. In photography the host matrix permits the halogen to diffuse away from the original crystal site and the halogens are no longer available for recombination when irradiation ceases. Hence, after suitable chemical treatment a permanent photographic image is formed by the stable silver particles. Photochromic glass, on the other hand, is impervious, chemically inert, and rigid. These characteristics prevent loss of the halogen by diffusion, and exclude atmospheric oxygen and moisture which might cause side reactions. Thus the released halogens are confined to the immediate vicinity of the crystal site and a recombination of silver and halogen can take place after active irradiation ceases. The recombination process is enhanced by exposure to heat (thermal bleaching) or visible radiation of relatively long wavelength (optical bleaching).

In both photographic film and photochromic glasses, silver ions and electrons resulting from the dissociation of silver halide regroup to form opaque silver colloids (specks) of appreciable size. These opaque silver colloids (specks) are the entities which absorb *visible* light in much the same manner as any metal and which causes the photochromatic glass to appear dark.

When the activating light is removed, the silver colloids in the glass begin to split apart into silver ions again. They recombine with the halide ions which are still trapped within the molecular structure of the glass and form light-sensitive silver halide microcrystals again.

A major factor in determining speeds of response, range of densities, and base colors of photochromic glass is the mixture of the host glass material. Corning Glass Works describes their host photochromic glass as *alkali borosilicate* crown glass, while Chance Pilkington Ltd. (England) label their photochromatic host glass as an *alumino phosphate* glass.

There are three manufacturers of "wide-range,"

swift-acting photochromic glass in the western world: Corning Glass Works, Chance Pilkington, and Deutsche Spezialglas AG (Desag). Confusion exists in marketing trade names of photochromic glass. Corning glasses are marketed under Corning's registered names, but are also sold under Rodenstock and Hoya glass trade names. Both Zeiss and Rodenstock use the same Desag photochromic glass, but sell it under different brand names. Chance Pilkington Ltd. markets its own photochromic glass but has issued a license for the production of its glass in the United States by Schott Optical. Bausch and Lomb uses both Corning and Chance Pilkington photochromic glasses.

The wavelength that induces darkening depends on the chemical composition of the glass. Glasses containing only silver chloride are sensitive to wavelengths from approximately 300 to 400 nm with optimum activation around 350 nm.[31] The longer wavelength limit of spectral sensitivity for darkening is *higher* for glasses containing the heavier halogens such as silver bromide (up to 550 nm) and silver iodide (up to 650 nm).[32] The addition of other "sensitizers" in trace quantities enhances the sensitivity and photochromic darkening. These include the oxides of arsenic, antimony, tin, and copper. Copper oxide is an especially effective sensitizer and is believed to enter the silver halide crystals as a cuprous ion.

Crystal *size* and *spacing* determine the resolution and transparency in the unactivated state. Silver halide crystals are 50 to 100 Å in diameter, and the average spacing is from 500 to 1000 Å.[31] In general, glasses with crystals less than 50 Å in diameter are not photochromic. Above 300 Å the glass becomes translucent. For particles with an average diameter of 100 Å and an average spacing of 600 Å between particles, the concentration will be about $4 \times 10^{15}$ particles per cubic centimeter.[31]

The photochromic glasses differ from organic photochromic materials in that they are immune from fatigue or deterioration of their photochromic performance with extended use. Specimens exposed to thousands of cycles of darkening and fading show no deterioration of photochromic properties. However, new photochromic lenses need to go through a "breaking-in" period. Exposure of the glass to sunlight for several full darkening-fading cycles is necessary before reaching the normal operating performance. If removed from exposure for long periods of time, the "breaking-in" cycles need to be repeated.

The first photochromic lens produced commercially in the United States was the *Bestlite* lens, introduced by Corning in 1965. The Bestlite lens was superseded by the *Photogray* lens in 1968, which had an increased darkening with sunlight exposure and a more uniform spectral recovery. In 1971 Corning introduced the *Photosun* lens. In sunlight it darkened to a transmittance of 25% but faded to a transmittance of only 68% overnight. Corning's second generation of lenses includes the *Photogray Extra* lens (1978) and the *Photobrown Extra* lens (1981). These lenses differ from the original Photogray lens in both darkening and fading more quickly. Within 60 seconds in sunlight the Photogray Extra darkens to 22% transmittance while the Photobrown Extra darkens to 24% transmittance. When removed from activating radiation for 5 minutes the Photogray Extra fades to 62% transmittance and the Photobrown Extra fades to 64% transmittance. Overnight the Photogray Extra fades to 87% transmittance while the Photobrown Extra fades to 86% transmittance.

Both Photogray Extra and Photobrown Extra respond more to visible light than other Corning photochromic glasses. Each will darken slightly in a brightly lighted room. Photogray Extra and Photobrown Extra are available in one-piece and fused multifocal construction. The segment used in fused multifocal construction is made of clear glass; therefore, when the lens is fully darkened, the segment area will be lighter than the distance portion of the lens.

The optical and physical properties of Photogray Extra and Photobrown Extra are similar to those of other Corning photochromic lenses. The index of refraction is 1.523, the same as ophthalmic crown glass. However, the specific gravity of Photogray Extra and Photobrown Extra is less than that of other Corning photochromatic glass and ophthalmic crown glass. Photogray Extra and Photobrown Extra weigh 2.41 g/cc compared to 2.54 g/cc for ophthalmic crown glass.

*Photosun II* was introduced in 1983. It is darker in both the faded and the darkened states than other Corning photochromic lenses. The transmittance is 41% in the faded state and 12% in the darkened state. After 1 minute of exposure to direct sunlight it darkens to 16% transmittance and within 1 hour it reaches the maximum darkness of 12%. In the fading process, Photosun II lightens to 29% transmittance within 5 minutes. Overnight it will fade to 41% transmittance.

The Photogray II was introduced, in 1984, as a "second-generation" replacement for the Photogray lens. Fully faded, the Photogray II offers the highest transmission of any Corning photochromic lens, transmitting 87% of incident visible light. When fully darkened, this lens transmits 42% of visible light. It will darken over 85% of its range within the first 60 seconds of activation, and within 1 hour it will reach its maximum darkness of 42% transmission. Photogray II lenses fade the fastest of any current Corning photochromic lens, recovering 70% of its fully faded transmittance in the first 5 minutes after being re-

moved from sunlight, then fading overnight to 87% transmittance.

Transmission curves for Photogray Extra, Photobrown Extra, Photosun II, and Photogray II lenses (in both the faded and darkened states) are shown in Figures 7-33 through 7-36; the transmission proper-

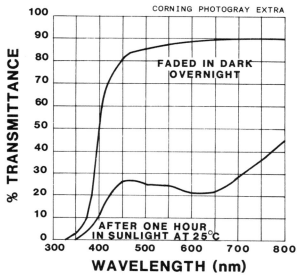

FIGURE 7-33. Spectral transmission curves for Corning Photogray Extra, in faded and darkened states. (From Corning Glass Works.)

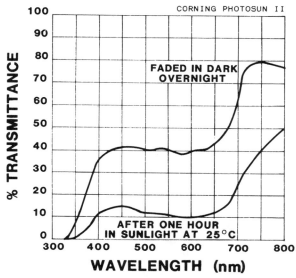

FIGURE 7-35. Spectral transmission curves for Corning Photosun II, in faded and darkened states. (From Corning Glass Works.)

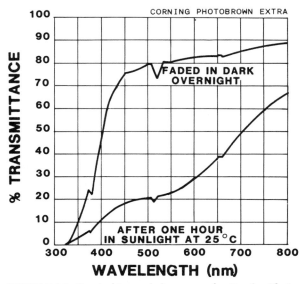

FIGURE 7-34. Spectral transmission curves for Corning Photobrown Extra, in faded and darkened states. (From Corning Glass Works.)

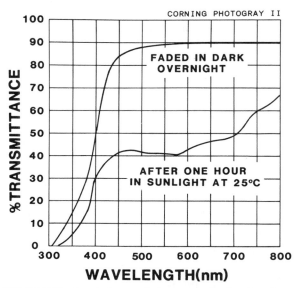

FIGURE 7-36. Spectral transmission curves for Corning Photogray II, in faded and darkened states. (From Corning Glass Works.)

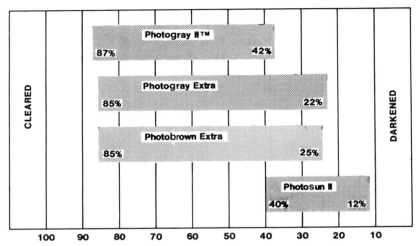

CLEARED

Photogray II™
87%    42%

Photogray Extra
85%    22%

Photobrown Extra
85%    25%

Photosun II
40%    12%

DARKENED

100  90  80  70  60  50  40  30  20  10

FIGURE 7-37. **Chart comparing the transmission properties of currently available Corning photochromic lenses. (From Corning Glass Works.)**

ties of currently available Corning photochromic lenses are compared in Figure 7-37.

Although many factors influence the transmittance of a specific photochromic glass, the principal factors are

1. The intensity of the incident radiation
2. The wavelength of the incident radiation
3. The temperature of the glass
4. The thickness of the glass
5. The previous heat treatment
6. The exposure history

At controlled temperatures the transmittance decreases as the intensity of the sunlight increases. When the transmittance of a photochromic lens is recorded over an entire day, it shows a characteristic pattern of decreasing transmittance beginning at dawn, continuing to decrease until saturation is achieved, then beginning to fade before sunset and continuing to fade until the next morning. On overcast or cloudy days, the lenses will not be as dark as in full sunlight. The closer the wavelength of incident radiation is to the optimum wavelength for activation, the darker the lens becomes.

As a general rule photochromic lenses are temperature dependent. The lenses will be *darker* in cool temperatures, and *lighter* in warm temperatures. The rate of fading in the dark increases with the temperature (thermal bleaching). The darkening rate is relatively insensitive to the temperature.

The silver halide microcrystals are distributed uniformly throughout the glass. However, photochromic glass can be uniformly activated throughout its total depth only if sufficient energy intensity is applied. Intense darkening occurs near the incident

surface while with increasing depth the glass becomes more and more transparent, depending on the exposure energy. The predominance of surface darkening means that the lenses function essentially as *uniform-density* lenses (see Section 7.24). Although the greatest darkening occurs near the exposed surface of the glass, some darkening occurs at deeper levels. Therefore, thicker lenses may darken somewhat more than thinner ones, and will take longer to fade to the near-clear state.

While photochromic lenses may be toughened by either chemtempering or air tempering, Corning recommends *chemtempering*. Lenses that have been chemtempered usually exhibit greater impact resistance than air-tempered lenses at the same thickness. Air tempering will produce a lens which is darker at high temperatures, retains more tint in the overnight faded state, and fades more slowly than a chemtempered lens. Chemically tempered lenses will fade more rapidly and have a wider range of transmittance between the fully faded and darkened states.

Up to a point, both the degree and speed of darkening increase with the number of darkening and fading cycles a photochromic lens undergoes. People who spend a large amount of time outdoors will generally have darker lenses in the fully activated state than those who spend only a moderate amount of time outdoors.

All photochromic lenses darken faster than they fade. For example, Photogray Extra lenses will darken in sunlight to 22% transmittance in 60 seconds, but will fade to only 62% transmittance in 5 minutes.

The darkening of photochromic lenses is triggered mainly by long ultraviolet radiation and short visible radiation. Automobile windshields and side windows will absorb and reflect much of the activating ultraviolet radiation, thereby reducing the amount of ultraviolet radiation reaching the photochromic lenses. Hence, the lenses will not darken completely when used for driving an automobile. People who drive during daylight hours may be disappointed by the performance of the lenses.

Replacement of a broken photochromic lens can sometimes present a problem, as one lens may be darker than the other or the two lenses may not darken or lighten at the same rate. In such cases it is recommended that the original lens be chemically tempered in the same bath as the new lens. If tempering the original lens along with the new lens is not feasible, the original lens may be placed in boiling water or in an oven at 212°F (100°C) for at least an hour.

When *coating* (antireflective or tinted) is applied to photochromic lenses it must be applied within a critical temperature range of approximately 230 to 375°C. Temperature controls in this range are critical to the performance of the photochromic lenses. Therefore the time, temperature, and cooling must be closely controlled. Any mismatch in color or transmittance that arises from the coating operation can usually be corrected by placing the lenses in a 400 to 425°C oven for approximately 25 minutes and then allowing the lenses to cool at room temperature.

Any coating that absorbs the near ultraviolet radiation should be applied only to the *back* surface of the lens. Otherwise it may reduce the photochromic action which causes the lens to darken on exposure to sunlight. An antireflection coating produces little change in the behavior of photochromic lenses, apart from the slightly higher transmission as expected.[33]

Although many people believe that a single pair of photochromic lenses can act as a substitute for two pairs of glasses (sunglasses for daytime outdoor wear and clear glasses for indoor use) there are two possible hazards involved in using photochromic lenses in this manner:

1. Wearing *any* photochromic lens for night driving is questionable, since even the lightest of these tints (such as Photogray II, Photogray Extra, or Photobrown Extra) fades to a transmittance of only 87, 85, or 85% (respectively), but this level is reached only after *overnight* fading. However, the wearing of Photosun or Photosun II lenses for night driving is completely unacceptable, since these lenses fade to a transmittance of only 68 and 41%, respectively.

2. Because of the high transmittance of photochromic lenses in the infrared region of the spectrum, Chase[34] has questioned the suitability of these lenses as *sunglass* lenses. In his investigation of the transmission properties of four types of photochromic lenses, Chase found that transmittance for the infrared wavelength of 1400 nm was approximately 90% for the Photobrown lens, 87% for the Photogray lens, 82% for the Photosun lens, and 75% for fusible photochromic glass (all of which have been discontinued and replaced by "second-generation" lenses). Pointing out that infrared radiation in the region of 1400 nm is heavily absorbed by the eye, he stated that while photochromic lenses may provide comfort, they do not provide the protection from infrared radiation that many other sunglasses provide. He suggested that they might be worn for comfort under those conditions not requiring protection from infrared radiation.

### Corning CPF Lenses

In 1983 the Medical Optics Department of Corning Glass Works introduced three special photochromic lenses intended to relieve the symptoms of glare discomfort and/or reduced visual acuity associated with certain eye disorders. These lenses eliminate virtually all of the ultraviolet energy and, in addition, the blue end of the visible spectrum. The lenses were not intended as a treatment or cure for ocular disorders, but were developed to reduce some of the side effects associated with these disorders. Patients having the following conditions are likely candidates for the CPF lenses: developing cataracts, aphakia and pseudophakia, diabetic retinopathy, macular degeneration, corneal dystrophies, retinitis pigmentosa, optic atrophy, albinism, aniridia, and glaucoma.

One or more of the following symptoms indicate that CPF filters may be helpful: reduced vision due to light scattering, sensitivity to glare, severe photophobia, reduced vision due to loss of contrast, and prolonged adaptation time.

As with other Corning photochromic lenses, the CPF lenses are manufactured from borosilicate glass containing silver halide microcrystals. The product designations CPF 511, CPF 527, and CPF 550 refer to the spectral cutoff characteristics of the lenses (in nanometers). The chart below lists the properties of each lens, and the spectral transmittance curves for each of the three lenses (in the darkened and lightened states) are shown in Figure 7-38.

| | | Transmission | |
|---|---|---|---|
| | Color in Faded State | Faded | Darkened |
| CPF 511 | Yellow-amber | 47% | 16% |
| CPF 527 | Orange-amber | 37% | 12% |
| CPF 550 | Reddish-amber | 21% | 5% |

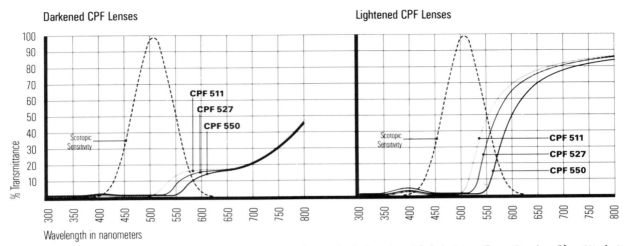

FIGURE 7-38. **Spectral transmittance curves for Corning CPF lenses, in darkened and faded states. (From Corning Glass Works.)**

The lenses are available only from Corning Medical Optics Department and can be obtained in plano power and by prescription (single vision, bifocals, trifocals, and aphakic lenticular).

It is recommended that the lenses be mounted in a Corning side-shield frame, but they can be mounted into most frames. Since the lenses absorb blue light, it may be difficult for some people to identify traffic signals and warning signs. They should therefore be worn with caution when driving. CPF lenses should not be worn at night. The lenses can be chemically tempered.

Lynch and Brilliant[35] conducted an evaluation of the CPF 550 lens on sixteen patients with retinitis pigmentosa. The CPF lenses improved visual acuity in the majority of subjects by 1/4 of a line. While contrast sensitivity and adaptation time were not improved, the subjects reported a decreased sensitivity to glare.

### Plastic Photochromic Lenses

In 1982, American Optical introduced a *plastic* photochromic lens called the *Photolite* lens. Although photosensitive plastic is not new, photosensitive plastic ophthalmic lenses are. In the faded state the Photolite lens exhibits 90% transmittance and in the darkened state has 45% transmission. The lens darkens to 45% of its darkened state within 2 minutes. As with other photochromic materials, the cooler the temperature the darker the lens. There is no variation in transmittance across the lens as the lens coloration is confined to a relatively thin region of the surface. The lens is manufactured not by a dye-pot process but by chemical impregnation.

When fully activated the Photolite lens turns blue. It can be tinted, however, to different colors. A Photolite lens that is tinted brown or tan in its faded state will become grayish in color when fully darkened. The lens has a life expectancy of approximately 2 years: After this period of time the lens is capable of only about 50% of the reaction that it exhibited when new. The Photolite lens is not intended as a sunglass lens, rather as a fashion tinted product.

### The Corlon Lens

In 1982 Corning introduced a new type of lens, called the *Corlon*, that combines both glass and plastic. The Corlon lens is made of two layers. The front (convex) layer is a thin glass lens of white crown glass or photochromatic glass (Photogray Extra); the back layer is a very thin layer of special polyurethane plastic bonded to the glass lens (Figure 7-39). The

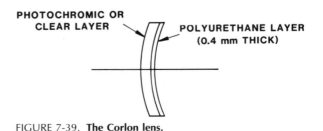

FIGURE 7-39. **The Corlon lens.**

glass layer has a central thickness of about 1.3 mm (in minus power prescriptions) in white crown glass and a center thickness of 1.5 mm in photochromic material. The polyurethane layer is only 0.4 mm thick. Hence the Corlon lenses are lighter in weight than ordinary glass lenses and will have thinner edges than either CR-39 plastic or ophthalmic crown glass lenses in minus prescriptions.

Because the front surface is glass, the front surface will not scratch as easily as the surface of a plastic lens. The two-layer construction of the lens makes it resistant to penetration of an impacting object. While the glass layer may be fractured by the impact of an object, the bonded polyurethane layer normally remains intact, preventing the glass particles or the object from striking the eye. The unique construction of the lens eliminates the need for tempering. The Corlon lens is more resistant to rear surface penetration than either chemtempered white crown glass or CR-39 plastic. Statistical drop-ball testing is performed by the manufacturer.

The glass used to make the photochromic Corlon lens is Photogray Extra. Since the photochromic layer is not as thick as a regular Photogray Extra lens, it will not darken as much. The polyurethane layer of the Corlon lens (made with either white crown or photochromic glass) can be tinted in solid colors or gradient tints, using special water-based dyes, by the dispenser.

*Younger PLS Filter Lenses*

In 1984, Younger optics introduced a series of four CR-39 plastic lenses designed to selectively filter out ultraviolet and visible blue radiation. These lenses are known as the Protective Lens Series (PLS). These lenses are not photochromic lenses, nor are they tinted (dyed, and therefore having the dye present only at the surface), but are manufactured in such a way that the protective additives are present throughout the lens material and therefore cannot be bleached or removed.

The PLS filter lenses cut off virtually all of the ultraviolet and blue visible radiation below the wavelengths designated by the product name: The PLS 400 lens is pale yellow in color, and eliminates all radiation shorter than 400 nm; the PLS 530 lens is orange-amber in color and will block out 95 to 97% of ultraviolet and blue radiation up to 530 nm; the PLS 540 lens is brown in color and blocks 95 to 97% of the ultraviolet and blue radiation up to 540 nm; and the PLS 550 lens is red in color and blocks out 95 to 97% of ultraviolet and visible light radiation up to 550 nm.

The natural color of any of the PLS lenses can be changed by the use of a cosmetic tint, using standard tinting methods, without affecting the performance of the lens.

The manufacturer recommends these lenses for protection against ultraviolet and visible blue radiation, since it has been suggested by Ham, Mueller, and Sliney[18] that short-wavelength radiation (including both ultraviolet and blue) can be hazardous. The lenses are also recommended for patients with pathological conditions such as cataracts, corneal dystrophies, macular degeneration, and retinitis pigmentosa.

## 7.22. Lens Thickness and Transmission

It is apparent from the discussion in Section 7.11 that when a prescription lens is made in an absorptive material, a variation in transmission across the lens will occur due to the variation in thickness. Normally, absorptive lenses are rated in terms of transmission for a lens of a constant thickness of 2 mm. However, the difference between a rated transmission for a 2-mm-thick lens and the actual transmission, at various points on a lens of high power, can be large. In addition, many patients find the variation in density across the lens cosmetically unattractive since it draws attention to the high power of the lens. Since tint identification by the practitioner is based on samples that are approximately 2 mm thick, tint recognition and identification become more difficult with large variations in thickness.

It is obvious that a high plus lens made from absorptive glass will have lower actual transmittance than the rated transmittance, and that the transmission increases from the center to the edge of the lens, while a high minus lens may have approximately the rated transmittance at the center of the lens with a decrease in transmission toward the edge. A diagram may be drawn which illustrates the transmission as a function of lens thickness (see Figure 7-40). For a lens of a given diameter, center thickness or edge thickness may be estimated on the basis of Figure 7-40 and Table 7-5. Using this procedure, the center-to-edge transmission of a particular prescription can be estimated. The same figures are also useful if one wishes to order a lens that will provide a desired transmission at the center or the edge of the lens.

As the visual axis passes through various areas of a strong absorptive lens, the eye must adapt to different levels of brightness. If an individual is adapted to a certain brightness level, the question arises as to

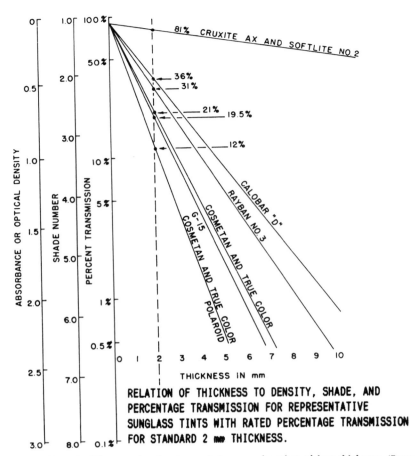

FIGURE 7-40.  **Diagram showing transmission as a function of lens thickness. (From Kors and Peters,** *Amer. J. Optom. Physiol. Opt.,* **Vol. 49, p. 727–735 (c) The American Academy of Optometry, 1972).**

## TABLE 7-4
**Chart for Estimating Thickness for Various Lens Diameters, Assuming (1) Edge Thickness Is a Minimum of 1.0 mm and (2) Average of Center Thickness and Edge Thickness Is a Minimum of 2.0 mm**

| | Diameter (mm) | | | | | | | | |
|---|---|---|---|---|---|---|---|---|---|
| *Power* | *58* | *56* | *54* | *52* | *50* | *48* | *46* | *44* | *42* |
| +7.00 | 7.1 | 6.7 | 6.2 | 5.8 | 5.4 | 5.0 | 4.6 | 4.3 | 4.0 |
| +6.50 | 6.7 | 6.3 | 5.8 | 5.4 | 5.1 | 4.7 | 4.4 | 4.1 | 3.8 |
| +6.00 | 6.1 | 5.8 | 5.4 | 5.0 | 4.7 | 4.4 | 4.1 | 3.8 | 3.5 |
| +5.50 | 5.7 | 5.4 | 5.0 | 4.7 | 4.4 | 4.1 | 3.8 | 3.6 | 3.3 |
| +5.00 | 5.3 | 5.0 | 4.7 | 4.4 | 4.2 | 3.9 | 3.6 | 3.4 | 3.1 |
| +4.50 | 4.9 | 4.6 | 4.3 | 4.1 | 3.8 | 3.6 | 3.3 | 3.1 | 3.0 |
| +4.00 | 4.4 | 4.2 | 4.0 | 3.7 | 3.5 | 3.3 | 3.1 | 2.9 | 2.8 |
| +3.50 | 4.0 | 3.8 | 3.6 | 3.4 | 3.2 | 3.0 | 2.9 | 2.8 | 2.7 |
| +3.00 | 3.5 | 3.3 | 3.2 | 3.0 | 2.9 | 2.8 | 2.8 | 2.7 | 2.6 |
| +2.50 | 3.1 | 3.0 | 2.8 | 2.8 | 2.7 | 2.6 | 2.6 | 2.5 | 2.5 |
| +2.00 | 2.8 | 2.7 | 2.5 | 2.5 | 2.5 | 2.5 | 2.5 | 2.4 | 2.4 |
| +1.50 | 2.6 | 2.6 | 2.3 | 2.3 | 2.3 | 2.3 | 2.3 | 2.3 | 2.3 |
| +1.00 | 2.4 | 2.3 | 2.1 | 2.1 | 2.1 | 2.1 | 2.1 | 2.1 | 2.1 |
| +0.50 | 2.2 | 2.1 | 2.0 | 2.0 | 2.0 | 2.0 | 2.0 | 2.0 | 2.0 |

Source: J.K. Davis and H.G. Fernald, *Amer. J. Optom. Arch. Amer. Acad. Optom.,* Vol. 46, No. 8, p. 574. © The American Academy of Optometry, 1969.

how much this level can be changed instantly without impairing visual performance or requiring a significant period of adaptation. Davis and Fernald[36] believe that a ±0.25 to ±0.50 log unit difference in illumination appears reasonable. The 0.25 log unit difference means a change in percentage transmission of approximately 1.8 times, while a 0.50 log unit difference means a change in percentage transmission of approximately 3.2 times.

For lenses having a cylindrical component, the two principal meridians will behave differently, so both meridians must be considered. For strong absorptive prescriptions, constant adaptation activity is required as various zones of the lens are used. Fatigue or impaired visual performance is likely to result. The solution for the problem of variations in transmission across a lens is to provide lenses designed so as to result in essentially uniform transmission across the entire lens: Such lenses are known as *uniform-density* lenses, and are discussed in Section 7.24.

When an anisometropic patient wears absorptive lenses, the transmittance of one lens (the more minus or less plus lens) will be greater than that of the other. The resulting difference in retinal illumination for the two eyes can give rise to spatial distortions in the presence of a moving object. In 1922, Pulfrich demonstrated that if a filter is held in front of one eye of a normally sighted subject, a pendulum swinging in the frontal plane will be perceived as moving in an elliptical path. The classical explanation for this phenomenon is that there are differential arrival times (latency differences) at the visual cortex for the neural impulses for the two eyes.

Hall[37] found that depth perception was diminished when the binocular visibility relationship is altered, and concluded that balanced binocular visibility is essential to good depth perception. Hence, binocularity may be compromised when a person wears two lenses which vary significantly in transmission. A discrepancy or a perceptual conflict between stereoscopic and monocular clues to spatial arrangements may result in ophthalmic vertigo and difficulties in orientation.[38] When driving an automobile, apparent displacements of moving objects would be likely to constitute a hazard.

In Section 7.20 it was stated that for occupational use the luminous transmittance of lenses with shade numbers 1.5 to 3.0 (61.1 to 13.9% transmission) must be within 10% of each other. For shade numbers 4.0 to 14.0 (transmission 5.18 to 0.0027%) the luminous transmittance of a pair of lenses should be within 20% of each other. Following these guidelines, the transmittance of a pair of sunglass lenses should be within 20% of each other. Lenses may be matched by alteration of thickness, by the use of coatings, or by using uniform-density lenses.

## 7.23. Prescribing Absorptive Lenses

With the large number of absorptive lenses available it is almost impossible for the optometrist to be familiar with all of them. Furthermore, the task of selecting absorptive lenses is complicated because lens composition, color, and cost are not suitable criteria for predicting the amount of near ultraviolet, visible, or near infrared radiation that a specific lens will transmit. While dark-appearing tints generally filter out more of the visible spectrum than the lighter tints, differences occur among lenses that appear to be the same. Generally, plastic lenses (CR-39) filter more near ultraviolet radiation than glass lenses, but exceptions occur. Within a range of similar dark tints there is no way to predict by appearance which of the lenses effectively filter significant quantities of near ultraviolet or near infrared radiation. Effective filtering in a specific region of the spectrum is no guarantee of effective absorption in another region of the spectrum.

An alarming number of *nonprescription* sunglasses have been found to transmit bands of ultraviolet radiation—to have ultraviolet "windows."[39] Studies suggest that an inappropriately large number of tinted lenses transmit large amounts not only of ultraviolet radiation but also of visible and infrared radiation.[40] Lenses that attenuate visible radiation much more than they attenuate ultraviolet radiation may be harmful on two counts. First, wearing lenses that attenuate the visible radiation while transmitting much of the ultraviolet radiation enables the eye to function in a high-intensity environment for longer periods of time than without the filter, thereby increasing the exposure to ultraviolet radiation. Second, the attenuation of the visible radiation by the filter causes the pupil of the eye to dilate, which in turn causes the incident radiation on the lens to be reduced by a greater amount than the retinal illumination. If the ultraviolet radiation is not appreciably reduced, the dilated pupil may pass more ultraviolet radiation with the lens than when *no lens* is worn.

Patients are subjected to a wide range of wavelengths and intensities of illumination in their work and in recreational activities. The optometrist must have available a range of absorptive lenses that would be suitable for the environment in which the wearer works or intends to use the glasses.

In order to prescribe intelligently and to explain to the patient why a lens or lenses is (are) recommended, the optometrist must be fully aware of the transmission properties of the lenses available for use. Other characteristics such as the lens composition (glass or plastic), method of obtaining the tint (solid or coating, fixed or variable tint), durability, and appearance should be considered when recommending a particular lens. Many patients are particularly interested in the cosmetic appearance of the lenses, and tend to select sunglasses on the basis of appearance. Since color affects different people differently, the subjective reaction to the appearance of the environment as seen through the lenses is also an important consideration.

Tints may be demonstrated by having plano lenses glazed *binocularly*, in plastic spectacle fronts. If feasible, the patient should be given the opportunity to assess the color and density of an absorptive lens in a natural sunlight environment. Allowance must be made for differences in the thickness of the sample and the finished prescription lenses.

Patients often depend on eye practitioners to advise them concerning the *need* for absorptive lenses, and concerning the color or shade to be worn. In order to assist practitioners in determining whether or not absorptive lenses are needed, an instrument known as the *Alpascope* was developed. The Alpascope target consists of two parallel bars of light. As the patient views the target, beginning with a low level of illumination, the illumination level is gradually increased until the patient reports that the two bars of light have fused into one. The lower the level of illumination required for the two bars to fuse into one, the greater the patient's sensitivity to light.

In a study conducted by Becnel, Fruge, and Coullard,[41] an extremely high correlation was found between Alpascope findings and the patient's complaint (or lack of complaint) regarding sensitivity to light. Although the use of such an instrument may provide the practitioner with information that will be helpful in advising the patient, the results of this study indicate that the patient's own complaint (or lack of complaint) concerning light sensitivity usually provides sufficient evidence: That is, patients' complaints of light sensitivity should be taken seriously.

In advising a patient concerning absorptive lenses, the practitioner should question the patient concerning not only his or her sensitivity to light, but also concerning the activities (and levels of illumination) for which the lenses are to be used. For outdoor activities in bright daylight, most patients who are at all sensitive to light will benefit most from one of the lenses having transmittances in the 10 to 20% range.

On the other hand, older patients who have small pupils (and perhaps early crystalline lens opacities) will require a somewhat *lighter* tint.

As for the lighter, indoor tints (Softlite, Cruxite, and others), even though the transmittance of these lenses is very little less than that of clear crown glass, some patients have worn such lenses for several years and are completely convinced that the lenses provide greater visual comfort than clear lenses. Experience shows that in these cases it is wise to *continue* the use of the tinted lenses. However, if a medium-dark shade is worn (Softlite B, Cruxite AX, etc.), the practitioner should probably recommend switching to the next lighter shade.

## 7.24. Miscellaneous Absorptive Lenses

Absorptive lenses falling in the "miscellaneous" category include uniform-density lenses, gradient-density lenses, and cosmetic tints.

**Uniform-Density Lenses.** As already pointed out, when an absorptive lens is made of a glass having a "through-and-through" tint, the amount of light absorbed is a function of lens thickness (Figure 7-40) and therefore varies from the center to the edge of the lens: A high minus lens will be noticeably lighter at the center, a high plus lens will be noticeably darker at the center, and a lens containing a cylinder will have a lighter streak running across the lens corresponding to the meridian of the minus cylinder axis. This problem can be avoided if a thin layer of tinted glass is fused to a clear lens. The tinted glass is fused to the *flatter* surface of the lens, usually the back surface of a plus lens (Figure 7-41A) or the front surface of a minus lens (Figure 7-41B). Such a lens is known as a *uniform-density* lens.

The problem of nonuniform density does not occur with dyed plastic lenses, with photochromatic lenses, or with coated lenses. In the case of *dyed plastic* lenses, the dye penetrates the lens surface uniformly throughout the surface, therefore resulting in a tint of uniform density. With *photochromatic* lenses the darkening involves mainly the silver halide crystals on or near the surface (as pointed out in Section 7.19), resulting in a uniform appearance. *Coated* lenses also have a uniform-density appearance, since the coating is applied uniformly over the lens surface.

**Gradient-Density Lenses.** A gradient-density lens is one that is designed so that light transmission gradually increases from the top to the bottom of the lens. Such a lens may be made either by coating a glass lens or by dying a plastic lens.

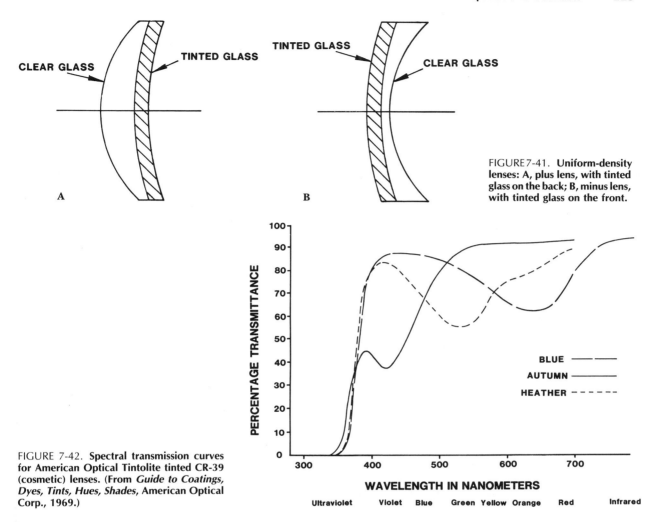

FIGURE 7-41. **Uniform-density lenses: A, plus lens, with tinted glass on the back; B, minus lens, with tinted glass on the front.**

FIGURE 7-42. **Spectral transmission curves for American Optical Tintolite tinted CR-39 (cosmetic) lenses. (From** *Guide to Coatings, Dyes, Tints, Hues, Shades,* **American Optical Corp., 1969.)**

**Cosmetic Lenses.** Plastic lenses differ from glass lenses in that they can easily be tinted by the ophthalmic dispenser, using a dyeing process. One factor in the increasing popularity of plastic lenses, beginning in the 1970s, was the availability of the tinting process. Although these dyed lenses are intended strictly as *cosmetic* lenses, they can be responsible for some of the same problems already described for photochromic lenses: (1) If the lenses are heavily dyed, they may be too dark for night driving and (2) there is a tendency for many of these tints to have high transmittances in the infrared region, with the result that they should not be used as sunglass lenses. Transmission curves for some of the tinted plastic lenses are shown in Figure 7-42.

## LENS REFLECTIONS AND COATINGS

### 7.25. Reflections from Spectacle Lens Surfaces

When light reaches the boundary between two transparent media having different indices of refraction, most of the light is refracted but a small amount is reflected. This reflected light may be troublesome to the wearer of spectacle lenses, because it can produce ghost images, falsification of image position, haze, and loss of contrast. In addition, the reflected light can be annoying to observers because it may obscure the wearer's eye with a veiling glare and accentuate the fact that glasses are being worn.

Reflected images may become troublesome to a spectacle wearer when

1. the reflected image is located near a fixated object;
2. the reflected image is sufficiently intense to stand out from its background;
3. the reflected image is in focus or capable of being made clear by accommodation.

For light of normal or nearly normal incidence, the intensity of light reflected from a surface separating two media of different indices is given, as shown in Section 7.11, by Fresnel's equation:

$$I_R = \frac{(n' - n)^2}{(n' + n)^2} (I) .$$

Also as shown in Section 7.11, for ophthalmic crown glass having an index of refraction of 1.523, surrounded by air, the intensity of light lost by reflection from the first surface is 4.3% of the original intensity; and if we assume that no light is lost by absorption, the loss by reflection from the back surface of the lens is 4.1%, with the result that the intensity of the transmitted beam is 91.6% of the intensity incident upon the lens.

By inspection of Fresnel's equation, we may conclude the following:

1. The intensity of the reflected beam is unaffected by the direction of the incident light, that is, whether it traverses from air to glass or from glass to air.
2. The intensity of the reflected beam increases as the index of refraction increases. Hence, the reflection intensity is greater for lenses made with high index glass such as High-Lite than for lenses made with ophthalmic crown glass or with CR-39 plastic.
3. Since the index of refraction increases toward the blue end of the spectrum, the reflected image will contain more blue light than red light. Therefore the reflected image may appear slightly bluish in color while the transmitted light, deficient in blue, may appear slightly yellowish.

## 7.26. Types of Surface Reflections

The surfaces that are primarily involved in forming reflected images are the front and back surfaces of the spectacle lens and the front surface of the wearer's cornea. While these reflecting surfaces may participate in the formation of an infinite number of reflected images which may or may not enter the eye, only a few of these images have any significance. In Figure 7-43, five common sources of spectacle lens reflections are shown. They may be described as follows:

1. *Image Formed by Light Coming from Behind the Spectacle Lens and Reflected by the Back Surface of the Lens into the Eye* (Figure 7-43A). An image formed by reflection from the back surface of the lens has the highest intensity of any possible spectacle lens reflection: As already shown, for crown glass having an index of refraction of 1.523 this reflection will have an intensity of 4.3% of the intensity of the incident light. When lens sizes were small this type of reflection was seldom a problem, but with the larger lens sizes now in use it may prove to be a problem for some wearers. For example, when the face is strongly illuminated, reflected images of the eye or the adnexa may be seen by the wearer. Fortunately, the concave back surfaces used for most ophthalmic lenses tend to limit the directions from which incident light may be reflected into the eye: Fitting the lens as close to the eye as possible and slightly "face-forming" the frame will tend to reduce the likelihood of reflections entering the eye.

2. *Image Formed by Light Coming from Behind the Lens, Undergoing Internal Reflection by the Front Surface and Then Refracted Again by the Back Surface into the Eye.* This reflection, shown in Figure 7-43B, is somewhat less intense than the reflection described above. For clear ophthalmic crown glass the intensity of this image will be 3.94% of that of the incident light. Neither of these two types of images, formed by light reflected from the back surface of the lens, is ordinarily troublesome to the wearer.

3. *Image Formed by a Light Source in Front of the Lens, Formed by Two Internal Reflections Initially by the Back Surface and Then by the Front Surface of the Lens and into the Eye.* This is a still less intense reflection (see Figure 7-43C) often referred to as the "double internal reflection" ghost image. If we assume no loss of light by absorption, the intensity of this image is only 0.17% of the intensity of the incident light, for ophthalmic crown glass.

Though of relatively low intensity, this reflected image is one of the most troublesome of the ghost images arising from a light source in front of the eye. Every lens presents ghost images of this type, but for most powers of lenses the images are not in focus and

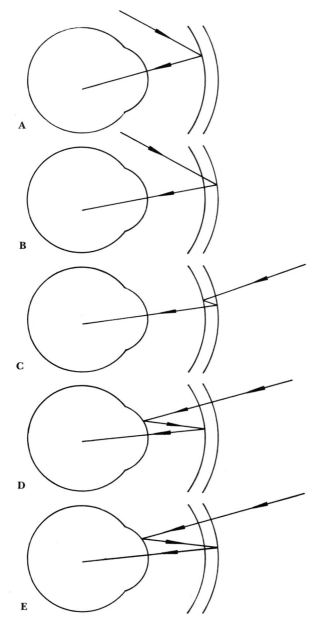

FIGURE 7-43. **Types of surface reflections: Note that in A and B the light is incident upon the back surface of the lens, while in C, D, and E the light is incident upon the front surface of the lens. See text for explanation.**

the intensity is so low that the images are not seen. However, if there happens to be a small, bright source of light against a dark background, the ghost image may be sufficiently bright to be noticeable, as, for example, a street light at night or lights in a darkened theater.

For weak minus lenses, the images of bright objects peripheral to the wearer's line of sight are reflected *toward the center* of the lens. For example, bright overhead lights are reflected downward, and the ghost image is formed between the source and the center of the lens. When the wearer moves his or her head, the ghost image moves in the same direction as the head. On the other hand, for the wearer of a plus lens, the ghost images are displaced farther from the optical center of the lens, so they usually cause little or no trouble. However, if the wearer looks at a light source directly through the optical center of the lens, the ghost image is superimposed on the source and is not seen.

The position and intensity of this ghost image are essentially independent of the base curve of the lens, so changing the base curve will prove to be of no help. Changing the pantoscopic tilt of the lens also has little effect on this ghost image. For lenses of low minus power, with the same power for both eyes, and if the source of annoyance is in one direction only, the lenses can be decentered (or can be ordered with prism) so that the optical center of the lens is moved toward the source forming the reflected image. For example, if overhead lights are a source of annoyance to a patient, the optical centers of the lenses can be raised, or base-down prism power can be ordered, so that the ghost image lies farther away from the patient's line of sight.

4. *Image Formed when Light Passes through the Lens, Is Reflected by the Cornea, and Then Is Reflected into the Eye by the Back Surface of the Lens.* This reflected image is illustrated in Figure 7-43D. If no loss of light occurs by absorption, the intensity of this reflected image is 0.09% of that of the incident light. If the front surface of the cornea has a radius of curvature of 8 mm, the catoptric (reflecting) power of the cornea is about -250.00 D, with the result that the cornea will form an image of the real source at a distance of 4 mm behind the apex of the cornea. This small image becomes a source of reflection at the back surface of the lens, and hence is reflected back into the eye. Several solutions to the problem present themselves: The position of the ghost image can be displaced by angling the lens, or a sharp image can be thrown out of focus either by changing the lens form (base curve) or by changing the vertex distance.

5. *Image Formed by Light Reflected from the Cornea and Then Internally Reflected by the Front Surface of the Lens and into the Eye* (see Figure 7-43E). As with ghost image number 4, the cornea forms an image of a real source approximately 4 mm behind the vertex of the cornea, which then becomes a source of reflection at

the front surface of the lens. The intensity of this ghost image is 0.08% of that of the incident light, for ophthalmic crown glass. This ghost image also may be displaced by angling the frame, by changing lens form, or by changing vertex distance.

6. *Other Reflections.* The wearer of a flat-top bifocal or trifocal lens may be subjected to ghost images and flare due to reflections from the segment top. The manufacturer may address this potential problem by angling the top of the segment and by placing an absorptive coating on the top of the segment before the segment is fused into the carrier lens. The coating will absorb the light that would normally participate in the ghost image or create flare.

## 7.27. Methods of Controlling Surface Reflections

The methods available for minimizing the annoyance of surface reflections include base curve selection; alteration of pantoscopic tilt; changing the vertex distance; selection of small lens sizes; patient education and counseling; and the use of antireflective coatings.

**Base Curve Selection.** Depending upon the source of the reflection, the location, size, and clarity of reflections may be altered by using a steeper or flatter base curve than would normally be used. However, the change in the base curve must usually be large, in order to correct the problem, and in making such a base curve change the advantages of "corrected-curve" lens construction would be compromised. The use of a special base curve would mean the grinding of a "custom" lens, which would add to the cost of the patients' lenses. Unless the source of the reflection has been analyzed correctly, a change in base curve may not achieve the desired result. For these reasons, base curves are seldom changed for the purpose of minimizing reflections. However, it is true that by choosing the flattest base curve that is consistent with acceptable optical performance, the object field of view of the two reflecting surfaces of the lens is reduced, thereby reducing the potential number of reflected images visible to the wearer or to an observer facing the wearer.

**Alteration of Pantoscopic Tilt.** When a constantly recurring reflection is determined to be produced by a given light source, a change in pantoscopic tilt of the lens has the effect of rotating the object fields of view of the reflecting surfaces away from troublesome light sources such as artificial lights or outdoor lighting coming through windows. However, the re-

sultant object field may prove to contain new light sources, with the result that little may be gained.

**Changing the Vertex Distance.** A change in the vertex distance may have the advantage of focusing the ghost image in a different plane. However, the vertex distance of a lens can be altered only to a limited extent: Both from the optical and cosmetic points of view, this alteration is practical only within no more than 2 or 3 mm in either direction. Alteration of vertex distance is particularly difficult with frames having fixed nose pads.

**Selection of Small Lens Sizes.** Selection of a smaller lens size than usual will also reduce the object fields of view of the two reflecting surfaces of the lens, by reducing the area available for reflection.

**Patient Education and Counseling.** When a patient complains of the presence of ghost images, he or she may need only to be informed that the reflected images are normal occurrences. If basic information is provided concerning the nature and origin of the ghost images, the patient may "accept" them and learn to ignore them.

## 7.28. Antireflective Coatings

If the methods discussed above (particularly alteration of the pantoscopic tilt and patient education) do not provide relief from reflected images, an antireflective coating should be placed on *both* surfaces of the lens. Although an antireflective coating does not completely eliminate the ghost images caused by reflection, it can reduce their intensity enough so that they will not be bothersome to the wearer or to an observer.

According to the principle of *interference*, whenever two waves of equal intensity or amplitude produced by a monochromatic source are superimposed, neutralization (cancellation) occurs if the path difference is an odd number of half wavelengths, while *reinforcement* occurs if the path difference is equal to an even number of half wavelengths. In addition, it should be understood that when light is reflected from a surface whose index of refraction is higher than the index of the medium in which the light has previously traveled, a phase change of 180 degrees takes place. When the reflecting surface has a lower index of refraction than the incident medium, no phase change occurs.

When a transparent coating has been applied to the surface of an ophthalmic lens, two reflection waves will occur, one from the front surface of the coating and one from the interface between the coating and the glass. According to the principles of interference, these two reflection waves will cancel one another provided that two conditions are met: The first, called the *amplitude* condition, requires that the two reflection waves must be of equal amplitude; the second, called the *path* condition, requires that they be one-half wavelength out of phase. Examination of the amplitude condition will tell us the index of refraction of the coating and the path condition will reveal the proper thickness of the coating.

In order to satisfy the amplitude condition, the intensity of the reflection from the air-coating interface, $R_1$, must be equal to the intensity of the reflection from the coating-glass interface, $R_2$. Using Fresnel's equation, where $n_c$ specifies the index of refraction of the coating and $n_g$ specifies the index of refraction of the glass,

$$R_1 = \left(\frac{n_c - 1}{n_c + 1}\right)^2 = R_2 = \left(\frac{n_g - n_c}{n_g + n_c}\right)^2,$$

or

$$\left(\frac{n_c - 1}{n_c + 1}\right) = \left(\frac{n_g - n_c}{n_g + n_c}\right),$$

and

$$(n_c + 1)(n_g - n_c) = (n_c - 1)(n_g + n_c)$$

or

$$n_c n_g + n_g - n_c^2 - n_c = n_c n_g - n_g + n_c^2 - n_c$$

$$2n_g = 2n_c^2,$$

and therefore

$$n_c = \sqrt{n_g}.$$

Thus in order to satisfy the amplitude condition the index of refraction of the coating should be equal to the square root of the index of refraction of the glass. When applied to ophthalmic crown glass ($n = 1.523$) the index of refraction of the coating should be

$$\sqrt{1.523} = 1.234.$$

In addition to having the ideal index of refraction, the coating must have good adhesive properties, it must be sufficiently hard, and it must be insoluble in water. Of the known materials having an index of refraction of 1.234, all are either too soft or are too soluble in water. A compromise material must therefore be accepted in order to obtain a sufficiently hard coating that is water resistant. *Magnesium fluoride* has the optimum combination of optical and physical properties, and is used almost universally for antireflective coatings of glass lenses. Its refractive index,

approximately 1.38, is too high to completely satisfy the amplitude condition when applied to ophthalmic crown glass: This results in an incomplete neutralization of reflections with the result that the reflectance of a surface cannot be reduced to zero for any wavelength.

The ideal glass to be used with magnesium fluoride would have an index of refraction of $1.38^2$, or 1.904: Therefore, more efficient results will be obtained when magnesium fluoride is applied to glasses with a high index of refraction. When applied to High-Lite glass, having an index of 1.70, the intensity of the reflections is reduced from 6.8% of the incident light (for ophthalmic crown glass) to 0.30% of the incident light.

In order to satisfy the *path* condition the two reflected waves must be out of phase by one-half wavelength at the point of interference. Since the wave reflected from the coating-glass interface will traverse the coating *twice*, this wave will be one-half wavelength out of phase with the reflected wave at the air-coating interface providing that the coating is made with an *optical* thickness of *one-quarter wavelength*. It should be understood that if two wavefronts are out of phase by any *odd* number of wavelengths, and if the wavefronts are generated by any coating having an optical thickness of the same odd number of quarter wavelengths the path condition is satisfied. However, the minimum optical thickness (one-quarter wavelength) produces the highest quality coating and is the thickness that is usually applied.

It is important to distinguish between *optical* thickness and *physical* (actual) thickness. The optical thickness is the product of the physical thickness and the index of refraction of the coating. Hence, the optical thickness is $\lambda/4$, and the physical thickness is $\lambda/4n$.

The stipulation that the optical path difference should be one-half wavelength is correct only for a particular wavelength of monochromatic light. Although the eye is sensitive to a wide range of wavelengths, the path-condition can be satisfied only for monochromatic light. The wavelength generally selected is 555 nm, the wavelength of maximum visibility. The reflection for this wavelength is essentially eliminated, but reflections gradually increase toward the red and violet ends of the spectrum. The combination of red and violet reflected light causes the lens surface to appear to have a *purplish*, or *magenta*, color. Early workers noted that the magenta color resembled the "bloom" of ripened fruit, and the coating process was spoken of as "blooming" the lenses.

The magenta color of a coated lens is so unique that, during the coating process, it is used to assess the *thickness* of the coating. If the coating is too thin, the lens surface will have a pale "straw," or amber, appearance; but if the coating is too thick, the surface will have a pale blue appearance.

To produce antireflective coatings that are effective for more than one wavelength requires the application of *several layers* of coating material. The refractive index and the optimal thickness of each layer are chosen so that the combination of the amplitudes of all the waves reflected from boundaries sums to zero across many wavelengths. Whereas a camera lens may have as many as seven or more layers of an antireflective coating, multicoated ophthalmic lenses are given only three or four layers. The Zeiss Super ET and the Hoya 99 Multicoat lenses (Figure 7-44) are examples of multicoated glass ophthalmic lenses.

### 7.29. Optical Principles of Antireflective Coatings

The optical principles involved in antireflective coatings may be understood with the help of the diagram shown in Figure 7-45. In this diagram, $N_g$ represents an ophthalmic crown glass lens, and $N_c$ represents a

coating of one-quarter wavelength applied to the lens. AP and BO are parallel rays, in air, incident upon the surface, XX′, of the coating. All angles of incidence, refraction, and reflection are considered to be very small. Some of the incident light designated by AP (solid line) as well as by BO (dashed line) is *reflected* (in directions PM and ON) while some is *refracted* (in directions PK and OL). When ray APK reaches the boundary, YY′, between the coating and the glass, some of the light is reflected (in the direction KO) while some is refracted (in the direction KQ). At point O on the surface of the coating, a portion of the light is reflected internally (in the direction OL) and a portion is refracted in the direction ON.

Inspection of the diagram indicates that ray APKO has traveled a distance PKO *farther* than ray BO. Since the angles of incidence and refraction are very small and since the coating has an optical thickness of one-quarter wavelength, ray APKO has traveled *one-half wavelength* farther than ray BO upon arrival at point O. Since the index $N_c$ is greater than $N_a$ (of air) and since $N_g$ is greater than $N_c$, there is a change in phase of 180° in the reflected ray KO, at K,

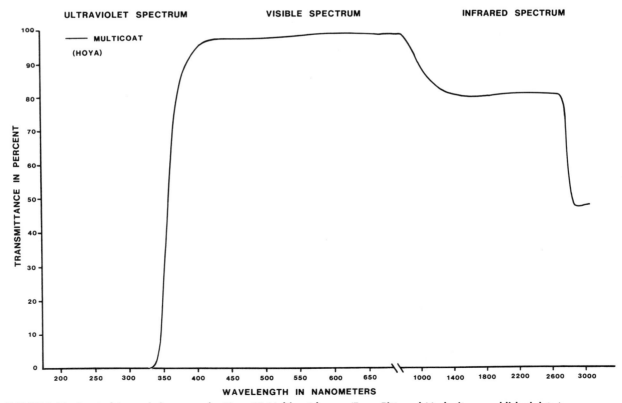

FIGURE 7-44. **Spectral transmission curve for Hoya 99 Multicoat lenses. (From Pitts and Maslovitz, unpublished data.)**

and in the reflected ray ON, at O. Therefore, as a result of the phase changes in the reflected rays at K and O and the difference in path lengths of APKO and BO, the reflected ray ON (dashed) and the refracted ray ON (solid) are out of phase by 180° at point O, and neutralization (cancellation) occurs.

The use of an antireflective coating results not only in the attenuation of bothersome reflections but also in an increase in the *transmission* of the lens. It is apparent in Figure 7-45 that at point O the refracted ray (dashed line) and the internally reflected ray (solid line) are *in phase*, with the result that reinforcement occurs, therefore increasing the transmission of the lens.

As pointed out by Knoll[3], *tinted* glass lenses will often reduce the intensity of reflected images. The effect of a tinted lens on each of the types of reflections obviously depends upon the number of times the reflected light traverses the lens before entering the eye. The table below indicates (see Figure 7-43) how many times the light involved in producing each type of reflection traverses the lens:

| Type of Reflection | Number of Times Traversed |
| --- | --- |
| A | Zero |
| B | Twice |
| C | Thrice |
| D | Once |
| E | Thrice |

When multiple traverses occur, the effective transmittance can be found by raising the nominal transmission to a power equal to the number of traverses.

For example, if the nominal transmittance, for a medium, of a tinted lens is 90%, the effective transmittance after two traverses is 81% and after three traverses is 72.9%. Therefore, for a lens medium having a nominal transmission of 90%, image B in Figure 7-43 will be reduced to 81%, image C and image E will each be reduced to 73%, and image D will have a transmission of 90% of the intensity produced when an untinted lens is used.

Even a lightly tinted lens having an absorbance of only 4% per 2 mm of lens thickness is effective in reducing the C or E type of reflection by about 12%; and a slightly darker tint, having an absorbance of 8% per 2 mm of thickness, has the effect of reducing the intensity of these reflections by 22%.

## 7.30. Production of Antireflective Coatings

The first step in the production of a magnesium fluoride coating is a thorough cleaning of the lens to be coated. The lens is then placed in a bell-shaped chamber in which a vacuum is created, to remove all of the air which might interfere with the deposition of magnesium fluoride. The lens is then heated to about 300°C, which contributes to the hardness of the coating. Magnesium fluoride pellets are then heated to about 2500°C in order to produce vaporization. As the magnesium fluoride evaporates, it is deposited on the surface of the lens. During the deposition, an operator observes the color of the light reflected from a test lens of high refractive

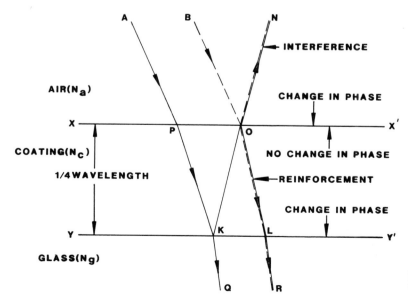

FIGURE 7-45. **Optical principles of an antireflective coating.**

index. The eye is particularly sensitive to the development of the characteristic magenta color, so this method of thickness control is satisfactory. However, the eye may be replaced by a photoelectric cell screened by a green filter transmitting light of wavelength 555 nm. The output of the photoelectric cell diminishes as the coating thickens, reaching a minimum when the correct thickness is reached.

## 7.31. Reflections Annoying to an Observer

There are reflections from the surface of an ophthalmic lens that are not noticed by the lens wearer but are annoying or distracting to an observer. These reflections cause the lenses to produce "flashes" of light, coming and going with the movement of the wearer's head. These reflections enhance the "glassiness" of the appearance of the lenses, calling attention to the lenses rather than to the eyes. They conceal the eyes from the observer, making eye contact difficult. Since the eyes serve as highly expressive aids to communication, this form of communication can be strongly impaired by the wearing of spectacles. This is detrimental enough in daily personal contacts with individuals or with small groups of people, but for entertainers, teachers, ministers, and others who wear glasses when appearing before large audiences, these reflections can be a serious impediment to effective communication. Spectacle wearers who must appear before television or movie cameras or before press photographers are subject to brilliant light sources which increase the number and intensity of these reflections.

An antireflective coating provides a distinct improvement in the appearance of the lenses, giving a cleaner, fresher, and warmer appearance to the wearer's eyes. In this image-conscious age, in which spectacles can detract from one's overall image, antireflective coatings are of particular benefit to those whose appearance receives careful scrutiny by the public.

**The Multiple-Ring Effect.** In addition to the reflections formed by the surfaces of the lenses, the *edges* of the lenses are responsible for the multiple rings seen just inside the edges of high minus lenses (see Figure 7-46). These are true images of the edge of the lens, and are caused by internal reflections which begin at the lens edge and reflect their way toward the center of the lens until they strike the surface at an angle which permits refraction by the lens surface and into the eye of the observer. This process is then repeated, and multiple ring reflections may be seen.

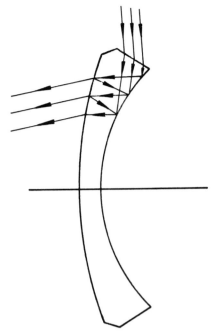

FIGURE 7-46. **The multiple-ring effect.**

The rings are more pronounced for oblique angles than for straight-ahead viewing. The ring reflections also depend upon such factors as the power of the lens, the curvatures of the surfaces, and the angle and distribution of the bevel. These reflections will be the same color as that of the edge of the lens.

An antireflective surface coating is helpful in attenuating the intensity of the multiple-ring effect for straight-ahead viewing or for viewing at moderately oblique angles, but is less helpful for extremely oblique angles of view. However, various forms of *edge treatment* can be helpful in this regard. *Edge coating* makes the rings less noticeable by reducing the granularity of the edge. *Painting* the edge a neutral gray color also tends to make the rings less noticeable. A third method of edge treatment is the use of a *semitransparent* edge. Such an edge can be produced by using an edging wheel that results in a fine, semitransparent grain rather than a coarse white grain on the bevelled lens edge. The same effect can be achieved, for a plastic lens, by slightly *buffing* the edge. Another effective procedure for a plastic lens is that of tinting (dyeing) the lens after it has been edged. The tint can be so light that the transmission of the lens is effected very little for straight-ahead viewing. The dyeing procedure is effective because of the fact that the dye penetrates a short distance beyond the lens surface: For multiple reflections, the reflected beam within the lens traverses this

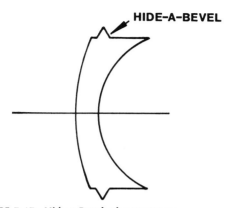

**HIDE-A-BEVEL**

FIGURE 7-47. **Hide-a-Bevel edge treatment.**

thin dyed zone several times at an oblique angle. The reflected beam is thus attenuated a significant amount.

One of the most effective ways of coping with the multiple-ring effect involves the use of the *Hide-a-Bevel* edge treatment. As shown in Figure 7-47, this edge treatment consists of a flat edge with a small, narrow bevel protruding from the flat edge. With such an edge there is a smaller area of ground glass from which reflections can arise, and the angle the flat edge makes with the lens surfaces is such that the rays tend to be reflected in such a direction that they do not enter the observer's eye.

## References

1. Hecht, S., Hendley, C.D., Ross, S., and Richmond, P.M. The Effect of Exposure to Sunlight on Night Vision. *Amer. J. Ophthal.*, Vol. 31, p. 1533, 1948.
2. Richards, O.W. Sunglasses for Eye Protection. *Amer. J. Optom. Arch. Amer. Acad. Optom.*, Vol. 48, pp. 200–203, 1971.
3. Knoll, H. Ophthalmic Lens Reflections. *Optom. Weekly*, Vol. 53, No. 31, pp. 517–518, 1962.
4. Pitts, D. G., and Cullen, A.P. *Ocular Ultraviolet Effects from 295 nm to 400 nm in the Rabbit Eye.* DHEW (NIOSH), Publication No. 75-175, 1977.
5. Hiller, R., Gracometti, L., and Yuen, K. Sunlight and Cataract, an Epidemiological Investigation. *Amer. J. Epidemiol.*, Vol. 105, pp. 450–459, 1977.
6. Hollows, F., and Moran, D. Cataract—The Ultraviolet Risk Factor. *Lancet*, pp. 1249–1250, Dec. 5, 1981.
7. Zigman, S. The Role of Sunlight in Human Cataract Formation. *Surv. Ophthalmol.*, Vol. 27, pp. 317–326, 1983.
8. Taylor, H. The Environment and the Lens. *Brit. J. Ophthalmol.*, Vol. 64, pp. 304–310, 1980.
9. Buckingham, R.H., and Pirie, A. The Effect of Light on Lens Proteins in Vitro. *Exp. Eye Res.*, Vol. 14, pp. 297–299, 1972.
10. Van Heyningen, R. What Happens to the Human Lens in Cataract? *Sci. Amer.*, Vol. 233, pp. 70–81, 1976.
11. Lerman, S. *Radiant Energy and the Eye.* Macmillan, New York, pp. 131–186, 1980.
12. Parrish, J., Anderson, R., Urback, F., and Pitts, D. *Biological Effects of Ultraviolet Radiation with Emphasis on Human Responses to Longwave Ultraviolet.* Plenum Press, New York, pp. 177– 219, 1978.
13. Lerman, S. Lens Fluorescence in Aging and Cataract Formation. *Doc. Ophthalmol. Proc. Series 8* , pp. 241–260, 1976.
14. Lerman, S., and Borkman, R. A Molecular Model of Lens Aging, Nuclear and Cortical Cataract Formation. *Metabol. Pediatr. Ophthalmol.*, Vol. 3, pp. 27–35, 1979.
15. Pitts, D. Threat of Ultraviolet Radiation to the Eye—How to Protect Against It. *J. Amer. Optom. Assoc.*, Vol. 52, pp. 949–957, 1981.
16. Boettner, E., and Wolter, J. Transmission of the Ocular Media. *Invest. Ophthalmol.*, Vol. 1, pp. 766–783, 1962.
17. Ham, W., Mueller, H., Ruffolo, J., Guerry, D., III, and Guerry, R.K. Action Spectrum for Retinal Injury from Near UV Radiation in the Aphakic Monkey. *Amer. J. Ophthalmol.*, Vol. 93, pp. 299–306, 1982.
18. Ham, W., Mueller, H., and Sliney, D. Retinal Sensitivity and Damage from Short Wavelength Light. *Nature*, Vol. 260, pp. 153–155, 1976.
19. Parrish, J., Anderson, R., Urback, F., and Pitts, D. *Biological Effects of Ultraviolet Radiation with Emphasis on Human Responses to Longwave Ultraviolet.* Plenum Press, New York, p. 142, 1978.
20. Luria, S.M. Preferred Density of Sunglasses. *Amer. J. Optom. Physiol. Opt.*, Vol. 61, pp. 379–402, 1984.
21. Pitts, D.C., Cullen, A.P., and Dayhaw-Barker, P. Determination of Ocular Thresholds for Infrared Radiation Cataractogenesis. DHHS (NIOSH) Publ. No. 80-121, June 1980.
22. Clark, B.A.J. Infrared Transmission Limits for Sunglasses. *Austr. J. Optom.*, Vol. 52, pp. 167–186, 1969.
23. Clark, B.A.J. Color in Sunglass Lenses. *Amer. J. Optom. Arch. Amer. Acad. Optom.*, Vol. 46, pp. 825–840, 1969.
24. Wright, W.D. *Photometry and the Eye.* The Hatton Press, London, 1949, p. 127.
25. Luckiesh, M., and Holladay, L.L. Penetration of Fog by Light from Sodium and Tungsten Lamps. *J. Opt. Soc. Amer.*, Vol. 31, pp. 528–530, 1941.
26. Richards, O.W. Yellow Glasses Fail to Improve Seeing at Night-driving Luminances. *Highway Res. Abstr.*, Vol. 23, pp. 32–36, 1953.
27. Blackwell, H.R. The Effect of Tinted Optical Media upon Visual Efficiency at Low Luminances. *J. Opt. Soc. Amer.*, Vol. 43, p. 815, 1953.
28. Davey, J.B. Seeing Times with Yellow Driving Glasses. *Optician*, Vol. 136, p. 651, 1959.
29. Bierman, E.O. Tinted Lenses in Shooting. *Amer. J. Ophthal.*, Vol. 35, pp. 859–860, 1952.
30. The American National Standards Institute Z87.1-1979, Practice for Occupational and Educational Eye Protection, p. 8.
31. Megla, G.K. Optical Properties and Application of Photochromic Glass. *Appl. Optics*, Vol. 5, No. 6, pp. 945–960, 1966.

32. Armistead, W.H., and Stookey, S.D. Photochromic Silicate Glasses Sensitized by Silver Halide. *Science*, Vol. 144, pp. 150–154, April 10, 1964.

33. Douthwaite, W.A., and Elliott, P. A Comparison of Photogray Glass with Surface Tints. *The Optician*, pp. 15–16, Sept. 24, 1971.

34. Chase, W.W. Radiation Transmission of Photochromic Lenses. *Optom. Weekly*, pp. 43–47, Nov. 6, 1975.

35. Lynch, P.M., and Brilliant, R. *Optom. Monthly*, pp. 36–42, June 1984.

36. Davis, J.K., and Fernald, H.G., Prescription Sunglasses. *Amer. J. Optom. Arch. Amer. Acad. Optom.*, Vol. 46, No. 8, pp. 572–577, 1969.

37. Hall, R.M. Theoretical and Practical Considerations of the Pulfrich Phenomenon. *Amer. J. Optom. Arch. Amer. Acad. Optom.*, Vol. 15, No. 2, pp. 31–46, 1928.

38. Lederer, J. Anisopia. *Australas. J. Optom.*, Vol. 40, No. 1, pp. 13–38, 1957.

39. Anderson, W.J., and Gebel, R.K.H. Ultraviolet Windows in Commercial Sunglasses. *Appl. Optics*, Vol. 16, pp. 515–517, 1977.

40. Borgwardt, B., Fishman, G.A., and Vander Meulen, D. Spectral Transmission Characteristics of Tinted Lenses. *Arch. Ophthal.*, Vol. 99, pp. 293–297, 1981.

41. Becnel, L., Fruge, P., and Coullard, G. Validation Studies of the Alpascope. *J. Tex. Optom. Assoc.*, Vol. 24, pp. 7–12, Oct. 1968.

## Questions

1. A lens is made of glass with an index of 1.69. What is the index of refraction for the ideal antireflection coating for the lens?

2. An ideal antireflection coating is to be placed on glass that has a refractive index of 1.621. What is the minimum physical thickness of the coating required for light of wavelength 590 nm?

3. What is the percentage of perpendicularly incident light reflected from a *surface* between air and glass with a refractive index of 1.62?

4. Assuming there is no absorption, what is the percentage of perpendicularly incident light transmitted by a lens of index 1.70?

5. What is the total amount of light transmitted through a clear lens of index 1.50 coated on both sides with a substance of index 1.30?

6. A lens is 4 mm thick at its center. The index of refraction is 1.50. The transmission factor is 0.5 per mm. Ignoring reflection, what is the transmission at its center?

7. A lens is 3 mm thick at its center. The index of refraction is 1.50. The transmission factor is 0.2 per mm. Ignoring reflection, what is the density at its center?

8. A lens is 2 mm thick at its center. The index of refraction is 1.50. The transmission factor is 0.2 per mm. Ignoring reflection, what is the opacity at its center?

9. What would be the combined optical density of two absorptive lenses each having a light transmission of 20%?

10. A lens having a transmission of 50% is placed next to a lens having a transmission of 20%. What is the transmission of the combination?

11. A +4.00 lens is 5.6 mm thick at its pole. The index of the glass is 1.5 and the transmission factor is 0.5/mm. Assuming no loss of light by reflection, at what distance from the pole do we have 25% transmission?

12. A spectacle or crown lens of index of refraction of 1.523 has a total transmission of 50% for a thickness of 2 mm. What is the total transmission for a lens of this material made 5.0 mm thick?

13. An ophthalmic crown lens with an index of refraction of 1.523 has a transmission of 15% for a thickness of 2.2 mm. What is the transmission for a lens of this material made 3.6 mm thick?

14. A tinted contact lens of index 1.49 has a transmission of 20% at a thickness of 1.0 mm. When made with a center thickness of 0.2 mm, what will be the total transmission?

15. A welder decides that when welding, the proper illumination occurs when he places a shade number 8 and a shade number 3 together. If he wished to replace the two lenses by a single lens, what would be the shade number of the single lens?

16. What would be the luminous transmission of a welding lens with a shade number of 8.0?

17. A glass filter of index 1.523, which is 1 mm thick, transmits 80% of the incident light upon it. If a transmission of 50% is desired, what would be the thickness of a single glass filter of the same material?

18. A −10.00 D tinted lens is 1 mm thick at its center. The index of refraction is 1.50. The transmission factor is 0.5 per mm. Assume no loss by reflection.
    (a) Plot a graph illustrating transmission as a function of its thickness.
    (b) Plot a graph illustrating transmission as a function of the *distance (h) from the optical center*.

# CHAPTER EIGHT

# Multifocal Lenses

The crystalline lens continues to grow throughout life. As additional layers of lens fibers are laid down underneath the lens capsule, the older layers of fibers are trapped in the nucleus of the lens where they become compressed, with the result that they gradually lose their elasticity. The loss of elasticity of the lens nucleus is responsible for the gradual loss of the ciliary muscle's ability to bring about an increase in the vergence of light when accommodation is attempted. The near point of accommodation therefore gradually recedes with age, so that by about the age of 45 it becomes difficult if not impossible for most people to see clearly for near-vision tasks without the assistance of plus lens power in addition to any lens power required for distance vision.

Although the near "add" may be supplied in the form of a pair of glasses for near work only, most people who require lenses for distance vision find it more convenient to wear only one pair of glasses having two or more focal powers. Whereas the early multifocal lenses provided only two focal powers, later developments included trifocal lenses, occupational multifocal lenses, and progressive addition lenses.

In this chapter we will first discuss the physical characteristics of multifocal lenses, beginning with the history and development of these lenses. This will be followed by discussions of the physical characteristics, manufacturing processes, optical design principles, performance characteristics, and the clinical application of multifocal lenses. The chapter will conclude with a discussion of invisible bifocals and progressive addition lenses.

# PHYSICAL CHARACTERISTICS

## 8.1. History and Development of Multifocal Lenses

Benjamin Franklin is credited with the invention of the bifocal lens. In a letter to a friend written in 1785, Franklin commented that he had formerly worn two pairs of spectacles, one for distance vision and one for reading, and finding the arrangement sometimes inconvenient he had the lenses cut in half, and half of each kind mounted in the same frame (Figure 8-1). Franklin's bifocal lens was similar in appearance to the Executive one-piece bifocal that is available today, having a dividing line going all the way across the lens.

Franklin's bifocal had excellent optical properties with sharp imagery over the entire reading field. Since the distance and reading portions were separate lenses, the optical centers could be placed at the same level as the dividing line or could be separated as desired. Because both parts of the lens were made of the same material, there was little difference in the chromatic aberration of the two portions. However, the Franklin bifocal suffered from two disadvantages. First, the dividing line produced annoying reflections, it tended to collect dust and dirt, and it was conspicuous and unsightly. Second, the two portions of the lens were held in position by the eyewire of the frame, providing a weak structure that could easily come apart.

The next bifocal lens was the *Solid Upcurve* bifocal, invented by Isaac Schnaitmann of Philadelphia in the year 1838. This was the first commercially successful one-piece bifocal lens. Starting with a lens ground for the wearer's reading prescription, this lens was made by grinding a flatter curvature on the upper portion of the back surface of the lens, thereby reducing the plus power sufficiently for distance vision (Figure 8-2). The desirable attributes of this lens included

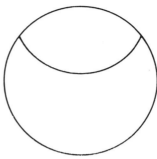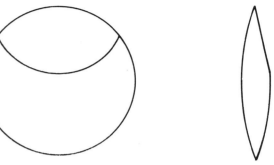

FIGURE 8-2. **The Solid Upcurve bifocal, showing base-down prismatic effect in the distance portion.**

invisibility, structural strength, little chromatic aberration, and a wide field of view for reading. However, there were serious disadvantages. Due to the limited choice of surface powers, significant amounts of aberrations were usually present in the distance portion, which restricted the field of vision. In addition, the resurfacing of the upper portion of the back surface moved the optical center of the distance portion of the lens below its original position. Consequently the distance portion contained a strong base-down prismatic effect. The combination of aberrations and strong prismatic effects in the distance portion produced intolerable optical results.

In 1888 (more than 100 years after Franklin's invention of the bifocal lens) August Morck reinvented the Franklin bifocal in a modified form. Called the *Perfection* bifocal, this lens had a curved dividing line, the two pieces of glass having bevelled edges, making the lens more stable in the frame (Figure 8-3). The lens performed much like the

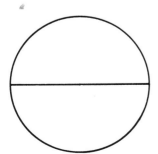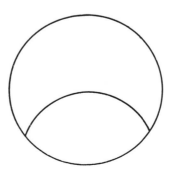

FIGURE 8-1. **Benjamin Franklin's bifocal.**

FIGURE 8-3. **The Perfection bifocal**

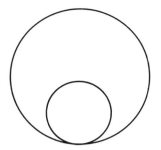

FIGURE 8-4. **The cemented bifocal.**

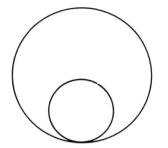

FIGURE 8-5. **The Cemented Kryptok bifocal.**

Franklin bifocal, but the distance portion was larger than the near portion. It possessed most of the disadvantages of the Franklin bifocal: The dividing line was conspicuous and tended to collect dust and dirt, and the lenses were difficult (and expensive) to manufacture.

Morck also invented the *Cemented* bifocal, filing a patent for this lens in 1888 (Figure 8-4). A thin glass wafer, having the same index of refraction as the major lens, was cemented onto the back surface of the major lens. The adhesive used was Canada balsam, whose index of refraction was essentially the same as that of glass. The front surface of the wafer had the same curvature as the back surface of the major lens and hence no power change occurred between the major lens and the wafer. The back surface of the wafer was made less concave than the back surface of the major lens, and the power of the addition was simply the difference in power between the back surface of the major lens and the back surface of the wafer.

This lens was satisfactory both optically and cosmetically, and was widely used until well after the turn of the century. However, its defects became well known. The shoulder around the dividing line collected dirt; the adherence of the wafer was affected by changes in temperature; the cement had a tendency to darken with use; and the wafer had a tendency to fall off. Cemented bifocals are still used occasionally, for special purposes such as a temporary bifocal or an experimental correction, or for a bifocal used as a low-vision aid. Epoxy resin (Araldite) is used as a cement, rather than Canada balsam.

The next step in the development of the bifocal lens was made by John Borsch who, in 1889, invented the *Cemented Kryptok* bifocal. Borsch ground a countersink curve, in the form of a depression in the front surface of the lens, then cemented a wafer made of flint glass ($n = 1.67$) into the countersink area, and covered the surface of the entire lens with a thin meniscus of glass which was cemented into place

(Figure 8-5). The name "Kryptok" was derived from the Greek word meaning "hidden." This was the first bifocal lens in which the reading addition was created by using a material having a higher index of refraction than that of the major lens. This lens was difficult to manufacture, since six surfaces had to be ground and polished. The cover plate was thin and fragile, the cemented surfaces tended to darken, and the lens would easily come apart.

In 1908 John Borsch, Jr., developed the first *fused* bifocal lens, which was also called the *Kryptok* (Figure 8-6). In making this lens, a flint bifocal segment in the form of a button was fused into a countersink area in the front surface of the major lens, under high temperature, and hence a cover glass was not needed. After the segment button was fused to the major lens, the same curvature was ground on the entire front surface of the lens. The front surface curvature was spherical, because a toric front surface would have produced a different cylindrical power in the distance portion than in the near portion of the lens. Hence, any cylindrical component had to be ground on the back surface of the lens. The Kryptok bifocal was the precursor of all modern fused bifocal lenses, and is still in use today.

The Kryptok provided many advantages over the prior bifocals. The segment edges did not collect dust

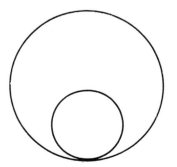

FIGURE 8-6. **The fused Kryptok bifocal.**

and dirt; the segment did not fall out, become discolored, or chip; and the segment was relatively invisible. Lenses could be produced in large quantities on a commercial scale, at low cost. The shortcomings of this lens were the result of the use of two kinds of glass and the temperature (600 to 700°C) required for fusing the segment to the major lens. In order to prevent strain from developing along the fused surface, the coefficients of expansion of the two glasses must be matched from the fusing temperature down to room temperature, and high-speed polishing of the surfaces to be fused often produces tension on the surfaces. During heating for fusion, this tension may be released, producing an irregular, undulating surface.

When competitively priced forms of this lens are made, the surfaces to be fused are deliberately not matched in curvature, with the surface of the button somewhat *steeper* than that of the countersink curve to allow air bubbles to escape during the fusing process. This procedure tends to produce a fused surface which is not perfectly spherical after the fusing is completed. If the fusing temperature is not rigidly controlled and becomes too high, the countersink surface of the major lens will collapse and become distorted. If the temperature rises too high, the flint and crown glasses will mix along the lines of fusion, forming an alloy layer which produces a hazy image.

An additional problem with the Kryptok bifocal is that *transverse chromatic aberration* is often a problem. The reading portion of a Kryptok bifocal, consisting of two optical elements having different indices of refraction, is in the form of a *doublet*. In order for a doublet to be achromatic, the dioptric powers of the two elements must be opposite in sign and the nu ($v$) values must be in the same ratio as the dioptric powers. It follows, therefore, that in order to be achromatic, a doublet having a resultant refractive power that is *positive* will consist of a convex element made of crown glass and a concave element made of a higher

index (flint) glass. Such a doublet is shown in Figure 8-7A. However, a Kryptok bifocal having a positive resultant refractive power has a form just the *opposite* of the form required for an achromatic lens—it has a convex element of flint glass and a concave element of crown glass (Figure 8-7B).

The reading portion of a Kryptok bifocal will have a positive resultant power whenever (*a*) the power of the distance portion is positive or (*b*) the power of the distance portion is negative but is weaker than the power of the addition. In these instances, noticeable amounts of chromatic aberration may be present with the result that the wearer may complain of seeing color fringes around objects; in addition the chromatism may reduce the depth of focus at the reading point.

On the other hand, in order for a doublet having a resultant *minus* power to be achromatic, it must consist of a concave crown glass and a convex glass of higher index of refraction: This is indeed the case for any Kryptok bifocal lens in which the minus power of the distance portion of the lens is higher than the plus power of the add; so, for such lenses, chromatic aberration is not likely to present a problem.

An important improvement in the design of fused bifocals was made by Henry Courmettes, a French national living in New York. In 1915, Courmettes patented the idea of fusing into a major lens a button (segment) that was made of two kinds of glass. The upper portion of the button was identical in index of refraction to that of the major lens, while the lower portion had a higher index of refraction. The two pieces of glass were ground so that they had perfectly flat edges, resulting in a straight dividing line after they had been fused along the flat edges. As shown in Figure 8-8, the composite button was subsequently fused into the countersink depression of the major lens. After the fused composite button was ground and polished, the upper half of the button could no longer be seen (being made of glass having the same index as the major lens), leaving visible only the boundary between the crown glass and the higher index glass in the lower part of the composite but-

FIGURE 8-7. **A,** an achromatic doublet, consisting of a convex crown glass element and a concave flint glass element; **B,** a Kryptok bifocal, whose reading portion consists of a concave crown glass element and a convex flint glass element.

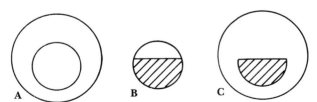

FIGURE 8-8. **The Courmettes fusing process. A,** major lens with countersink curve; **B,** button made of two kinds of glass; **C,** the finished straight-top bifocal lens.

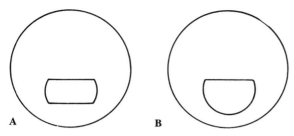

FIGURE 8-9. **A, the first straight-top fused bifocal, having a B segment; B, the D-style straight-top fused bifocal.**

ton. Using the Courmettes process, all of the fused straight-top bifocals and trifocals now in use were designed by other manufacturers subsequent to 1915.

The first straight-top bifocal to be patented was the "B" or bar segment, shown in Figure 8-9A. It was designed by two Englishmen, Watson and Culver, and was patented in 1931. In an unusual development, four men (Clement, Price, N.H. Stanley, and V. Hancock) working independently almost simultaneously filed substantially identical patent applications for the "D"-style segment (which looks like a capital "D" lying on its back) as shown in Figure 8-9B. The patent was finally issued to N.H. Stanley in 1933. The "R" or ribbon segment was developed by Silverman and patented in 1932. The *R-Compensated* series of lenses (designed for the compensation of vertical prismatic effects at near and consisting of 7 segments designated R-4 through R-10) was patented by Silverman in 1933.

Many other designs of fused bifocals, using the basic Courmettes process, have been developed since 1915, with only a few actually reaching the manufacturing stage. Modifications of the D style have included the *Panoptik*, having rounded corners, designed by Hammon; and the curved-top version, called the *Widesite*, designed by Price. Both of these lenses are shown in Figure 8-10.

The first *one-piece* bifocal to have any commercial success was the Solid Upcurve bifocal, already de-

scribed, introduced in 1836. In the following 70 years many patents were issued for other one-piece bifocals, but none of these was produced successfully on a commercial basis. In 1910 Charles Conner, an optician from Indianapolis, succeeded in producing a bifocal ground from a single piece of glass having a uniform index of refraction, which he called the *Ultex* bifocal. Patents were granted to Conner for both the lens and the grinding process. In 1910 the One Piece Bifocal Company, predecessor of the Continental Optical Company, was organized for the purpose of manufacturing the Ultex bifocal.

For the Ultex lens, the additional plus power for the reading segment is obtained by a change in curvature, usually on the back surface. Using a large lens blank, the back surface is ground in the form of a saucer having a round disk at the center (which forms the near portion of the lens) and a flatter or less concave curvature in the periphery (forming the distance portion). This is shown in Figure 8-11. While it is possible to grind a one-piece segment of any size, certain stock diameters are produced. For example, the *Ultex A* is ground with a central disk 38 mm in diameter. After grinding, the blank is cut into two lenses, each having a segment height of 19 mm (Figure 8-12A). If a segment height greater than 19 mm is required, only one lens (having a segment height of 32 mm) is cut from the blank. The *Ultex AA*, also called the AL (Figure 8-12B), is made in this manner.

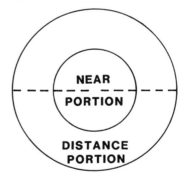

FIGURE 8-11. **Large lens blank used for grinding the Ultex one-piece bifocal, resulting in two lenses.**

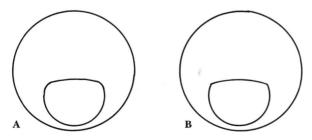

FIGURE 8-10. **Modified straight-top fused bifocals: A, the Panoptik; B, the Widesite.**

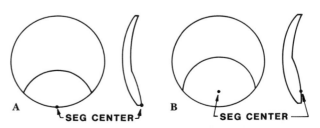

FIGURE 8-12. **Ultex one-piece bifocals: A, Ultex A; B, Ultex AL.**

Segments as small as 22 mm round (the *Ultex B*) can be produced. The Ultex-style bifocal is currently available in the A, AA (or AL), and B segments.

The next step in the development of the one-piece bifocal took place almost 50 years later when, in 1954, the Americal Optical Company developed the *Executive* bifocal. The front surface of this lens has a more convex curvature in the lower portion than in the upper portion, creating a ledge that one can easily feel, and having a straight dividing line going all the way across the lens (Figure 8-13). The Executive-style bifocal has proven to be a popular lens, and is now manufactured by a large number of companies under a variety of trade names.

The invention of the *trifocal* lens can be traced to an English inventor, John Isaac Hawkins. In 1826 Hawkins, at the age of 54, published a description of a pair of trifocal lenses that he had designed for his own use. The trifocals were made of three separate

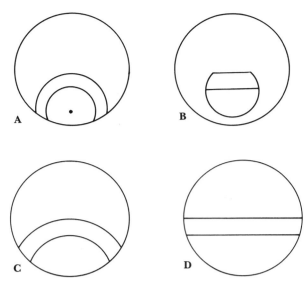

FIGURE 8-14. **Glass trifocal lenses: A, round-top fused; B, straight-top fused; C, Ultex one piece; D, Executive one piece.**

pieces of glass, and Hawkins provided detailed specifications for their manufacture. The first commercially available fused trifocal was the round-top trifocal (Figure 8-14A) which was developed simultaneously, but independently, by a number of individuals. The developers of this lens were L.W. Bugbee, Clement, Hammon, V. Hancock, and Price, with the patent being issued to V. Hancock in the year 1932. The first straight-top fused trifocal (Figure 8-14B) was developed by V. Hancock in 1936. Shortly after the development of the Ultex bifocal in 1910, Conner developed the curved-top Ultex trifocal (Figure 8-14C); and in 1954 American Optical designed the Executive trifocal (Figure 8-14D) to accompany the Executive bifocal.

## 8.2. Fused Bifocal Lenses

Fused bifocal lenses are currently available in a large variety of segment styles, all of which can be categorized as either round segments, straight-top segments, or modified straight-top segments.

### Round Segments

The original Kryptok bifocal is the least expensive bifocal available, not being a "corrected-curve" lens, and having a flint segment which, because of its small

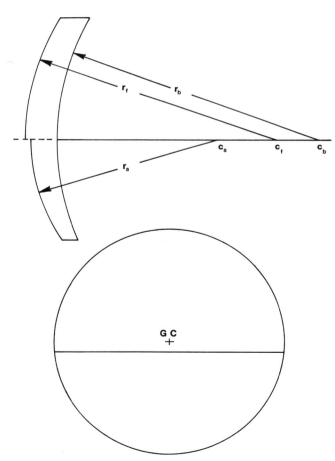

FIGURE 8-13. **The Executive one-piece straight-top bifocal. GC = geometrical center.**

nu value, often results in a large amount of chromatic aberration. The word "Kryptok" is no longer protected by a patent, so the lens can be manufactured by any company. In the past, the Kryptok was available in many segment sizes but it is now available only as a 22-mm segment.

In addition to the Kryptok, each manufacturer has developed a round segment bifocal, using the company's particular corrected curves. These lenses all use barium crown glass (rather than flint) in the segment. Because of its high nu value, barium crown has no more chromatic aberration than ophthalmic crown glass. Segments of these bifocals are routinely 22 mm in diameter with the segment optical center located at the geometrical center of the segment in the uncut form (11 mm below the segment top), as shown in Figure 8-15A. Lenses in this category are the American Optical *Tillyer D*, the Univis *Unachrome*, the Shuron-Continental *Kurova D*, and the Vision-Ease *CRF*. Representative round fused bifocals are listed in Table 8-1.

## Straight-Top Segments

Undoubtedly the most commonly used bifocal is the straight-top fused bifocal, first developed by the Univis Lens Company of Dayton, Ohio. The original Univis *Sentinel D* was made, as already described, by fusing together a truncated round high-index segment and a small crown glass segment, and fusing the resulting segment into the countersink area. The upper edge of the straight top segment is located 6 mm below the 22-mm circle, with the result that the segment center is located 5 mm below the segment top (Figure 8-15B).

When the Univis patent expired, other lens manufacturers introduced straight-top bifocal lenses. These lenses include the American Optical *Tillyer Masterpiece S*, the Shuron-Continental *Kurova D*, and the Vision-Ease *D*. All of these lenses are now available in a 22 × 16 mm segment, a 25 × 17.5 mm

**TABLE 8-1**
**Representative Glass Bifocals, Fused**

| Shape | Name | Segment Size | Seg Top to Seg Pole |
|---|---|---|---|
| Round | Kryptok | 22 | 11 |
| | Tillyer D | 20 | 10 |
| | | 22 | 11 |
| | Orthogon F | 22 | 11 |
| | Unachrome | 22 | 11 |
| | Kurova O | 22 | 11 |
| | Vision-Ease CRF | 22 | 11 |
| Straight top | Univis Sentinal D | 22 × 16 | 5 |
| | | 25 × 17.5 | 5 |
| | | 28 × 19 | 5 |
| | Kurova D | 22 × 16 | 5 |
| | | 25 × 17.5 | 5 |
| | | 28 × 19 | 5 |
| | B&L | 22 × 16 | 5 |
| | Straight Top | 25 × 17.5 | 5 |
| | | 28 × 19 | 5 |
| | Tillyer | 20 × 15.5 | 5.5 |
| | Masterpiece S | 22 × 16.5 | 5.5 |
| | | 25 × 18 | 5.5 |
| | | 28 × 19.5 | 5.5 |
| | Vision-Ease D | 22 × 16 | 5 |
| | | 25 × 17.5 | 5 |
| | | 28 × 19 | 5 |
| | | 35 × 22.5 | 5 |
| Modified straight top | Tillyer Sovereign | 20 × 14.4 | 4.5 |
| | | 22 × 15.5 | 4.5 |
| | | 25 × 17.5 | 5.5 |
| | Kurova CT | 22 × 16 | 4 |
| | B&L Panoptik | 22 × 15 | 3.5 |
| | | 24 × 16.5 | 4.0 |
| | AO Panoptik | 24 × 16.4 | 4.5 |
| | Univis F | 22 × 16 | 4 |
| | Vision-Ease C | 22 × 16 | 4 |
| | | 25 × 17.5 | 5 |
| Ribbon segments | Univis B | 22 × 9 | 4.5 |
| | Univis R | 22 × 14 | 7.0 |
| | Kurova B | 22 × 9 | 4.5 |
| | Kurova R | 28 × 14 | 7.0 |
| | Vision-Ease B | 22 × 9 | 4.5 |
| | Vision-Ease R | 22 × 14 | 7 |
| | Vision-Ease R-Compensated | 22 × 14 | 4–10 |

segment, and a 28 × 19 mm segment. In addition, the Tillyer Masterpiece S bifocal is available in a 20 × 15.5 mm segment and the Vision-Ease D is available in a 35 × 22.5 mm segment. In all of these lenses the segment center is located 5 mm below the segment top. Representative straight-top fused bifocals are listed in Table 8-1.

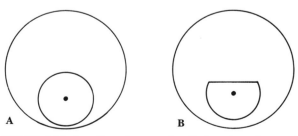

FIGURE 8-15. **Location of segment optical centers: A, round fused segment; B, straight-top segment.**

## Modified Straight-Top Segments

Following the successful introduction of the Univis D bifocal, other manufacturers developed modified forms of the straight-top bifocal. The earliest of these lenses were the Bausch and Lomb *Panoptik* and the Shuron *Widesite*, already mentioned and illustrated in Figure 8-10. The Panoptik bifocal has a 23 × 15 mm segment with slightly rounded corners, while the Widesite has a slightly curved top. More recently the Univis *F* and the Vision-Ease *C* have been introduced, similar in shape to the Panoptik; as well as the American Optical *Tillyer Sovereign* and the Shuron-Continental *Kurova CT*, similar in shape to the Widesite. The Sovereign is available in 20-, 22-, and 25-mm widths, while the Kurova CT is available only in a 22-mm width. See Table 8-1 for a list of representative modified straight-top bifocals.

## Ribbon Segments

The ribbon segment is essentially a straight-top segment with the lower part cut off (as shown in Figure 8-9A) and is designed so that the wearer will have distance vision below the segment as well as above. These lenses are available in 22 × 14 mm segments as the Univis *R* and the Vision-Ease *R*, and in a 28 × 14 mm segment as the *Kurova B*. See Table 8-1 for representative ribbon segment bifocals.

All fused bifocal segments, whether round, straight top, modified straight top, or ribbon segments, have the bifocal segment located on the *front* surface of the lens. Since the segment side of the lens must have a spherical surface, a fused bifocal lens requiring a cylindrical correction must be made in a negative toric form.

## 8.3. One-Piece Bifocal Lenses

One-piece bifocal lenses are available in both round and straight-top styles. For most round one-piece bifocals, the segment is located on the *back* surface of the lens, so if a cylindrical correction is needed it must be ground on the front surface. However, for those few one-piece bifocals in which the segment is on the front surface, a negative toric form is used (as with fused bifocals).

## Round Segments

The original Ultex A and AL, described in Section 8.1, have large, round segments, the lower parts of

which have been cut off. These are sometimes referred to as *hemispherical* segments. Both the A and AL segments are 38 mm in diameter, the A being 19 mm high and the AL being 33 mm high (in uncut form), and both having the segment on the back of the lens. An additional hemispherical one-piece bifocal is the Robinson Houchin *Hydray*, which is available in either a 40 × 20 mm or a 38 × 33 mm segment, with the segment on either the front or the back surface of the lens. Also, both the Ultex and the Hydray are available in 22-mm round segments (the Ultex having the segment on the back of the lens and the Hydray having it on the front). These lenses look very much like a Kryptok or other round fused bifocal, but the fact that they are one-piece bifocals is disclosed by feeling the change in curvature from the segment to the major lens.

## Straight-Top Segments

Straight-top one-piece bifocals now available, in addition to the Executive, are the Univis *E*, the *Kurova M*, the Vision-Ease *Bifield*, and the *Hydray EX*. The standard height (in the uncut form) for all of these segments is 25 mm, with the exception of the Hydray, which is 29 mm high. Representative one-piece bifocals are listed in Table 8-2.

TABLE 8-2
**Representative Glass Bifocals, One Piece**

| Shape | Name | Segment Size | Seg Top to Seg Pole |
|---|---|---|---|
| Round | Kurova Ultex B | 22 | 11 |
| | RH Hydray CX B (seg on front) | 22 | 11 |
| Hemispheres | Kurova Ultex A | 38 × 19 | 19 |
| | Kurova Ultex AL | 38 × 33 | 19 |
| | RH Hydray Cx (seg on front) | 40 × 20 | 20 |
| | RH Hydray CC (seg on back) | 40 × 20 | 20 |
| | RH Hydray ALX (seg on front) | 38 × 33 | 19 |
| | RH Hydray AL | 38 × 33 | 19 |
| Straight top | AO Executive | 25 high | 0 |
| | B&L Dualens | 25 high | 0 |
| | Univis E | 25 high | 0 |
| | Kurova M | 25 high | 0 |
| | RH Hydray EX | 28 high | 0 |
| | Vision-Ease Bifield | 25 high | 0 |

## 8.4. Double-Segment Bifocals

A double-segment bifocal lens has two segments, one below and the other above eye level, and is intended for use by electricians, painters, and others whose occupations require close work above eye level. Most of these lenses are of the straight-top variety, and are available in both fused and one-piece styles. In almost all of these lenses, the separation between the upper and lower segments is 13 mm.

The first double-segment bifocal to be introduced was the Univis *Double D*, which is now available in either 22- or 25-mm segment widths. Vision-Ease also makes both 22- and 25-mm Double D bifocals; American Optical makes both the Tillyer S and the Executive in double bifocal form; Robinson Houchin makes the Hydray DS (a one-piece bifocal similar in appearance to the Double Executive); and Shuron-Continental makes the Kurova double bifocal in a large number of combinations (for example, double D, lower D segment with upper B segment, and lower D segment with upper R segment).

With one exception, the power of the addition in the upper segment is equal to that of the addition in the lower segment. The exception is the Tillyer Double Executive, in which the upper segment has approximately two-thirds the power of the lower segment. For example, for a lower segment power of +1.50 D add, the upper segment has an addition of +1.00 D, and for a lower segment having an add of +2.50 D the upper add is +1.75 D. Representative double-segment bifocals are shown in Figure 8-16 and are listed in Table 8-3.

## 8.5. "Minus Add" Bifocal

The "minus add" bifocal is a lens designed predominantly for near work but with a relatively small distance window at the top. Although not widely used, this lens should be considered when prescribing for a

TABLE 8-3
**Representative Glass Bifocals, Double Bifocals**

| Name | Shape | | Separation |
|------|-----|--------|------------|
| | Top | Bottom | |
| Kurova D-D | D | D | 13 |
| Kurova D-B | D | B | 13 |
| Kurova D-R | D | R | 13 |
| Kurova D-O | D | O | 13 |
| Kurova O-O | O | O | 13 |
| Kurova Double M | M | M | 13 |
| Univis Double D-22 | D | D | 13 |
| Univis Double D-25 | D | D | 13 |
| RH Hydray DS | (Double straight top) | | 14 |
| Tillyer S Double Seg 22 | S | S | 13 |
| Tillyer Double Executive | (Double straight top) | | 14 |
| Tillyer D Sovereign | D | D | 13 |
| Vision-Ease Double D-22 | D | D | 13 |
| Vision-Ease Double D-25 | D | D | 13 |

presbyope whose occupation requires a large field for near work, such as a barber or postal clerk. The first lens of this kind was the Solid Upcurve bifocal, described in Section 8.1. A minus add lens is currently available, in an Ultex one-piece form, as the *Rede-Rite* bifocal. Instead of the segment top being located at the upper edge of the wearer's lower eyelid (as in most bifocal fitting), when fitting this lens the top of the segment is usually located above the center of the pupil.

Additional methods of obtaining an unusually large reading field include the use of a 28-mm-high straight-top one-piece (Executive-style) bifocal, and the use of an Ultex AL bifocal which can be as high as 33 mm. Even the "standard" Executive-style bifocal can be fitted as high as 25 mm, and for most wearers this will put the segment top well above the center of the pupil. See Figure 8-17 and Table 8-4 for representative minus add lenses.

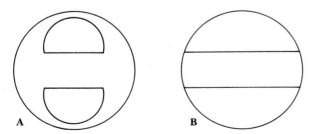

FIGURE 8-16. **Double-segment bifocals: A, Univis fused double-D; B, Tillyer Double Executive bifocal.**

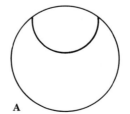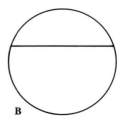

FIGURE 8-17. **"Minus add" bifocals: A, Ultex Rede-Rite bifocal; B, 28-mm-high Executive-style bifocal.**

TABLE 8-4
**Representative Glass Bifocals, Rede-Rite Style**

| Name | Segment Size | Seg Top to Seg Pole |
|---|---|---|
| Kurova Ultex Rede-Rite | 38 × 19 | 19 |
| RH Clear Vue | 40 × 20 | 20 |

## 8.6. Trifocal Lenses

A presbyope who requires a near addition of +1.75 D or more is likely to have difficulty seeing at "arm's length" through *either* the distance or the near portion of his bifocal lenses: For arm's length vision the near point of accommodation through the *distance* portion of the lens is beyond the object of regard, whereas through the *near* portion of the lens the far point of accommodation is too close for the object of regard. Clear vision at the arm's length "gap," however, can be provided by the use of a trifocal lens. First introduced by Univis as *Continuous Vision* lenses, trifocal lenses have an intermediate segment located just above the near segment (as described in Section 8.1) whose power is normally one-half that of the reading addition.

In the original Univis Continuous Vision lens, the height of the intermediate segment was 6 mm. Later, an 8-mm "occupational" segment was introduced; but at the present time almost all trifocals have a 7-mm-high intermediate segment. Trifocal lenses are available in straight-top styles, in both fused and one-piece forms. Univis, American Optical, Vision-Ease, and Shuron-Continental all make a fused straight-top trifocal lens; Americal Optical, Vision-Ease, and Robinson Houchin all make an Executive-style one-piece trifocal. Representative trifocal lenses are illustrated in Figure 8-14 and are listed in Tables 8-5 and 8-6.

## 8.7. Plastic Multifocal Lenses

In recent years plastic multifocal lenses have taken an increasing share of the ophthalmic lens market. As demand has increased, additional styles have become available. At the present time all plastic multifocals are of the *one-piece* construction. The multifocal segment is usually located on the front side of the lens. The lenses are produced either in the finished form (both sides finished) or in the semifinished form, with only the front side molded by the manufacturer. When only the front surface has been finished, the

TABLE 8-5
**Representative Glass Trifocals, Fused**

| Name | Segment Width | Height of Intermediate |
|---|---|---|
| AO Sovereign | 24 | 7 |
| AO S-22 | 22 | 6 |
| AO S-724 | 24 | 7 |
| B&L Straight Top | 23 | 7 |
| B&L Panoptik | 24 | 6.8 |
| Kurova Straight Top | 22 | 6 |
|  | 23 | 7 |
|  | 25 | 7 |
|  | 28 | 7 |
| Univis D | 23 | 7 |
|  | 25 | 7 |
|  | 28 | 7 |
| Univis D Vocational[a] | 25 | 7 |
| Vision-Ease ST | 22 | 6 |
|  | 28 | 6 |
|  | 23 | 7 |
|  | 25 | 7 |
|  | 28 | 7 |
|  | 35 | 7 |
|  | 22 | 8 |
|  | 28 | 8 |
|  | 28 | 10 |
|  | 35 | 10 |
| Vision-Ease C | 24 | 7 |

[a]Intermediate addition is 40% for +1.50 and +1.75 D adds, 60% for +2.00 and +2.25 D adds, and 70% for +2.50, +2.75, and +3.00 D adds.

TABLE 8-6
**Representative Glass Trifocals, One Piece**

| Name | Total Width | Width of Near Area | Height of Intermediate |
|---|---|---|---|
| AO Executive | — | — | 7 |
| B&L Dualens | — | — | 7 |
| Kurova M | — | — | 7 |
| Univis E | — | — | 7 |
| Vision-Ease Trifield | — | — | 7 |
| Kurova Ultex A | 47–49 | 32 | 7 |

laboratory has the job of finishing the back surface to the required prescription.

The traditional *thermosetting* plastic (such as CR-39) multifocals include those having round segments, which are available in 22-, 24-, 25-, 38-, and 40-mm diameters. Straight-top segments are available in 22-, 25-, 28-, 35-, and 40-mm widths. Full-field Executive-style bifocals are made available by several manufacturers.

A wide variety of thermosetting plastic *trifocals* are also available. The most common types are the straight-top, round-bottom trifocals, available in 7 × 22 mm, 7 × 28 mm, and 8 × 35 mm segments. In addition, the Executive-style trifocal is available with a 7-mm intermediate segment. Special trifocals include the concentric 40-mm round, 8-mm intermediate segment trifocal by Younger, the CRT trifocal by American Optical and Orcolite, and the ED trifocal, made by Sola.

The CRT trifocal (Figure 8-18A) has a 14-mm-high intermediate segment, and was designed for the growing number of video computer terminal operators. Because of the very generous vertical field of view, this lens is suitable for any occupation in which a high percentage of near work must be done at an intermediate distance. The ED trifocal (Figure 8-18B) is an Executive-style lens with a D segment. This lens has the advantage that the intermediate field extends not only 8 mm above the bifocal segment but also below the bifocal and on either side.

Polycarbonate *thermoplastic* multifocal lenses have recently been introduced. Gentex manufactures a lens having a straight-top 28-mm-wide bifocal segment and one having an Executive-style segment. American Optical and Vision-Ease provide a 28-mm straight-top bifocal segment. Gentex also makes a

7 × 25 mm straight-top, round-button trifocal. Representative plastic bifocal, double bifocal, and trifocal lenses are listed in Tables 8-7, 8-8, and 8-9.

**TABLE 8-7**
**Representative Plastic Bifocals**

| Material | Style | Manufacturer | Width |
|---|---|---|---|
| CR-39 | Round | American Optical | 22 |
| | | Armorlite | 22 |
| | | Coburn | 22, 40 |
| | | Robinson-Houchin | 40 |
| | | Signet | 22, 24, 38 |
| | | Silor | 22 |
| | | Sola | 22, 24, 38 |
| | | Titmus | 22 |
| | | Vision-Ease | 22, 25, 40 |
| | | Younger | 22, 40 |
| CR-39 | Straight top (D style) | American Optical | 25, 28 |
| | | Armorlite | 22, 25, 28, 35 |
| | | Coburn | 22, 25, 28, 35 |
| | | Orcolite | 25, 28, 35 |
| | | Signet | 22 28 |
| | | Silor | 25 28, 35 |
| | | Sola | 25 28, 35 |
| | | Titmus | 25 28, 35 |
| | | Vision-Ease | 22, 25 |
| | | Younger | 25, 28, 35, 40 |
| CR-39 | Modified straight top | Vision-Ease | 25, 28 |
| CR-39 | Executive style | American Optical | |
| | | Armorlite | |
| | | Orcolite | |
| | | Sola | |
| | | Titmus | |
| | | Vision-Ease | |
| Polycarbonate | Straight top | American Optical | 28 |
| | | Gentex | 28 |
| | | Vision-Ease | |
| Polycarbonate | Executive style | Gentex | |

**TABLE 8-8**
**Representative CR-39 Plastic Double Bifocals**

| Style | Manufacturer | Width, Upper Seg/ Lower Seg | Separation |
|---|---|---|---|
| Double executive style | American Optical | | 14 |
| | Orcolite | | 14 |
| | Sola | | 10 |
| Double straight top | Sola | 25/28 | 12 |
| | Sola | 25/28 | 15 |
| | Vision-Ease | 25/25 | 13 |
| | Younger | 28/28 | 14 |

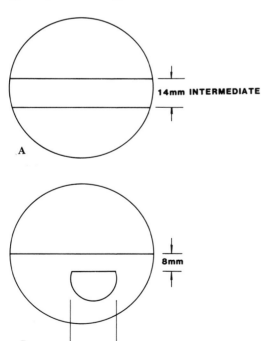

FIGURE 8-18. **Unusual plastic trifocal lenses: A, CRT, having a 14-mm intermediate segment; B, ED, an Executive-style lens with a D segment.**

TABLE 8-9
**Representative Plastic Trifocals**

| Material | Style | Manufacturer | Size |
|---|---|---|---|
| CR-39 | Straight top | American Optical | 7-25, 7-28, 8-35 |
| | | Armorlite | 7-25, 7-28 |
| | | Coburn | 7-25, 7-28, 8-35 |
| | | Orcolite | 7-25, 7-28 |
| | | Signet | 7-25, 7-28, 8-35 |
| | | Silor | 7-25, 7-28, 8-35 |
| | | Sola | 7-25, 7-28, 8-35 |
| | | Titmus | 7-25, 7-28 |
| | | Vision-Ease | 7-25, 7-28 |
| | | X-Cel | 7-25, 7-28, 8-35 |
| | | Younger | 7-25, 7-28 |
| CR-39 | Executive style | American Optical | 7 intermediate |
| | | Armorlite | 7 intermediate |
| | | Orcolite | 7 intermediate |
| | | Sola | 7 intermediate |
| CR-39 | CRT (Exec.) | American Optical | 14 intermediate |
| | | Orcolite | 14 intermediate |
| CR-39 | ED (Exec./D) | Sola | 8 intermediate |
| | | | 25 D near seg |
| CR-39 | Round top | Aire-O-Lite | 7-22 |
| | | Younger | 8-32 |
| Polycarbonate | Straight top | Gentex | 7-25 |

# MULTIFOCAL LENS MANUFACTURING PROCESSES

## 8.8. Glass Multifocals

The manufacturing process for multifocal lenses made of glass will be discussed in terms of processes for lenses of (1) fused and (2) one-piece construction. These processes are illustrated in Figures 8-19 through 8-22.

1. *Fused multifocals* (Figure 8-19) are constructed of two or more separate pieces of glass which are fused together at a high temperature. A blank is selected for the major portion of the lens, and the concave side is ground to match the curvature of the fusing slab upon which it rests. This prevents sagging of the blank's convex surface during fusion, and prevents changes in curvature of the countersink surface. A countersink or depression curve is then ground and polished over the reading area, usually on the convex surface of the major lens. The curvature, depth, and position of the countersink are pre-

cisely determined, to provide a segment with the desired amount of power in the reading addition. The countersink is usually made slightly larger than the finished segment diameter.

The segment button is made of a single piece of glass for *round* bifocals, or two truncated pieces of glass for *straight-top* bifocals, and of three truncated pieces for *ribbon* bifocals and *straight-top trifocals*. If the segment button is composed of more than one piece

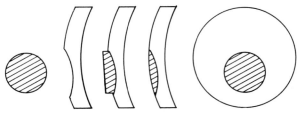

FIGURE 8-19. **Manufacturing process for a round fused bifocal lens such as the Kryptok.**

of glass, the truncated pieces are fused along their thin edges.

In making round fused bifocals, the segments are usually made of barium crown or flint glass, having a higher index of refraction than the major lens. In straight-top bifocals, the upper part of the segment button has the same composition and index of refraction as the major lens, and the lower portion is of barium crown (Figure 8-20). In trifocals the upper portion of the segment button is identical to the major lens, the intermediate portion has a higher index of refraction than the major lens, and the lower portion has a still higher index. In making ribbon segments, the upper and lower portions are identical to the major lens, and the middle portion has a higher index than the major lens.

If the segment button is composed of more than one piece of glass, the truncated pieces are fused at their edges to form the composite segment button. Some manufacturers coat the truncated edges with pigmented coatings before fusion, to help reduce the reflections from the segment line. The segment button is then ground and polished to precisely match the curvature of the countersink surface. The segment button and countersink surfaces of the major lens are meticulously cleaned, and are then assembled for fusing. The assembled blank then goes to the furnace room where it is heated until fusion occurs, and is then slowly annealed and cooled to room temperature.

A generator is used for the front surface, to remove excess glass from the protruding button, to obtain the desired front surface curvature, and to reduce the segment to its approximate finished size. The front surface is then finely ground, smoothed, and polished to a precise finish. A final inspection is conducted, to verify the accuracy of all prior procedures.

The curvature placed on the front surface is always spherical, and if a cylinder is required it is placed on the back surface, in negative toric form. If the front surface were to be made toric, not only would there be a difference in the cylinder power in the distance and near portions (due to the difference in refractive indices) but for larger cylinders a toric front surface would alter the *shape* of the segment.

Once the front surface has been finished, the blank is in the *semifinished* form, which is the common form supplied to optical prescription laboratories. The back surface is then finished, by the laboratory, to the requirements of the prescription. Since the front surface is *fixed*, on a semifinished blank, the curvature of that surface is often referred to as the *base curve*. Upon completion of the front surface of a *straight-top* fused multifocal lens, the ophthalmic crown portion of the segment becomes indistinguishable from the major lens. As already described in the discussion of the Courmettes design, it blends into the major lens because it has the same index of refraction. The border between the major lens and the higher index portion of the major lens is visible because of the difference in the *reflectivity* of the two types of glass.

2. *One-piece glass multifocals* are made from a single homogeneous piece of glass, usually ophthalmic crown glass having an index of refraction of 1.523 and a nu value near 59. The increase in plus power in the reading portion is provided by a change of curvature on either the front or the back surface of the lens. Ingenious machinery has been designed to achieve the desired accuracy.

*Round*, or Ultex-style, one-piece bifocals are made by starting with a rough blank in the form of a large saucer-shaped disk molded to the approximate curves to be ground (Figure 8-21). The molded saucers are then ground and polished on the surface opposite the segment side, to aid final inspection of the dividing line between the distance and near portions of the lens. The blanks are then prepared for grinding the segment side. The surface containing the segment is ground by the use of special abrasive wheels consisting of two adjoining curves which grind the peripheral (distance) and the central (reading) areas on the same surface. The difference in curvature between these two areas determines the power

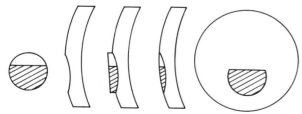

FIGURE 8-20. **Manufacturing process for a straight-top fused bifocal lens.**

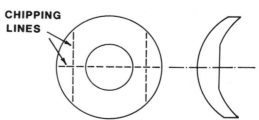

FIGURE 8-21. **Manufacturing process for an Ultex-style one-piece bifocal lens.**

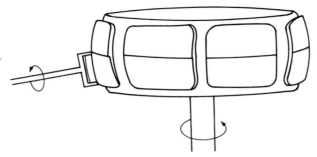

FIGURE 8-22. **Manufacturing process for an Executive one-piece bifocal lens (re-drawn from Int. Ophthal. Clin., Vol. 11, No. 1, p. 234, 1971).**

of the bifocal addition. The abrasive wheel used for trifocals consists of three curves, each cutting its own area automatically. Thus, the abrasive wheel simultaneously cuts the two fields of a bifocal lens or the three fields of a trifocal lens, as it rotates abouts its center.

Each area on the front surface is then smoothed, fined, and polished. When the segment is placed on the back surface of the lens, the segment area has a longer radius of curvature (less minus power) than the distance portion. However, when the segment is on the front surface, the segment has a shorter radius (more plus power) than the distance portion. Any cylinder is, of course, ground on the side opposite the segment. The base curve is the surface power of the distance area on the segment side of the lens.

As for the *straight-top*, or Executive-style, one-piece multifocals, the production of these lenses is initiated by mounting molded blanks on a truncated sphere (Figure 8-22). A rotating cup tool generates and polishes the reading segment surface as the block and the mounted lens rotate in a circular path around a fixed axis of revolution. Since the generating and polishing tool is restricted to the lower portion of the front surface of the lens, a straight dividing line is created between the distance and reading portions of the lens. The reading area will have a shorter radius (more plus power) than the distance area. Any cylindrical power will be ground on the back surface, with the base curve of the lens designated as the front surface power in the distance portion of the lens.

## 8.9. Plastic Multifocals

Most plastic multifocal lenses in the United States are made of either CR-39 (thermosetting) or polycarbonate (thermoplastic) material. Lenses made from these two materials are made by the use of somewhat different methods.

*Thermosetting* multifocal lenses are commonly made by pouring liquid monomer, to which a suitable catalyst has been added, into glass casting molds. When heated, polymerization takes place. To obtain precise lens curvatures and smooth surfaces, the molds are usually made of glass, with their mold surfaces highly polished—polished more meticulously than the glass surfaces of spectacle lenses.

The process of manufacturing the segment surface of such a mold is unusual. The surface of the distance portion is generated first, on an ophthalmic crown glass blank which then undergoes chemical tempering. Ordinarily, toughening would follow the completion of the lens, but if the blank were to be toughened after the reading surface had been finished, the distance surface would likely become distorted. After polishing the distance surface, the depressed segment surface is cut and polished. The mold retains the properties of a toughened lens, because the toughened layer extends beyond the segment depression surface.

For economic reasons, molds usually are not made for lens prescriptions that fall outside the normal usage range. For these prescriptions, a cast CR-39 blank is individually surfaced by conventional methods. In making semifinished blanks, the surface containing the segment (usually the front surface) is cast and the second surface is cut and polished at the prescription laboratory by the use of methods similar to those used in finishing glass lenses. For thermosetting lenses that are finished by the manufacturer, *both* surfaces of the lens are cast.

*Thermoplastic* polycarbonate lenses are made by means of a process known as *injection molding*. Beginning with the material in solid form, it is melted down and injected into a steel mold at a temperature of approximately 400°C. In the injection process the polycarbonate will assume, under pressure, the same configuration as the highly polished steel mold. The mold is made to squeeze the lens in order to prevent shrinkage and to ensure accuracy of the surfaces. Each lens-forming cycle requires approximately 90 to 130 seconds.

The lenses are then inspected and sent through a coating machine. The surfaces of polycarbonate material are soft and are usually susceptible to scratches—far more so than the surfaces of CR-39 material. Therefore, all polycarbonate lenses are coated with an *organosilicon* material which greatly increases the surface hardness. Although the first coatings used on polycarbonate lenses could not be tinted, the coatings currently in use may be tinted.

# OPTICAL PRINCIPLES OF MULTIFOCAL LENS DESIGN

## 8.10.  Powers of the Distance and Reading Portions

The total (or resultant) refracting power in the reading portion of a bifocal lens consists of the power of the distance prescription combined with the power of the bifocal addition, or "add." This is shown in the following examples:

> Power of distance portion: +1.00 DS −0.50 DC x 180.
> Power of reading addition: +2.00 DS
> Power of reading portion: +3.00 DS −0.50 DC x 180.

The formula for a bifocal lens is written so as to represent the formula for the distance portion of the lens combined with the power of the reading addition, as shown below (for the right lens):

> OD +1.00 DS −0.50 DC x 180, +1.00 D add.

When this formula appears in a lens prescription (without further instructions) it may be filled with a bifocal lens or with two single-vision lenses, one for distance vision and one for near vision. The power of the lens for near vision would correspond to the power of the near portion of the bifocal prescription. In writing an order for a pair of lenses, the power of the distance portion is referred to as the *distance* prescription, and the power of the reading portion is referred to as the *near* prescription.

## 8.11.  One-Piece Bifocals

### Cemented and Ultex-Style Bifocals

The powers of the distance and reading portions of cemented and Ultex-style one-piece bifocals may be analyzed in the following manner. In a sagittal section of a cemented or Ultex-style one-piece bifocal, the lens can be considered as being made of two components: (1) the major lens and (2) the segment. For a cemented bifocal, the *wafer* is obviously the segment. In the case of a one-piece bifocal, both the major lens and the segment are constructed from a solid piece of glass with the segment usually on the back surface. Of course, in such a bifocal, the surface dividing the major lens and the segment is purely *imaginary*, but it can be represented by a continuation

FIGURE 8-23. **An Ultex-style bifocal, showing the imaginary surface dividing the segment and the major lens in the near portion of the lens.**

of the back surface of the distance portion of the lens (Figure 8-23).

The major lens of a cemented or Ultex-style one-piece bifocal encompasses both the distance field and the reading field, but the segment encompasses only the reading field. The glass in the segment has the same index of refraction (1.523) as the glass in the major lens.

For a cemented bifocal, the front side of the segment (which is cemented onto the back of the major lens) has a power which corresponds to the power of the back surface of the major lens but is opposite in sign. The power of the major lens is that of the distance prescription. If $F_1$ is the front surface power of the distance portion of the lens, the approximate power of the distance portion is given by the formula

$$\text{approximate power of distance portion} = F_1 + F_2. \tag{8.1}$$

If $F_3$ is the front surface power of the segment and $F_4$ the back surface power of the segment, the power of the reading addition is equal to the power of the segment and is given by the formula

$$\text{power of reading addition} = F_3 + F_4. \tag{8.2}$$

However, since

$$F_3 = -F_2,$$

the power of the reading addition $= F_4 - F_2$,
$$\tag{8.3}$$

and hence the power of the reading addition can be determined directly by measuring $F_4$ and $F_2$ with a lens measure. This can be done with either a one-piece bifocal or a cemented bifocal (even after the wafer has been cemented to the lens).

In the case of a one-piece bifocal, the segment is ground in such a way that it has a thin edge, and the edge therefore represents the intersection of the two spherical surfaces and is consequently circular. The *upper* portion of the Ultex A segment therefore has a thin edge which is circular, but at the *lower* portion of the lens the edge of the lens does not extend all the way to the (imaginary) circle represented by the segment, with the result that the lower edge of the segment has a finite thickness (Figure 8-23). However, a one-piece *prism* bifocal has a round prism segment which is thick on one side and thin on the other, and the amount of prism is specified in terms of the prismatic effect at the geometrical center of the segment.

The center thickness of a round feather-edge one-piece segment can be calculated by the use of the following formula, in which $h$ represents one-half the diameter (or one-half the width) of the segment:

$$\text{center thickness} = \frac{h^2 F_{add}}{2(n' - n)} . \qquad (8.4)$$

### The Executive Bifocal

This lens can be considered as a single-vision distance lens upon which has been placed a reading segment with a power of the required addition, and whose back surface curvature matches the front surface curvature of the single-vision distance lens (see Figure 8-24). It follows that the segment could have been placed toward either the nasal or the temporal side of the lens, creating a *decentered* segment. In practice, decentration is achieved by grinding the distance center where it is required *after* the back surface of the lens has been finished. In this manner, two different curvatures are ground on the front

surface of the blank. The distance portion of the blank is ground with a preselected curvature, and a steeper curve (more plus power) is ground on the lower portion (reading area) of the blank. The difference between the powers of these two surfaces determines the reading addition.

The power of the upper (distance) portion of the lens corresponds to the power of the distance portion in ordinary bifocal terminology, and the power of the lower portion of the lens is the *total power* (distance power plus add) through the reading portion.

The centers of curvature ($C_s$ and $C_u$) of the two front surface radii ($r_s$ and $r_u$) fall on the common axis which intersects the lens surface at the point of minimum thickness of the *ledge* (see Figure 8-25). The optical center falls on this, the common axis, at the point of minimum ledge thickness. The position of the segment optical center is fixed, as in the case of any semifinished bifocal blank. When the back surface of the reading portion of the lens is finished (surfaced), the position of the optical center of the segment is not changed, remaining as the point of minimum ledge thickness on the front surface. It can, however, be decentered horizontally or vertically, as with any bifocal lens, by specifying its position in relation to the major reference point on the distance portion of the finished lens.

The *base curve* of the lens is the surface power of the distance portion on the segment side of the lens.

## 8.12. Fused Bifocals

In the case of the fused bifocal, there are three surfaces to be considered: (1) the front surface of the major lens, which is continuous with the front sur-

FIGURE 8-24. **An Executive-style bifocal considered as a single-vision distance lens with a reading segment.**

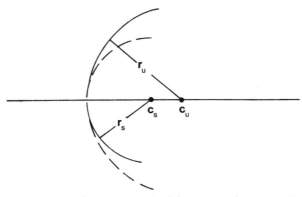

FIGURE 8-25. **The common axis of the centers of curvature ($C_s$ and $C_u$) of the two front curves ($r_s$ and $r_u$) of an Executive-style bifocal.**

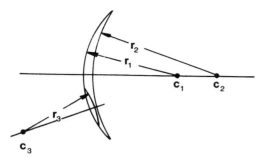

FIGURE 8-26. **The three radii of curvature in the reading portion of a fused bifocal.**

face of the segment and has a radius of curvature $r_1$; (2) the back surface of the major lens, whose radius of curvature is denoted by $r_2$; and (3) the back surface of the segment which contacts and becomes identical with the countersink surface of the major lens, and whose radius of curvature is $r_3$ (see Figure 8-26).

If we define the index of refraction of the glass comprising the major lens as $n'$, the index of the glass comprising the segment as $n''$, the index of air as $n$, and if we assume for the moment that all three surfaces are spherical, the powers of the distance and reading portions of the lens can be represented by the following equations:

power of the distance portion

$$F_D = \frac{n' - n}{r_1} + \frac{n - n'}{r_2}. \qquad (8.5)$$

power of the reading portion

$$F_A = \frac{n'' - n}{r_1} + \frac{n' - n''}{r_3} + \frac{n - n'}{r_2}. \qquad (8.6)$$

Fundamentally, the power of the reading addition is equal to the power of the reading portion minus the power of the distance portion, and is given by the following formula:

power of reading addition

$$F_A = \frac{n'' - n}{r_1} + \frac{n' - n''}{r_3} + \frac{n - n'}{r_2} - \frac{n' - n}{r_1} - \frac{n - n'}{r_2}$$

$$= \frac{n'' - n}{r_1} + \frac{n' - n''}{r_3} - \frac{n' - n}{r_1}$$

$$= \frac{n'' - n}{r_1} - \frac{n' - n}{r_1} + \frac{n' - n''}{r_3}$$

$$F_A = \frac{n'' - n'}{r_1} + \frac{n' - n''}{r_3}. \qquad (8.7)$$

The term $(n'' - n')/r_1$ is equivalent to the power of the front surface of the segment covered by the material of the major lens, and the term $(n' - n'')/r_3$ is the power of the fused surface between the segment and the major lens.

This condition actually existed for the *cemented Kryptok* bifocal in which the segment (instead of being fused into the countersink) was cemented onto the front surface of the lens, and the whole front surface was covered with a concentric shell of glass having the same index as the major lens.

**Important Note:** Caution must be taken not to assume that the power of the segment of a fused bifocal in air is equal to the power of the reading addition, as in the case of the wafer of a cemented bifocal or the imaginary segment of a one-piece bifocal.

It is apparent that the reading addition of a fused bifocal lens is obtained by the *gain in power* on the front surface of the segment as compared to that of the front surface of the distance portion *plus* the power of the fused surface between the segment and the major lens. Equation (8.7) means that the power of the reading addition would correspond to the power of the segment if it were completely embedded in glass having an index equal to $n'$. Clearly, the gain in power on the front surface of the segment over that of the front surface of the distance portion is equal to the front surface power of the segment embedded in the major lens.

If every term of Eq. (8.7) is multiplied by the term

$$\frac{n' - n}{n'' - n'},$$

we obtain

$$\frac{n' - n}{n'' - n'} F_A = \left(\frac{n' - n}{n'' - n'}\right)\left(\frac{n'' - n'}{r_1}\right)$$

$$+ \left(\frac{n' - n}{n'' - n'}\right)\left(\frac{n' - n''}{r_3}\right)$$

$$\frac{n' - n}{n'' - n'} F_A = \frac{n' - n}{r_1} + \left(\frac{n' - n}{n'' - n'}\right)(-1)\left(\frac{n'' - n'}{r_3}\right)$$

$$\frac{n' - n}{n'' - n'} F_A = \frac{n' - n}{r_1} - \frac{n' - n}{r_3}.$$

The quantity $(n' - n)/r_1$ represents the power of the front surface as measured by means of a lens measure, and may be designated as $F_1$; the quantity $(n' - n)/r_3$ represents the power of the countersink curve (in air) as measured with a lens measure, and is designated as $F_C$. Therefore,

$$\frac{n' - n}{n'' - n'} F_A = F_1 - F_C. \tag{8.8}$$

Solving for $F_C$,

$$F_C = F_1 - \frac{n' - n}{n'' - n'} F_A, \tag{8.9}$$

and solving for $F_1$,

$$F_1 = F_C + \frac{n' - n}{n'' - n'} F_A. \tag{8.10}$$

Equation (8.9) is useful in calculating the countersink curve required to obtain a given reading addition when $F_1$ has been specified; Eq. (8.10) is used to calculate $F_1$ when the countersink curve and the reading addition are known.

### EXAMPLE 1

A 20-mm round bifocal segment is fused into a concave countersink surface having a radius of curvature of 50 cm, on the front surface of the lens. The index of refraction of the distance portion of the lens is 1.51 and the index of refraction of the segment is 1.68. In order to obtain a $+2.50$ D add, what must the front surface power be?

$$F_1 = F_C + \frac{n' - n}{n'' - n'} F_A$$

$$= F_C + \frac{1.51 - 1.00}{1.68 - 1.57} (2.50)$$

$$F_C = \frac{n' - n}{r_C} = \frac{1.51 - 1.00}{-0.5} = -1.02$$

$$F_1 = -1.02 + \frac{0.51}{0.17} (2.50)$$

$$= -1.02 + 7.50$$

$$= +6.48 \text{ D}.$$

### EXAMPLE 2

The refractive index of a fused front surface segment is 1.60 and the refractive index of the distance portion (carrier lens) is 1.50. The power of the front surface is $+6.00$ D. and the power of the countersink surface (in air) is $-1.50$ D. What is the power of this add?

$$\frac{n' - n}{n'' - n'} F_A = F_1 - F_C$$

$$\frac{1.50 - 1.00}{1.60 - 1.50} F_A = 6.00 - (-1.50)$$

$$5 F_A = +7.50$$

$$F_A = \frac{+7.50}{5}$$

$$= +1.50 \text{ D}.$$

For a *finished lens* received from the laboratory, it is obviously no longer possible to obtain a measurement of the countersink curve with a lens measure. In this case, in order to find the power of the add it is necessary (*a*) to measure the power of the reading portion by an optical method; (*b*) to measure the power of the distance portion either with a lens measure or by an optical method; and (*c*) to subtract the power of the distance portion from the power of the reading portion. In practice, the spherical or spherocylindrical formula of the major lens is determined by the use of a *lensometer*, following which the lensometer is used to determine the spherical power of the reading portion. In order to find the power of the add, the spherical power of the distance portion is subtracted from the spherical power of the reading portion.

Both the front and back surfaces of the segment are spherical, with the result that, for a *round* fused bifocal, a feather-edge segment can be obtained by letting these two surfaces intersect at the boundary of the segment (see Figure 8-27A). The optical center lies at the geometrical center of the segment, which is also the segment reference point.

The upper edge of the segment of a *straight-top* fused bifocal such as the Univis D is relatively *thick*, but the remainder of the segment boundary is circular and progresses to a feather edge at the bottom of the segment (Figure 8-27B). The Panoptik, Widesite, and Fulvue bifocals also have a thick edge at the top and a thin edge all the way around the circular portion of the contour. In each of these cases, the segment optical center lies at the center of the circular portion of the segment, which is also the reference point of the segment.

As stated above, equations (8.9) and (8.10) indicate that the curvature of the countersink curve is used to obtain the desired amount of add in the finished lens. Since the power of the add is independent of the power of the back surface of the lens, the front surface of a bifocal blank can be finished at the factory; the grinding and polishing of the back surface, to meet the requirements of the patient's prescription, can be left to the laboratory. Semifinished bifocal blanks with adds from $+0.50$ to $+3.00$ D or more, in 0.25 D steps, are stocked by prescription laboratories, and a number of manufacturers will supply blanks with higher adds on special order.

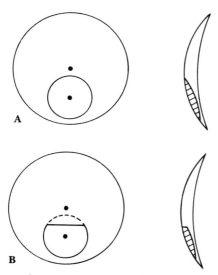

**A**

**B**

FIGURE 8-27. **(A) With a round fused bifocal, a feather edge can be obtained; (B) with a straight-top fused bifocal, the upper edge of the segment is relatively thick but a feather edge can be obtained at the bottom of the segment.**

The term

$$\frac{n' - n}{n'' - n'}$$

is sometimes referred to as the "ratio." The numerical value of this ratio is determined by the selection of glasses to be used in the two parts of the lens: For ophthalmic crown glass of index 1.523 and flint glass of index 1.69, the ratio is approximately 3/1. This means that the add will be changed by 1 D by changing the value of $F_1$ by 3 D. Rearranging Eq. (8.9),

$$F_A = \frac{F_1 - F_C}{ratio}. \qquad (8.11)$$

If the ratio = 3/1, and if $F_1 = +4.50$ D, and if $F_C = -1.50$ D, the power of the add is +2.00 D. But if $F_1$ is increased by 3.00 D to +7.50 D, the power of the add will be +3.00 D (being increased 1 D in power). Using this relationship it will be seen that if $n'' < n'$, a "minus add" will be obtained.

If enough information is known to calculate the power of the add, the thickness of the segment can easily be found by considering the segment to be completely imbedded in the distance portion of the lens. The thickness sometimes needs to be known in order to avoid grinding into the segment while finishing the back surface of the lens. If the segment is considered to be completely embedded, as in the cemented Kryptok bifocal, the center thickness can

be found by the equation

$$\text{center segment thickness} = \frac{h^2 F_A}{2(n'' - n')}. \qquad (8.12)$$

**EXAMPLE 3**
A 22-mm round fused-front-surface bifocal has a distance power of +2.00 D and an add of +3.00 D. The index of refraction of the distance portion of the lens is 1.54 and that of the segment is 1.63. What is the center thickness of the segment?

$$t_C - t_P = \frac{h^2 F_A}{2(n'' - n')}$$

$$t_C - 0 = \frac{(0.011)^2(+3.00)}{2(1.63 - 1.54)}$$

$$t_C = \frac{(0.000121)(3)}{2(0.09)} = \frac{0.00363}{0.18}$$

$$= 0.002 \text{ m} = 2 \text{ mm}.$$

Some segments are available with thick edges all around, permitting the production of *prism segments*. The amount of prism is specified in terms of the prismatic effect at the segment reference point.

As shown in Figure 8-28, the *vertical* position of the segment of a finished lens is specified in terms of the height of the top of the segment above the bottom of the lens, or in terms of the distance from the top of the segment to the major reference point of the distance portion of the lens. The *horizontal* position of the segment is specified in terms of the displacement of the segment center to the right or the left of the distance major reference point.

Segments having *toric* countersink surfaces could be produced, but since there is seldom an occasion for using such a lens, bifocals with segments of this type are not maintained as stock items.

For a fused bifocal lens having the segment em-

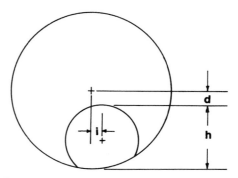

FIGURE 8-28. **Specification of the position of a fused bifocal segment, both vertically and horizontally.**

bedded in the front surface (the usual situation), the front surface of the major lens has a spherical curvature which is identical and continuous with that of the segment, whereas for a fused bifocal having the segment embedded in the back surface, the back surface is spherical and its curvature is continuous with that of the segment. As already stated, for a fused bifocal a cylinder, if required by the prescription, is ground on the side of the lens opposite the segment side.

# PERFORMANCE CHARACTERISTICS

## 8.13. Vertical Placement of the Optical Center of the Segment

In order to evaluate the performance of a given bifocal lens, we must know (in addition to the distance and near powers) the positions of the optical centers of the distance portion of the lens and the segment, and the point in the lens through which the wearer's line of sight will pass during reading or other close work.

As discussed in Section 2.17, all lenses function at their best when the optic axis of the lens passes through the center of rotation of the eye, and lens designers assume this to be the case when they specify the performance in the periphery of a lens. It follows that the level of the distance optical center of a bifocal lens, as with a single-vision lens, should be determined on the basis of the amount of *pantoscopic tilt* of the lens, the distance optical center being lowered 1 mm for every 2° of pantoscopic tilt. Since a pantoscopic tilt of 6° is usually considered to be cosmetically desirable, it follows that the distance optical center should normally be located *3 mm* below the point directly in front of the center of the pupil when the eyes are in the primary position (as shown in Figure 8-29). The position of the distance optical center of a bifocal lens is usually specified, both laterally and vertically, in relation to the *geometrical center* of the lens.

An important respect in which various bifocal styles differ from one another has to do with the distance from the segment pole to the segment top. For fused and one-piece *round* knife-edge segments, the optical center of the segment lies at the geometrical center of the segment: For example, the distance from the segment optical center to the segment top in the case of a 22-mm round fused segment is 11 mm, whereas the corresponding distance for a 38 × 19 mm Ultex A-style segment is 19 mm. These and other segment optical center locations are illustrated

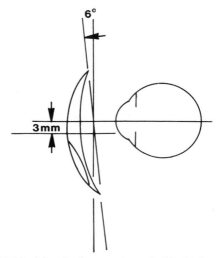

FIGURE 8-29. **A lens having a pantoscopic tilt of 6 degrees, with the distance optical center located 3 mm below the point directly in front of the center of the pupil.**

in Figure 8-30. For regular or modified *straight-top* fused bifocal segments, having thick (rather than knife-edge) segment tops, the segment optical center is located at the center of the circular contour of the segment: For example, for a 22 × 16 mm Univis D-style segment, the distance from the segment optical center to the segment top is 5 mm. For the *Executive-style one-piece* bifocal, the segment optical center is located *at* the segment top. For prism segments, having thick edges all the way around, the contour of the segment is of no assistance in locating the optical center of the segment.

For an individual who does not wear spectacles or who wears single-vision lenses, when the eyes are lowered during reading the *head* is lowered also. Consequently, for a wearer of single-vision lenses, the lines of sight for the two eyes pass through the lenses at only a slightly lower level for reading than

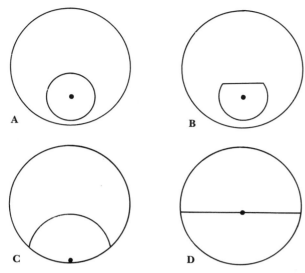

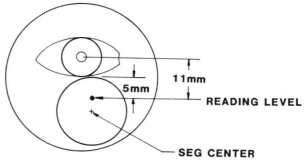

FIGURE 8-32. **A round fused bifocal, showing segment top at lower border of the iris, with reading center located 5 mm below the segment top and 11 mm below the center of the pupil.**

FIGURE 8-30. **Positions of the segment optical center: A, Kryptok segment, 11 mm below segment top; B, straight-top fused, 5 mm below segment top; C, Ultex style, 19 mm below segment top; D, Executive style, segment center at segment top.**

for distance vision. However, for a bifocal wearer, the head must be kept more or less erect, in order to see through the bifocal segments (some practitioners tell novice bifocal wearers to "point with the chin, rather than with the eyes," when reading). Fry and Ellerbrock[1] found that, during reading, the foveal line of sight passes through a point in the lens (which they called the "reading center") approximately 11 mm below the point directly in front of the center of the pupil when the eyes are in the primary position (see Figure 8-31). If the lens has a pantoscopic tilt of 6° and the distance optical center is 3 mm below the pupillary level (in order for the optic axis of the lens to pass through the center of rotation of the eye), the *reading center* will be located approximately 8 mm below the distance optical center. If we assume that the "visible iris diameter" in the vertical meridian

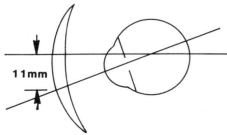

FIGURE 8-31. **The reading center, located approximately 11 mm below the point directly in front of the pupil.**

averages 12 mm and that the segment top falls at the lower border of the iris (or 6 mm below the center of the pupil), the reading center would then fall 5 mm below the top of the segment as shown in Figure 8-32. If it is considered desirable for the wearer to read through the segment, at the level of the segment optical center, the segment center should be located *5 mm* below the segment top.

## 8.14. Lateral Placement of the Optical Center of the Segment

When determining the specifications for a pair of bifocal lenses, the segment optical centers should be decentered inward from the major reference points of the two lenses, so that they will coincide with (or fall directly above or below) the reading centers. In reality, the segment optical centers seldom fall exactly at the reading centers, and may be displaced horizontally as well as vertically.

The inward displacement of the segment optical center from the distance major reference point is referred to as *segment inset*. Segment inset is always *in* unless the segment position is manipulated to provide a horizontal prismatic effect. Segments are inset for two purposes: (1) to assure that the fields of view through the two segments coincide (or overlap) for the two eyes and (2) to prevent the segment from producing any horizontal prismatic effects at the reading center. When the optical center of the segment is placed directly above or below the reading center, the segment itself produces no horizontal prismatic effect at the reading center.

The *field of view* is the area of useful vision that can be obtained with the eye held stationary, the center of the entrance pupil being the reference point: It may also be thought of as the solid angle of vision, for the stationary eye, subtended at the center of the entrance pupil. A distinction must be made between the *field of view* (as just defined) and the *field of fixation*, which is the area of object space within which direct foveal vision can be obtained by moving the eye with the head stationary. The reference point for the field of fixation is the center of rotation of the eye. The field of view and the field of fixation are illustrated in Figure 8-33.

The correct amount of segment inset depends upon:

1. The distance interpupillary separation, usually referred to as the *distance PD*
2. The distance from the back pole of the lens to the center of rotation of the eye
3. The fixation distance
4. The power of the distance correction, in the horizontal meridian

Segment inset is normally determined by measuring the distance and near interpupillary distances (using the method to be described in Chapter 9), and then insetting each of the segments one-half of the difference between the distance and near interpupillary distances. For example, if the distance PD is 64 mm and the near PD is 60 mm, the segment inset would be 2 mm for each lens.

A near PD taken without the distance-correcting lenses before the eyes, or calculated on the basis of geometrical relationships, ignores the *prismatic effects* of the distance correction (in the horizontal meridian) on the positions of the segment optical centers. For example, when a hyperope, wearing plus lenses centered for the distance PD, converges upon a near object, the visual axes move inward and pass through points in the lenses located nasally from the distance optical centers. At such points, *base-out* prismatic effects are produced, and the eyes must converge *to a greater extent* than would be necessary without the glasses. On the other hand, for a myope, wearing minus lenses centered for the distance PD, converging the visual axes produces a *base-in prismatic effect*, with the result that the eyes are required to converge *less* than would be necessary without the glasses.

Scott Sterling[2] has mathematically determined the amount of segment inset required (for a 40-cm reading distance) to make the reading fields coincide. His calculations depended upon the powers of the distance portions of the lenses, and were based on assumptions that the visual axes pass through the segment optical centers and that the distance from the lens to the center of rotation of each eye (the "stop distance") was 25 mm. The amounts of segment inset for various lens powers and interpupillary distances, as determined by Sterling, are shown in Table 8-10. The use of this table is recommended.

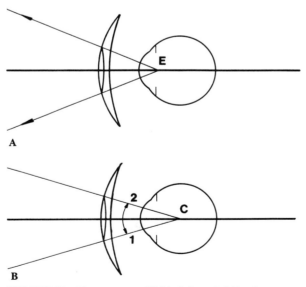

FIGURE 8-33. **The concepts of field of view of a bifocal segment: A, peripheral field of view, subtended at the center of the entrance pupil, E; B, field of fixation, or macular field of view, subtended at the center of rotation, C.**

TABLE 8-10
**Segment Inset Required (for Each Bifocal Segment) to Make the Two Reading Fields Coincide at a 40-cm Reading Distance**[a]

| Distance Lens Power in 180 Meridian | Distance PD | | | | | | | | |
|---|---|---|---|---|---|---|---|---|---|
| | 56 | 58 | 60 | 62 | 64 | 66 | 68 | 70 | 72 |
| +14.00 D | 2.5 | 2.5 | 2.6 | 2.7 | 2.8 | 2.9 | 3.0 | 3.1 | 3.2 |
| +12.00 D | 2.3 | 2.3 | 2.4 | 2.5 | 2.6 | 2.7 | 2.8 | 2.9 | 3.0 |
| +10.00 D | 2.2 | 2.2 | 2.3 | 2.4 | 2.5 | 2.5 | 2.6 | 2.7 | 2.8 |
| + 8.00 D | 2.0 | 2.1 | 2.2 | 2.2 | 2.3 | 2.4 | 2.5 | 2.5 | 2.6 |
| + 6.00 D | 1.9 | 2.0 | 2.1 | 2.1 | 2.2 | 2.3 | 2.3 | 2.4 | 2.5 |
| + 4.00 D | 1.8 | 1.9 | 2.0 | 2.0 | 2.1 | 2.1 | 2.2 | 2.3 | 2.3 |
| + 2.00 D | 1.7 | 1.8 | 1.9 | 1.9 | 2.0 | 2.0 | 2.1 | 2.1 | 2.2 |
| 0.00 D | 1.6 | 1.7 | 1.8 | 1.8 | 1.9 | 1.9 | 2.0 | 2.1 | 2.1 |
| − 2.00 D | 1.6 | 1.6 | 1.7 | 1.7 | 1.8 | 1.9 | 1.9 | 2.0 | 2.0 |
| − 4.00 D | 1.5 | 1.5 | 1.6 | 1.7 | 1.7 | 1.8 | 1.8 | 1.9 | 1.9 |
| − 6.00 D | 1.4 | 1.5 | 1.5 | 1.6 | 1.6 | 1.7 | 1.8 | 1.8 | 1.9 |
| − 8.00 D | 1.4 | 1.4 | 1.5 | 1.5 | 1.6 | 1.6 | 1.7 | 1.7 | 1.8 |
| −10.00 D | 1.3 | 1.4 | 1.4 | 1.5 | 1.5 | 1.6 | 1.6 | 1.7 | 1.7 |
| −20.00 D | 1.1 | 1.2 | 1.2 | 1.2 | 1.3 | 1.3 | 1.4 | 1.4 | 1.4 |

[a]For distances from 30 to 40 cm add 0.2 mm to the amount of inset for each 5 cm of reading distance nearer than 40 cm.
Source: Adapted from S. Sterling, Ophthalmic Lenses, Bausch and Lomb Optical Co., 1935.

*Total inset* is defined as the lateral displacement of the segment optical center from the geometrical center of the lens. When the distance portion of the lens has no power or contains only prismatic power (in the form of a plano-prism), total inset may be inward, zero, or outward, depending upon the position of the major reference point of the distance lens in relation to the geometrical center of the lens. If the major reference point falls *inward, on,* or *not farther than 1.5 mm outward* from the geometrical center of the lens, the total inset will be *in*. However, if the major reference point falls *2 mm outward* from the geometrical center, the total inset will be *zero*. If the major reference point falls *farther than 2 mm* outward from the geometrical center, the total inset will be *out*. These relationships are illustrated in Figure 8-34.

The following examples will illustrate the concepts of segment inset and total inset.

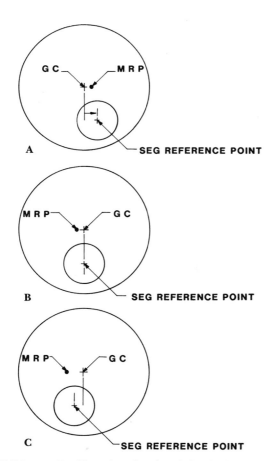

FIGURE 8-34. **Total inset, based on the relationship between the geometrical center (GC) and the major reference point (MRP) of the lens: (A) Total inset is inward (the usual situation); (B) total inset is zero; (C) total inset is outward.**

### EXAMPLE 1

Given the distance prescription −2.00 DS with +2.00 D add, OU. The frame size is 50 mm with a distance between lenses (DBL) of 18 mm, and the patient's PD is 64 mm at distance and 60 mm at near. What will be the segment inset and the total inset?

The *frame PD* will be 50 + 18, or 68 mm. Since the patient's distance PD is 64 mm, the major reference point must be decentered 2 mm inward from the major reference point, or:

$$\text{Frame PD} - \text{patient's PD} = 68 \text{ mm} - 64 \text{ mm}$$
$$= 4 \text{ mm}.$$
$$\text{Major reference points} = 2 \text{ mm in, each lens.}$$
$$\text{Segment inset} = (64 - 60)/2$$
$$= 2 \text{ mm in, each lens.}$$
$$\text{Total inset} = (68 - 60)/2$$
$$= 4 \text{ mm in, each lens.}$$

### EXAMPLE 2

Given the distance prescription +2.00 DS combined with 1 prism diopter base in with +2.00 D add, OU. The frame size is 46 mm, with a DBL of 18 mm. The patient's PD is 70/66. What will be the segment inset and the total inset?

$$\text{Frame PD} = 46 \text{ mm} + 18 \text{ mm}$$
$$= 64 \text{ mm}.$$
$$\text{Frame PD} - \text{patient's PD} = 64 \text{ mm} - 70 \text{ mm}$$
$$= -6 \text{ mm}.$$
$$\text{Major reference points} = 3 \text{ mm } out, \text{ each lens.}$$
$$\text{Segment inset} = (70 - 66)/2$$
$$= 2 \text{ mm in, each lens.}$$
$$\text{Total inset} = (64 - 66)/2$$
$$= 1 \text{ mm } out, \text{ each lens.}$$

## 8.15. Differential Displacement (Image Jump)

The optical performance of a bifocal lens can be considered in terms of:

1. Differential displacement at the segment top ("image jump")
2. Differential displacement at the reading level
3. Total displacement at the reading level
4. Chromatic aberration

Fundamentally, any modern bifocal lens can be considered as the equivalent of a cement bifocal, having a segment wafer cemented to a single-vision lens (as shown in Figure 8-35). While reading through the segment, the patient may be considered as looking through two lenses, the distance lens and the added segment wafer. In effect, a bifocal lens may have three centers: (1) the distance optical center; (2) the segment optical center; and (3) the resultant optical center, or the point of zero prism in the segment position. The resultant optical center may actually fall outside the segment area.

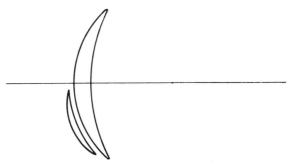

FIGURE 8-35. **A bifocal lens considered as the equivalent of a cement bifocal, made up of a single-vision lens and a segment wafer.**

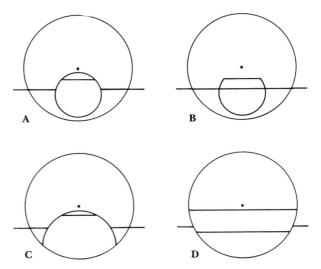

FIGURE 8-36. **Differential displacement of various bifocal lenses, when viewing a horizontal line through a bifocal lens held about 8 inches in front of the eye. The greater the vertical distance between the line in the distance portion and that in the segment portion, the greater the differential displacement.**

As discussed in Chapter 4, prismatic effects exist for a spherical lens at all points on the lens except for the optical center of the lens. Prentice's rule provides an easy formula for the calculation of the magnitudes of these prismatic effects. When the wearer looks through the segment area of a bifocal lens, it is apparent that in addition to the prismatic effect caused by the distance lens there is a prismatic effect caused by the segment itself (unless the patient looks through the optical center of the segment). The prismatic effect at any point in the segment is equal to the sum of the prismatic effects of the distance lens and the segment.

The prismatic power of the segment alone determines the amount of *image jump*, or *differential displacement*. Image jump is due solely to the segment. An effective way of demonstrating image jump at various parts of a bifocal segment is to observe a distant horizontal line through a bifocal lens held about 8 inches or more in front of the eye. Differential displacement manifests itself by the fact that the parts of the line seen through the distance portion of the lens and seen through the segment are *not continuous*. Examples of differential displacement are shown in Figure 8-36. By comparing the level of the image seen through the distance portion with that seen outside the field of the lens, it is possible to determine the prismatic effect of the distance portion. In addition, the total prismatic effect at a specified level in the reading area can be determined by comparing the level of the part of the line seen through the segment with the part seen outside the field of view of the lens.

If the prismatic power of the distance portion of the lens at a point on the segment boundary differs from the prismatic effect of the segment, the image "jumps" as the line of sight moves from the distance portion into the segment area. This "jump" is mani-

fested to the wearer by an apparent alteration in the position of an image of any viewed object, as the line of sight passes from the distance portion of the lens to the segment area, or vice versa. Image jump may occur at any point on the boundary of the segment; however, the term "jump" is usually used to describe the phenomenon occurring at the *top* of the segment, where it is particularly bothersome. The amount of jump present at the top of the segment depends entirely upon the prismatic effect of the segment at that point. It is simply the differential displacement at the top of the segment. As shown in Figure 8-37, this is a *base-down* prismatic effect, resulting in an apparent scotomatous (blind) area.

The amount of jump is independent of the power of the distance portion of the lens but, as indicated by Prentice's rule, it is equal to the distance, $d$, from the segment top to the segment pole times the power of the add, $F_A$

$$\text{image jump} = dF_A. \qquad (8.13)$$

For a bifocal addition of a given power, image jump depends *solely* upon the position of the segment optical center in relation to the segment top.

### EXAMPLE
A patient requires the distance prescription +3.00 DS with a +2.00 D add. What will be the amount of image jump (*a*) for a 22-mm round fused bifocal segment; (*b*) for a 22-mm-wide fused straight-top

segment; (c) for a 38-mm round Ultex segment; and (d) for an Executive-style segment?

(a) For the 22-mm round fused bifocal segment (segment optical center 11 mm below the segment top),
    jump = 1.1(+2.00)
        = 2.2 prism diopters, base down.

(b) For the 22-mm-wide fused straight-top bifocal segment (segment optical center 5 mm below the segment top),
    jump = 0.5(+2.00)
        = 1.0 prism diopter, base down.

(c) For the Ultex bifocal segment (segment optical center 19 mm below the segment top),
    jump = 1.9(+2.00)
        = 3.8 prism diopters, base down.

(d) For the Executive-style segment (segment optical center at the top of the segment)
    jump = 0(+2.00),
        = 0 prism diopters.

**Size of the Scotoma.** For a 40-cm reading distance, the linear height of an apparent scotoma subtending an angle of 1 prism diopter would be 40/100 cm, or 0.4 cm. Therefore, the scotomas created for each of the lenses described here would have the following heights:

(a) For the 22-mm round segment,
    2.2(0.4) = 0.88 cm.

(b) For the straight-top segment,
    1(0.4) = 0.4 cm.

(c) For the Ultex segment,
    3.8(0.4) = 1.52 cm.

(d) For the Executive segment,
    0(0.4) = 0 cm.

As shown in Figure 8-37, jump would be expected to cause a blind area at the top of the segment. Whether or not the wearer actually experiences a scotoma depends not only upon the amount of jump, but upon the *pupil size* and the *vertex distance* of the

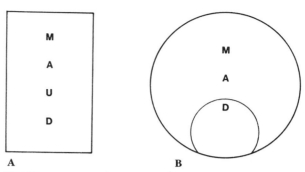

FIGURE 8-38.  **A, Reading card used to demonstrate a "no-jump" bifocal; B, only the letters *MAD* are seen, unless a "no-jump" bifocal is used.**

lenses. For an extremely small pupil, or for a very large vertex distance, it would be *impossible* to see any object which would fall within the angle between the two beams (Figure 8-37). However, with a normal pupil, none of the common types of bifocals worn in the usual position before the eyes would be capable of producing the effect of a blind area.

A bifocal lens having the segment optical center located at or close to the segment top is sometimes referred to, by the manufacturer, as a "no-jump" bifocal. When these lenses were first developed, some optical dispensers demonstrated the phenomenon of jump with the help of a reading card such as the one shown in Figure 8-38. The patient was asked to look at the letters on the card while changing fixation from the distance portion to the near portion of the lens. Depending on the position of the line of sight, the patient would see either the word MAD or the word MUD. Then, changing to a "no-jump" bifocal, the word MAUD would be seen.

A phenomenon which consists of an *overlapping* of the distance and near fields at the top of the segment is also sometimes referred to as a "blind area," but the term "zone of confusion" more aptly describes this phenomenon.

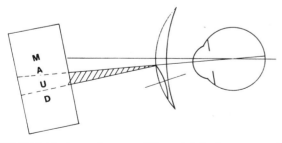

FIGURE 8-37.  **Image jump, or differential displacement at the segment top: Note the base-down prismatic effect, resulting in a blind area. The letter *U* is in the blind area, with the result that the word *MAD* is seen.**

## 8.16. The Zone of Confusion

When a bifocal wearer attempts to look at an object through the top edge of the segment, he sees two images, because a portion of the bundle of rays entering the pupil passes through the distance lens and a portion passes through the segment. The images formed by these bundles of rays may differ from each other in three ways:

1. In direction (differential displacement)
2. In focus (differences in power)
3. In size (differences in magnification)

The difference in direction depends upon the differential prismatic effects at the top of the segment, and can be controlled by varying the position of the segment optical center. It is possible to completely eliminate the difference in direction by placing the segment center at the top of the segment. However, this is of questionable value, since the differences in image focus and in image size would still be present.

The term "zone of confusion" may be applied to any portion of the lens where the line of sight traverses the dividing line, since the phenomena of change in image vergence and change in image size occur simultaneously. The depth of the zone of confusion is dependent solely upon the *size of the pupil* and the *vertex distance*, and will occur relatively independently of the differential displacement at the top of the segment. Consequently, if the goal is to improve the wearer's vision at the top of the segment, there is probably little advantage in attempting to eliminate prismatic effects in that area.

## 8.17. Differential Displacement at the Reading Level

Referring to Figure 8-39A, if the reading level corresponds to the position of the segment center (as shown by the solid line), the segment will cause no prismatic effect so the displacement of the object of regard will be the same as it would be through the distance portion of the lens, with the result that there is no differential displacement at the reading level. However, if the reading level is *above* the segment center, a *base-down* prismatic effect (due to the segment) will be present (Figure 8-39B); if the reading level is *below* the segment center, the segment would cause a *base-up* prismatic effect (Figure 8-39C).

If the vertical distance between the segment optical center and the reading level is indicated by $d_A$, and the power of the add by $F_A$, the differential displacement at the reading level is given by

$$\text{differential displacement} = d_A F_A. \quad (8.14)$$

If the reading level is above the segment center, the segment contains base-down prism; if the reading level is below the segment center, the segment contains base-up prism.

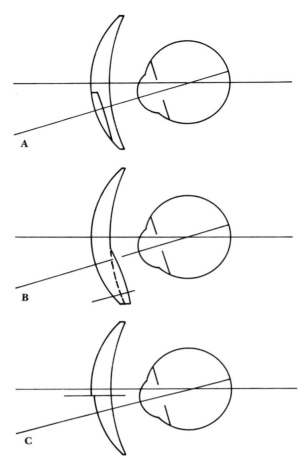

FIGURE 8-39. **Influence of segment type on differential displacement at the reading level: A, no displacement; B, base-down prismatic effect; C, base-up prismatic effect.**

We can compare differential displacement at the reading level to image jump, by using as examples the same prescription used in the previous examples, and also using the same bifocal segments.

### EXAMPLE

Given the distance prescription +3.00 DS with a +2.00 D add. What will be the amount of differential displacement at the reading level, if the reading level is 8 mm below the distance optical center of the lens (*a*) for a 22-mm round fused bifocal segment, whose segment top is 2 mm below the distance center, with a reading level 6 mm below the segment top (Figure 8-40A); (*b*) for a 22-mm-wide straight-top fused bifocal segment, whose segment top is 3 mm below the distance center, with a reading level 5 mm below the segment top (Figure 8-40B); (*c*) for a 38-mm round

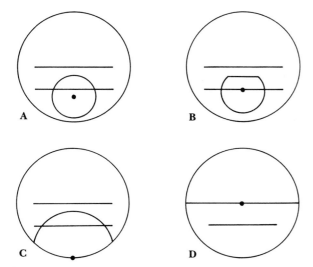

FIGURE 8-40. **Differential displacement: A, 22-mm round fused bifocal; B, 22-mm-wide straight-top fused bifocal; C, 38-mm round Ultex A bifocal; D, Executive-style bifocal.**

Ultex A bifocal, whose segment top is 2 mm below the distance center, with a reading level 6 mm below the segment top (Figure 8-40C); and (d) for an Executive-style bifocal, whose distance and near optical centers are both located at the segment top, with a reading level 5 mm below the segment top (Figure 8-40D)?

(a) For the 22-mm round fused segment, differential displacement
= 0.5(+2.00)
= 1.00 prism diopter, base down.
(b) For the straight-top fused segment, differential displacement
= 0(+2.00)
= 0 prism diopters.
(c) For the Ultex A bifocal, differential displacement
= 1.3(+2.00)
= 2.6 prism diopters, base down.
(d) For the Executive-style bifocal, differential displacement
= 0.5(+2.00)
= 1.00 prism diopter, base up.

It is interesting to note that the Executive-style bifocal, which induces no image jump at all, is responsible for 1 prism diopter of differential displacement at the reading level, and that the Ultex bifocal, having the largest amount of jump of any of the four bifocal types, has considerably more differential displacement at the reading level than the Executive-style bifocal (but opposite in direction).

## 8.18. Total Displacement at the Reading Level

Whereas image jump and differential displacement at the reading level are independent of the power of the distance portion of the lens, total displacement at the reading level depends upon both the power of the distance lens and the power of the add. Inspection of Figure 8-41 indicates that total displacement is equal to the prismatic effect caused by the distance lens plus that caused by the bifocal add. If we use $F_D$ to indicate the power of the distance portion of the lens, $F_A$ to indicate the power of the add, $d_D$ to indicate the distance from the distance optical center of the lens to the reading level, and $d_A$ to indicate the distance from the segment pole to the reading level,

$$\text{total displacement} = d_D F_D + d_A F_A. \quad (8.15)$$

It should be understood that the displacement due to the distance portion of the lens, $d_D F_D$, is in all

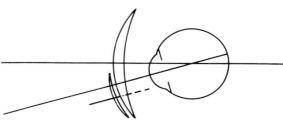

FIGURE 8-41. **Total displacement at the reading level, due both to the power of the distance lens and to the power of the segment.**

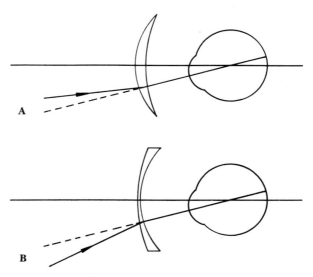

FIGURE 8-42. **Displacement at the reading level due to the power in the distance portion of the lens: A, base-up effect for a plus lens; B, base-down effect for a minus lens.**

cases base up for a distance lens of plus power, and base down for a distance lens of minus power. This is shown in Figure 8-42. A comparison of total displacement for a positive as opposed to a negative distance prescription can be made by calculating the total displacement for our four types of segments, each having a +2.00 D add, for (1) a distance prescription of +3.00 DS, as previously used, and (2) a distance prescription of −3.00 DS.

**EXAMPLE 1**

For the distance prescription of +3.00 DS (with all other values the same as in the example in Section 8.17):

(a) For the 22-mm round fused segment,
total displacement = 0.8(+3.00) + 0.5(+2.00)
= 2.4 base up + 1.0 base down
= 1.4 prism diopters, base up.

(b) For the straight-top fused segment,
total displacement = 0.8(+3.00) + 0(+2.00)
= 2.4 prism diopters, base up.

(c) For the Ultex A segment,
total displacement = 0.8(+3.00) + 1.3(+2.00)
= 2.4 base up + 2.6 base down
= 0.2 prism diopter, base down.

(d) For the Executive-style segment,
total displacement = 0.5(+3.00) + 0.5(+2.00)
= 1.5 base up + 1.0 base up
= 2.5 prism diopters, base up.

**EXAMPLE 2**

For the distance prescription of −3.00 DS (again, with all other values the same as in the previous example):

(a) For the 22-mm round fused segment,
total displacement = 0.8(−3.00) + 0.5(+2.00)
= 2.4 base down + 1.0 base down
= 3.4 prism diopters, base down.

(b) For the straight-top fused segment,
total displacement = 0.8(−3.00) + 0(+2.00)
= 2.4 base down + 0
= 2.4 prism diopters, base down.

(c) For the Ultex A segment,
total displacement = 0.8(−3.00) + 1.3(+2.00)
= 2.4 base down + 2.6 base down
= 5.0 prism diopters, base down.

(d) For the Executive-style segment,
total displacement = 0.5(−3.00) + 0.5(+2.00)
= 1.5 base down + 1.0 base up
= 0.5 prism diopter, base down.

These examples show that for the +3.00 D lenses the Ultex A provides the smallest amount of total displacement, since the base-down effect of the segment opposes the base-up effect of the distance lens. On the other hand, for the −3.00 D distance lens, the Executive style provides the smallest amount of total displacement, since the base-up effect of the segment opposes the base-down effect of the distance lens.

Note also that for the straight-top fused bifocal, having no differential displacement at the reading level (since the segment pole is at the reading level), total displacement is entirely due to the distance lens. This means that when a wearer of single-vision lenses first begins to wear bifocals, the use of a straight-top bifocal will mean that the total displacement at the reading level will be no different than it was while wearing single-vision lenses. (However, if we consider the fact that the wearer of bifocal lenses is forced to lower the lines of sight of the eyes more than the wearer of single-vision lenses, the total displacement at the reading level will be greater than with single-vision lenses.)

## 8.19. Transverse Chromatic Aberration

As discussed in Chapter 6, the chromatic aberration of a lens can be considered in terms of either *longitudinal* (axial) or *transverse* chromatic aberration. It was pointed out in that discussion that since the eye itself has longitudinal chromatic aberration, any longitudinal chromatic aberration caused by a lens will only add to or subtract from the eye's chromatic aberration. However, for lenses of high power, transverse chromatic aberration can sometimes be a problem. A lens having a large amount of transverse chromatic aberration ("lateral color") could conceivably result in the wearer complaining of the presence of "color fringes" when viewing an object through the peripheral portion of the lens.

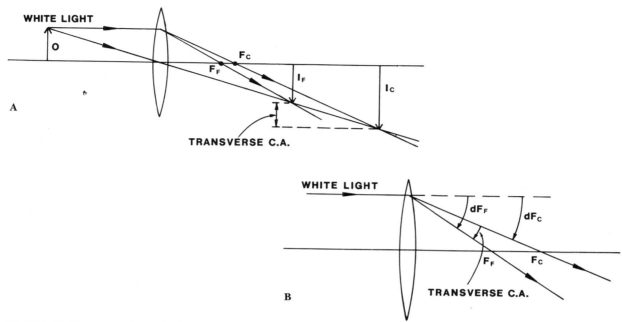

FIGURE 8-43. **Transverse chromatic aberration (C.A.): A, expressed in terms of the difference in image sizes for the red and blue rays; B, expressed in angular terms, as the difference in the angles of refraction for the red and blue rays.**

Transverse chromatic aberration, as noted in Chapter 6, may be expressed in linear form as a difference in image size of the red and blue images (Figure 8-43A) or in angular terms as the difference between the refracted red and blue rays (Figure 8-43B). As noted in Section 6.6, transverse chromatic aberration in angular terms is given by Eq. (6.6):

$$\text{transverse chromatic aberration} = \frac{dF}{v},$$

where $d$ represents the distance from the optical center of the lens to the point in question, $F$ represents the power of the lens, and $v$ represents the nu value. When $d$ is expressed in centimeters, the numerator of the equation will be the equivalent of Prentice's rule, and the transverse chromatic aberration in angular terms will be expressed in prism diopters. Therefore, transverse chromatic aberration is dependent upon the point in a lens through which the foveal line of sight passes.

Although transverse chromatic aberration exists for any point in the periphery of a lens and, for a given power, increases as the eye rotates away from the optical center, it does *not* exist at the optical center.

For a *bifocal lens*, the total transverse chromatic aberration for any point within the segment area is found by combining the transverse chromatic aberration due to the distance lens with that due to the segment. If the subscript D is used to indicate the distance lens and the subscript A is used to indicate the add, the transverse chromatic dispersion at any point within the segment is given by the expression

$$\text{transverse chromatic aberration} = \frac{d_D F_D}{v_D} + \frac{d_A F_A}{v_A}.$$

$$(8.16)$$

If we are interested in the transverse chromatic aberration at the reading level, the parameters represented by $d_D$ and $d_A$ are those given in Sections 8.15 and 8.17: The transverse chromatic aberration at the reading level is related to the total displacement at the reading level, as described in Section 8.17. For a one-piece bifocal, a comparison of equations (8.15) and (8.16) indicates that the transverse chromatic aberration at the reading level is due simply to the total displacement (expressed in prism diopters) divided by the nu value.

Since the nu value of crown glass is 59, the transverse chromatic aberration at the reading level for an Ultex A bifocal having a distance power of +3.00 DS and a +2.00 D add would be

$$0.2/59 = 0.003 \text{ prism diopter},$$

while for an Ultex A having a distance power of −3.00 DS and a +2.00 D add, the transverse chromatic aberration at the reading level would be

$$5.0/59 = 0.08 \text{ prism diopter.}$$

For the Executive bifocal having a distance power of +3.00 DS and a +2.00 D add, the transverse chromatic aberration would be

$$2.5/59 = 0.04 \text{ prism diopter,}$$

while for an Executive bifocal having a distance power of −3.00 DS and a +2.00 D add, the transverse chromatic aberration at the reading level would be

$$0.5/59 = 0.008 \text{ prism diopter.}$$

For a fused bifocal segment made of flint glass, having a nu value of only 36, the transverse chromatic aberration due to the segment itself would be approximately twice that due to the segment of a one-piece bifocal, for a given value of $d_A$. However, in the case of a straight-top fused bifocal the segment causes little or no displacement at the reading level, so the lower nu value of the flint glass proves to be no problem. Even for the Kryptok bifocal, the value of $d_A$ is sufficiently small that (for the examples used here) the displacement due to the +2.00 D add, with a distance power of +3.00 D, is only 1.4 prism diopters. This would result in chromatic dispersion at the reading level of only

$$1.4/36 = 0.04 \text{ prism diopter.}$$

### Transverse Chromatic Aberration at the Segment Periphery

Although the transverse chromatic aberration due to a 22-mm flint-segment fused bifocal is negligible at the reading level, it is possible that in a given combination of distance powers and reading additions the transverse chromatic aberration at the *periphery* of the segment could prove to be a problem. We might, for example, consider two extreme cases, as shown in Figure 8-44: one in which the wearer of a plus-powered distance lens looks through the lower edge of the segment, and one in which the wearer of a minus-powered distance lens looks through the upper edge of the segment.

If the power of the plus lens is +3.00 DS and the power of the add is +2.00 D, and the segment top

is located 3 mm below the distance optical center, the transverse chromatic aberration occurring 1 mm above the lower edge of the segment would be

$$\frac{2.4(3)}{59} + \frac{1(2)}{36} = 0.122 \text{ base up} + 0.056 \text{ base up}$$

$$= 0.178 \text{ prism diopter, base up.}$$

For a distance power of −3.00 DS and an addition of +2.00 D, the segment top again being located 3 mm below the distance optical center, the chromatic dispersion at a point 1 mm below the segment top would be

$$\frac{0.4(3)}{59} + \frac{1(2)}{36} = 0.020 \text{ base down} + 0.056 \text{ base down}$$

$$= 0.076 \text{ prism diopter, base down.}$$

Even in these extreme cases the transverse chromatic aberration at the edge of the segment is less than 1 prism diopter. This explains why very few bifocal wearers complain of seeing color fringes.

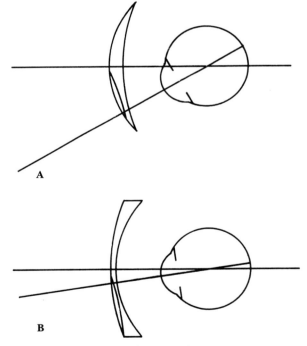

FIGURE 8-44. **Transverse chromatic aberration: A, wearer of a plus-powered distance lens looking through the lower edge of the segment; B, wearer of a minus-powered distance lens looking through the upper edge of the segment.**

_____ CLINICAL CONSIDERATIONS _____

## 8.20. Theories of Bifocal Selection

The problems of image jump, differential displacement at the reading level, and total displacement at the reading level have led to three theories of bifocal selection:

1. The bifocal segment should be selected so that "jump" is eliminated. This is done by choosing a bifocal whose segment pole is located at the dividing line. Such lenses have been referred to as "no-jump" bifocals. The straight-top one-piece (Executive-style) bifocal satisfies this criterion.

2. The bifocal segment should be selected so that differential displacement at the reading level is eliminated. This is done by choosing a bifocal whose segment pole is located at the reading level. The straight-top fused bifocal (having the segment pole 5 mm below the segment top) satisfies this criterion.

3. The bifocal segment should be selected so that the total displacement at the reading level is zero, or as near zero as possible. This can be done by selecting a bifocal whose segment provides prismatic effect opposite to that provided by the distance lens. Two extreme approaches are available here: (1) For a minus distance prescription, the *base-down* prismatic effect at the reading level can be opposed by the *base-up* prismatic effect of an Executive-style segment (Figure 8-45A); (2) for a plus distance prescription, the *base-up* prismatic effect of the distance power can be opposed by the *base-down* prismatic effect of an Ultex A segment (Figure 8-45B). This was shown quantitatively by Examples (*c*) and (*d*) in Section 8.18.

A number of arguments, pro and con, can be made concerning each of these three theories. As for the idea of eliminating *image jump*, it will be recalled that, because of the finite size of the pupil and the normally used vertex distance, the patient may not be *aware* of the predicted image jump; and because of the change in vergence of light and the change in image size when the line of sight crosses the dividing line, there exists a *zone of confusion*, even for a bifocal (such as the Executive) that has no jump. One can make the argument, therefore, that image jump is not a severe problem.

As for the theory of attempting to eliminate the *total displacement* at the reading level, wearers of

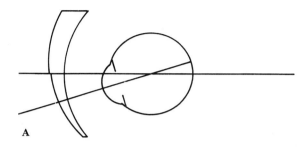

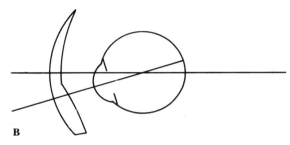

FIGURE 8-45. **Selection of a bifocal segment to eliminate total displacement at the reading level: A, minus distance lens with Executive-style segment; B, plus distance lens with Ultex-style segment.**

single-vision lenses are accustomed to displacement at the reading level: The only way it could be avoided would be for the wearer to tip the head forward sufficiently during reading so that the foveal line of sight passes through the optical center of the lens. Since wearers of single-vision lenses *are* accustomed to this displacement, one can argue that there is no point in trying to *eliminate* it, once the wearer switches to bifocals.

Therefore, the goal should not be to reduce the prismatic effect at the reading level to zero, but to avoid changing the effect to which the patient has already become adapted.

In a situation in which bifocals will be the first pair of glasses to be worn by a patient, he or she would not have become accustomed to prismatic effects at the reading level, so the easiest type of bifocal for this patient to become adjusted to would be one which has *no* prismatic effects at the reading level. In such a case the distance correction is not likely to be very strong, and in order to avoid prismatic effects at the reading level, the segment optical center would have to be placed very close to the reading center.

On the other hand, when a patient requires a strong distance correction, glasses will very likely have previously been worn, and hence the patient will have become accustomed to the prismatic effect at the reading center due to the power in the distance portion of the lens. In order to provide the same prismatic effect that was present for single-vision lenses, the segment optical center must be placed at the reading center.

One may conclude that, of the three problems, the problem of *differential displacement at the reading level* is most likely to be a source of annoyance to the bifocal wearer. Whereas image jump, along with change in vergence, occurs fleetingly and can, with practice, be ignored (or can be eliminated by *blinking* as the eye crosses the dividing line), differential displacement at the reading level is more difficult to ignore. Fortunately, it can be easily eliminated (as already stated) by prescribing a straight-top fused bifocal having the segment center located approximately 5 mm below the top of the segment.

### 8.21. Segment Size and Shape

The selection of a bifocal style for a given patient involves a number of considerations in addition to image jump and displacement, one of the most important of which is the selection of the segment size and shape that most nearly meet the patient's needs.

### Segment Width

Just as spectacle lenses have progressively grown larger and larger in recent years, bifocal segments, too, have grown. The fused Kryptok bifocal, invented early in this century, was for many years made in a diameter of 17 mm, but it gradually grew to 19 mm, 20 mm, and finally to 22 mm in diameter, and the "standard" width of the fused straight-top bifocal segment has grown from 19 to 22 mm, with segments as wide as 35 mm available, whereas the epitome of segment width is expressed by the Executive-style bifocal, whose segment width equals the width of the lens. The early, smaller, bifocal segments were designed to infringe as little as possible on the wearer's distance field of vision (and, incidentally, to make the segment as *invisible* as possible), whereas the larger segments now in use were designed mainly with the *desk worker* in mind.

By the use of very simple mathematics it may be shown that, for a 22-mm-wide segment, the width of the macular field of view (or field of fixation) for a working distance of 40 cm, neglecting the effect of lens power, is approximately 35 cm (or 14 inches). Referring to Figure 8-46, for a distance of 2.7 cm from the spectacle plane to the center of rotation of the eye, we find by the use of similar triangles that

$$\text{field width} = \frac{(42.7)(2.2)}{2.7}$$
$$= 34.8 \text{ cm}$$
$$= \text{approx. } 14 \text{ inches.}$$

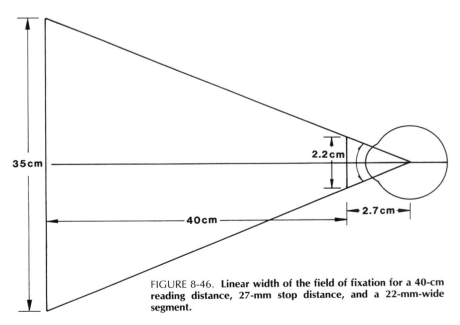

FIGURE 8-46. **Linear width of the field of fixation for a 40-cm reading distance, 27-mm stop distance, and a 22-mm-wide segment.**

The field given here is based on the head remaining in a stationary position, and can, of course, be widened by moving the head. For the Executive-style bifocal, having a lens width of 50 mm, the width of the field of view (without moving the head) would be

$$\frac{(42.7)(5)}{2.7} = 79 \text{ cm} = \text{approx. 31 inches.}$$

Even though a 22-mm-wide segment is considered to be wide enough for most everyday near tasks such as reading a magazine or a book, writing, or sewing, many patients (whose activities would not seem to require them) request to be fitted with the Executive-style bifocal. One can make the argument that, with a narrower segment, *distance vision* is available on either side of the segment. However, a counterargument to this is that, for a fused (22- or 25-mm-wide) segment, the wearer is aware of the dividing line when the eyes are moved to one side or the other (when looking through the lower part of the lens), whereas with a segment that goes all the way across the lens, the field of view (although blurred, for distance vision) presents no dividing line.

In any event, when a patient requests a wide segment, there is usually nothing to be gained by attempting to change his or her opinion.

### Segment Height

Many patients (and even some eye practitioners and optical dispensers) believe that a bifocal segment should be placed as low as possible in the lens, so as to infringe on the distance field as little as possible. Fitting a bifocal in this manner is almost always a mistake, and results in the patient getting a *stiff neck* as a result of throwing the head back to find the bifocal segment. It is undoubtedly true that errors are more often made by fitting bifocal segments *too low* than by fitting them too high!

The reference point most often used in specifying segment height is the *ciliary line* at the top edge of the lower lid. For most people the ciliary line is on the same level as the lower limbus (corneoscleral junction), as shown in Figure 8-47. However, if a patient's lower lid is very much lower or higher than the lower limbus, the *limbus* should be used as the reference point. Since the vertical extent of the cornea (the visible iris diameter) is usually from 11 to 12 mm, the vertical distance from the ciliary line to the center of the pupil is usually very close to 6 mm.

One disadvantage of a *round* bifocal segment is that the widest (and the most useful) part of the segment is a considerable distance below the dividing

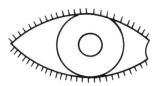

FIGURE 8-47. **Normally the ciliary line for the lower lid is at the same level as the lower limbus.**

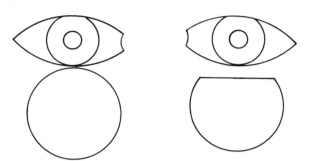

FIGURE 8-48. **A round segment is normally fitted at the level of the ciliary line (A), whereas a straight-top segment is fitted 1 or 2 mm below the ciliary line (B).**

line: Because of this, the segment top must be fitted higher than would be the case for a straight-top bifocal, whose widest portion is very close to the dividing line (see Figure 8-48). As a general rule a round segment should be fitted at about the level of the ciliary line, whereas a straight-top segment is usually fitted 1 or 2 mm below the ciliary line.

Occupation, head position, and many other factors are of importance in determining the segment height. In general, the more time spent in reading or other close work, the higher the segment should be. In the extreme case of an individual whose entire working day involves near work at or near eye level (as a barber or postal clerk, as mentioned earlier), a "reading lens with a distance window," such as a very high Executive style or an Ultex AL, should be considered. As for head position, an individual who customarily walks with the head held back may require a somewhat lower segment, whereas one who walks with the head slouched forward will require a somewhat higher segment. It is a good idea to get into the habit of observing a patient's posture, both in walking and in reading, prior to deciding upon the bifocal segment height.

### Segment Shape

Probably the only advantage of a *round* fused segment (with its knife-like edges) is the fact that it is *less visible* than a straight-top fused segment. The least visible segment is a "no-chrome" round segment (made of barium crown glass, having a nu value very close to that of crown glass) in a flesh-colored tint such as Softlite A or Cruxite A. In years past, a very small fused round segment, about 12 mm in diameter (called a "spot" or "button" segment), was sometimes recommended for a patient whose occupation required mainly distance vision. Although such segments are no longer manufactured, a small round segment of any size may be obtained by resurfacing the front surface of a 22-mm round fused segment until the desired segment size is reached.

A *ribbon* segment (in particular, a B segment, 9 mm high) should be considered for a worker, such as a construction worker, who may have to do a large amount of climbing on a ladder or other structures, but who must have occasional near vision available. However, in order for the distance field below the segment to be of any value to the wearer, the segment must be fitted sufficiently high (and with a sufficiently large frame) so that the distance portion of the lens below the segment is at least 7 or 8 mm high. This means that the segment height should be at least 16 or 17 mm. There is little advantage in fitting the 14-mm-high R segment for such a patient, since it would have to be fitted *very* high in order for any distance field to be available below the segment. The most appropriate use of the R segment is for the compensation of differential vertical prismatic effects. This will be discussed in Section 10.3.

A final factor that must be taken into consideration is the style of bifocal that the patient is *currently wearing*. In most cases, if the patient is completely satisfied with the type of bifocal that he or she has been wearing, the best procedure is *not* to change the bifocal type. For example, a patient having a moderate amount of astigmatism who has worn Ultex bifocals (which are usually in positive toric form) may not be happy if he or she is switched to an Executive-style or straight-top fused bifocal (in negative toric form), even though the differential displacement at the top of the segment is greatly decreased. An analogous situation arises when a wearer of Orthogon single-vision lenses (having a positive toric form) is first fitted with bifocals: The patient may have difficulty adjusting to an Executive-style bifocal made in negative toric form. However, because of the important advantages of negative toric lenses (discussed in Chapter 6) and the increased differential displacement at the segment top often caused by an Ultex bifocal, it is probably best to switch these people to a negative toric bifocal and to warn them of the possible adaptation problems.

## 8.22. Horizontal Prismatic Effects

When bifocal lenses are prescribed, unintentional horizontal prismatic effects may be induced in the reading portion of the lens, due either to the *distance* power or to the power of the *add*. In order to understand these prismatic effects, the bifocal segment should be thought of as existing *behind* the distance lens, as if it were a cement bifocal (as already discussed).

If the segments have been correctly inset for the wearer's near PD (as determined by Sterling's table, here Table 8-10), the fields of view for the two eyes will coincide. However, the *distance* power of the lens *does* induce a horizontal prismatic effect at near, unless the distance prescription has no power in the horizontal meridian. For example, for a distance correction of +3.00 DS and an interpupillary distance of 66/62, the prismatic effect at the near PD, due to the distance power, will be

$$0.2(+3.00) = +0.6$$
$$= 0.6 \text{ prism diopter, base out,}$$

and for a −3.00 DS with the same PD, the prismatic effect at the near PD due to the distance power will be

$$0.2(-3.00) = -0.6$$
$$= 0.6 \text{ prism diopter, base in.}$$

If the wearer's right and left lenses both have the same power (both +3.00 DS or both −3.00 DS), the total prismatic effect would be 1.2 prism diopters (base out in the first case and base in in the second).

Note that this prismatic effect, due to the distance power, is independent of the power of the add: As long as the segments are inset to correspond to the near PD, the addition itself causes no prismatic effect.

Since the addition causes no prismatic effect, when a wearer of single-vision lenses is first fitted with bifocals, the prismatic effect at near will be no different than it would have been with single-vision lenses (assuming that the single-vision lenses had been centered for the distance PD.) Therefore, if the patient had no complaints of asthenopia due to the prismatic effect induced by the single-vision lenses, one would think that he or she would have no complaints due to the prismatic effect when wearing bifocal lenses. However, with the application of the plus addition for reading, the patient becomes more exophoric at near, due to the loss of accommodative convergence, and this may precipitate complaints (although Sheedy and Saladin[3] found that many presbyopic patients had no complaints in spite of the exophoria at near). Of course if the patient had been *esophoric* at near before becoming presbyopic, the plus addition would reduce the esophoria. In any event, most presbyopes are exophoric through the reading addition.

The *myopic* presbyope will obtain some relief from the induced exophoria due to the plus addition, because of the *base-in* effect of the distance correction, but the *hyperopic* presbyope is more likely to be bothered by symptoms of exophoria, due to the *base-out* effect of the distance lenses.

If a presbyope has symptoms accompanying near work which appear to be caused by exophoria at near, some relief may be obtained by increasing the segment inset in order to obtain a base-in prismatic effect due to the segment. For example, for a +2.50 D addition, if each segment is inset an *additional* 3 mm, the additional inset would produce a horizontal prismatic effect of 0.75 prism diopters base in for each lens, or a total of 1.5 prism diopters base in.

Care must be taken that the additional segment inset designed to provide a base-in effect through the segment does not create a new problem. Additional inset of small segments should be avoided. The outline of the segment acts as a field stop, which determines the monocular field of fixation; if the segment is small, the temporal limits of the monocular fields of fixation could be significantly reduced. Hence, the overall binocular field of fixation would be appreciably contracted. Large straight-top segments such as the 28- and 35-mm fused segments can undergo additional inset without reducing the binocular field of fixation below that of an average-sized segment. However, these segments have an unusual appearance because of the cutoff appearance on the nasal side of the lens (Figure 8-49). For this reason, the Executive-style bifocal is very suitable when additional segment inset is necessary, because the monocular fields of fixation are maintained with no noticeable change in the appearance of the lenses.

Occasionally, a practitioner may decide to incorporate prismatic power in the *segment only*. This may be done for the purpose of compensating for a lateral

phoria at near, or possibly for the purpose of counteracting the prismatic effect induced at near by the distance prescription. Some bifocal manufacturers have available *prism segment* bifocals, which can incorporate as much as 2 or 3 prism diopters of base-in or base-out prism. However, prism segment bifocals are always "factory orders," involving increased cost and a very long delay. Accordingly, their use is recommended only as a last resort.

Another solution to providing horizontal prism in the segment only is by the use of Fresnel *Press-on prisms* (discussed in Chapter 4). The Fresnel prism may be cut from a sheet or may be purchased in precut form. These Press-on prisms unfortunately have the disadvantages of poor appearance and a slight loss of visual acuity.

It is fortunate that, as already pointed out, presbyopes usually have no difficulty compensating for a high exophoria at near.

### 8.23. Differential Vertical Prismatic Effects at the Reading Level

If the distance powers of a patient's lenses differ for the two eyes in the vertical meridian, the two eyes will experience differential prismatic effects at the reading level. For a wearer of single-vision lenses this may not constitute a problem, since the head is normally tilted downward in reading with the result that the reading level may be very little different than the level used for distance vision. However, a bifocal wearer is forced to read at a level approximately 10 mm below the distance optical centers of the lenses, so differential prismatic effects associated with reading may create a problem.

A number of methods of compensating for differential vertical prismatic effects are available. These will be discussed in Chapter 10.

### 8.24. Ordering and Dispensing Bifocals

A laboratory order for a pair of bifocal lenses must include the following information:

1. The *distance prescription*, as ordinarily specified for single-vision lenses, including for each eye the sphere power, cylinder power and axis, the distance PD, the vertical level of the major reference point, and (if prism is prescribed) the amount of prism and the direction of its base.

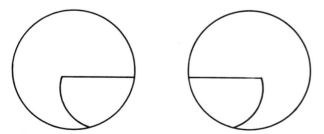

FIGURE 8-49. **Wide fused bifocal segments, inset sufficiently to provide a base-in prismatic effect.**

2. The *power of the add*, specified as a plus sphere addition, for example, +1.50 D add. The total power through the segment area would be the sum of the distance prescription and the add.

3. The *bifocal style*, or type. It should be understood that if a trade name (for example, "Executive," or "Univis") is specified, the laboratory is obligated to provide the specific trade-name lens. However, if a descriptive term is used, such as "straight-top one-piece bifocal," "straight-top fused bifocal," or even "Executive style" or "Univis style," the laboratory is free to fill the order with any manufacturer's lens that fulfills the description.

4. The *segment size*. As with lens sizes in general, the horizontal dimension (measured where the segment is at its widest) is specified first. The vertical dimension of the segment is the vertical measurement of the *unfinished* lens, before cutting and edging are done. For example, a Kryptok bifocal would be specified as 22-mm round, and a Univis D would be specified as 22 by 16 mm, even though the lenses are to be mounted with the top of the segment only 14 mm high in the frame. Obviously it is not necessary to specify the *width* of an Executive-style segment.

5. The *segment height*. The height of the segment is specified as the perpendicular distance from a line tangent to the lowest part of the lens bevel to a line tangent to the highest point on the top of the segment (see Figure 8-50). The segment height is simply specified, for example, as *19 mm high*.

6. The *segment inset*. The segment inset is specified as the distance nasally (inward) from the major reference point of the distance portion of the lens to the optical center of the segment (see Figure 8-50). When the horizontal meridian of the distance lens contains no power or has only prismatic power, the *total inset* must be specified.

7. The *frame specifications*. The size of the frame is specified on the basis of the *box system*, such as 50 by 42

mm. The DBL (distance between lenses), shape of the frame, and temple style and length must also be specified.

8. The *lens material and tempering process*. Most prescription order forms have boxes for indicating whether glass or plastic lenses are desired. If *glass* is specified, the method of tempering (heat or chemical) must also be specified. If *plastic* is specified, the order will almost always be filled with CR-39 plastic, having an index of refraction of 1.498. If polycarbonate or another plastic material is desired, it must be specified. The surfaces of polycarbonate lenses are routinely coated with a scratch-resistant coating. If a scratch-resistant coating is desired for CR-39 lenses, it must be specified in the order.

9. *Tint*. As with single-vision lenses, the order must indicate either clear lenses or the specific tint, if a tint is desired.

When the lenses have been received from the laboratory, they should be carefully verified. Verification will be discussed in the following section.

At the *dispensing visit*, the patient should be instructed in the use of the bifocals. This is particularly true for the patient's *first pair* of bifocals. The patient should be told to lower the eyes, rather than the head, when reading, and should be told to try to ignore the dividing line as the line of vision sweeps across it. One way to ignore the dividing line is to *blink* when lowering the eyes.

Some new bifocal wearers tend to worry about whether they are looking through the upper or the lower portions of the lenses. A way to take care of this problem is to suggest to the patient that he can make sure he is looking through the upper part of each lens by tilting his head *downward* slightly; and he can make sure that he is looking through the lower part by tilting his head *upward* slightly.

In addition to the mechanical problems of learning how to hold the head and to move the eyes for near vision, some new bifocal wearers have difficulty in adapting to such optical factors as the difference in magnification, difference in reading distance, and differences in the stimuli to accommodation and convergence.

Most new bifocal wearers are pleased to find that "everything looks larger" through their bifocals, although some may have difficulty learning to hold their reading material at a sufficiently close distance. It is sometimes disappointing, after prescribing bifocal lenses that definitely improve the patient's near acuity, to have the patient complain that he "has to hold his reading too close." Fortunately, patients typically adapt to this "problem" within just a few days.

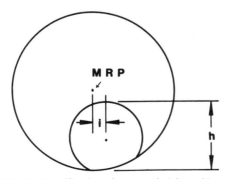

FIGURE 8-50. **Specification of segment height and inset. MRP = major reference point.**

The fact that bifocal lenses reduce the *accommodative demand* means that less accommodative convergence is available, causing the near phoria to change in the *exophoric* direction. It is fortunate that most presbyopes have little difficulty compensating for this exophoria. Sheedy and Saladin[3] have hypothesized that presbyopes compensate for exophoria at near by virtue of *unrestricted use of accommodative convergence*: Since accommodative convergence is not accompanied by the usual "dose" of accommodation, a presbyope may freely make use of accommodative convergence to compensate for the exophoria.

Fortunately, if the patient persists in wearing the bifocals (assuming that they were prescribed and fitted correctly), complete adaptation usually takes place within 2 or 3 weeks. For new bifocal wearers, it is a very good idea to schedule the patient for a *progress visit* at the end of 2 or 3 weeks.

## 8.25. Verification of Bifocals

When the practitioner receives a finished pair of glasses from the laboratory, they must be evaluated in terms of accuracy of workmanship and quality of materials. Procedures for the verification of *single-vision* lenses will be discussed in Chapter 9—only the aspects of verification that apply to bifocal lenses will be discussed here.

The operations involved in the layout, marking, surfacing, and cutting of bifocal lenses are more complicated than are those for single-vision lenses, and the lenses must be carefully examined to ensure that they have been made according to specifications. In addition to determining the accuracy of the placement of the distance centers, bifocals must be evaluated for segment type, segment width, segment height, segment inset, and power of the add. In verifying segment height each lens must be measured, and for straight-top segments a straight edge should be used to assure that both segment tops lie in the horizontal meridian.

As described in Chapter 4, when a lensometer is used to determine the back vertex power of a single-vision lens (or of the distance portion of a bifocal lens) the lens is placed in the lensometer with its back pole against the lens stop and the target is moved until the image of the target on the reticle is in sharp focus. Using this procedure, one actually measures the *back focal length* of the lens—the distance from the back pole of the lens to the secondary focal point. Lenses used in refractors or as trial lenses in trial frames are calibrated in terms of back vertex power,

and the use of such lenses is appropriate when determining a patient's *distance* lens prescription. It will be recalled that the *effectivity*, or effective power, of a lens depends upon the distance from the back pole of the lens to the vertex of the cornea; so whatever the form or thickness of a lens, lenses having the same back vertex power for parallel light will have the same effectivity for axial rays as long as their back poles are placed at the same distance from the corneal vertex.

A different situation exists, however, when a refractor or trial lens is used to determine a patient's *near* prescription. In this situation the light rays emerging from the test object (a near-point card) are *divergent* rather than parallel, with the result that the concept of back vertex power does not apply. Unless the spectacle lens for the near prescription is made in the same form and thickness as the test lens, it will not have the same effectivity as the test lens—that is, it will not focus the image of a near object at the same position as with the test lens. For example, if a +5.00 D lens is needed for distance vision and the total power of the lenses in the refractor or trial frame for a near object is +7.50 D, a single-vision spectacle lens measuring +7.50 D (back vertex power) in a lensometer will not have the same effect as the test lenses unless the spectacle lens and the test lenses have the *same form*. This fact accounts for the discrepancy often found between the clearest working distance obtained with the refractor and that found when the finished spectacle lenses are worn.

It is, therefore, difficult to duplicate the exact *effectivity* achieved under the test conditions for a near object, with either single-vision lenses or bifocal lenses. The difference is usually small, and it isn't necessary in most cases to have the exact correspondence in effectivity of the test lenses and the patient's lenses: Most patients can make a slight alteration in the near working distance without difficulty. It is important, however, that the powers of the lenses fall within the individual patient's tolerance.

Standards published by the American National Standards Institute (ANSI Z80.1-1979)[4] recommend, as do manufacturers of bifocal lenses, that the power of the near portion of a bifocal lens should be measured with the *segment side* of the lens placed against the lensometer stop.

1. For a *back surface segment*, such as an Ultex-style bifocal, the power of both the distance and near portions of the lens should be measured by placing the *back* surface of the lens against the lensometer stop, therefore measuring back vertex power (see Figure 8-51). The power of the bifocal addition will

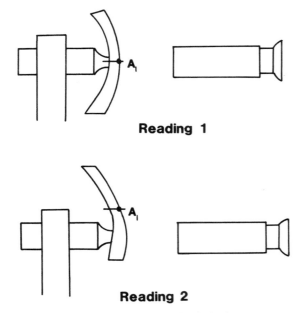

**Reading 1**

**Reading 2**

FIGURE 8-51. **For a back surface bifocal,** *back* **vertex power should be measured: First lensometer reading is the back vertex power of the distance portion; second reading is the back vertex power of the near portion. The power of the add is equal to the second reading less the first reading. $A_1$ is the position of the front pole or vertex of the distance portion.**

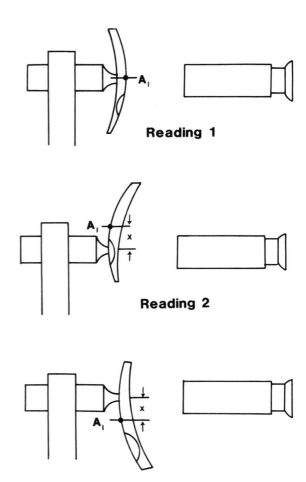

**Reading 1**

**Reading 2**

**Reading 3**

FIGURE 8-52. **For a front surface bifocal, $A_1$ denotes the position of the front pole or vertex. The first lensometer reading is the back vertex power of the distance portion, at the distance pole; second reading is the front vertex power of the reading portion; third reading is the front vertex power of the distance portion. Note that the distance $x$ is the same in the second and third diagrams, in order to balance the effect of thickness. The power of the add is equal to the difference between the reading 2 and reading 3. The back vertex power of the distance portion is reading 1.**

be equal to the back vertex power of the near portion of the lens minus the back vertex power of the distance portion.

2. For a *front surface segment*, such as a fused bifocal, the back vertex power of the distance portion is measured in the usual manner, with the back surface of the lens against the lens stop. To obtain the power of the add, both the distance and near portions of the lens should be measured by placing the lens in the instrument with its *front* surface against the lensometer stop (Figure 8-52): Using this procedure, front vertex power is measured. The power of the bifocal addition will then be the front vertex power of the near portion of the lens minus the front vertex power of the distance portion.

Confusion exists in the minds of many optometrists concerning the recommended method of measuring the power of the addition of a front surface bifocal segment.[5] With few exceptions optometrists use refractors with lenses calibrated in terms of back vertex power, and therefore it seems logical that the finished lens should conform to the back vertex power determined by the use of the refractor. Furthermore, the patient looks through the lens from

the back, rather than the front, so why must the power of the addition be verified by reversing the lens in the lensometer? Even though the use of front vertex power does not appear to be logical, there are limitations in the back vertex power notation.

Whether the lensometer is used to measure back vertex power (with the back pole of the lens placed against the lensometer stop) or front vertex power (with the front pole placed against the stop), in order

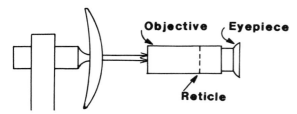

FIGURE 8-53. **In order for the image of the lensometer target to be seen sharply focused on the reticle, the light between the test lens and the eyepiece must be parallel.**

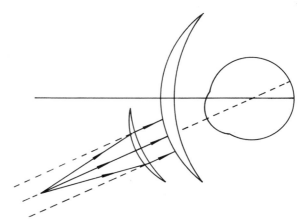

FIGURE 8-54. **For an absolute presbyope, the bifocal add should render diverging rays from a near object parallel, as if they came from the far point of the eye as corrected by the distance lens.**

for the image of the lensometer target to be seen sharply focused on the reticle, the light between the test lens and the eyepiece must be *parallel* (see Figure 8-53).

A bifocal lens can be considered as a distance lens with a specified back vertex power, having a plus spherical add placed on the front or back surface. The power of the distance lens is correctly verified in terms of back vertex power by placing the back pole of the lens against the lens stop. For a front surface bifocal, it is apparent that the back surface of the lens provides the same change in vergence through both the distance and the near portions of the lens. The difference in vergence of the wavefronts between the distance and near portions is generated solely by the *segment* on the front surface. To measure the difference in effectivity occurring at the front surface of the lens, for a near object, the readings must be made with the front surface of the lens against the lens stop.

It should be understood that while neither of the two front vertex powers (for the distance and near portions of the lens) is actually experienced by the wearer, the important consideration is that the *difference* between these two powers determines the power of the add.

The purpose of the bifocal addition is to change the vergence of the light from a near object so that it apparently originates from a greater distance: In fact, for an absolute presbyope the add must render diverging rays from a near object *parallel*, as if they came from infinity. The distance lens then directs the rays toward the far point of the eye, and the optical system of the eye focuses the rays on the retina. This is shown in Figure 8.54.

There will be a slight difference in effectivity for near refraction when placing an add on the front surface as compared to the back surface of a lens. Even though the two adds will have the same power when each is measured on the segment side of the lens, the effectivity of the back surface segment is slightly less than that of a front surface segment.

For most lenses the measurement of a front surface segment add will result in essentially the same power whether the front or the back surface of the lens is placed against the lensometer stop: Any difference will not be large enough to be significant. However, for lenses having an *appreciable center thickness* (high plus lenses or industrial safety lenses) the difference in the two methods of measurement will often exceed 0.25 D.

The difference will be smaller for strong minus lenses than for strong plus lenses, for two reasons: (1) Strong minus lenses have smaller center thicknesses than strong plus lenses, and (2) strong minus lenses have flatter front surfaces than strong plus lenses, creating smaller vergence changes at the back surface of the lens when measured with the back pole of the lens against the lens stop.

In summary, it appears to be almost impossible, for the viewing of a near object, to provide exactly the same effectivity with a spectacle lens (either for a single-vision reading lens or for a bifocal lens) as with the test lenses calibrated in back surface power. The manufacturer places the addition either on the front surface or on the back surface, and if measurements are made through the distance portion and the near portion with the *segment surface* against the lensometer stop, the difference between the two readings will be equal to the addition specified in the prescription. It may appear that the recommended procedure for measuring bifocal additions exists only for the purpose of standardization—to convince all parties to agree with, rather than to argue about, the procedure. This is true, up to a point, but it should be understood that the recommended procedure *is accurate.*

One other point should be understood. Most manufacturers make multifocal lenses in *semifinished* form, only the segment side being finished. The add is unalterably fixed before the other parameters (such as distance power, back surface power, and thickness) are known. This applies to all multifocals, whether bifocals, trifocals, or progressive addition lenses, and to segments whether on the front or back surface.

## 8.26. Prescribing and Fitting Double-Segment Bifocals

It is possible that the double-segment bifocal is the most *underutilized* bifocal style. Experience shows that, when a double-segment bifocal is prescribed (and correctly fitted) for a patient who must make use of near vision above eye level on a routine basis, wearers have little difficulty adjusting to them and usually request them again when the prescription must be changed. A probable reason for their underutilization is that many practitioners fail to question their patients concerning the need for above-eye-level vision.

The correct method of fitting a double-segment bifocal is illustrated in Figure 8-55. It will be recalled that most double-segment bifocals have a separation of *13 mm* between the upper and lower segments. Since the vertical extent of the cornea is about 11 or 12 mm, this means that if the top edge of the lower segment is fitted in the usual manner (from 1 to 2 mm below the lower limbus), the lower edge of the upper segment will be just about on the level with the upper limbus (which is usually slightly *above* the ciliary line of the upper lid).

It should be understood that the double-segment bifocal is *not a trifocal*. Since, for most people, near vision above eye level involves about the same working distance as near vision below eye level, the power

of the add in the upper segment should in most cases be the same as that in the lower segment. As noted in Section 8.4, all double-segment bifocals have the same power in both the upper and lower segments with the exception of the Tillyer Double Executive, whose upper segment has a power approximately two-thirds of that of the lower segment.

## 8.27. Prescribing and Fitting Trifocals

When a patient gets to the point that a +1.75 or +2.00 D add is required (which usually occurs at about 50 years of age), the practitioner should seriously consider the possibility of recommending *trifocal* lenses. Like double-segment bifocals, trifocals are generally underutilized. One reason for this is that many practitioners think of trifocals as vocational lenses. However, almost *any* advanced presbyope will have difficulty with many arm's-length tasks (such as seeing merchandise and prices in a supermarket) through either the upper or lower portion of a bifocal lens.

The suggestion that the use of trifocals should be considered for a patient who requires a +1.75 or +2.00 D add is based on the following rationale. As a rule of thumb, a patient should have to use no more than half of his or her amplitude of accommodation on a sustained basis. This being the case, a patient having 2.00 D of accommodation should have to use only 1.00 D of accommodation and therefore should be given (for a 40-cm reading distance) a +1.50 D add. Having 2.00 D of accommodation, the near point of accommodation (through the *distance* lenses) would be 50 cm; the far point through a +1.50 D segment, or the most remote distance that would allow clear vision through the *segment*, would be

$$\frac{1}{1.50} = 67 \text{ cm},$$

and the closest distance of clear vision through the *segment* would be

$$\frac{1}{2.00 + 1.50} = \frac{1}{3.50} = 29 \text{ cm}.$$

Figure 8-56A shows that there is an area of *overlap* between the distances at which objects may be clearly seen through the distance and near portions of the lens. Moreover, the depth of field of the eye makes this area of overlap even larger.

For a patient who has an amplitude of accommodation of only 1.00 D, only 0.50 D of accommodation

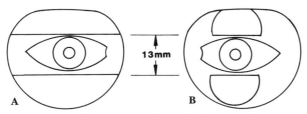

FIGURE 8-55. **The correct method of fitting a double-segment bifocal: Double Executive (A) and double straight-top fused (B). The top edge of the lower segment is fitted in the usual manner, from 1 to 2 mm below the lower limbus.**

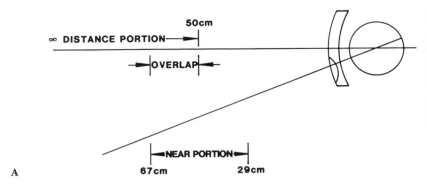

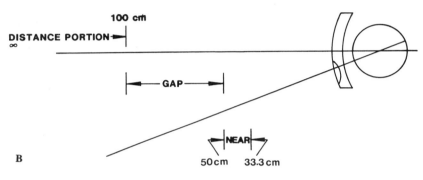

FIGURE 8-56. **(A) For a patient having 2.00 D of accommodation, there is an area of overlap (from 50 to 67 cm) in which objects may be seen clearly through either the distance or near portion of a bifocal lens; (B) for a patient having only 1.00 D of accommodation, there is a gap (from 44 cm to 1 m) where vision will be blurred with both the distance and near portions of the lens.**

should be used on a routine basis, with the result that a +2.00 D add will be required for a 40-cm working distance. Having 1.00 D of accommodation, the near point of accommodation through the *distance* portion of the lens is 1 m, and the range through the near portion of the lens would extend outward to

$$\frac{1}{2.00} = 50 \text{ cm}$$

and inward to

$$\frac{1}{+1.00 + 2.00} = \frac{1}{3.00} = 33.3 \text{ cm.}$$

As shown in Figure 8-56B, there will now be a *gap* from a distance of 1 m to a distance of 50 cm, where vision will be blurred through both the distance and near portions of the lens. If we consider that the depth of field would add approximately 0.25 D to the far end of the near range (through the segment) and the near end of the distance range, there would still be a gap extending from 80 to 57 cm. It should be understood that when a practitioner increases a patient's bifocal addition from +1.50 to +2.00 D, as far as the patient is concerned it is the *practitioner's fault* that the gap exists, since it wasn't there with the previous glasses! For this reason, it is the practitioner's responsibility to warn the patient, when increas-

ing the add to +2.00 D (or even to +1.75 D), that a gap will be present. Better still, the existence of the gap can be *demonstrated* to the patient, either by the use of the refractor or by the use of trial lenses.

Many presbyopic patients are unaware of the existence of trifocals or do not understand their function, and are therefore unaware of the possibility of improved vision at intermediate distances. A careful analysis of the patient's visual status should consider not only the age and amplitude of accommodation but the occupation, hobbies, and the requirements for clear vision for specific intermediate tasks. As a *routine* matter, the practitioner should demonstrate to any patient requiring a +1.50 D add or greater that clear vision is available at distance, at 40 cm, and at intermediate ranges with appropriate lens powers. Through education and demonstration, the inclusion of this procedure will substantially increase the use of trifocals and improve the visual performance of many presbyopes.

Unfortunately, many practitioners are hesitant to suggest trifocals unless the patient specifically complains of blur at an intermediate distance. This hesitancy may be due partially to the increased cost of the lenses, but, more important, to the practitioner's failure to take an adequate case history and to take the time necessary to inform the patient of the problem that will result when the power of the addition is increased (including the demonstration described above). Practitioners who wear trifocals themselves are in a position to inform their patients in a manner not open to younger practitioners—that of offering personal testimony concerning their benefits.

Most patients adjust to trifocals at least as easily as to bifocals, and in many cases more easily. Upon dispensing it is helpful to instruct the beginning trifocal wearer in the proper head movements for best vision through the intermediate and near segments.

It will be recalled that trifocals are available in both fused and one-piece styles. For the fused styles, the intermediate segment is made of a kind of glass having an index of refraction intermediate between that of the distance portion and that of the segment; for the one-piece styles the intermediate segment has a curvature that is halfway between that of the distance portion of the lens and the bifocal segment. Most trifocals, whether fused or one piece, are of straight-top construction. The standard power of the intermediate segment is 50% of that of the lower segment, although some manufacturers of "vocational" trifocals vary the power of the intermediate segment from 40% for weaker adds to 70% for higher adds.

### Field of View

The vertical extent of the field of view through the intermediate segment depends upon the following parameters:

1. The vertical dimension of the intermediate segment
2. The pupil size
3. The vertex distance
4. The viewing distance
5. The total power through the intermediate segment

1. As shown in Figure 8-57A, the field of view in the vertical meridian is directly proportional to the vertical dimension (or *depth*) of the intermediate segment. The most commonly used depth for an intermediate segment is 7 mm, although 6-, 8-, 10-, and even 14-mm segments are available. Although one may imagine that "bigger is better," it must be remembered that increasing the vertical extent of the intermediate segment causes an infringement of either the near or the distance portion of the lens: The deeper the intermediate segment, the lower the upper boundary of the *near* portion will be (or the higher the lower boundary of the *distance* portion will be).

2. With a given vertical dimension, the smaller the patient's pupil the larger the vertical field of view

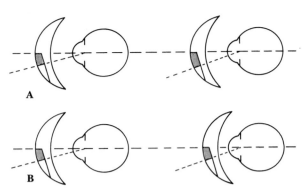

FIGURE 8-57. **(A) The vertical extent of the field of view for an intermediate segment is directly proportional to the vertical dimension (depth) of the intermediate segment. (B) the vertical extent of the field of view for an intermediate segment of a given depth depends on the distance of the segment from the eye.**

will be before doubling of the image occurs. If the edge of the intermediate field splits the pupil, two images will be seen, varying in position, size, and focus. Fortunately, most presbyopic patients have small pupils.

3. The vertical extent of the field of view for an intermediate segment of a given depth depends upon the distance of the segment from the eye (as shown in Figure 8-57B), the field of view decreasing as the vertex distance increases. For this reason the field of view is also influenced by the pantoscopic tilt, the curvature of the segment surface, and whether the segment is placed on the front or the back surface of the lens.

4. For a constant lens power, the *linear* size of the field of view increases as the viewing distance increases, even though the *angular* field of view remains constant. For a pupil size of 3.5 mm, a vertex distance of 12 mm, and a viewing distance of 56 cm (equal to 22 inches), Holmes, Jollife, and Gregg[6] have shown that the absolute field of view (the field in which there will be no diplopia) will be as follows:

6.0 cm for a 6-mm intermediate segment
8.3 cm for a 7-mm intermediate segment
10.6 cm for an 8-mm segment

For a viewing distance of 69 cm (27 inches), the absolute field of view will be:

7.6 cm for a 6-mm intermediate segment
10.6 cm for a 7-mm intermediate segment
13.5 cm for an 8-mm intermediate segment

5. The vertical extent of the field of view for the intermediate segment also varies with the *power of the lens* in the intermediate portion of the lens, decreasing with increasing plus lens power and increasing with increasing minus lens power.

## Segment Height

The segment height of a trifocal lens is specified, in the laboratory order, as the distance of the *upper edge* of the intermediate segment from the lower edge of the lens.

The specification of no other trifocal parameter has been more controversial than that of segment height. It must be kept in mind that although the purpose of a trifocal lens is to provide the wearer with intermediate vision, with few exceptions the wearer will use the lenses much more often for the normal near-point (reading) range than for the intermediate range. Comfortable near-point vision, therefore, should not be sacrificed for the sake of intermediate vision.

The trifocal segment can be considered as "sitting on top" of the bifocal segment, in the position in which the bifocal segment would be *without* the intermediate segment, as shown in Figure 8-58. For example, if a patient would normally require a 17-mm-high bifocal, the height of the trifocal segment should be specified as 23 mm for a 6-mm segment or as 24 mm for a 7-mm segment. The result of this method of fitting is that the area occupied by the intermediate segment is taken *entirely* from what would otherwise have been the distance portion of the lens. Any substantial departure from this practice, such as placing the top of the intermediate segment at the customary bifocal height, will make the near segment inaccessible for close work. On occasion, the intermediate segment may be placed in a unique position for a specific vocational purpose, but this is rarely done.

For the example given above, Figure 8-58 shows that if the top of the bifocal segment is 2 mm below the lower limbus (and if the vertical diameter of the cornea is 12 mm with a 3-mm pupil) the top of a 7-mm segment will fall slightly above the lower margin of the pupil.

When a trifocal lens is fitted for occupational use (that is, for a person who must routinely work at an intermediate distance), as a general rule the top of

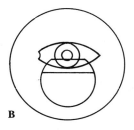

FIGURE 8-58. **(A) Segment height for a bifocal lens (B) Trifocal segment height: The top edge of the near segment of the trifocal lens (B) should be in the same position that it would be in without the intermediate segment.**

the intermediate segment should be placed at the lower edge of the pupil. The usual variations in head position and in physical stature must, however, be taken into consideration. For nonoccupational use, the top of the intermediate segment should normally be placed no more than 1 or 2 mm below the lower margin of the pupil. For a short, stout person or for a chronic "line-noticer," the top of the segment may be placed near the midpoint between the lower limbus and the lower edge of the pupil. However, as with bifocal lenses, many more errors are made in setting segments *too low* than in setting them too high! The problem resulting from fitting a trifocal lens too low is that the use of the reading segment becomes very inconvenient: It requires the wearer to tilt the head back excessively, just as he or she would have to do while wearing a bifocal segment fitted too low.

Several devices are available for measuring segment height, for both near and intermediate segments. For example, glazed frames may be used for marking trifocal segment height. However, it should be understood that the *most accurate* measurement of segment height is made with the spectacle frame to be worn by the patient in place, with the actual dimensions of the frame to be ordered. For a patient's first trifocal prescription it is advisable to use a frame or mounting with *adjustable nose pads*, since such a frame will allow some latitude in adjusting the segment height after dispensing.

# INVISIBLE BIFOCALS AND
## PROGRESSIVE ADDITION LENSES

### 8.28. Invisible Bifocals

Upon reaching the age at which a reading addition is required, the majority of ametropic patients will readily (although reluctantly) accept the practitioner's advice that bifocal lenses will offer the best solution to their problems. Although the use of two pairs of glasses—one for distance vision and one for reading—may be considered as an alternative, most people who try this system eventually find (as Benjamin

Franklin did, 200 years ago) that the constant switching from one pair to the other becomes very tiresome. However, some people, if given a choice, will accept bifocals only on the condition that the bifocal segments are as *invisible* as possible.

As discussed earlier, an inexpensive method of making bifocal segments *almost* invisible is to use a fused no-chrome round segment with a flesh toned tint such as Softlite A or Cruxite A. Such a segment has a feather edge all the way around, and its invisibility is enhanced by the fact that the chromatic dispersion is almost the same (having almost the same nu value) for the barium crown segment as for the ophthalmic crown major lens. Although a round-top segment is a little more difficult for most people to become adapted to than a straight-top segment, this lens, on the whole, provides very good optical performance. One reason for the slower adaptation is the fact that the widest portion of the segment is at the center (rather than at or near the top, as with a straight-top segment) with the result that the eyes must be lowered farther into the segment for useful near vision; another reason is the larger amount of *image jump*, when compared to either a fused or a one-piece straight-top segment.

An often overlooked advantage of the round segment fused bifocal is its very small amount of *differential displacement at the reading level.* If a 20-mm round segment is fitted with the segment top *2 mm above the ciliary line*, there will be only a small amount of differential displacement at the reading level (if we assume a reading level 10 mm below the position of the visual axis in the primary position). Although fitting the segment top at this height may result in a segment that is too high for many patients (depending upon factors such as the lens size and shape, pantoscopic tilt, and anatomical factors) it is nevertheless a good idea to fit this bifocal relatively high—*at least* at the level of the ciliary line. This will result not only in a minimum of differential displacement at the reading level but also (and perhaps more important) will make it unnecessary for the patient to throw his head back to use the wider portions of the bifocal segment.

It should be recalled that while the Kryptok bifocal is a holdover from the era when all lenses had a "6.00 D base curve," no-chrome round fused bifocals (such as the Tillyer D and the Vision-Ease CRF) are high-quality corrected-curve lenses. In summary, although a straight-top segment has obvious advantages over a round segment, when a patient requests an "invisible" bifocal the practitioner should seriously consider a tinted, fused, no-chrome, round segment bifocal.

Multifocal lenses designed for the specific purpose of being *invisible* are available in two cate-gories: (1) blended bifocals and (2) progressive addition lenses. A *blended bifocal* is a one-piece round segment bifocal in which the transition between the major lens and the bifocal segment has been blended over a relatively small zone, in order to reduce its visibility. A *progressive addition lens* is also a one-piece lens with no visible segment area, but it differs from a blended bifocal in that there is a progressive increase in plus power over a relatively large area between the upper (distance) and lower (near) portions of the lens. The portion of the lens within which the power of the addition varies is known as the *progressive zone:* The power of the addition in this corridor varies, in a continuous manner, from zero in the distance portion of the lens to that of the full add in the near portion. Although progressive addition lenses are advocated for the purpose of providing a continuity of vision from infinity to the near point, and hence satisfying the needs of the patient who would otherwise wear trifocals, often the predominant factor motivating the use of this lens (on the part of practitioners as well as patients) is the fact that the "segment" is completely *invisible.*

## 8.29. Blended Bifocals

When a blended bifocal lens is made, the procedure by means of which the one-piece segment is blended results in an annular band, from 3 to 5 mm in width, encircling the segment (see Figure 8-59A). The width of the annular band varies from one manufacturer to another, and depends upon the base curve of the lens and the power of the add. It should be understood that although the power in a blended bifocal changes gradually (rather than abruptly) from the distance to the near portion of the lens, the blended area is *unusable,* as far as the wearer is concerned: There is no usable "progressive zone," as with a progressive addition lens. It should also be understood that none of the blended bifocal segments is completely invisible, especially when viewed obliquely: Each has a *bump* on the front surface, and the higher the power of the add, the larger and the more noticeable the bump.

The blended bifocal was first introduced in 1946, by Howard Beach, as the *Beach Blended* bifocal. This lens is no longer available. The *Younger Seamless* bifocal was introduced in 1954, and is currently available in ophthalmic crown glass, in CR-39 plastic, in photochromic glass, and in High-Lite (high-index) glass. In 1979 Coburn Optical began marketing the Coburn *E-Z 2 Vue* lens. It is available in CR-39 plastic only. Shortly afterward, many other manufacturers intro-

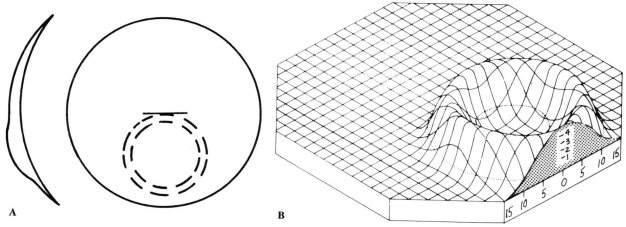

FIGURE 8-59. (A) A blended bifocal lens has an annular transition area, 3 to 5 mm wide, which is unusable because of unwanted astigmatism. (B) The annular-shaped zone of unwanted astigmatism in a blended bifocal lens. (From American Optical Truvision Progressive Lenses, American Optical Corp., 1984.)

duced their own blended bifocal lenses. American Optical introduced the *Ultravue Blended* bifocal, also made only in CR-39 plastic; Sola Optical followed with the *Sola Blend 2000*, also a CR-39 product; Silor Optical introduced the *Super Blend 25*, another CR-39 plastic lens; and Robinson Houchin designed the *Ultimate* blended lens, still another CR-39 plastic lens.

In the manufacture of a blended bifocal, when the line of intersection between the major lens and the segment is blended, the rate of change of curvature in the radial meridians (those meridians passing through the segment center in all directions) is substantially more rapid than the rate of change of curvature in the circumferential meridians. The difference between these two curvatures, at any point in the transition zone, represents *cylinder power*, usually referred to by lens designers simply as *astigmatism*. This unwanted astigmatism (see Figure 8-59B) is an inevitable consequence of manufacturing a "blended" bifocal, and cannot be avoided. The amount of astigmatism is a function of the rate of change in curvature, which in turn depends upon the width of the blended area and the power of the add. The amount of astigmatism decreases as the width of the blend decreases; so the manufacturer may choose either a wide blended zone having a smaller amount of astigmatism, or a narrow blended zone having a larger amount of astigmatism. And, of course, the narrower the blended area, the narrower the *unusable* portion of the lens.

Knoll[7] found that the magnitude of the cylindrical component in the transition zone of a blended bifocal was roughly equal to the power of the add. For a given distance from the center of the segment, the amount of astigmatism will be the same, but the

principal meridians of the unwanted cylinder will vary with the direction of the patient's gaze (since one of the principal meridians will be parallel to the direction of the radial meridian and the other principal meridian will be parallel to the direction of the corresponding circumferential meridian).

In fitting and dispensing the blended bifocal, the top of the blended zone should normally be placed 1 or 2 mm above the lower ciliary line. This will place the astigmatism-free area of the segment in the most favorable position. The manufacturer of a blended bifocal places an ink mark on the lens, at the upper end of the blended zone (Figure 8-59), to assist the practitioner in arriving at the segment height. In order to provide as large a vertical field of view through the segment as possible, it is important to select a frame with a deep vertical dimension. It is best to use a corneal reflection method of measuring the PD, and to fit the segments to the monocular PD in order to ensure maximum use of the entire segment width.

## 8.30. Progressive Addition Lenses

A progressive addition lens is a one-piece lens having distance and near portions that are relatively stable in power and free of aberrations, with a "progressive zone," extending across the entire width of the lens, and connecting the distance and near portions. As shown in Figure 8-60, the central portion, which is the usable area of the progressive zone, is known as the "progressive corridor." Within the progressive corridor, the power increases continuously from the

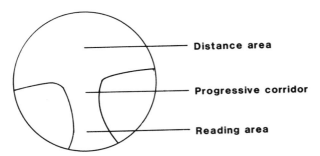

**FIGURE 8-60. Progressive addition lens, showing progressive corridor connecting the distance and near portions.**

distance to the near portion of the lens. All powers between the distance and near prescriptions are present in the progressive corridor. There is no visible reading segment and there are no dividing lines; hence there is an absence of "image jump."

The advantages of clear vision at all distances and the absence of a visible reading segment are partially offset by areas of the lens where the aberration of *astigmatism* is present. The astigmatism occurs as a result of the fact that a progressive addition surface (normally the front surface of the lens) is produced by generating an *aspherical* curvature, a process which produces local (and gradual) variations in refractive power as well as astigmatism. It is impossible, both mathematically and physically, to design and construct a progressive addition lens which does *not* produce astigmatism in the lateral portion of the progressive zone.

The principal parameters of a progressive addition lens are interrelated, and include (*a*) the size of the distance and near areas, (*b*) the types and intensity of the aberrations, and (*c*) the depth and usable width of the progressive corridor. Differences in lens design reflect differences in the designer's priorities. Compromises must be made between areas of high image performance and the severity of the aberrations.

When an aspherical surface having a variable radius of curvature is generated, the inherent astigmatism found to the right or to the left of the line at the center of the progressive corridor (a line known as the "umbilicus") is proportional to the *rate of change in curvature*. Hence, there are two basic approaches to the design of progressive addition lenses. These are sometimes referred to as "hard" and "soft" designs, and refer to the amount and distribution of the astigmatism on the convex lens surface.

1. Designs that result in substantial amounts of astigmatism and distortion that are restricted to small regions of the lens surface are known as "hard" designs. Such a lens provides relatively large areas of high-quality imagery (the distance portion, the progressive corridor, and the near portion), but at the expense of a large amount of astigmatism in the lateral portions of the progressive zone.

2. Designs in which astigmatism and distortion are minimal but are widely distributed throughout the lens are known as "soft" designs. Such a lens provides smaller areas of high-quality imagery (due to the fact that astigmatism encroaches into the distance portion, the progressive corridor, and the near portion) but has less astigmatism in the lateral portions of the progressive zone.

When the visual axes pass through areas of the lenses where astigmatism is present, the wearer's vision is not only blurred but objects in the visual field tend to "rock" or "swim," especially under conditions in which relatively large, and repeated, eye movements are made. These subjective symptoms are directly related to the astigmatism and to changes in magnification resulting from the gradual increase in power toward the bottom of the lens; the more severe the symptoms, the longer the adaptation period.

It is possible to calculate the approximate amount of astigmatism adjacent to the progressive corridor.[8] Assume that no astigmatism exists along the center line (umbilicus) of a symmetrical aspherical surface, and that a refractive change of 1 D occurs along the center line over a vertical distance $Y$ (see Figure 8-61). Along a vertical line parallel to the center line and at a lateral distance $Y$ from the umbilical line, the astigmatism will be approximately 2.00 D.

### The Omnifocal Lens

The first commercially successful progressive addition lens to be introduced in this country was the

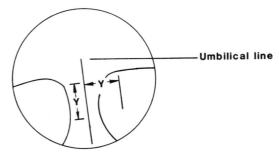

**FIGURE 8-61. The amount of astigmatism at a lateral distance $Y$ from the umbilical line is equal to twice the change in power for the vertical distance $Y$ in this diagram.**

Omnifocal, invented by David Volk and Joseph Weinberg in 1961. It was a glass lens, manufactured by Robinson-Houchin, and marketed to the profession in 1965 by Univis Lens Co. The *front* surface of the lens was the aspherical (progressive) surface. On this surface, the radius of curvature in the *vertical* meridian gradually decreased, from the top to the bottom of the lens, while the radius of curvature in the *horizontal* meridian did not vary. The power of the cylinder (in the horizontal meridian) was +0.75 D greater than the power of the desired addition: For example, for a +1.50 D add, the cylinder on the front surface would have a power of +2.25 D.

Since the spherical power on the front surface of the lens increased gradually from top to bottom with a constant amount of astigmatism, a *plus cylinder, axis 90* was created, whose power was constant in the horizontal meridian. In order to eliminate this plus cylinder and to produce the correction for the eye's astigmatism, a compensating toric surface had to be ground on the back surface of the lens. This resulted in a *bi-toric* lens, with the axis of the back surface cylinder normally at an oblique angle to the 90 degree axis of the front surface cylinder. Calculation of the amount (and axis) of the back surface cylinder for this lens, as well as the manufacturing process, were complicated procedures. These and other problems contributed to the slowness with which this lens was accepted.

The distance and near centers of the Omnifocal lens are separated by a vertical distance of 25 mm. Since the power in the vertical meridian varies over the entire extent of the lens, the total amount of plus power actually *decreases* from the distance optical center upward, while increasing from the distance optical center downward (see Figure 8-62).

The increase in plus power, as the line of sight moves downward through the lens, is *nonlinear*, resulting in a maximum addition of +1.75 D for a lowering of the line of sight by 25 mm, but only one-fourth this amount of added plus power is cre-

ated by lowering the line of sight 12.5 mm. The increase in astigmatism, too, is nonlinear, increasing ever more rapidly as the visual axis approaches the bottom of the lens. Since the progression occurred over the entire front surface of the lens (from top to bottom) this lens was the epitome of a "soft" design. The Omnifocal is no longer available, but a large number of other progressive addition lenses have been introduced.

### The Varilux Lens

Developed by Bernard Maitenaz, the original Varilux lens was introduced by Essel Optical of France in 1959. This lens enjoyed considerable success in Europe before being introduced in the United States, by Titmus Optical Co., in 1967. The original Varilux lens, now called the Varilux 1, differed from the Omnifocal in that the upper half of the lens had no progression in power. The progression was confined to a 12-mm-deep zone in the center of the lens, throughout which the power increased in a linear fashion. Below the progressive corridor there was a zone of maximum addition having a constant power, and having a width of about 22 mm.

The astigmatism-free progressive corridor of the original Varilux lens was approximately 5 mm wide. Outside the hourglass-shaped area containing the distance area, the progressive corridor, and the reading area, the amount of unwanted astigmatism increased toward the bottom of the lens. In the areas lateral to the progressive corridor, the axis of the astigmatism varied in such a manner as to cause a sensation of blurring and vertigo. Unfortunately, the usable width of the progressive corridor decreased with increasing power of the add.

The design of the original Varilux lens evolved over a period of 8 years, and took place in four principal stages.[9] Although the theoretical problems of surface design were challenging, the design and construction of the surfacing machinery proved to be even more difficult.

As shown in Figure 8-63, the distance portion of this lens has a conventional spherical surface. At the beginning of the progressive zone (without any visible line of demarcation) the radius of curvature begins to shorten and uniformly decreases over a distance of 12 mm until the full power of the addition is reached. At that point, the progressive surface merges into a second portion that is essentially spherical, which has the curvature necessary to produce the power necessary for the add. The boundaries of this stabilized near area are, of course, invisible. No

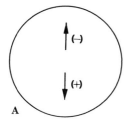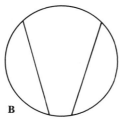

FIGURE 8-62. **The Omnifocal lens: (A) The plus power in the distance portion decreases from the distance optical center upward; (B) the useable area of the lens is wide at the top, gradually narrowing toward the bottom.**

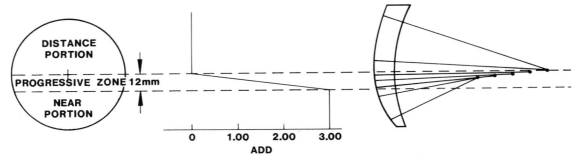

FIGURE 8-63. **The original Varilux lens: The distance portion has no progression of power, but in the progressive zone the radius of curvature decreases over a distance of 12 mm until the full power of the addition is achieved. From** *All About Varilux: The Progressive Addition Lens* **(Titmus Optical, Petersberg, Virginia).**

image jump occurs, either between the lower portion of the distance area and the beginning of the progressive zone, or between the end of the progressive zone and the stabilized near area.

Since the progressive surface is confined to a 12-mm-deep zone (rather than extending from top to bottom) the Varilux lens is considered as a "hard" design.

The umbilical line (the line of symmetry at the center of the progressive corridor) has the following property: For small distances around each point on this line, the vertical and horizontal radii of curvature are essentially equal. Hence, for any point along the line of symmetry, quasi-spherical surface properties exist. With the determination of the curvature along the vertical meridian, and with the requirement that the line of symmetry forms an umbilical line, the progressive surface and its image properties are largely predetermined.

Varilux lenses are manufactured differentially for the right and left eyes, with the line of symmetry inclined nasally toward the bottom of the lens so that it will conform to the convergence of the visual axes for any point in the progressive zone. This inclination provides a segment inset of 2.5 mm from the distance optical center. As a result of this design, equal horizontal eye excursions to the right or the left of the line of symmetry provide uniform power changes (Figure 8-64A). Other progressive addition lenses achieve segment inset by rotation of the segments nasally, in which case equal horizontal eye movements to the right or left of the line of symmetry do not result in uniform power changes (Figure 8-64B). Laboratory finishing to the prescribed distance power, whether spherical or sphero-cylindrical, is accomplished on the *back* surface in the same manner as for any front surface multifocal lens. In the United States, the original Varilux lens was available only in glass.

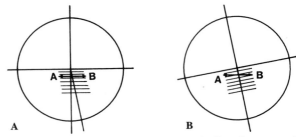

FIGURE 8-64. **(A) With the Varilux lens, the line of symmetry is inclined nasally toward the bottom of the lens, so that equal horizontal eye movements provide uniform power changes. (B) if segment inset is provided by nasalward rotation, equal horizontal eye movements fail to provide uniform power changes. (From Essilor Optical.)**

In 1969, Silor Optical became the distributor for Varilux in the United States. In the same year, Titmus Optical introduced the *Progressor*, manufactured by Benoit Berthiot of France but apparently similar to the Varilux lens. The Progressor lens was known as the *Zoom* lens in France, and as the *Progressive* lens in Germany. The Progressor lens was discontinued by Titmus in 1971.

Essel's patent on the original Varilux lens expired in 1973. In the same year, Essel patented the *Varilux 2* lens, a lens in which the entire front surface (not just the progressive zone) was of aspherical design. As shown in Figure 8-65, the cross sections of curves that form the front surface of the Varilux 2 lens are a family of *conic sections*. Starting in the upper portion of the lens, the conic section is first an oblate ellipse, gradually becoming a circle at the optical center of the lens, below which the section gradually changes to increasingly flatter prolate ellipses, then to a parabola, and finally (in the near segment area) the sections become increasingly flatter hyperbolas. In any horizontal section of the *upper portion* of the lens, the power (plus spherical power) is at a minimum at the midline and increases toward either side;

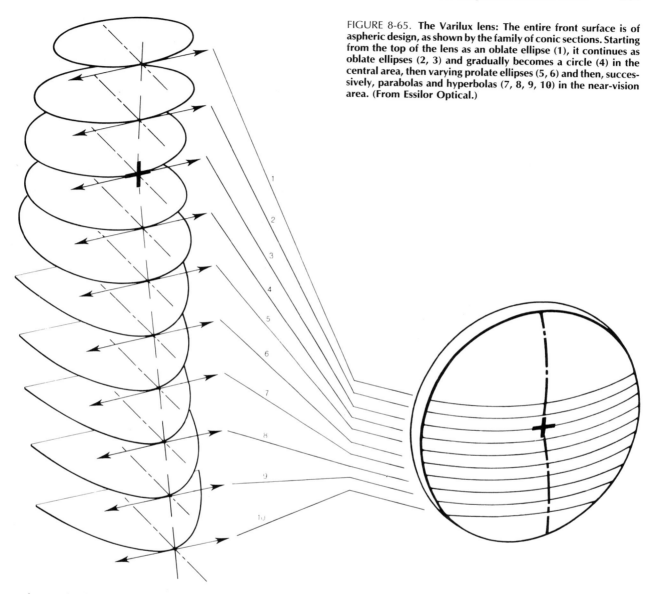

FIGURE 8-65. **The Varilux lens: The entire front surface is of aspheric design, as shown by the family of conic sections. Starting from the top of the lens as an oblate ellipse (1), it continues as oblate ellipses (2, 3) and gradually becomes a circle (4) in the central area, then varying prolate ellipses (5, 6) and then, successively, parabolas and hyperbolas (7, 8, 9, 10) in the near-vision area. (From Essilor Optical.)**

whereas in the *lower portion* of the lens, the plus power is at a maximum at the midline and decreases toward either side.

This design reduces the intensity of the surface astigmatism and distortion in the peripheral areas of the lens, as compared to the original Varilux and therefore can be considered as a "softer" design than the original Varilux. At the same time, the distance area and the segment area, where high image resolution is present, are reduced in size. This modification in design not only reduced the amount of unwanted astigmatism in the peripheral portions of the progressive zone, but controlled the *axis* of the unwanted

astigmatism in such a way that the rectangularity of peripherally viewed objects was preserved, thus reducing the rocking and swaying and the disturbing deformation of the image.

The Varilux 2 is available in CR-39 plastic, ophthalmic crown glass, photochromic glass, and high-index (1.70) glass.

Multi-Optics Corporation, the American distributor of the Varilux 2 lens, has recently introduced a new version of the Varilux progressive addition lens called the *Overview* lens. This lens features a 40-mm round one-piece segment, placed 9 mm above the fitting cross, to allow overhead viewing. Originally designed for airline pilots, it can also be used by anyone else who requires a near correction for overhead viewing. This lens, therefore, is an alternative to the conventional double-segment bifocal.

### The Ultravue Lens

When the patent on the original Varilux lens expired, other manufacturers quickly began to introduce their own versions of progressive addition lenses. In 1973, American Optical introduced the *Ultravue* lens, a CR-39 plastic lens. This lens had a segment area width of 25 mm, and it was later renamed the *Ultravue 25* lens. This was followed in 1978 by the *Ultravue 28* lens, having a segment area 28 mm wide. The Ultravue lens differs from the Varilux 2 by having a well-defined distance portion, free of surface astigmatism, and a relatively wide astigmatism-free segment area, but at the expense of a higher rate of progression in the 10- to 12-mm progressive corridor and larger amounts of astigmatism adjacent to the progressive corridor and the segment area. This lens, therefore, was of the "hard" design category.

A second difference between the Ultravue and Varilux lenses is that the progressive corridor of the Ultravue lens blank lies vertically rather than being angled nasally toward the bottom of the lens. In finishing this lens, the laboratory must rotate the axis of symmetry 10 degrees from the vertical, in order to provide 2.4 mm of segment inset. If a correction for the eye's astigmatism is required, it is incorporated into the back surface of the lens (taking the 10 degree rotation into account).

### The Younger 10/30 Lens

Younger optics introduced the Younger 10/30 lens, a progressive addition CR-39 plastic lens, in 1978. The designation "10/30" is based on the fact that the lens has a 10-mm-deep progressive corridor and a 30-mm-wide usable segment area (having no more than 1.00 D of astigmatism). The Younger 10/30 lens epitomizes the "hard" progressive addition lens design. Not only is the unaberrated distance area larger than that of other progressive addition lenses, but image distortion is eliminated, both nasally and temporally, for a short distance below the mid-horizontal meridian of the lens. The near area is spherical, and although designated as 30 mm wide, the usable width depends on the base curve, the power of the add, and the segment height (i.e., the vertical position of the

segment area): The effective segment width is usually no greater than about 25 mm.

The Younger 10/30 lens differs from the Varilux and Ultravue lenses in that the progression is *nonlinear*, as with the Omnifocal lens. The first one-third of the power change occurs in the upper 5 mm of the progressive corridor, whereas the remaining two-thirds occurs in the lower 5 mm. This means that the amount of unwanted astigmatism, outside the progressive corridor, increases progressively in the downward direction.

Since the lens has large unaberrated distance and near areas, the aberrated areas are small (as typified by the "hard" design). Therefore, the intensity of the aberrations, confined to small areas, is relatively high, making these areas unusable.

### The Super NoLine Lens

Silor Optical, which had marketed the original Varilux lens since 1969, renamed the Varilux lens, calling it the *NoLine* lens, in 1978. In early 1980, Silor Optical began marketing the *Super NoLine* lens, made by Benoit Berthiot of France. This is an improved version of the original NoLine progressive addition lens, having a progressive corridor 12 mm deep. It has wide distance and segment areas, the segment area being about 25 mm wide and is therefore in the "hard" design category. Although surface astigmatism and distortion are present in the peripheral portions of the progressive corridor, the lens is designed to control the axes of the unwanted astigmatism in order to prevent rotation of vertical and horizontal lines of peripherally viewed objects. The Super NoLine lens is made with the line of symmetry inclined nasally, as is the Varilux lens, with the result that separate blanks are required for right and left lenses. This lens is available in CR-39 plastic, in ophthalmic crown glass, and in photochromic glass.

### The Unison Lens

In mid-1980, Univis Lens Co. introduced the *Unison* lens. This lens was initially made only in CR-39 plastic but is now available also in ophthalmic crown and photochromic glass. In 1983, the lens division of Univis was sold to Vision-Ease.

The Unison lens falls into the "hard" design group. It has a completely spherical distance area, a 12-mm-deep but relatively narrow progressive corridor, and a relatively wide segment area. Astigmatism and distortion are concentrated adjacent to the segment area, with little rotation of vertical or horizontal lines of peripherally viewed objects.

### Other Progressive Addition Lenses

In early 1982 Younger Optics introduced the Cosmetic Parabolic Sphere (CPS) progressive addition lens. This lens, made of CR-39 plastic, represents the counterpart to the Younger 10/30 lens: Whereas the Younger 10/30 lens falls into the "hard" design group, the CPS lens is a lens of "soft" design. This lens has an unaberrated distance area, a 12.5-mm-deep progressive corridor of reasonable width, and a small reading area. The orientation of vertical and horizontal lines of peripherally viewed images is fairly well controlled.

In 1982, American Optical introduced the *Truvision* lens. This lens was designed by John Winthrop, who was the designer of the Ultravue lens. While the Truvision lens may be placed in the "hard" design group, it has less astigmatism in the periphery of the progressive corridor than other lenses in this group, because the progressive corridor has been lengthened to 15 mm. It has an unaberrated distance area and a fairly wide segment area, but the segment area is positioned lower in the lens blank because of the increased depth of the progressive corridor. The Truvision lens is available in CR-39 plastic, ophthalmic crown glass, and in Photogray Extra. The areas and distribution of astigmatism for the Truvision, Varilux 2 and the CPS progressive addition lenses are shown in Figure 8-66.

Titmus Optical, in 1983, began marketing the *NeuVue 75* lens, which in 1984 was renamed the *NaturalVue* lens. This lens is currently made in CR-39 plastic but will soon be available in ophthalmic crown glass and in Photogray Extra. It has an aberration-free distance area, a 15-mm-deep progressive corridor, and a fairly wide segment area, and can be considered as a compromise between the "hard" and "soft" design groups. Due to the increased depth of the progressive corridor, the aberrations in the periphery of this portion of the lens are attenuated: In this respect, it resembles the Truvision lens.

In 1984, Coburn Optical Industries began marketing the *Progressiv R*, a lens designed by Rodenstock in Germany in 1981. The lens is available in CR-39 plastic, ophthalmic crown glass, Photogray Extra, and high-index glass. This lens fits into the "hard" design category, having a virtually aberration-free distance area, a 12-mm-deep progressive corridor, and a relatively narrow segment area. While the peripheral portions of the progressive zone con-

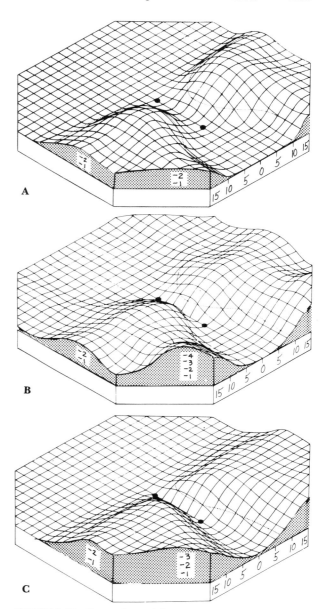

FIGURE 8-66. **Comparison of the areas of astigmatism for the Truvision (A), Varilux (B), and Younger CPS (C) progressive addition lenses. The height of each contour represents the amount of astigmatism. (From American Optical Truvision Progressive Lenses, American Optical Corp., 1984.)**

tain astigmatism, the orientation of vertical and horizontal lines of peripherally viewed objects is largely retained. The lens must be rotated 8 degrees, to obtain 2 mm of segment inset.

Sola Optical, in late 1984, introduced a CR-39 plastic lens called the *VIP* lens. This is a lens of "soft" design, having a large distance area and a relatively large near area, with the unwanted astigmatism widely distributed in the peripheral portions of the lens. Orientation of vertical and horizontal lines of peripheral objects has been well controlled.

In 1986 Seiko Optical Products introduced two progressive addition lenses, the P-2 and the P-3. The P-2, intended for general wear, was of the "soft" design, reducing the magnitude of astigmatism but increasing the area containing the astigmatic error. The P-3 is recommended for active patients and resembles the "hard" design, having larger areas free of astigmatism but with more intense levels of astigmatism concentrated in small regions of the lens.

Polarite, in 1986, developed a plastic *polarized* progressive addition lens, called the *Progressive "M."* This lens may be tinted.

### The Effect of Pupil Size

For a ray of light passing through any point on the line of symmetry in the progressive corridor, the power is essentially spherical and free from astigmatism: But when we consider the situation in which the wearer's eye, having a *pupil of finite size*, looks through the progressive zone, the rays passing through the pupil's upper margin will have a different dioptric value than those passing through the lower margin. With a progressive corridor 12 mm deep and a +2.50 D add, the dioptric difference in power for the upper and lower margins of a 6-mm pupil would be +1.25 D. In addition, the larger the pupil, the smaller the usable area of the progressive corridor. Fortunately, the pupil tends to become smaller with age, so the majority of progressive addition lens candidates have relatively small pupils.

### Edge Thickness

Because of the increase in curvature toward the bottom of the lens necessary to obtain the required increase in power, the lower edge of the lens is substantially *thinner* than the upper edge. This thickness difference creates a *base-up* prismatic effect. The amount of prism created is dependent upon the power of the add, and imposes a limitation on the minimum thickness of the finished lens.

In order to equalize edge thickness, some manufacturers work *base-down* prism over the entire back surface of the lens. The amount of prism required is determined by the difference in the edge thickness between the top and bottom of the lens. Since the thickness difference is the same for both the right and left lenses, the prism is the same for each lens, and no residual vertical imbalance is created. However, both distance optical centers would be lowered for plus lenses, and would be raised for minus lenses. The equalization of the edge thicknesses results in a substantial reduction in overall lens thickness.

## 8.31. Patient Selection and Dispensing Considerations

Patient selection is an important consideration in the fitting of progressive addition lenses. Although these lenses have been designed to provide continuous vision for all distances, many practitioners and their patients consider these lenses simply as "invisible bifocals." If a patient is interested *only* in invisibility, the practitioner should consider the possibility of prescribing either a "no-chrome" round fused bifocal, which has the advantages of being inexpensive and optically sound (as described in Section 8.25), or a "blended" one-piece bifocal (as described in Section 8.26). However, a blended bifocal has the disadvantage of a blurred transition area, and lacks the advantage of vision for intermediate distances. Progressive addition lenses should therefore be considered *both* for patients who require vision for intermediate distances (who would otherwise require trifocal lenses) and for those who request an invisible bifocal.

Progressive addition lenses are normally *not* recommended for (a) patients who are currently wearing bifocals or trifocals (in particular, large, Executive-style trifocals) and are completely satisfied with them; (b) patients who are accustomed to wearing (and are satisfied with) single-vision lenses for reading only; (c) patients requiring vertical prism (which is not available in progressive addition lenses); and (d) patients who are "nervous" or "high-strung." An important consideration, in adapting to progressive addition lenses, is the wearer's ability to become a "head-mover" rather than an "eye-mover." When the wearer successfully learns to move the head (in viewing objects to one side or the other) rather than moving the eyes, he or she no longer complains of the sensation of blurring or "swimming."

Lewis[10] has recently devised a scoring system designed to rule out patients who are unsuited for progressive addition lens wear. A specified number of points is given for each of three factors: age refractive error, and various motivational considerations. If a patient has a score of less than 20 out of a total of 40 possible points, he or she is considered not to be suitable for progressive addition lens wear.

Although each progressive addition lens has a different theory of construction, they all share common design problems: Those of (1) producing distance and near areas of high image performance, (2) merging these zones gradually, and (3) placing the inevitable aberrations and distortions where they will interfere as little as possible with comfortable vision. As discussed previously in this section, the available lenses vary from one another in terms of the size of the aberration-free distance area; the power and width of the usable near area; the depth and usable width of the progressive corridor; and the dimensional orientation, intensity, and location of the aberrations and distortions. Due to these differences, the practitioner should become familiar with the characteristics of the available lenses. After determining that the patient is motivated and has the emotional makeup conducive to wearing progressive addition lenses, the practitioner should analyze the patient's visual characteristics and visual requirements for both work and leisure, taking into consideration:

1. The power of the distance correction
2. The power of the add
3. The pupil size
4. Prior experience with multifocals
5. Habitual eye and head movements
6. Height of commonly viewed near visual objects
7. The relative usage of near and intermediate ranges

Matching the design to the individual patient will increase the potential for patient acceptance. However, appropriate design characteristics can only be selected if the practitioner is provided with the information necessary for making a choice. Most manufacturers of progressive addition lenses have provided qualitative descriptions or comparisons of lens design rather than *quantitative data*. Quantitative data provided by various investigators have not been reported in a standardized format that would permit comparisons of the relevant parameters of lens design. In 1982 the American Optometric Association Commission on Ophthalmic Standards recommended a format for providing a graphical representation of the optical properties of progressive addition lenses that would allow rapid comparison of lens design. Two diagrams would be used, one to display the iso-spherical equivalent lines and the other to portray iso-cylindrical lines and the induced cylinder axis orientation. The selection of the most suitable lens design could be made more intelligently if manufacturers would furnish information in this format.

### Essential Fitting Measurements

The measurements necessary for fitting progressive addition lenses differ from those for the fitting of conventional "segment" multifocal lenses in two respects:

1. Interpupillary distance is measured *monocularly*, as opposed to a single measurement taken with a millimeter ruler.
2. The reference point in the vertical meridian (analogous to "segment height" for a conventional multifocal lens) is the *center of the pupil* rather than the ciliary line.

Because of the relatively narrow width of the progressive corridor, the fitting of progressive addition lenses must be done very precisely. Monocular distance PD measurement is *essential*, in order to make sure that the line of sight of each eye remains in the progressive corridor as the eyes turn downward. This may be done by means of a corneal reflection pupillometer or by the use of a special device made available by the manufacturer of the lens.

Once the monocular distance PD measurement has been made, it is necessary to measure the vertical distance from the center of the pupil to a horizontal line tangent to the lowest point on the bottom edge of the lens (while the patient's head is held erect and the visual axes are in a horizontal plane), as shown in Figure 8-67A. The manufacturer will then place on the finished lens a temporary marking called a *fitting cross* at a point on the lens corresponding to the center of the pupil.

This measurement should be made by using the exact frame (in the correct eye size and bridge size) in which the patient's lenses will be mounted. One method of making this measurement, for an unglazed frame, consists of placing a strip of clear tape on the frame with the adhesive side toward the fitter, the tape being approximately centered across the pupil. The fitter holds a penlight in a position level

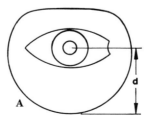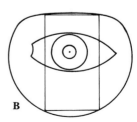

FIGURE 8-67. **Necessary measurements for fitting progressive addition lenses. See text for explanation.**

with the patient's eye, instructs the patient to look at the penlight, and observes the corneal reflex. Using a felt pen, a dot is placed on the tape, directly in front of the corneal reflex (Figure 8-67B). Other methods of making the measurement include the use of segment measuring devices, and special templates supplied by lens manufacturers.

Two systems designed especially for progressive addition measurements are the Grolman Fitting System, supplied by American Optical, and the Magna/Mark System. The Grolman Fitting System attaches directly to the patient's frame, and is provided with movable horizontal and vertical scales for use in determining the necessary measurements. The Magna/Mark System employs magnetically held translucent target spots, which move over a plastic lens that is mounted into the eyewire of the frame. When the target spots are centered before each pupil, the frame is removed and the target spot location is transferred to the plastic lens by a finepoint pen.

For most lenses, the progressive corridor begins about 2 mm *below* the fitting cross. Although the manufacturer's recommendations should be kept in mind, there are times when this fitting technique should be modified. For many patients, this technique will result in the reading area being *too low in the lens* to be of much use. For a wearer who would like to make considerable use of vision at intermediate distances, the fitting cross will need to be placed 1 or 2 mm *above* the center of the pupil. In any event, factors such as head posture, location of near visual tasks (in regard to the vertical level of the lenses), and pupil size are important considerations.

### Frame Selection

The frame selected should provide a minimum vertical distance of *22 mm* from the center of the pupil to a horizontal line tangent to the lowest point on the bottom edge of the lens. For a lens having a 12-mm progressive zone beginning 2 mm below the pupil, the reading area (having the full power of the addition) would then be 8 mm deep. As shown in Figure 8-68, this generally means that the "B" (vertical) measurement must be a minimum of 38 mm. If the distance below the pupil were to be less than 22 mm, the effective depth of the full-power reading area would be so small that the lower rim of the frame would intrude into the patient's visual field when reading (particularly for a patient having large pupils).

Lens shapes having cutaway nasal portions, such as goggle shapes that rapidly slope away from the

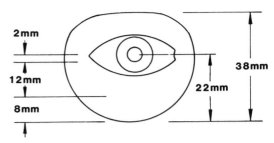

FIGURE 8-68. **The vertical box measurement of a frame for a progressive addition lens should be a minimum of 38 mm.**

nose, should be avoided, particularly if a large amount of inward decentration is necessary.

In order to provide a wide lateral field of view through the progressive corridor and the stabilized near area, it is recommended that the patient's frame should be fitted at as close a vertex distance as possible (no greater than 12 to 13 mm) and should have a pantoscopic tilt of 12 to 15 degrees with a slight amount of face forming. As with other multifocals, adjustable nose pads will allow flexibility in positioning the frame, both at dispensing and afterward.

### Verification of Progressive Addition Lenses

Most progressive addition lenses have two sets of markings, a temporary set and a permanent set, to assist in verification of power and in fitting the lenses to the patient's face. The temporary markings usually consist of a fitting cross which should fall at the center of the pupil, together with a distance reference center and a near reference center for use in checking the powers of the distance and near segment areas (Figure 8-69A). The PD and the vertical height of the fitting cross are confirmed by placing the lens on the manufacturer's centering or verification chart. The temporary markings should be left on the lens until the position of the fitting cross can be verified on the patient's face.

The permanent markings are "semi-invisible," and consist of two engraved circles that establish the horizontal line at the beginning of the progressive corridor (Figure 8-69B). Some manufacturers also place their own identification mark and the power of the add near the engraved circle on the temporal side of the lens. In the event that the temporary markings have been removed, they can be reconstructed by locating, marking, and placing the engraved circles over the manufacturer's centration chart. If the lens is placed against a dark background and viewed with an intense light, the circles can be seen more easily.

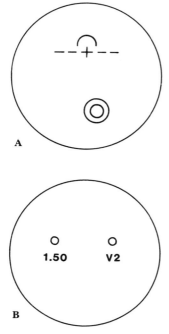

**FIGURE 8-69. Progressive addition lens markings: A, temporary markings; B, permanent "semi-invisible" markings.**

*Patient Communication*

Patient instruction, demonstration, and follow-up are important aspects of progressive addition lens fitting. The patient must be told exactly how to make use of the lenses and should be scheduled for at least one return visit after the lenses have been dispensed, in order to evaluate his or her progress in adapting to the lenses. The patient should understand the importance of maintaining the lenses in correct adjustment: Return visits for readjustment will need to be made on a more frequent basis than for wearers of conventional multifocal lenses.

*Patient Acceptance of Progressive Addition Lenses*

Numerous reports[11–15] indicate success rates for progressive addition lenses (based on patient acceptance) as high as 90% or even higher. The report by Jan[15] indicates essentially the same rate of success (over 90%) with Varilux 1, Varilux 2, Ultravue, and Younger 10/30 lenses. Given the differences in the design philosophies of these lenses, Jan's results would suggest that lens design is not a critical factor in patient acceptance, but that other factors (such as

patient selection and meticulous fittings, verification, and dispensing procedures) together with patient instruction may be of great importance.

In spite of these reports of very high success rates, progressive addition lenses have failed to penetrate the multifocal market in the United States to any significant extent. Whereas progressive addition lenses account for an estimated 30% of the European multifocal market,[16] despite aggressive marketing campaigns by some manufacturers (to the public as well as to the eye professions) progressive addition lenses account for about 16% of the multifocal market in this country.[17]

Although the reasons for their lack of use in the United States are hard to determine, the perceived advantages of progressive addition lenses are apparently not yet great enough to override the accepted benefits of conventional multifocal lenses. However, the market for progressive addition lenses will undoubtedly expand due to consumer awareness; an increase in the knowledge, skill, and confidence of many practitioners; the positive recommendations of practitioners currently wearing the lenses; and referrals by successful progressive addition lens wearers.

## References

1. Fry, G.A. and Ellerbrock, V.J. Placement of Optical Centers in Bifocal Lenses. *Optom. Weekly*, Vol. 37, No. 35, pp. 989–996, 1941.
2. Sterling, S. Ophthalmic Lenses: Their History, Theory, and Selection, pp. 53–81 in *Modern Bifocal Lenses*, Bausch and Lomb Optical, Rochester, N.Y., 1935.
3. Sheedy, J.E., and Saladin, J.J. Exophoria at Near in Presbyopia. *Amer. J. Optom. Physiol. Opt.*, Vol. 52, pp. 474–481, 1975.
4. American National Standard Recommendation for Prescription Ophthalmic Lenses, Z80.1-1979. American National Standards Institute, Inc., New York, 1979.
5. Field, K. The "Correct" Method of Checking Bifocals. *Opt. J. Rev. Optom.*, pp. 19–20, Feb. 1, 1966.
6. Holmes, C., Jollife, E., and Gregg, J. *Guide to Occupational and Other Visual Needs*, Vol. 1. Silverlake Lithographers, Los Angeles, 1958.
7. Knoll, H. The Optical Characteristics of Beach Blended Bifocals. *Amer. J. Optom. Arch. Amer. Acad. Optom.*, Vol. 29, pp. 150–154, 1952.
8. Davis, J.K. Aspheric Lenses: What's Possible and What Isn't. *Rev. Optom.*, Vol. 115, No. 5, pp. 68–74, 1978.
9. Maitenaz, B. Four Steps That Led to Varilux. *Amer. J. Optom. Arch. Amer. Acad. Optom.*, Vol. 43, pp. 441–450, 1966.
10. Lewis, R. *Conquest of Presbyopia*, pp. 42–45. R & M Lewis, Sydney, Australia, 1983.

11. Wittenberg, S. Field Study of New Progressive Addition Lens. *J. Amer. Optom. Assoc.*, Vol. 49, pp. 1013–1021, 1978.

12. Chapman, D.T. One Clinic's Experience with Varilux 2—The First 400 Patients. *Optom. Monthly*, Vol. 69, pp. 946–949, 1978.

13. Betouray, A. A Study of 250 Varilux 2 Prescriptions. *Optom. Monthly*, Vol. 70, pp. 885–888, 1979.

14. Augsburger, A., Cook, S., Dietrich, R., Shackleton-Hardenstein, S., and Wheeler, R. Patient Satisfaction with Progressive Addition Lenses in a Teaching Clinic. *Optom. Monthly*, Vol. 75, pp. 67–72, 1984.

15. Jan, D. The Results of Fitting Progressive Addition Lenses. *Opt. Index*, Vol. 55, pp. 77–78, 1980.

16. Young, J.M. The Progress of Progressives. *Ophthal. Optician*, Vol. 24, pp. 300–306, 1984.

17. Vinciguerra, S. The Progressive Boom. *20/20*, Vol. 12, No. 3, p. 100, 1985.

## Questions

1. A bifocal lens is to be made for a patient to the following specifications:

   OD +2.00 DS

   Add +3.00 D (22-mm round fused seg, 16 mm high)

   Lens size 46 × 40 mm

   Assume that the bifocal inset is zero and that the distance optical center is located at the geometrical center of the lens. What is the "jump" at the top of the segment?

2. A bifocal lens is to be made for a patient to the following specifications:

   OD +2.00 DS

   Add +2.00 D (22-mm round fused seg, 16 mm high)

   Lens size 44 × 40 mm

   The patient reads through a point 9 mm below the distance optical center. Assume that the bifocal inset is zero and that the distance optical center is located at the geometrical center of the lens. If the reading distance is 50 cm, what is the "jump" at the top of the segment *expressed in centimeters?*

3. A bifocal lens is to be made for a patient to the following specifications:

   OD +4.25 DS −1.75 DC x 180

   Add +2.00 D (22-mm round fused seg, 16 mm high)

   Lens size 44 × 40 mm

   The patient reads through a point 9 mm below the distance optical center. Assume that the bifocal inset is zero and that the distance optical center is located at the geometrical center of the lens. What is the differential displacement at the reading level?

4. A bifocal lens is to be made for a patient to the following specifications:

   OD +3.25 DS +1.75 DC x 180

   Add +2.00 D (22-mm round fused seg, 16 mm high)

   Lens size 44 × 40 mm

   The patient reads through a point 10 mm below the distance optical center. Assume that the bifocal inset is zero and that the distance optical center is located at the geometrical center of the lens. What is the total vertical displacement at the reading level?

5. The prescription is −2.00 DS; add +2.00 D; Ultex A bifocal; lens size 46 × 40 mm; seg 17.0 mm high; distance optical center at the geometrical center of the lens. What is the vertical prismatic effect everywhere in the segment?

6. The prescription is OD +1.50 DS, OS +1.50 DS; +2.00 D add. The top of the seg is 6 mm below the distance optical center. The reading level is 6 mm below the top of the seg. What size of round bifocal would give *zero vertical prismatic effects* at the reading level?

7. A bifocal lens is to be made for a patient to the following specifications:

   OD +2.00 DS

   Add +1.50 D (Ultex A, 17 mm high)

   Lens size 42 × 38 mm

   Assume that the bifocal inset is zero and that the distance optical center is located at the geometrical center of the lens. Where is the combined optical center or point of zero vertical prismatic effect of the segment located?

8. In a front surface fused bifocal the power of the distance portion is −6.00 D and the power through the near portion is −4.00 D. The power of the fused surface formed by the rear surface of the segment and the countersink surface is +2.00 D. What is the power of the back surface of the distance portion?

9. What is the dioptric value of the countersink surface in air of a front surface fused bifocal whose specifications are as follows: −4.00 DS, +2.00 D add, +10.00 D base curve, index of refraction of glass for distance lens = 1.50, index of refraction of segment glass = 1.60?

10. A 20-mm round Kryptok bifocal is made with a +10.00 D curve on the front side of the distance

portion. The power of the seg in air is +18.00 D. The index of the glass of the major lens is 1.5, and the index of the seg is 1.6. What is the power of the add?

11. A fused front surface bifocal has a carrier (distance) portion made with an index of refraction of 1.50 and the segment is made with an index of refraction of 1.60. The power of the add is +2.50 D. What is the power of the segment in air?

12. A fused trifocal is made with a +6.00 D front surface on the distance portion with a countersink surface of −2.00 D. The index of refraction of the distance portion is 1.48 and the near add is +2.00 D. If the intermediate add is to be half of the near add, what are the indices of refraction of near and intermediate segments?

13. The index of refraction of the segment flint glass is 1.69; the index of refraction of the carrier crown glass is 1.53. If the reading add is +2.00 D, and the seg is 22 mm round, what is the center thickness of the segment?

14. A 20-mm round fused bifocal has a distance prescription of −4.00 D and a +3.00 D add. The surface power on the distance front side is +6.00 D; index of refraction of carrier is 1.52; index of refraction of seg button is 1.65. After the segment of this lens has been fused to the carrier lens, what powers need be ground on the front and back surfaces of the distance portion so as to obtain a +7.00 D add and a distance prescription of −9.00 D?

15. A 20-mm round front surface fused bifocal has a distance power of −4.00 D and a +4.00 D add; +6.00 D base curve; index of refraction of the carrier is 1.52; index of refraction of the seg is 1.65. The lens size is 52 × 42 mm and the bifocal height is 19 mm. If the center of the back of the seg is covered by 0.6 mm of the distance portion, what would be the center thickness of the distance lens?

16. A golfer requires a low-segment fused 20-mm round bifocal. The round bifocal lens blank is 56 mm in size. The lower edge of the segment is 6 mm above the lower edge of the blank and the finished lens size is 48 × 40 mm. What is the minimum segment height obtainable?

17. A lens measure reads as follows on an Ultex B lens: +6.00 D along the 180° meridian, +7.50 D along the 90° meridian; the spherical curve is −6.50 D, and the segment surface power is −4.25 D. What is the approximate prescription of this lens?

18. Executive bifocal prescription, +2.00 DS −1.50 DC x 180, add +2.00 D, is made on +6.50 D base curve. What are the surface powers on the back surface in the principal meridians?

19. The pupil of a patient wearing Varilux 2 lenses with an add of +2.75 D is 6 mm in diameter. It is assumed that the progressive zone is 14 mm deep and that the power increases linearly in the zone. When the entire pupil is in the progressive zone, what is the approximate difference in power of a ray passing near the upper edge of the pupil and near the lower edge of the pupil?

20. A 20-mm round fused front surface bifocal is made with a +8.50 D front surface of the distance portion. The distance power is +4.50 D and the add is +2.00 D. The index of refraction of the major lens is 1.55, nu = 60. The index of refraction of the glass of the segment is 1.66, nu = 40. What is the total longitudinal chromatic aberration through the reading portion of the lens?

21. A fused bifocal is made with a tinted carrier glass with an index of refraction of 1.54 and a white (clear) segment with an index of refraction of 1.63. The distance portion has a power of +3.00 D with 2.7 prism diopters base up; the add is +3.00 D. The front surface power of the distance portion is +9.00 D. The round segment is 22 mm in diameter. At the geometric center, the lens is 4 mm thick. The top of the segment is located 1 mm below the geometric center. If the transmission is 10.0% through the geometric center, what is the *difference* in transmission through the geometric center and through the optical center of the segment?

# CHAPTER NINE

# Eyewear Design and Dispensing

No matter how careful the practitioner has been in his or her examination, and no matter how expertly a pair of ophthalmic lenses have been prescribed, they will be of little value to the intended wearer unless they are mounted in a well-fitting and cosmetically pleasing frame or mounting. Attention to the design of eyewear—including the frame or mounting as well as the lenses—is an important aspect of the vision care practitioner's daily work.

Although optometrists have historically performed the functions of eyewear design and dispensing themselves, in recent decades this task has largely been delegated to assistants. However, this work is sufficiently important so that the practitioner should be fully conversant with it: This includes being able to evaluate (and to criticize, if necessary) the work of the assistant and—since relatively few optometric assistants have had prior training—being able to train newly hired assistants.

Whether eyewear design and dispensing are done by the optometrist or by an assistant, extreme care must be taken in all phases of the work, from the physical measurements such as interpupillary distance and segment height to frame selection, verification of the finished eyewear, and adjustment of the eyewear to the patient's face.

This chapter describes in some detail the various types of spectacle frames and mountings, beginning with a brief historical introduction, and then goes on to present the principles involved in the measurement of interpupillary distance, frame selection, the ordering of materials, and the verification and dispensing of eyewear.

# SPECTACLE FRAMES AND MOUNTINGS

## 9.1. Historical Introduction

Whereas an optical device intended for use by only one eye is known as an *eyeglass*, or a *spectacle*, such a device intended for use by both eyes may be referred to as a pair of *eyeglasses*, a pair of *spectacles*, or simply as a pair of *glasses*.

As discussed in Chapter 1, the art of glassmaking was developed, and glass was used for a variety of purposes, hundreds of years before the invention of spectacles. Evidence suggests that the first visual aid was a simple magnifying glass, or *eyeglass*, mounted in a rim of wood, metal, bone, horn, or leather, and held in front of the eye by means of a short handle. Eventually, two such devices were hinged in such a way that one eyeglass was placed before each eye (see Figure 9-1), constituting a rudimentary pair of eyeglasses or spectacles.

The true inventor of spectacles is not known, but the honor has been attributed to several individuals including Roger Bacon, a monk at Oxford who died in 1294; Alexandria de Spina, a monk at Pisa who died in 1313; and Salvino d'Armati of Florence who died in 1317. Bacon, in his *Opus Magnus*, written in 1267, described the magnification of letters or small objects brought about by a strong plano-convex lens held on a page of manuscript. He concluded that such an instrument could be useful for those having weak eyes. No evidence exists that Bacon moved the lens to the eye, or mounted lenses in any frame-like structure. The reference to the invention of glasses by de Spina can be found in a manuscript belonging to the monastery of St. Catherine's, in Pisa, whereas the claim of d'Armati is made on an epitaph in the church of Santa Maria Maggiore, in Florence. The inscription (translated from the Italian) reads:

HERE LIES
SALVINO D'ARMATI OF THE ARMATI
OF FLORENCE
INVENTOR OF SPECTACLES
MAY GOD FORGIVE HIM HIS SINS
IN THE YEAR OF OUR LORD 1317

Owing to a lexigraphic error on the epitaph and a discrepancy between the age of the bust on his tomb and the date of the epitaph, investigators believe that d'Armati's claim is spurious.

FIGURE 9-1. **An early spectacle frame.**

The earliest written evidence of the invention of spectacles was given in a sermon by the monk Giordano da Rivalto, who stated in 1305 that it was not quite 20 years since there was found the art of making eyeglasses.[1] This statement places the invention of spectacles at about 1285. His sermon was recorded and is still in existence in the Medicio-Laurenziana Library, in Florence.

From their inception, early spectacles, because of their weight and construction, were difficult to mount securely before the eyes. Initially, hinged or riveted spectacles, clamped to the nose, were uncomfortable and often interfered with breathing. Later, various devices in the form of head bands were used. Late in the sixteenth century, Spanish spectacle makers were using loops of silk or cord that were attached to the outer edge of the frames and extended to loop securely over the ears. Around 1730 Edward Scarlett, an English optician, is credited with perfecting rigid side pieces (temples), which became the precursors of the temples of modern spectacles.

## 9.2. Modern Frames and Mountings

The various types of devices used for supporting spectacle lenses before the eyes can be classified under two categories: *frames* and *mountings*. Although these terms are often used interchangeably, the two terms have different meanings.

A *frame* has three parts: a *front*, which encircles the lenses, a *bridge*, and a pair of *temples*, and may be made of either metal or plastic (Figure 9-2). A *combination*

A

B

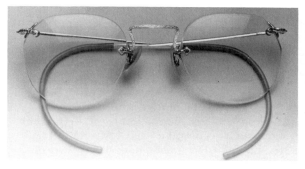

C

FIGURE 9-2. **Modern spectacle frames: A, metal; B, plastic; C, combination.**

*frame* combines both metal and plastic, the upper part being made of plastic and the lower part (and in some cases the entire "eyewire") being made of metal. Typically a metal frame or a combination frame is provided with adjustable *pads*, designed to fit against the sides of the nose, while a plastic frame has stationary pads or none at all.

A *mounting* is a device which holds the two lenses in front of the eyes, but does not completely encircle the lenses. Mountings are typically made of gold-filled materials, and are classified as either rimless or semi-rimless.

A *rimless mounting* (Figure 9-3) has five parts: a *center*, or bridge, which holds the two lenses together nasally; two *endpieces*; and two *temples*. Two holes must be drilled in each lens, one nasally (to attach to the center) and one temporally (to attach to the end-piece). The center, or bridge, is provided with two

adjustable pads, fitting against the sides of the nose and carrying the weight of the mounting and the lenses. Each pad has an extension, called the *pad arm*, holding it to the center. The *strap* is a device that butts up against the lens, at both the endpiece and the center.

A *semi-rimless mounting* has three parts: a *front*, which includes the bridge and two arms, and two *temples*. There are two versions of the semi-rimless mounting. The first version to be developed, the American Optical *Numont* (Figure 9-4), attaches only to the nasal side of the lens, requiring only one hole per lens. This is a very lightweight mounting, making use of a *tri-flex spring*, located where the lens attaches to the mounting. Although the Numont mounting looks fragile, this springs tends to keep the lenses from breaking in the presence of bumps and shocks.

A later version of the semi-rimless mounting was

FIGURE 9-3. **A rimless mounting.**

FIGURE 9-4. **A Numont semi-rimless mounting.**

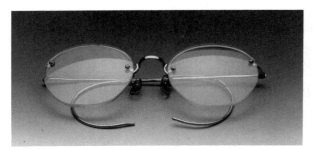

FIGURE 9-5. **A Rimway semi-rimless mounting.**

the American Optical *Rimway* (Figure 9-5) or the Shuron *Shurset*, which attaches to both the nasal and the temporal sides of each lens (requiring two holes per lens). Although appearing to be less fragile than the Numont mounting, this mounting has the problem that the temporal corner of the lens may break if the glasses are bumped, with the result that it may be more fragile than the Numont.

In addition to conventional frames or mountings (having bridges and temples), other devices are available for holding a pair of lenses in front of the eyes (or a single lens in front of one eye). The term *pince-nez*, as well as the term *eyeglasses*, is used for a pair of lenses that are attached to the head by pinching the nose (having no temples). The term *lorgnette* indicates either a pair of lenses or a single lens (usually a pair of lenses) held in front of the eyes by means of a handle. The term *monocle* is used to designate a single lens (looking very much like a trial case lens), usually used only for occasional wear, and held in front of the eye by means of pressure exerted by the facial and brow muscles.

## 9.3. Metal Frame and Mounting Materials

If there is no marking (engraving) on a metal frame, the frame can be made of *any* material, but it is illegal to represent the frame as "gold" unless the gold content is specified. In order for a spectacle frame or mounting to be said to contain gold, it must meet specific government standards.

The chemically pure, unalloyed metal is known as *fine gold*. A spectacle frame or mounting made of fine gold would be too soft to be of any use, so other metals are added to improve gold's hardness and durability. According to the *karat* system of designation, *fine gold* is 24 karat gold. In any gold alloy, the proportion (by weight) of fine gold to the weight of other metals is expressed in terms of karats, that is, as a multiple of 1/24 times the total weight. Thus, "10 karat gold" means 10 parts of gold combined with 14 parts of other materials.

According to the Federal Trade Commission, the word "gold" used alone (without a karat designation) indicates fine gold of 24 karat quantity. The term "solid," when used with respect to fine gold or gold alloy, refers not to the fineness or the karat quality of the gold but to the fact that the article is "solid" rather than "hollow" (or rather than containing a core of a base metal, such as a wire-core temple). The term "solid gold" (or "solid 10 karat gold") may be properly used only as a descriptive term for an article that is made entirely of fine gold (or of 10 karat gold) and is solid rather than being hollow or having a core made of a base metal. The term "solid," used with respect to a gold alloy article, must immediately *precede* (not follow) the karat marking, that is, "solid 12 karat gold." If an alloy having a fineness of *less than 10 karat* is used, it cannot be represented as "karat gold."

The majority of metal frames and rimless or semi-rimless mountings are categorized as *gold-filled*. Gold-filled material consists of a base metal (usually a special steel or copper alloy) onto which a layer of gold has been bonded by the use of heat and pressure. The term gold-filled actually has two meanings: It is used to specify the *process* by means of which the material is made (as just described) and also as a *mark of quality*.

In an article that is marked "gold-filled," the weight of the gold must be at least 1/20 of the total weight of the item, and the fineness must be at least 10 karat. Gold-filled may either be spelled out or may be abbreviated "GF." The fraction does not have to be included in the marking as long as the gold material is at least 1/20 of the total weight: In the marking "10 K GF," 1/20 is implied. If an article is made by the gold-filled process and the weight of the gold is *less than* 1/20 of the total weight, the term gold can be used together with the correct descriptive term, such as *gold plate*, *gold overlay*, or *rolled gold plate*. However, the fraction (karats) must be specified. For example, the abbreviation "1/30 10 K GP" means 1/30 10 karat gold plate.

Gold may also be applied to the base metal by means of an *electroplating* process. The layer of gold must be at least 7 millionths of an inch thick, but *heavy gold electroplating* is 100 millionths of an inch thick. Abbreviations cannot be used for gold applied by electroplating: The words must be spelled out, to avoid any misunderstanding.

Finally, if the word "gold" is used alone, it implies *pure* gold; and the use of terms such as "embodying gold," "dura-gold," "mira-gold," "gold-craft," and "gold-effect" are misleading if not illegal.

The responsibility for the correct marking of gold can be placed at any point in the distribution of a spectacle frame or mounting, including the manufacturer, the importer (if a foreign frame), the distributor, or the dispenser. Mislabeling or misrepresentation is a misdemeanor, punished by a fine of not more than $5,000 or imprisonment of not more than one year, or both, for *each article* improperly represented. The criminal provisions of the National Stamping Act are enforced by the U.S. Department of Justice. In addition, competitors injured by mislabeling practices in the jewelry trade associations may bring civil action.

## 9.4. Plastic Frame Materials

As discussed in Chapter 1, there are two types of plastic materials, thermosetting and thermoplastic. A *thermosetting* material begins as a liquid and becomes solid during the manufacturing process, due to the application of heat and pressure. Once the product has been manufactured, it will never again soften to any significant extent, even at elevated temperatures or with the application of pressure: Under these circumstances it will simply decompose. Examples of thermosetting plastics are *melanines* (used, for example, for Melmac dishes); *phenolics* (Bakelite); *polyesters* (used for clothing and other materials); and *allyls*. It will be recalled that the plastic lens material CR-39 is *allyl diglycol carbonate*.

A *thermoplastic* material has the property of softening when heated and hardening when cooled. No matter how often the process is repeated, the basic structure of the material remains unchanged. Examples of thermoplastic materials are *acrylics* (Plexiglas, Perspex, and Lucite), including *polymethyl methacrylate*, used to make "hard" contact lenses; *polycarbonate*, the toughest spectacle lens material available; *vinyl*, used for seat covers and upholstering; and two materials used for the manufacture of spectacle frames, *cellulose nitrate* and *cellulose acetate*.

Although cellulose nitrate and cellulose acetate are similar in appearance, when used for spectacle frames they have very different properties. Cellulose nitrate, also called *zylonite* (and called *celluloid* by the film industry), is superior to cellulose acetate in the following respects:

1. Cellulose nitrate is tougher than cellulose acetate, and because of its inherent toughness, frames

can be made thinner with it than with cellulose acetate.
2. The surface of cellulose nitrate is harder, allowing it to take a better polish, which is easily maintained.
3. Cellulose nitrate is somewhat easier to work, as it can be stretched when heated and shrunk when cooled without measureable deterioration.
4. The softening point of cellulose nitrate is higher than that of cellulose acetate and its water absorption is lower, so that it has better dimensional stability in warm and humid climates.

On the other hand, cellulose acetate is superior to cellulose nitrate in three respects:

1. Cellulose acetate can be produced in a much shorter period of time than cellulose nitrate.
2. Cellulose acetate frames are more *colorfast* than those of cellulose nitrate.
3. More important, cellulose acetate frames are far *less combustible* than frames made of cellulose nitrate.

Although cellulose acetate frames may burn, the propagation of flames is slow, whereas frames made of cellulose nitrate are highly inflammable, igniting very easily (almost with an explosive force), with the flames propagating rapidly. Because of its flammability cellulose nitrate has been banned for the manufacture of spectacle frames in the United States, although some foreign manufacturers still use it.

Upon casual inspection it is difficult to differentiate a frame made of cellulose acetate from one made of cellulose nitrate. However, *camphor* is commonly used as a plasticizer in the manufacture of cellulose nitrate (but not cellulose acetate), and if the frame is rubbed briskly with a cloth, an odor of camphor will be produced. Both of these materials are soluble in ketones such as *acetone*, but neither is very soluble in alcohol. Acetone is frequently used to repolish or repair frames made from both materials.

*Acrylic* is the common name for the family of thermoplastic materials which includes polymethyl methacrylate. In addition to its use for hard contact lenses, this material has been used, on occasion, for the manufacture of spectacle frames. It has many desirable characteristics for this purpose, including dimensional stability, clarity, colorfastness, surface hardness, good wear resistance, lightness in weight, and flame resistance; and if heated properly, frames made of this material can be easily adjusted. However, its hardness is offset by brittleness and low impact resistance, both of which have been found to be major handicaps. This material is therefore not often used for spectacle frames.

*Nylon* is a generic name given to a class of thermoplastic polymers known as *polymides*. The nylon material is too expensive if produced in sheet form, so an injection molding technique is used. Nylon is tough and hard, but its usefulness is limited by its inherent handicaps of brittleness, lack of transparency, poor color selection, and a high rate of water absorption. Nylon frames are currently available but are not often used.

*Optyl*, a modified thermosetting material, is an epoxy resin introduced in 1968 by an Austrian firm. An Optyl frame is made by casting a liquid at high temperature, followed by a curing process. It is colorless when molded, and is dyed (either partially or successively) with various colors on different parts and at various angles. The advantages of Optyl include hardness, high luster, noninflammability, dimensional stability, and lightness in weight.

Unlike traditional thermosetting materials, Optyl becomes soft and pliable when heated above 87°C, and can be bent or shaped with the fingers to any desired form. When cool, it retains the adjustment indefinitely. It has a "locked-in" chemical memory; that is, when reheated, the frame will return to its original shape. Because the frame is cast in the final form, the plastic eyewires do not stretch or shrink to any extent, as do those made of cellulose nitrate and cellulose acetate, so the lenses must be edged to the *exact* size to fit the frame. Optyl has the disadvantage that it must be heated to a higher temperature than a cellulose acetate frame, and the frame will break if an attempt is made to adjust it while it is cold.

In recent years another ester of the cellulose family, *cellulose propionate*, has entered the plastic frame market. It has many attributes similar to Optyl, including the fact that it is made by a molding process. Frames made of this material are strong, light in weight, and can be made with a strong element of styling, including sculpturing effects. At present, no cellulose propionate frames are made in the United States.

In summary, although a great many materials have been used for the manufacture of plastic frames, the great majority of plastic frames are currently made of the thermoplastic material *cellulose acetate*.

## 9.5. Bridge and Temple Styles

The following discussion applies (with exceptions, which will be pointed out) to plastic frames rather than to metal frames or to rimless or semi-rimless mountings.

### Bridges

Whereas a metal frame or a rimless or semi-rimless mounting makes no contact with the nose (the contact being made by means of adjustable pads), a plastic frame is designed so that the bridge itself makes contact with the sides of the nose. Bridges of plastic frames can be classified as either saddle bridges or keyhole bridges, or as modifications of either of these forms.

A *saddle bridge* (see Figure 9-6) rests directly on the crest of the nose, having no pads to contact the sides of the nose. In order for a saddle bridge to fit correctly, the shape of the saddle must fit the contour of the nose. There are also *modified* saddle bridges, designed in such a way as to shorten the apparent length of the nose. A very few *metal* frames (mainly older styles that are now considered as antiques) have saddle bridges.

A *keyhole bridge* (Figure 9-7) makes contact only on the sides of the nose. The contact is made by means of fixed, nonadjustable pads, made of the same material of which the frame is made. Although usually fitting the wearer's nose more precisely than a saddle bridge, a keyhole bridge sometimes has the cosmetic disadvantage of accentuating the length of the nose. Some plastic frames have bridges that are, in effect, compromises between saddle and keyhole bridges.

FIGURE 9-6. **Plastic frame with a saddle bridge.**

FIGURE 9-7. **Plastic frame with a keyhole bridge.**

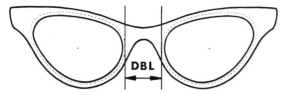

FIGURE 9-8. **Specification of bridge width: distance between lenses.**

*Bridge width* is specified, for rimless and semi-rimless mountings as well as for plastic and metal frames, as the shortest distance between the two lenses, or simply as DBL (distance between lenses), specified in millimeters (see Figure 9-8).

## Temples

The major types of temples that are available are shown in Figure 9-9. These are the skull temple, the library temple, the riding bow temple, and the comfort cable temple.

The *skull* temple is the most commonly used for plastic frames. It is bent downward behind the ear, and conforms to the contours of the skull. There are many modifications of the skull temple that vary from the "standard" skull temple mainly in terms of *width*. Most skull temples (and in particular the thinner styles) have a *wire core*, to provide added strength.

The *library* (Hollywood, spatula) temple extends straight back over the ear, having no bend whatsoever. This type of temple is difficult to fit, since the glasses must stay on the head on the basis of pressure exerted by the temples on the sides of the skull. However, this type of temple is convenient for a person who wears glasses only occasionally, usually wearing them only for brief periods of time.

The *riding bow* temple encircles the back and lower part of the wearer's ear. It is a plastic temple with a metal core, and is used for children's frames and safety frames.

The *comfort cable* temple also encircles the back and lower part of the ear, but differs from the riding bow temple in that all or part of the temple (particularly the part encircling the ear) is made of a coiled metal cable rather than plastic. The terms "riding bow" and "comfort cable" are to some extent used synonomously, a comfort cable being considered as a type of riding bow temple. Metal frames and mountings usually have comfort cable (riding bow) temples.

*Temple length* was formerly specified in terms of either "length to bend" or "overall length": However, temples are currently specified in terms of overall length. Although other frame measurements (lens size and distance between lenses) have always been specified in millimeters, temple length was formerly specified in inches. At the present time, almost all manufacturers specify overall temple length in millimeters.

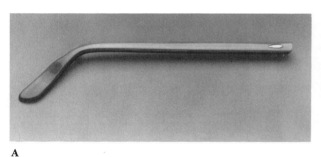

A

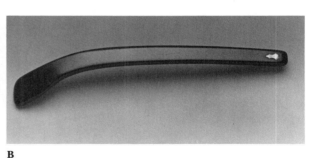

B

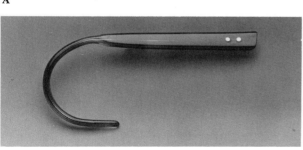

C

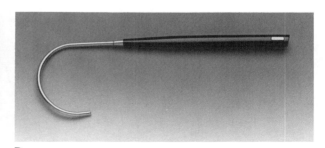

D

FIGURE 9-9. **Types of temples: A, skull; B, library; C, riding bow; D, comfort cable.**

# FRAME SELECTION AND ORDERING MATERIALS

## 9.6. Measuring Interpupillary Distance

The first step in the design and ordering of eyewear for a patient is the measurement of the interpupillary distance, usually referred to as the PD. Both the distance PD and the near PD are measured.

The desired measurements for distance and near vision, respectively, are the separation between the two visual axes where they intercept the spectacle plane. For distance vision the lines of sight are parallel, and their separation will be the same whether measured at the centers of rotation, the plane of the cornea, or the spectacle plane. However, when the eyes converge for near work, they rotate about their centers of rotation with the result that the separation of the lines of sight decreases from the centers of rotation to the corneal plane, and decreases even more at the spectacle plane (see Figure 9-10).

In order to measure the *distance PD*, the examiner faces the patient at a distance of approximately 40 cm and holds a millimeter ruler in the patient's spectacle plane. The patient is instructed to look at the examiner's left eye (see Figure 9-11A), and the examiner aligns the temporal edge of the patient's right pupil with the *zero* on the scale of the millimeter ruler. He then instructs the patient to look at his right eye, and notes the millimeter reading which is aligned with the nasal edge of the patient's left pupil. For patients having dark irides, it will usually be necessary to align the scale reading on the millimeter ruler with the temporal limbus (corneoscleral junction) of the right eye and the nasal limbus of the left eye.

In measuring the *near* PD, the examiner again faces the patient at a distance of approximately 40 cm and instructs the patient to fixate either his right or left eye (which is positioned on the patient's midline). He aligns the temporal edge of the right pupil with the zero on the scale, and notes the scale reading corresponding to the nasal edge of the left pupil (Figure 9-11B). For the 40 cm distance, the near PD will usually be about 4 mm less than the distance PD.

Rather than measuring the near PD, it is probably

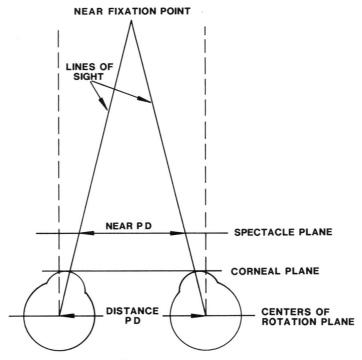

FIGURE 9-10. **Interpupillary distance: The distance between the visual axes of the two eyes where they intersect the spectacle plane.**

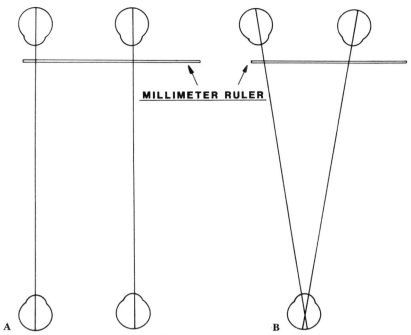

FIGURE 9-11. **Measurement of PD: A, distance; B, near.**

more accurate to carefully measure the distance PD and then to determine the near PD by calculation or by reference to a table. A simple formula for calculating the near PD may be derived with the aid of Figure 9-12. For *each eye*, the difference between the distance and near PD, $x$, is given by the equation

$$\frac{x}{27} = \frac{\frac{1}{2} \text{ distance PD}}{427},$$

$$x = \frac{27 \left(\frac{1}{2} \text{ distance PD}\right)}{427}.$$

**EXAMPLE 1**
For a distance PD of 60 mm,

$$x = \frac{27(30)}{427} = 1.9,$$

and

$$\text{near PD} = 60 - 2(1.9) = 56.2 \text{ mm}.$$

**EXAMPLE 2**
For a distance PD of 70 mm,

$$x = \frac{27(35)}{427} = 2.2,$$

and

$$\text{near PD} = 70 - 2(2.2) = 65.6 \text{ mm}.$$

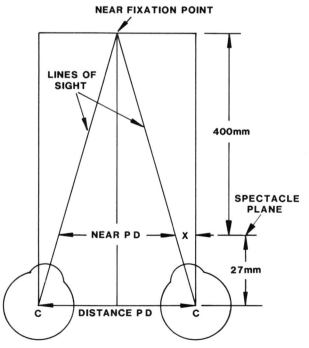

FIGURE 9-12. **Diagram used for deriving a formula to calculate the near PD.**

TABLE 9-1
**Calculated Near PD Values, as a Function of Distance PD, for a Near Working Distance of 40 cm**

| Distance PD | 58 | 60 | 62 | 64 | 66 | 68 | 70 | 72 |
|---|---|---|---|---|---|---|---|---|
| Near PD | 54.5 | 56 | 58 | 60 | 62 | 63.5 | 65.5 | 67.5 |

A compilation of near PD values, as a function of the distance PD, is given in Table 9-1.

### The Use of a Pupillometer

Many practitioners and dispensers prefer to use an instrument known as a pupillometer, rather than a millimeter ruler, to measure the interpupillary distance. The use of such an instrument has the advantage that the *monocular PD* can be measured: As discussed in Chapter 8, monocular PD measurement is a necessity in the fitting of progressive addition lenses.

### 9.7.  Frame Selection

The primary purpose of a spectacle frame is to hold the prescribed lenses in such a way as to provide the optimum visual efficiency. In addition, the frame should be physically comfortable and attractive in appearance, and meet the expectations of the patient.

A skilled dispenser may make sure that the requirements for optical performance and comfort are fulfilled by (1) specifying lens parameters meticulously as to refractive power, lens centration, base curve specification, lens material, multifocal type, and tint; (2) selecting a bridge design that will be stable, provide the proper weight distribution, and help maintain the glasses in the desired position in front of the face; (3) fitting the lenses as close to the face as possible; (4) providing the appropriate pantoscopic tilt and the corresponding vertical centration; (5) selecting the appropriate temple style and length, adjusted to conform to the contour of the ear and the shape of the mastoid process.

If these requirements are not met, the visual performance may not be as the prescriber intended, and the patient may be uncomfortable due to a sore nose, sore ears, or both.

Before frame selection is begun, circumstances which may influence the frame selection should be assessed. For example, the powers of the lenses together with the probable size of the lenses will affect the thickness and weight of the lenses; and multifocal or progressive addition lenses may require a particular minimum vertical frame dimension. The patient's PD as related to the frame PD indicates whether the eyes will appear to be approximately centered laterally in the frame, and whether the lenses can be cut from standard uncut blanks. The purpose or purposes for which the glasses are to be worn must be taken into consideration, to avoid unsuitable frame types.

### Fitting the Bridge

Although the percentage of weight of the spectacles carried by the nose varies both with the style of the frame and the patient's head position, most of the weight is normally borne by the nose when the head is held in an erect position. To prevent irritation of the nose, the bridge of the frame should distribute the weight of the spectacles over as large an area as possible. Since the fitting of the bridge is so important, the bony angular configuration of the nose should be determined by palpating the bridge of the nose with the thumb and forefingers. The pads of the frame should closely match both the frontal angle and the transverse angle, or splay angle of the nose. The frontal angle is the angle through which each side of the nose departs from a vertical line through the midline of the nose, as shown in Figure 9-13A, whereas the transverse or splay angle is the angle formed by the side of the nose with the median sagittal plane (the anterior−posterior plane through the midline of the nose), as shown in Figure 9-13B.

### Fitting the Temples

As the head is tilted forward, the weight of the spectacles shifts from the nose to the ears. This can create a problem if library or "skull" temples are worn, since they can hold the glasses in place only by pressure of the sides of the temples against the head in the region behind the ears. Riding bow temples, on the other hand, encircle the ears and secure the frame by making contact at the lower arc of the "crotch" of the ear.

The important features in the fitting of temples are (1) the shape of the "crotch" or angle of the external ear in relation to the side of the head, in terms of both the top of the ear and the back of the ear, and (2) the shape of the mastoid process. It is therefore necessary for the dispenser to get an unimpeded view of the back of the ear and the mastoid process before deciding upon the type of temple to be fitted.

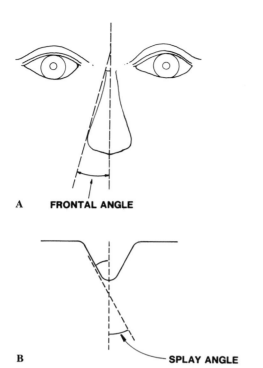

**A    FRONTAL ANGLE**

**B**

**SPLAY ANGLE**

FIGURE 9-13. **Bridge angles: A, frontal angle; B, transverse, or splay, angle.**

### Fashion in Eyewear Design

Once the optical and comfort criteria have been met, the features that satisfy the esthetic and cosmetic components are a matter of choice, taste, and fashion. Fashion is not new in eyeglasses. One needs only to glance through *Fashions in Eyeglasses*[2] to appreciate the variety of eyewear styles that have been introduced since the invention of glasses. Until the 1930s, developments in eyewear styles arose from the nature of the materials available and the skill of the optical artisans.

In the 1940s, designers of spectacle frames began to study fashion trends and based their designs on these trends. Since that time, eyewear has become an increasingly important component of the fashion picture, a component that is not likely to go away. Not surprisingly, the public is convinced that attractiveness in eyeglasses, as judged in relation to current fashion norms, is not an option but a necessity. This emphasis on fashion has served a useful purpose by making it easier for many patients to accept their practitioner's advice that corrective lenses are needed.

Although optometrists understandably feel competent concerning the visual performance and comfort the patient may expect with the lenses being prescribed and fitted, the lack of formal training in the fields of fashion and cosmesis may lead to a less secure feeling concerning the more esthetic and personal aspects of frame selection. However, most practitioners eventually come to grips with the situation, realizing that the patient's desire for attractive eyewear can be used as a justifiable means toward a desirable end.

Having undergone several years of professional training some newly licensed optometrists resent the fact that a spectacle frame, which they believe should be considered only as a "clinical appliance," is instead considered as an item of style and fashion. The emphasis on fashion does, after all, focus attention on products rather than on the services the optometrist has been trained to provide. In recent years this emphasis on fashion has progressed to the point that manufacturers consider fashion to be of greater importance than fitting requirements—as witnessed by the increasingly large number of "designer" frames that are available only in one size!

Since fashion in eyewear is here to stay, the practitioner who does not wish to deal with this aspect of optometric practice is well advised to delegate the task of frame selection to a well-trained assistant. Eyewear styling is a complex matter of color, texture, shape, size, and bridge and temple design. To a great extent it is an individualist undertaking, with the patient often having strong ideas concerning frame styling rather than seeking the advice of the optometrist or the dispenser. In matters such as esthetic suitability the patient must have the final say, but patients are receptive to the guidance of a dispenser if he or she demonstrates a strong grasp of cosmetic principles and a keen eye for facial structures. The patient can be assisted in making his or her choice if the dispenser is able to demonstrate how a particular facial shape is modified by eyewear that will create balance and will emphasize positive facial features while minimizing ones that may be considered as negative.

### Multiple Pairs

For many years, eyewear manufacturers and distributors have carried out promotional campaigns intended to convince the public of the necessity of owning more than one pair of glasses. Using slogans such as "The right eyewear at the right time" and "A second pair for outdoor wear," these campaigns have been only moderately successful. Whereas most members of the ametropic public have not been convinced that different occasions require different frame styles, many have become aware of the value of a second pair of glasses in the form of absorptive lenses for outdoor wear. However, during the past decade, with the availability of photochromic lenses, many patients are electing (often upon the suggestion of the optometrist or the dispenser) to order only one pair of glasses with the result that "second pair" campaigns tend to lose their momentum.

When the patient decides to obtain only one pair of glasses, compromises must often be made among such factors as color, comfort, weight, durability, and other factors. For example, the wearer of high minus lenses who would normally be a candidate for plastic lenses may decide upon a single pair of photochromic glasses—which are not widely available in plastic—thus avoiding the expense of a second pair of spectacles while putting up with the extra weight of glass lenses.

Detailed descriptions of frame selection are given by Sasieni,[3] Brooks and Borish,[4] and Stimson.[5]

## 9.8. Fitting Principles

Assuming that a spectacle frame or mounting has been properly *aligned* by the manufacturer or the laboratory, the correct fitting (to the patient's face) may be discussed in terms of the fitting triangle, the pantoscopic tilt, and the temple angle.

### The Fitting Triangle

A frame or mounting can be considered as fitting like a *triangle*, as shown in Figure 9-14, having points of contact at the crest of the nose and the tops of the ears. When the head is held erect, approximately two-thirds of the weight of the spectacles should be borne by the nose, and one-third by the ears. When

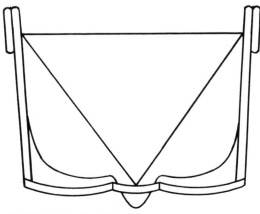

FIGURE 9-14. **The fitting triangle.**

the head is tilted downward, the weight of the frame is shifted to the ears. A common problem occurring with library temples is a tendency for the frame to slip down on the nose when the head is tilted downward: solving this problem requires a significant increase in the pressure of the temples against the sides of the head.

Since most of the weight of a pair of spectacles is borne by the nose, it is often necessary to *cushion* the load on the nose. For frames with adjustable pads, it is important to make sure that the *entire surface* of the pad makes contact with the nose. For patients who have sensitive noses, the area of contact of the pad may be increased by the use of *jumbo* pads. In selecting a plastic frame for a patient, since most of these frames lack adjustable pads, the frame selected must be one that provides as large an area of contact with the nose as possible. Since a saddle bridge must fit mainly on the *crest* of the nose, this frame style is appropriate for a patient who has a relatively wide, protruding bridge, or has a high bridge. For patients having very narrow bridges or flat, recessed bridges (often the case with blacks or Asians), a keyhole bridge or an adjustable pad bridge may be required.

### Pantoscopic Tilt

The pantoscopic tilt of a frame is defined as the amount of inward tilt of the front, away from the vertical (Figure 9-15). The provision of a pantoscopic tilt makes it possible to bring the lower edges of the lenses closer to the wearer's cheeks. Not only does this increase the field of view and afford greater protection from flying objects, but it also improves the appearance of the glasses. The optical principles relative to pantoscopic tilt were derived in Section 2.19.

When adjusting a pair of glasses, if one lens is higher on the patient's face than the other, the two lenses may be made level by increasing the pantoscopic angle for the lens that is low (or by decreasing the pantoscopic angle for the lens that is high). If it is necessary to tilt a lens so that the lower part of the lens tilts *away* from the face, this is called a *retroscopic* tilt. This should be used only if absolutely necessary to achieve a satisfactory fit on the wearer's face.

### Temple Angle

The temple angle is defined as the angle (in the horizontal plane) that the front makes with the temple, as shown in Figure 9-16. Although the temple angle depends upon a number of factors including

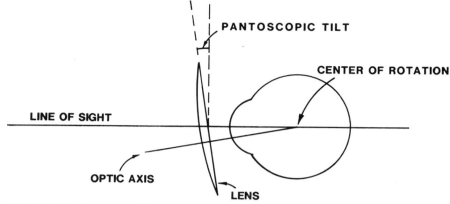

FIGURE 9-15. **Pantoscopic tilt.**

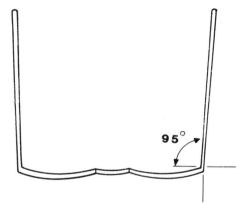

FIGURE 9-16. **Temple angle.**

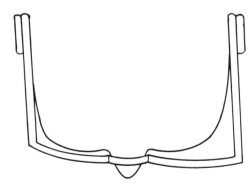

FIGURE 9-17. **Unequal temple angles.**

the width of the front and the width of the patient's head, in most cases the temples are bent *outward* a few degrees. However, with the popularity of large-size frames, it is sometimes necessary to bend the temples *inward* slightly.

Whether or not the two lenses are equally distant from the patient's brow (as determined by having the patient bend his head downward) depends upon the temple angles for the two lenses. If the temple angle on one side is *too small*, it will cause an excessive amount of pressure on the side of the head, as shown in Figure 9-17, causing the lens to extend outward in comparison to the other lens. This problem can be solved by increasing the temple angle on the protruding side (and, in some cases, also decreasing the temple angle on the other side).

### 9.9. Frame Alignment

The fitting principles described in the foregoing section apply both when selecting a frame or mounting for a patient, prior to ordering the spectacles from the laboratory, and when adjusting the completed spectacles to the patient's face. In either case it may be assumed that the frame has been placed in "standard alignment," or "trued," by the manufacturer or the laboratory. However, this is not always the case, so the task of alignment is often left to the dispenser.

In order to make sure that a frame or a pair of spectacles is properly aligned, the first procedure is to verify the alignment of the front itself, and the remaining procedure is to verify the alignment of the temples as they relate to the front.

### Alignment of the Front

This procedure involves two steps, each of which requires the use of a straightedge (usually a millimeter ruler). The straightedge is first used to verify the *horizontal* alignment of the front. As shown in Figure 9-18, the straightedge is placed against the back of

FIGURE 9-18. **Checking horizontal alignment of the front.**

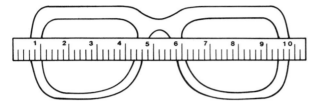

FIGURE 9-19. **Checking vertical alignment of the front.**

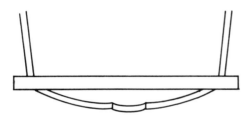

FIGURE 9-20. **Face-form.**

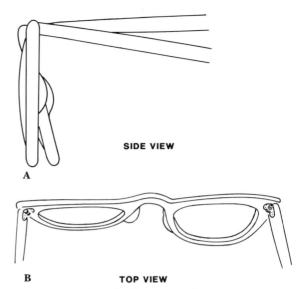

SIDE VIEW

A

B    TOP VIEW

FIGURE 9-21. **X-ing of a frame: (A) side view, (B) front view** (redrawn, from Brooks and Borish, System of Ophthalmic Dispensing, 1979, published by Professional Press Books, division of Fairchild Publications, New York).

the frame, below the endpieces (usually at the level of the tops of the pads), while the frame is held horizontally. When this is done, the left and right endpieces should be at equal distances above the straightedge. If not, the bridge must be adjusted to raise (or lower) one endpiece with respect to the other.

The straightedge is then used to verify the *vertical* alignment of the front, as shown in Figure 9-19. The frame is held vertically, with the temples extending upward, and the straightedge is placed along the back of the frame underneath (or possibly above) the pads or pad arms. Normally, a "four-point touch" is expected: the straightedge should touch the eyewire above each lens (or the back surfaces of each lens, for a rimless mounting) at the nasal and temporal edges. However, some spectacle frames are designed in such a way that the front is bowed *outward* slightly (known as "face-form," as shown in Figure 9-20) so that the nasal edges of the front will fail to make contact with the straightedge. This is particularly true of aviation-type goggles or sunglasses. If a four-point touch (or the desired amount of face-form) is not present, this can be corrected by appropriate adjustment of the bridge.

Occasionally, upon verifying the vertical alignment of the front, it will be found that the two lenses are not in the same vertical plane, one lens being tilted inward (or outward) with respect to the other. This condition, shown in Figure 9-21, is known as *X-ing*, and is corrected with an adjustment of the bridge.

### Alignment of the Temples

This procedure involves two of the adjustments described in the previous section: those of pantoscopic tilt and temple angle. Normally, a properly aligned frame should have a small amount of pantoscopic tilt (usually no more than about 10 degrees) and the angle of each temple should be somewhat greater than a right angle (on the order of 95 degrees). In order to make sure that both sides of the frame have

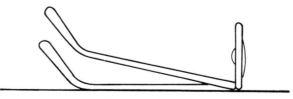

FIGURE 9-22. **Unequal pantoscopic angles.**

TABLE 9-2
**Spectacle Adjustments: Common Maladjustments and Their Remedies**

| Condition | Possible Causes | Remedies |
|---|---|---|
| Pads dig into sides of nose | Pads too tight<br>Temple tension too great<br>Pads too small for weight of glasses<br>Tender skin | Widen distance between pads<br>Reduce temple tension<br>Use large pads<br>Toughen skin with alcohol or other means |
| Pads dig into nose unevenly | Pads do not fit flatly against nose<br>Temple angle greater on one side | Correct pads to fit flatly against nose<br>Correct temple angles |
| Spectacles slide down on nose | Distance between pads too great<br>Temple tension too weak or temples too long | Bring pads closer together<br>Increase tension or replace with shorter temples |
| Lashes touch the lenses | Distance between pads too great<br><br>Temple tension too great or temples too short<br>Temples too short<br>Pad arms too short | Bring pads closer together or use deeply curved lenses<br>Reduce tension or replace with longer temples<br>Replace temples<br>Lengthen pad arms |
| Eyewires touch cheeks | Too great a distance between pads<br>Too great a pantoscopic tilt<br>Lenses too large in vertical dimension | Bring pads closer together<br>Reduce pantoscopic tilt<br>Refit with smaller eye size |
| Lenses touch brow | Lenses inset too far<br>Resting point on nose is too high<br>Too much retroscopic tilt | Lengthen pad arms<br>Lower spectacles by raising pad arms or pads<br>Reduce retroscopic tilt |
| One lens higher than the other | Temples not adjusted to different ear heights<br>One pad arm and pad lower than the other<br>Deformed nose | Correct pantoscopic tilt for one or both lenses<br>Adjust to balance<br>Compensate for difference with pad arm and pad |
| One lens out farther than the other | Temple angle uneven<br>One side of face larger than the other | Correct temples in reference to angle<br>Widen temple angle on side of face which is larger |
| Temples hurt behind ears | Tension too great<br>Uneven pressure<br>Area of contact between temple and ear too small<br>Temples too short<br>Temples ride too far out on top of ear<br>Temple tip rides too far at base of ear | Loosen temple tension or fit larger temples<br>Bend temples to conform to shape of ears<br>Use temple tubing to increase temple surface<br>Fit longer temples<br>Bend tip of temple to correct<br>Bend tip of temple in |
| Lenses too far from eyes | Pads improperly fitted<br>Pad arms too long | Spread pads<br>Shorten pad arms |
| Bifocal segments cause blurred area | Segments too high<br>Segments too low | Raise pad arms or use more pantoscopic tilt<br>Lower pad arms or use less pantoscopic tilt |

the same amount of pantoscopic tilt, the frame is placed on a tabletop with the temples downward, as shown in Figure 9-22. If one temple fails to make contact with the table, the endpiece on that side has a greater amount of pantoscopic tilt than the one on the other side. Both sides should, of course, have the same temple angle. This is easily verified by inspection. Specific procedures for adjusting spectacle frames or mountings, in order to remedy various maladjustments, are given in Table 9-2.

## 9.10. Frame and Mounting Specifications

When ordering a frame or mounting for a patient, the following parameters must be specified:

**Name.** The frame or mounting will have either a name or a number (usually a name). This must be specified, along with the name of the manufacturer.

**Color.** Metal frames or mountings are usually available in either yellow or white gold. However, some of the newer metal frames are available in darker colors, for example, bronze, brown, gray, or black. Plastic frames are typically available in an assortment of colors. Frame colors (as well as the other specifications, listed below) are available in manufacturers' catalogs and references such as the *Frame Fax* catalog.

**Size.** The "eye size" is usually specified in terms of the *boxing system*, described in Chapter 4, so the box measurement is normally specified. Bridge size is normally specified in terms of distance between lenses (DBL). Temple length is normally specified in terms of overall length. All specifications are given in millimeters.

**Temple Style.** This should be specified if, as sometimes occurs, the frame is available with more than one style of temple.

**Lens Shape.** This must sometimes be specified, since a frame or mounting may be available with more than one lens shape. This is particularly true of rimless and semi-rimless mountings.

## 9.11. Lateral Placement of Optical Centers

The correct specification of the optical center of each lens is of importance in order to (1) manage prismatic effects and (2) control the placement of the optic axis of each lens relative to the center of rotation of the eye. The management of prismatic effects involves both the lateral and vertical placement of the lens centers, whereas controlling the placement of the optic axis relative to the center of rotation of the eye depends not only on the lateral and vertical placement of the centers but also on the amount of pantoscopic tilt of the lenses.

The reader will recall that, in Chapter 4, the *major reference point* of a lens was defined as the point for which the specified prismatic effect is present. If no prismatic effect is specified, the major reference point coincides with the optical center, or *pole*, of the lens. If a prismatic effect is specified, the major reference point for each lens should be the point intersected by the line of sight of the eye, when the eyes are in the primary position, with reference to both the vertical and horizontal meridians. The optical center (if no prism is to be incorporated) or the major reference point for each lens is usually specified for *distance* vision, but when the correction is intended primarily for near visual tasks, specification for *near* vision may be considered. This will be discussed in Section 9.13. In any case, the position of the major reference point must be considered in terms of both its lateral and vertical placement.

Lateral placement of the optical centers, if no prismatic effects are called for in the prescription, depends simply on the relationship between the patient's PD and the frame PD. As discussed in Chapter 4 (and illustrated in Figure 9-23) the frame PD is equal to the horizontal box measurement plus the distance between lenses. With today's large lens sizes, the optical center of the lens must usually be decentered *nasalward*, in order for the separation of the optical centers of the lenses to be equal to the patient's PD. For example, if a patient having a distance

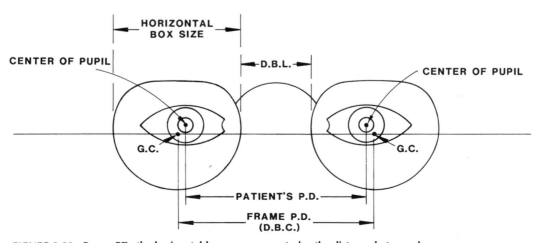

FIGURE 9-23. **Frame PD: the horizontal box measurement plus the distance between lenses.**

PD of 64 mm is fitted with a 50/20 frame and if the glasses are to be centered for distance vision, the optical center of each lens will have to be decentered inward 3 mm.

Lateral placement of the optical centers may be altered to help relieve a particular patient's binocular vision problem. For example, if a myope has exophoria at near, centering the lenses for the distance PD will induce *base-in* prismatic effect at near, reducing the amount of exophoria at near. Likewise, if a hyperopic patient is esophoric at near, centering the lenses for distance will induce a *base-out* prismatic effect at near, which will reduce the esophoria. The reader may recall that samples of such cases were given in Chapter 4.

If in the cases described above the patient also has a *distance* phoria, and if the phoria is in the same direction at both distance and near, consideration should be given to decentering the lenses with respect to the patient's distance PD. For the exophoric myope, decentering the lenses *outward* would provide base-in prism at both distance and near, whereas for the esophoric hyperope decentering the lenses outward would provide base-out prism for both distance and near (in both cases the amount of prismatic effect being greater at near than at distance). The same principle may be applied for the esophoric myope (whose lenses may be decentered *inward*, providing more base-out prism at distance than at near), and for the exophoric hyperope (whose lenses may also be decentered inward, providing more base-in prism at distance than at near).

In summary, when making the decision concerning lateral placement of the optical centers of the lenses, each case should be considered in terms of its own particular merits. However, if the frame selection is done by a dispenser or a technician (rather than by the practitioner who examined the patient) he or she will normally center the lenses for either distance or near, as stated in the prescription, not decentering the lenses inward or outward (in order to reduce a phoria) unless the prescription calls for such decentration.

## 9.12. Vertical Placement of Optical Centers

Although the horizontal position of the optical center receives close scrutiny, the vertical position is largely neglected.

Various authors have recommended what they believe to be the optimum vertical positions for optical centers. Waters[6] stated that the lenses should be "so positioned that the optical center of the area used is as close as possible to the normal level of the eyes"; whereas Maxwell[7] believed that the lenses should be mounted "so that the visual axes, when directed straight ahead, will pass through the lenses 3 mm above their centers." Laurance[8] suggested that the heights of the optical centers should be determined on the basis of the use that will be made of the glasses, listing three possible positions depending upon whether the glasses will be used for distance vision only, for constant wear (both distance and near), or for near only.

For glasses to be worn for distance vision only, Laurance suggested that the optical centers should be positioned so that they are pierced by the lines of sight when the eyes are directed toward a point on the ground at a distance of 12 m, or halfway to the ground at a distance of 6 m. For a person whose eye level is 5 feet 6 inches from the ground, and for a stop distance of 25 mm, those points would fall 3.5 mm below the centers of the pupils (see Figure 9-24). For

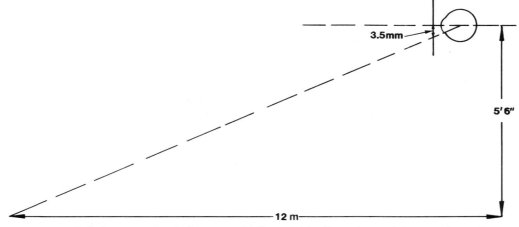

FIGURE 9-24. **Vertical placement of optical centers, with lines of sight directed toward the ground at a distance of 12 m.**

constant wear, the optical centers would be lowered about 1.5 mm or would be placed 5 mm below the centers of the pupils. For close work only, the optical centers would be placed in line with the lower borders of the irides, or about 6 mm below the centers of the pupils.

In general, it is desirable to vertically position the optical centers at the *most used* portions of the lenses. There are two reasons for this:

1. Prismatic effects are absent at the optical centers of the lenses. Therefore, when looking through the optical centers, objects are not subject to apparent displacement.
2. Chromatic aberration is absent at the optical centers, with the result that objects will not be affected by color dispersion.

In order for a lens of a given design to perform at its maximum potential, it must be worn as its designer intended. As discussed in Chapter 6, physiological approaches to lens design vary from one lens designer to another, and serve to differentiate one particular lens or lens series from another. However, in every lens design there is agreement on one basic assumption: The optic axis of the lens should pass through the center of rotation of the eye.

When the optic axis of the lens passes through the center of rotation of the eye, the optic axis and the visual axis coincide when the wearer looks through the optical center of the lens; that is, the visual axis is perpendicular to both lens surfaces. For any other viewing angle, a lens designer is able to anticipate the specific location in the lens through which the visual axis will pass. Therefore, by appropriate design, the designer can predict and can control image quality.

When the optic axis of the lens fails to pass through the center of rotation of the eye, the chief ray of the image-forming bundle, which essentially coincides with the visual axis, intersects the lens surfaces at locations and at angles of incidence not anticipated by the designer. This "mismatch" between the designer's intention and the actual use of the lens creates differences in the angles of incidence in the tangential and sagittal planes and therefore differences in the tangential and sagittal foci, resulting in the production of the aberration of *oblique astigmatism*. These differences in the angles of incidence in the tangential and sagittal planes occur even at the optical center, and hence the refracted bundle of rays entering the eye is astigmatic even though the lens is a spherical lens.

When looking through the optical center of a lens when the optic axis of the lens fails to pass through the center of rotation of the eye, the region of the optical center of the lens suffers both an increase in spherical power and a cylindrical power effect. A plus cylinder effect is created for a plus spherical lens, and a minus cylindrical effect is created for a minus spherical lens. Formulas for calculating these effects were given in Section 2.19. The axis of the induced cylinder is always *parallel* to the axis of rotation of the lens tilt: For example, the axis is 180° for a pantoscopic tilt, and 90° for a "face-form" tilt.

The formulas for calculating these effects (which occur when the optic axis of the lens fails to coincide with the visual axis) apply *only* to the area of the optical center of the lens: However, the lens also fails to perform as intended in the peripheral portions.

**A Note of Caution.** The student may conclude (because some textbooks are not clear on this subject) that, when any lens is tilted, power changes will occur at the optical center. This is incorrect. It should be understood that a lens *can be tilted* without bringing about power changes at the optical center *as long as the optic axis of the lens passes through the center of rotation of the eye* (see Figure 9-15). However, oblique astigmatism will occur in the peripheral portions of the lens, and the amount of oblique astigmatism will depend upon the particular lens design (that is, on the amount of "bend").

If we adhere to the principle that the optic axis of the lens should pass through the center of rotation of the eye, the required level of the optical center depends upon the pantoscopic tilt necessary to bring the lens into the proper relationship with the wearer's eyebrows and cheeks. In a given case it is best to provide the pantoscopic tilt that gives the *best appearance*, and to lower the optical centers of the lenses from the pupillary level (when the eyes are in the primary position) *1 mm for each 2° of pantoscopic tilt*, as derived in Section 2.20. One can estimate the pantoscopic tilt, or use a protractor to approximate the tilt of the front relative to the vertical plane when the head is held in the primary position.

It is a common laboratory procedure, unless the dispenser specifies otherwise, to place the optical center at the level of the *geometrical center* of the lens. Since the level of the geometrical center is usually in the neighborhood of 3 to 5 mm below the wearer's eye level, we have the happy coincidence that the placement of the optical center is consistent with the typical "cosmetic" pantoscopic angle of from 6 to 10°, with the result that the lens will function as the designer intended it to. An exception may occur, however, with *very large lens sizes*, for which the level of the geometrical center of the lens may be as far as 8 to 10 mm below the wearer's eye level. With these large lenses the dispenser should specify the position of the optical center of the lens, on the basis of the amount of pantoscopic tilt, so that it can fall some-what *above* the geometrical center.

Fortunately, the average levels of the lines of sight for various visual tasks tend to be consistent with the level usually selected for the optical center on the basis of the amount of pantoscopic tilt required to conform to the structure of the wearer's face.

Except in cases of large amounts of anisometro-pia, it is not a desirable practice to place the optical centers of the lenses higher than usual when glasses are to be worn for *distance only*, or to place the optical centers lower than usual when the glasses will be worn for *near only*: To do so would require the glasses to be tilted at an unattractive angle in order for the optical axes of the lenses to pass through the centers of rotation of the eyes.

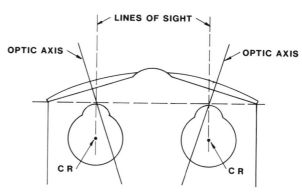

FIGURE 9-25. **Face-forming, with optical centers coinciding with the wearer's distance PD, causing the optic axes of the lenses to pass nasalward to the centers of rotation of the eyes.**

## 9.13. Centering Problems and Solutions

### Centering Lenses for Near Work Only

An interesting question arises in regard to the centration of lenses made solely for close work: Should the lenses be centered for the distance PD or for the near PD? The answer depends upon whether it is considered to be more important (*a*) to avoid inducing horizontal prismatic effects or (*b*) to assure that the optical axes of the lenses pass through the centers of rotation of the eyes.

Placing the optical centers of the lenses so that they coincide with the *near* PD produces zero prismatic effects when doing close work, but the optic axes will not pass through the centers of rotation. On the other hand, positioning the optical centers so that they coincide with the *distance* PD would result in the optic axes passing through the centers of rotation of the eyes, therefore controlling oblique astigmatism, but with prism induced as the eyes converge for close work. Although most practitioners would normally position the optical centers of the lenses for the near PD, both approaches should be given consideration, and the decision made on the basis of the alternative that will best serve the patient.

### Face-Forming

"Face-forming" refers to the process of curving a frame in such a way as to conform to the wearer's face. As shown in Figure 9-20, this procedure involves both an outward bowing of the bridge and a widening of the temple angles. Although face-forming improves the appearance of a large frame, it is responsible for two problems:

1. If the optical centers are placed so that they coincide with the patient's distance PD, face-forming causes the optic axes to pass *nasalward* to the centers of rotation of the eyes (see Figure 9-25), changing the effective performance of the lenses. This can be corrected by decentering both lenses outward (with the result that the optic axes of the lenses will now pass through the centers of rotation of the eyes), but unfortunately this procedure would induce unwanted horizontal prism.

2. The laboratory normally places the major reference points so that their distance from the nasal edge plus DBL is equal to the specified PD. If the frame has face-form, the shortest distance between the major reference points will be *smaller* than the PD. In practice, the shortest distance between the major reference points should of course correspond to the distance PD, but unless allowance is made for face-forming, the PD of the finished spectacles will not match the patient's PD.

Although the effect of pantoscopic tilt on the direction of the optic axis can be corrected by lowering the optical centers of the lenses 1 mm for each 2 degrees of tilt, the problems created by face-forming have no satisfactory solution. Therefore, face-forming a frame, especially one with lenses of high power, will result in inferior optical performance of the lenses for foveal vision.

### Hyper-Eye

Hyper-eye is defined as the condition in which one eye is positioned higher in the head than the other (like the little man on the cover of *MAD* magazine). It is not the same as *hyperphoria*, in which the visual axis

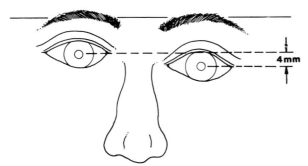

FIGURE 9-26. **Hyper-eye: The center of the pupil of the right eye is approximately 4 mm higher than that of the left eye.**

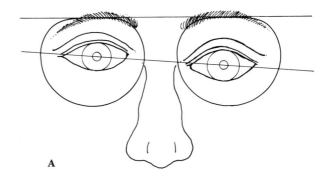

A

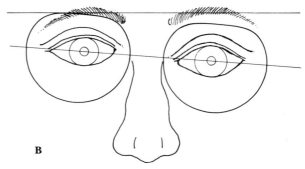

B

FIGURE 9-27. **Hyper-eye: A, with frame parallel to the eyebrow level; B, with frame parallel to the eye level.**

of one eye is aimed higher than that of the other. When writing an order for a pair of glasses for a patient having hyper-eye, the dispenser must decide where to specify the vertical positions of the optical centers of the two lenses. In making this decision, the relationship between the eyes and the eyebrows must be taken into consideration: specifically (as shown in Figure 9-26), the relationship between the imaginary line between the two eyebrows and that between the centers of the two pupils.

From the cosmetic point of view the appearance of a pair of glasses on a patient's face is usually more pleasing if the frame is roughly parallel to the *eyebrow level* (Figure 9-27A) rather than to the *eye level* (Figure 9-27B). However, in many cases a compromise is the best solution. In any case, once the most cosmetically acceptable solution has been arrived at, the dispenser should then determine the relative vertical positions of the centers of the patient's pupils when the frame (the same style and size as the frame in which the finished lenses will be mounted) is worn in the desired position on the patient's face. This may be done simply by measuring the vertical distance from the bottom rim of the frame to the center of the pupil, for each eye. Using the familiar rule that the optical center of each lens should be placed 1 mm below the center of the pupil for each 2° of pantoscopic tilt, the dispenser should then specify the vertical position of the optical center for each lens.

If the power of the prescription in the vertical meridian is plano or is very low (e.g., +1.00 DS −0.75 DC axis 180), it is not necessary to specify differential vertical positions for the optical centers. On the other hand, if *multifocal lenses* are being ordered, failure to specify differential positions of the centers will mean that one segment top will be farther below the eye than the other. However, for a hyper-eye patient

who has previously worn bifocals, the optometrist or dispenser should check the segment heights of the present bifocals. Some patients will be found to have no complaints of asthenopia in spite of the fact that one segment is 2 or 3 mm lower (with respect to the center of the pupil) than the other: If the patient is well adapted to this situation, it is probable that differential vertical placement of the segments in the new glasses *will* result in complaints of asthenopia.

## 9.14. Bifocal Segment Inset

As stated briefly in Section 8.16, there are two reasons for specifying *segment inset* in an order for a pair of bifocal lenses: to assure that the segments do not induce unwanted prismatic effects, and to assure that the reading fields of the two segments coincide.

The first step in deciding upon the amount of segment inset is the patient's *near PD*. The near PD may be measured, along with the distance PD, or it can be determined by reference to a table (see Table 9-1). It should be understood that even when the bifocal segments are inset for the exact near PD so that the segments themselves induce no prismatic effects, the *distance prescription* (unless it has zero power in the horizontal meridian) will induce an unwanted prismatic effect at near.

As discussed in Section 8.26, the distance lens may be considered as existing *behind* the segment (which actually is the case for a one-piece bifocal) and will induce horizontal prismatic effects at near due to the distance prescription. Since a bifocal add always has *plus* power, insetting the segments closer than required by the near PD will induce *base-in* prism. Therefore, for a hyperopic patient, whose distance lenses will induce a *base-out* prismatic effect at near, this effect can be counteracted to some extent by increasing the segment inset; for a myopic patient, whose distance lenses will induce a *base-in* prismatic effect at near, decreasing the segment inset will help to counteract this effect.

Since most presbyopes are exophoric at near, the *hyperopic* patient will be the one of most concern to the practitioner. For this patient, an additional 2 mm of segment inset, for a +2.00 D add, will induce a total base-in prismatic effect of 0.8 prism diopter (for the two lenses). It should be understood, however, that an excessive amount of segment inset will tend to cause the reading fields for the two eyes to fail to completely coincide, and therefore should be accompanied by the use of a reasonably wide segment (instead of the usual 22-mm-wide fused segment, a 28- or 35-mm-wide segment could be used, or even an Executive-style segment).

Fortunately, the reading fields will coincide reasonably well, even with excessive amounts of bifocal inset, unless the patient requires relatively high-powered lenses. This may be a problem particularly in aphakia. (The relationship between distance lens power and segment inset on the coincidence of the reading fields is illustrated by the data shown in Table 8-10.)

## 9.15. Prescription Order Forms

Prescription laboratories normally supply their "accounts" with their own prescription order forms. A typical laboratory order form is shown in Figure 9-28. Although one form differs from another in minor details, they all have provision for the lens and frame information necessary to order a pair of finished glasses. However, it should be noted that most prescription order forms *do not* have spaces for information concerning base curve, cylinder location (front or back surface), or thickness (center or edge).

Normally, a prescription laboratory will deal mainly with one or two lens suppliers. If, for example, the laboratory routinely supplies Bausch and Lomb Orthogon lenses, the optometrist or dispenser can expect base curves of the Orthogon series, with *front surface cylinders*; but if the laboratory supplies lenses from almost any other manufacturer, the lenses will have *back* surface cylinders (and whatever base curves are used in the particular series). Most laboratories will be pleased to supply customers with information concerning available series of lenses, together with charts showing the base curve ranges on the basis of lens powers.

The optometrist or dispenser may wish to specify, in the order, the "trade name" of the lens (e.g., Orthogon, Tillyer, Univis, Panoptik, Photogray Extra). When this is done, the manufacturer is obligated to supply the exact product ordered (although if the laboratory does not deal with the manufacturer of the specified lens, a substitute will probably be offered). On the other hand, many *bifocal* lenses made by different manufacturers are sufficiently similar so that the practitioner may wish to specify an "Executive-style" bifocal, a "Univis-style" bifocal, etc. When this is done, the laboratory is *not* required to supply an Executive or a Univis bifocal, but may supply a lens which looks (and performs) like the trade-name lens. In many cases, the availability of these "look-alike" lenses is due to the fact that the original patents have long been expired. This custom, however, does not apply to *progressive addition* lenses: Lens design in this field has not yet reached the point where one manufacturer's lens performs much the same as that of another.

As for *frame* specifications, if the optometrist or dispenser plans to order both the lenses and the frame from the same laboratory and to have the laboratory assemble the glasses, he or she will have to know which frame manufacturers the laboratory deals with, and, more important, which frame styles (and sizes and colors in each style) the laboratory carries *in stock*. Specifying a frame that the laboratory must order from the factory is sure to result in an unnecessary delay. The laboratory manager should be asked to supply this information. As already noted, there are a number of catalogs and indexes, such as *Frame Fax*, which list complete (and current) information on all available frames and mountings.

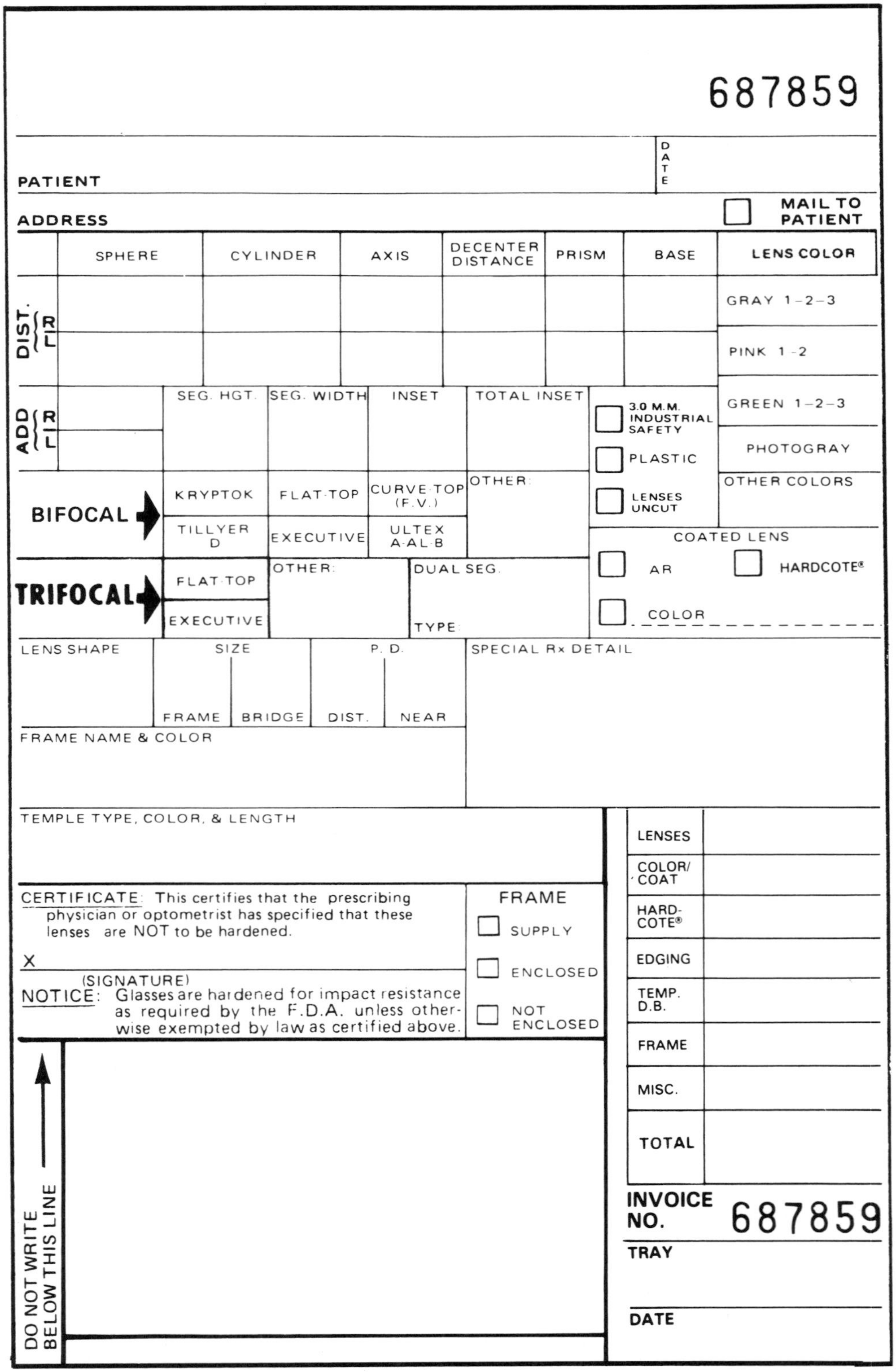

687859

| | SPHERE | CYLINDER | AXIS | DECENTER DISTANCE | PRISM | BASE | LENS COLOR |
|---|---|---|---|---|---|---|---|
| DIST. R | | | | | | | GRAY 1-2-3 |
| DIST. L | | | | | | | PINK 1-2 |

PATIENT

DATE

☐ MAIL TO PATIENT

ADDRESS

| | | SEG. HGT. | SEG. WIDTH | INSET | TOTAL INSET | | |
|---|---|---|---|---|---|---|---|
| ADD R | | | | | | | GREEN 1-2-3 |
| ADD L | | | | | | | PHOTOGRAY |

☐ 3.0 M.M. INDUSTRIAL SAFETY

☐ PLASTIC

☐ LENSES UNCUT

OTHER COLORS

**BIFOCAL** ➤

| KRYPTOK | FLAT-TOP | CURVE-TOP (F.V.) | OTHER: |
|---|---|---|---|
| TILLYER D | EXECUTIVE | ULTEX A-AL-B | |

COATED LENS

☐ AR    ☐ HARDCOTE®

**TRIFOCAL** ➤

| FLAT-TOP | OTHER: | DUAL SEG. |
|---|---|---|
| EXECUTIVE | | TYPE |

☐ COLOR _ _ _ _ _ _ _ _

| LENS SHAPE | SIZE | | P. D. | | SPECIAL Rx DETAIL |
|---|---|---|---|---|---|
| | FRAME | BRIDGE | DIST. | NEAR | |

FRAME NAME & COLOR

TEMPLE TYPE, COLOR, & LENGTH

CERTIFICATE: This certifies that the prescribing physician or optometrist has specified that these lenses are NOT to be hardened.

X _____
    (SIGNATURE)

NOTICE: Glasses are hardened for impact resistance as required by the F.D.A. unless otherwise exempted by law as certified above.

FRAME

☐ SUPPLY

☐ ENCLOSED

☐ NOT ENCLOSED

| LENSES | |
|---|---|
| COLOR/ COAT | |
| HARD-COTE® | |
| EDGING | |
| TEMP. D.B. | |
| FRAME | |
| MISC. | |
| TOTAL | |

↑ DO NOT WRITE BELOW THIS LINE

INVOICE NO. 687859

TRAY

DATE

312    FIGURE 9-28. **A typical prescription order form.**

# VERIFICATION AND DISPENSING

## 9.16. Verification

After the completed glasses arrive from the laboratory, they should be examined for accuracy of the prescription and for quality of the workmanship. To ensure that no phase of inspection is overlooked, a routine method of evaluation should be established. Furthermore, the accuracy and quality of workmanship must be judged against a preexisting list of tolerances, usually referred to as a *standard*.

### Ophthalmic Lens Standards

The American National Standards Institute (ANSI) is a nationally recognized nongovernmental body whose job is to act as a coordinator of voluntary standards development and a clearinghouse for information on national and international standards. ANSI approval procedures for standards strive for a consensus among those within its scope and affected by its provisions. Ophthalmic lens standards were first developed in 1964, revised in 1972, and further revised in 1979. The 1979 edition has wider tolerances than the 1972 Standard that it replaces. Table 9-3 provides a comparison of the refractive and prismatic requirements for the 1964, 1972, and 1979 standards.

The Z80.1-1979 Standard applies to all ophthalmic prescription lenses, either edged lenses or lenses mounted in a frame. It also applies to uncut lenses supplied by an optical laboratory in filling a specific prescription. However, all relevant optical specifications and tolerances included in the standard are considered as *recommendations*, not as specific requirements. The standard serves as a guide for tolerances, but is not restrictive. A lens may meet or exceed all specifications of the standard and be judged by the dispenser as unacceptable, or a lens not in conformance with the standard may be considered acceptable by the dispenser.

The optometrist should come to an agreement with the laboratory as to what his or her acceptable tolerances are. If the optometrist wishes to impose stricter tolerances than those specified in Z80.1-1979, this may be done as long as the optical laboratory agrees. A copy of the adopted standards should be posted, for reference, in the verification area of the optometric office.

The optical laboratory will include with the completed glasses a copy of the prescription order form (this will normally be one of the carbon copies of the form submitted by the optometrist, unless the glasses were ordered by telephone, in which case a laboratory employee will have completed the form). Since orders may contain clerical errors, the laboratory copy should be checked against both the optometrist's order form and the original examination record to make sure that all agree. In this way, any transcription errors will be found before the patient is notified that the glasses are ready for dispensing.

Verification of the completed eyewear should include the following procedures, performed in any sequence that seems appropriate.

### Lens Verification

1. The lens material (ophthalmic crown glass, high-index glass, CR-39 plastic, or polycarbonate) should be verified. High-index glass can be identified by the lower *approximate power* as determined by the lens gauge (it will be recalled that the lens gauge is calibrated for the index of refraction of ophthalmic crown glass). CR-39 and polycarbonate plastic can be identified on the basis of weight alone (being approximately half the weight of crown glass). One can differentiate polycarbonate from CR-39 plastic by the surface power found using the lens gauge, since CR-39 has a lower index of refraction than crown glass (1.498) while polycarbonate has a higher index (1.586).

2. The lens should be of the specified color and free of media defects such as stress, veins, bubbles, and foreign particles.

3. If the lens has been toughened by *air tempering*, it should exhibit a typical strain pattern when placed between the crossed polarizing filters of a Colmascope. Two modified versions of the Colmascope are available (the Tempr-a-Scope by Kirk and the Chem-Check by American Optical) to identify a lens that has been *chemically tempered*. These instruments are designed on the basis of the following principle: If the edge of a chemically tempered lens is immersed in a cell containing glycerine (between crossed polarizing filters) the edge of the lens will show a characteristic ring of light.

4. The lens surface should be free of defects that

TABLE 9-3
**Comparison of the Three ANSI Ophthalmic Lens Standards**

|  | *1964* | *1972* | *1979* |
|---|---|---|---|
| Refr. power (D) | 0.00 to 6.00 ±0.06<br>6.25 to 12.00 ±1%<br>>12.00 ±0.12 | 0.00 to 6.00 ±0.12<br>6.25 to 12.00 ±2%<br>>12.00 ±0.25 | 0.00 to 6.50 ±0.13<br>>6.50 ±2% |
| Cyl. power (D) | ±0.12 | ±0.12 | 0.00 to 2.00 ±0.13<br>2.12 to 4.50 ±0.15<br>>4.50 ±4% |
| Cyl. axis (Deg) | 0.12 to 0.37 ±3°<br>0.50 to 1.00 ±2°<br>>1.12 ±1° | 0.12 to 0.37 ±5°<br>0.50 to 1.00 ±3°<br>>1.12 ±2° | 0.125 to 0.375 ±7°<br>0.50 to 0.75 ±5°<br>0.875 to 1.50 ±3°<br>≥ 1.62 ±2° |
| Refr. power imbalance | No greater than single lens | No greater than single lens |  |
| Vert. prism (Δ) | ±0.25 each lens<br><br>0.25 imbalance | ±0.25 each lens<br><br>0.25 imbalance | ≤3.25 D ±0.33 each lens<br><br>0.33 imbalance<br>>3.37 D 0.1 $(F_v)$" imbalance |
| Horiz. prism (Δ) | ±0.25 each lens<br><br>0.50 imbalance | ±0.25 each lens<br><br>0.25 imbalance | ≤3.25 D 0.33<br>>3.37 D 0.1 $(F_v)$"<br>≤2.67 D 0.67 imbalance<br>>2.67 D 0.25 $(F_v)$" imbalance |
| Thickness | ±0.2 mm | ±0.2 mm | ±0.3 mm |
| Warpage | No tolerance specified | ±1.00 D<br>(not applicable to plastic lenses in metal frames) | ±1.00 D |

"$F_v$ refers to back vertex power of the lens—for imbalance it is the back vertex power of the weaker lens. The maximum amount of imbalance can be calculated from Prentice's rule.
Source: Reproduced, with permission, from V.M. King, Tolerance to Tolerances: Some Thoughts on the 1979 American National Standard Recommendations for Prescription Ophthalmic Lenses, *J. Amer. Optom. Assoc.*, Vol. 50, pp. 585–588, 1979.

would impair its function such as grayness, scratches, pits, abrasions, waves, cloth marks, and water marks.

5. The accuracy of the prescription specifications (spherical power, cylinder power and axis, prism power and base direction, and centration) should be verified, using a lensometer. The optical center (or the major reference point if prism is prescribed) should be marked and measured, in terms of its position both horizontally and vertically from the geometrical center.

6. For multifocal lenses, segment type, width, height, and inset should be verified.

7. The base curve (if specified in the order) should be verified, as well as the side (front or back) on which the cylinder is ground.

8. Center thickness (minus lens) or edge thickness (plus lens) should be verified with a caliper. If thickness was not specified in the order, the lens should be no thicker than required to provide mechanical stability.

*Frame Verification*

1. The frame should be verified as being the correct type (as indicated by the manufacturer's trade name or number), color, eye size (usually the box measurement), bridge size (distance between lenses), and temple style and length. The frame should be inspected to assure that it is in good condition with no scratches, plier marks, or other damage.

2. The quality of workmanship in mounting or glazing should be verified. The lenses should not wobble or feel loose in the frame. No air spaces should be present between the lens and the frame. The bevel should be examined for evenness and for the absence of chips. Screws used on hinges, on rimless mountings, or to attach trims on plastic frames should be checked to assure that the excess length has been properly cut off, filed smoothly, and peened over to prevent them from backing out.

3. For rimless mountings or metal frames the lenses should be checked, between crossed polarizers, for strain at the mounting points and around the metal rim. Excessive strain may cause the lenses to fracture.

4. Alignment of the frame should be checked, to make sure that the frame was in "standard alignment" when it left the laboratory. Temples should open easily, and adjustable pads should swivel freely.

In the interest of time, errors detected upon inspection should be corrected in the office if possible. If errors of workmanship not correctable in the office are present, or if the lens parameters do not meet the specified tolerances, the glasses should be returned to the laboratory for correction. Errors should be described precisely, so there will be no doubt of the validity of the cause for rejection. Although the patient will be disappointed by the delay, most patients will be pleased to know that their doctor requires accuracy on the part of the laboratory. Furthermore, the laboratory manager quickly learns whether a given practitioner demands accuracy and good workmanship or will accept substandard merchandise.

Augsburger[9] has described a method of ophthalmic material verification that results in both better service by the laboratory and a lowered rejection rate. He attributes this success to the fact that when glasses must be returned the laboratory is supplied with thorough documentation of the evaluation process. Much time can be saved and aggravation avoided by selecting a laboratory having high standards of accuracy and workmanship and capable of handling any prescription, no matter how complicated, and entrusting most (if not all) of the prescription work to this laboratory.

## 9.17. Dispensing and Adjusting

The most convenient method of dispensing and adjusting the finished glasses is for the practitioner or dispenser to face the patient, seated on the opposite side of a small table. Relatively little adjusting should be required if the frame fitting was done accurately and if the glasses were received from the laboratory in "standard alignment." In the standard alignment procedure, laboratory personnel will normally provide a temple angle slightly greater than 90° (although this should *not* be done with very large frames), a small amount of pantoscopic tilt, and (depending upon the type of frame) possibly a small amount of face-forming. The procedures for determining whether or not the glasses are in standard alignment have been described in Section 9.9. If it is found that the glasses, when received from the laboratory, are *not* in standard alignment, they should be aligned before they are dispensed to the patient. The procedures for doing this have also been described in Section 9.9.

### Alignment with Facial Features

One of the first things to be checked, when the glasses have been placed on the patient's face, is whether or not they are level with the patient's eyebrows and other facial features. A significant percentage of people will be found to have one ear higher than the other or one eyebrow slightly higher than the other, with the result that the glasses will appear to be "crooked" in relation to the face. The solution to this problem is to increase the pantoscopic angle of the temple on the "low" side or to decrease the pantoscopic angle on the "high" side. As a part of this process, pantoscopic angle for both lenses may have to be either increased or decreased, in order to maximize both the cosmetic effect and the optical performance. For example, the pantoscopic angle will have to be *increased* if the distance between the lower edges of the lenses and the cheekbones is too great (more than the "thickness of a finger") or if the upper edges of the lenses are too close to the eyebrows, whereas it will have to be *decreased* if the bottoms of the lenses are too close to the cheekbones or the tops are too far from the eyebrows. If the frame fitting and lens ordering were done correctly, the resulting pantoscopic tilt (in degrees) should be approximately twice the distance (in millimeters) between the optical center of each lens and the center of the patient's pupil.

Often, adjusting the glasses so that the front appears to be properly aligned with the patient's face is a matter of compromise: This is particularly true in cases of *hyper-eye*, or in cases where the two eyes are on the same level but one eyebrow is higher than the other. For a discussion of hyper-eye, see Section 9-13.

*Temple Adjustments*

In addition to angling the endpieces to make the glasses level on the patient's face and to provide the correct amount of pantoscopic tilt, it may be necessary to adjust the temple angles and to adjust the temples where they wrap around the patient's ears.

If the temple angles are correct, the temples should contact the sides of the head firmly but not tightly enough to make grooves in the skin. Whereas the temples of a "standard"-sized frame (e.g., less than 50 mm wide) will usually contact the sides of the head from just behind the temporal angle of the orbit (Figure 9-29A), those of an extra-large frame may contact the head only for an inch or so in front of the ear (Figure 9-29B). Such a frame may even require temple angles of *less than 90°*. Whatever the frame size, if the temple makes too little contact with the side of the head the temple angle must be decreased, whereas if there is too much contact the temple angle must be increased. As a part of the temple angle adjustment the patient should be asked to bend his or her head downward, and the dispenser should make sure that the top of the frame is equally close to both upper orbital ridges (in the region of the eyebrows).

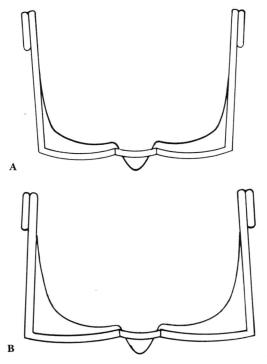

**A**

**B**

FIGURE 9-29. **Temple angle: A, desired temple angle for a "standard"-sized frame; B, temple angle for an extra-large frame.**

If this is not the case, the temple angles are unequal (as shown in Figure 9-17), and the temple angle should be *increased* on the side that is too far from the brow region.

The position of each temple where it bends over and behind the patient's ear should be carefully noted and, if necessary, the temple should be adjusted by bending it so that it will gently contact the skin in the area of the temple crotch (at the top of the ear) and in the region of the mastoid bone, without putting pressure on the outer ear. If the bend is found to be in front of or behind the crotch of the ear, this should be corrected. For most temples (particularly plastic temples) the best way to make this adjustment is to straighten the temple (almost completely removing the bend), and then to place the bend where it is needed.

Occasionally, it will be found that with all reasonable adjustment the "length to bend" of both temples is either too long or too short. In such a case the error was obviously made at the time of frame selection and ordering, and the temples should be replaced with ones of the correct length. Unfortunately, with some of the currently available "fashion" frames, only one or two temple lengths are available, so it is occasionally necessary to "make do" with temples that are really too short or too long for an optimum fit by judicious positioning of the bend and adjusting behind the ears.

*Bridge Adjustments*

If the glasses being dispensed incorporate a plastic frame with no adjustable pads, little can be done to improve the fitting of the bridge. However, if a pad bridge frame is used, the pads can easily be adjusted to move the glasses up or down on the patient's face or to move them closer or farther away from the eyes. Adjusting the pads downward will, of course, move the glasses upward on the patient's face, while moving them upward will allow the glasses to fit lower on the face; spreading the pads apart will tend to make the glasses fit closer to the eyes, while moving them closer together will tend to move them out a little. However, in any pad adjustment, care must be taken to ensure that the pressure of the pads is spread evenly over each side of the nose: A pad that digs in at the bottom or the top edge, or at the back or the front edge, will be sure to cause discomfort. If a patient has a tender nose (and particularly if heavy lenses or a heavy frame, or both, are being dispensed), it is a very good idea to replace the pads that come with the frame with *jumbo* pads.

## Verifying Visual Acuity

At the dispensing visit, the patient should be given an opportunity to read the letters on the distance Snellen chart with the new glasses; if the glasses are intended mainly for near vision (or if they are multifocals) the patient should be asked to read the letters on a reduced Snellen chart or a reading card. Although the optometrist may be very sure that the patient will have 20/20 vision or better with the new glasses, many patients will be disappointed if they are not given the opportunity to "try out" the new glasses by reading the eye chart. For this reason, some optometrists routinely do all of their dispensing in the refracting room, with the patient seated in the refracting chair (even though frame selection and other aspects of dispensing are done by a dispenser or technician with the patient seated at a fitting table). Not only can the glasses be conveniently adjusted in the refracting chair, but the patient does not have to be moved to another room for the visual acuity check.

## Checking Centration of Finished Spectacles

If the practitioner is concerned about the centration of the patient's lenses, this may be checked by the use of a retinoscope or ophthalmoscope in a semi-darkened room. With the spectacles in place, the patient is asked to fixate the instrument light while the examiner observes the patient's eye through the instrument's peephole. The examiner will see two images of the light, reflected from the front and back surfaces of the spectacle lens; and by moving about, the two images can be made to coincide (the student will recognize this procedure as reminiscent of that of lining up the reflected images from the "cornea" of a schematic eye when practicing retinoscopy).

At this point, the examiner's line of sight coincides with the optical axis of the lens. A third image, the corneal reflex, will also be seen. Since the corneal reflex lies near the patient's visual axis (which passes through the center of rotation of the eye) the separation of the two superimposed images and the corneal reflex indicates the discrepancy between the optical axis of the lens and the visual axis. By altering the pantoscopic tilt or by raising or lowering the spectacles, the two lens reflections may be brought into alignment with the corneal reflex. The optical axis of the lens will then pass through the center of rotation, so the fitting criteria will match the designer's assumption for optimal performance.

## Care and Maintenance of Glasses

Once the adjustment of the glasses has been completed, the dispenser should instruct the patient (particularly if this will be the first pair of glasses) concerning their care and maintenance. The patient should be cautioned not to spread the temples apart when putting the glasses on (it is surprising how many people do this) but to put the glasses on by tilting the temples over the ears. He or she should also be told that when the glasses are taken off they should be either put in the case, placed upside down on a table or other surface with the temples wide open, or folded and placed in such a way that they rest on the folded temples: In any event they should not be placed with the front surfaces of the lenses in contact with the table (again, it's surprising how many patients will do this, if not warned against it). The patient should understand that if the lenses are *plastic lenses*, extra care should be taken to avoid scratching them. Lenses (whether glass or plastic) should never be cleaned with a dry cloth. They should be cleaned by rubbing the lenses between the fingertips moistened by a liquid detergent while being held under running tap water. A soft cloth or tissue should be used to dry the lenses: While glass lenses may be wiped dry, plastic lenses should be *patted* dry.

## Advice Concerning Use of the Glasses

After the glasses have been properly adjusted, the patient should be instructed (even if this has been done during the examination) concerning what to expect from the new glasses. If a substantial change has been made in the prescription, the patient should be informed of the likely perceptual changes that will be encountered, and advised that these changes will disappear with time.

For a patient who will be wearing his or her first bifocals, trifocals, or progressive addition lenses, adequate instruction should be given in their use in walking (particularly in going up and down stairs) and in the positions of the head, eyes, and arms that result in comfortable reading posture.

## Subsequent Adjustments

When the glasses are dispensed, the patient should be told that the glasses may need further adjusting after they have been worn for several days, and should be encouraged to return for readjustment. Many optometrists routinely schedule each patient to return in approximately 2 weeks for a "progress check," so that he or she can make sure that the glasses are performing as they were intended to perform. At this time, clinical findings such as visual acuity may be verified, but, more important, the *adjustment* of the glasses should be verified. There is no doubt that more patients are "lost" to other practitioners by *poor dispensing or adjusting* than by poor refracting or prescribing! The reader's attention is again called to the common maladjustments and remedies listed in Table 9-2.

## References

1. Gasson, W. The Florentine Legend. *Ophthal. Optician*, Vol. 9, No. 17, pp. 924−928, 1969.
2. Corson, R. *Fashion in Eyeglasses*. Peter Owen Ltd., London, 1970.
3. Sasieni, L.S. *Principles and Practice of Optical Dispensing and Fitting*, 3rd ed. Butterworths, London, 1975.
4. Brooks, C.W., and Borish, I.M. *System for Optical Dispensing*. Professional Press, Chicago, 1979.
5. Stimson, R.L. *Ophthalmic Dispensing*, 3rd ed. Charles C. Thomas, Springfield, IL, 1979.
6. Waters, E.H. *Ophthalmic Mechanics*, Vol. 1, p. 262. Edward Brothers, Ann Arbor, 1964.
7. Maxwell, J.T. *Outline of Refraction*, p. 380. Medical Publishing Co., Omaha, 1937.
8. Laurance, L. *Visual Optics and Sight Testing*, p. 399. School of Optics, London, 1920.
9. Augsburger, A. Evaluation of Ophthalmic Materials. *Amer. J. Optom. Physiol. Opt.*, Vol. 55, pp. 700−705, 1978.

## Questions

1. What would be the pure gold content of a center stamped 1/10 12 K GF?

2. A patient has a distance PD of 70 mm. Assume the distance from the spectacle plane to the centers of rotation is 2.7 cm. What is the calculated near PD for a near distance of 25 cm from the spectacle plane?

3. Assume that the distance from the spectacle plane to the center of rotation is 27 mm. If the near PD is 56 mm for a point 25 cm from the spectacle plane, what is the calculated distance PD?

4. The prescription:
   OD +4.00 DS 1 prism diopter base in
   OS −2.00 DS 1 prism diopter base in
   Add = +2.00 D
   The prism is obtained by decentration,
   PD 62/58 mm
   DBL = 20 mm
   Horizontal box size = 46 mm
   What is the bifocal inset (amount and direction) of the right and left lenses, respectively?

5. The prescription:
   OD +4.00 DS 1 prism diopter base in
   OS −2.00 DS 1 prism diopter base in
   Add = +2.00 D
   The prism is obtained by decentration,
   PD = 62/58 mm
   DBL = 20 mm
   Horizontal box size = 46 mm
   What is the total inset (amount and direction) of the right and left lenses, respectively?

6. The distance PD is 64 mm and the near PD is 60 mm. The prescription is:
   OD +1.50 D
   OS +1.50 D
   +2.00 D add
   If we wish to have *zero horizontal prismatic effects* at the reading level, what must be the bifocal inset for each lens?

7. A frame is 52 × 44 mm in size and will be worn with a pantoscopic tilt of 10 degrees. The center of the pupil falls at a point 24 mm from the bottom of the frame. Assuming we use corrected-curve lenses and wish them to function as such, where is the vertical location of the "optical center" in the above frame?

8. An emmetropic presbyope requires an add of +2.50 D OU (both eyes) for normal close work and reading at 40.0 cm. He stands 25 cm from a plane mirror for shaving and experiences difficulty in seeing well enough to get a close shave. What prescription in single-vision lenses would be most satisfactory in solving his shaving problem?

9. The outside surface of a Kryptok bifocal is resurfaced with the same tool to remove scratches. The segment was reduced in size with the optical center of the segment remaining in the same position. The segment originally measured 18 mm high, 22 mm wide. If the seg height is reduced to 16.0 mm, what is the segment width?

10. Discuss the characteristics of cellulose acetate and cellulose nitrate as frame materials.

11. Give the variables that determine the near interpupillary distance.

12. Discuss the importance of the "fitting triangle" as a foundation for spectacle adjustments.

13. Discuss the merits of centering for the distance PD or the near PD in single-vision lenses for near use only.

14. Describe the adjustment procedure when the left lens is farther from the face than the right lens.

15. Describe the adjustment procedure when the right lens is higher in the face than the left lens.

16. Outline the procedures for evaluating an order for spectacles received from a laboratory.

# CHAPTER TEN

# Anisometropia and Aniseikonia

For the great majority of wearers of glasses or contact lenses, the refractive error differs little for the two eyes. However, for a small percentage of wearers the refraction differs by a significant amount, giving rise to the condition known as anisometropia.

Anisometropia—both in the uncorrected and corrected states—is responsible for a number of problems. For example, an individual having uncorrected anisometropia will not be able to have sharply focused images on the retinas of both eyes simultaneously. This, in extreme cases, may lead to suppression and even to amblyopia. And once anisometropia is corrected, it is responsible for base-in and base-out prismatic effects whenever the eyes are rotated horizontally, so that the visual axes fail to correspond to the optic axes of the lens, and for differential vertical prismatic effects when the visual axes are lowered for reading.

One of the most bothersome problems due to corrected anisometropia, however, is induced aniseikonia—a condition in which the retinal images for the two eyes differ in size or in shape. Whether or not aniseikonia will be induced, for a given patient, depends both upon the nature of the patient's ametropia (whether due mainly to the "axial" or "refractive" properties of the eyes) and upon the magnification characteristics of the correcting lenses.

# ANISOMETROPIA

## 10.1. Problems Resulting from Anisometropia

Anisometropia is considered to exist when the spherical equivalent refraction of the two eyes differs by 1.00 D or more. Anisometropia of this amount or greater can be responsible for a number of problems, both in the uncorrected state and in the corrected state.

### Uncorrected Anisometropia

As pointed out in Section 5.8, Hering's law of equal innervation tells us that the two eyes respond with equal amounts of accommodation. Therefore, when uncorrected anisometropia exists, the eyes may accommodate, in any given situation, so that the retinal image of one eye or of the other is in focus, or so that the two retinal images are more or less equally out of focus. For a child under the age of 5 or 6 years, uncorrected anisometropia may lead to the development of functional amblyopia or other binocular vision problems. If the child is emmetropic in one eye and has uncorrected *hyperopia* in the other, the more hyperopic eye will never have a sharply focused retinal image (since it will be easier for the child to use the emmetropic eye for vision at *any* distance) with the result that functional amblyopia may develop in the more hyperopic eye. On the other hand, if the child is emmetropic in one eye and has uncorrected *myopia* in the other eye, he or she will be likely to use the emmetropic eye for distance vision and the myopic eye for near vision, so amblyopia is not likely to develop. However, poor stereopsis or other binocular vision problems may result. If an older child or an adult is found to have uncorrected anisometropia, normal binocular vision seldom exists, but in many cases it can be restored by prescribing appropriate lenses.

### Corrected Anisometropia

Problems which may arise when lenses are worn for the correction of anisometropia may involve (1) the accommodative system; (2) the vergence system; and (3) the relationship between the retinal image sizes for the two eyes. Problems involving the accommodative system, as discussed in Section 5.8, occur as a result of the fact that lens *effectivity* differs for different fixation distances. Lenses which are equally effective for the two eyes of an anisometrope for distance vision are not equally effective for near vision. Problems involving the vergence system occur as a result of differential prismatic effects that are present when the visual axes pass through points in the lenses other than the optical centers. These effects can be considered in terms of either the horizontal or the vertical meridian (although they are present also for oblique meridians) and will be discussed in the following paragraphs. Problems involving retinal image size differences for the two eyes will be discussed later in this chapter.

## 10.2. Horizontal Prismatic Effects

When lenses of unequal powers are worn for the correction of anisometropia, any amount of movement of the visual axes away from the optical centers of the lenses brings about a prismatic imbalance. The amount and direction of the imbalance can easily be calculated with the aid of Prentice's rule.

### EXAMPLE
If a patient wears the prescription OD +3.00 D sphere and OS +1.00 D sphere, centered for the distance PD, we can easily calculate the prismatic effects resulting, during distance fixation, when the two visual axes pass through points in the lenses *(a)* 20 mm to the *left* of the optical centers and *(b)* 20 mm to the *right* of the optical centers.

*(a)* As shown in Figure 10-1, when the visual axes are turned toward the left, the right eye is subject to a *base-out* prismatic effect, and the left eye is subject to a *base-in* prismatic effect, the base-out prismatic effect being the greater of the two.

OD    $P = dF = 2(+3.00) = 6$ prism diopters, base out.

OS    $P = dF = 2(+1.00) = 2$ prism diopters, base in.

Resultant prismatic effect = 4 prism diopters, base out.

*(b)* As shown in Figure 10-2, when the visual axes are turned toward the right, the right eye is subject to a *base-in* prismatic effect and the left eye is subject to a *base-out* prismatic effect, the base-in effect being the greater.

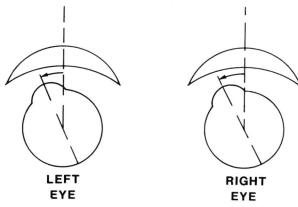

FIGURE 10-1. **Prismatic effects induced when the visual axes are turned toward the left while wearing the prescription OD +3.00 DS and OS +1.00 DS.**

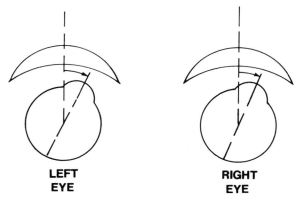

FIGURE 10-2. **Prismatic effects induced when the visual axes are turned toward the right while wearing the prescription OD +3.00 DS and OS +1.00 DS.**

OD   $P = dF = 2(+3.00) = 6$ prism diopters, base in.

OS   $P = dF = 2(+1.00) = 2$ prism diopters, base out.

Resultant prismatic effect = 4 prism diopters, base in.

These results indicate that when horizontal conjugate (versional) movements are made, the visual system must somehow compensate for an ever-increasing base-out prismatic effect when looking toward the left, and an ever-increasing base-in prismatic effect when looking toward the right. Even though our calculations show that the resultant prismatic effect is caused by the excess prismatic power of the *right* lens, it should be understood that in any direction of gaze the prismatic effect is considered to be shared equally by the two eyes.

Compensating for these horizontal prismatic effects seldom causes a problem for the anisometrope, since the visual system is capable of horizontal fusional vergence movements of reasonably large amplitude (*positive* fusional vergence movements to overcome base-out prism, and *negative* fusional vergence movements to overcome base-in prism).

## 10.3. Differential Vertical Prismatic Effects at the Reading Level

Since the eyes are capable of vertical fusional movements of only a small amplitude, vertical prismatic effects which differ for the two eyes are likely to cause symptoms of eyestrain. Such effects can occur when a corrected anisometrope either elevates or lowers the visual axes, looking through points in the lenses considerably above or considerably below the optical centers of the lenses. These effects are most likely to become a problem when the visual axes are lowered for reading. If the lenses are centered in front of the wearer's pupils for distance vision, lowering the visual axes for reading will induce a *base-up* prismatic effect for the more hyperopic eye of an anisometropic hyperope, or a *base-down* prismatic effect for the more myopic eye of an anisometropic myope (see Figure 10-3).

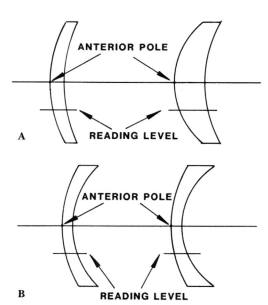

FIGURE 10-3. **Primatic effects induced when an anisometrope lowers the visual axes for reading: A, for an anisometropic hyperope: B, for an anisometropic myope.**

For a wearer of single-vision lenses these differential prismatic effects at the reading level may not constitute a problem, since the head is normally tilted downward in reading with the result that the reading level may be very little different than the level used for distance vision. However, a bifocal wearer must read through a level that is approximately 10 mm below the distance optical centers, with the result that differential vertical prismatic effects at near can become a problem.

For moderate amounts of anisometropia, the vertical imbalance at the reading level can be overcome by means of vertical fusional movements; but since the amplitude of vertical fusional movements is usually very small, a patient having a significant anisometropia may have to use virtually all of his or her vertical fusional vergence amplitude just to achieve single vision. For those anisometropic patients doing a great deal of near work, attention should be directed toward the possibility of a vertical imbalance at the reading level. Any asthenopia or other symptomatology accompanying near work should suggest an evaluation of an induced vertical imbalance.

### EXAMPLE
If a bifocal wearer is given the distance prescription

$$OD +1.50 DS -1.00 DC \times 180$$
$$OS +2.50 DS$$

the powers in the vertical meridian are $+0.50$ D for the right lens and $+2.50$ D for the left lens. If the patient reads at a level 10 mm below the distance optical centers, the vertical prismatic effects (see Figure 10-3) will be

OD    $dF = 1(+0.50) = 0.5$ prism diopter, base up

OS    $dF = 1(+2.50) = 2.5$ prism diopters, base up,

and the differential prismatic effect will be
2.0 prism diopters, base up, left eye.

Note that the power of the *add* does not enter into the situation, as long as it is the same for both eyes: In fact, it wasn't even necessary to specify the power of the add.

As shown in this example, when a patient's lenses are spherical or when the cylinder axes are horizontal or vertical, determination of the vertical imbalance at the reading level is easily done by the application of Prentice's rule. However, for cylinders of oblique axes, calculation of the vertical imbalance must be done by the method described in Chapter 4. On occasion, casual inspection of the patient's prescription may suggest that induced vertical imbalance at the reading level is a potential problem; but for prescriptions with strong oblique cylinders it may not be readily apparent that vertical imbalance is a potential problem. Fischer[1] has developed some aids that assist in making a rapid determination of the induced vertical imbalance.

Individual tolerance for induced vertical imbalance varies considerably from one patient to another. Ellerbrock and Fry[2] found that some anisometropic patients developed the capacity to "compensate" for prismatic inequalities at the reading level. This compensation was defined as the difference between a measured vertical phoria and the calculated vertical imbalance. Cusick and Hawn[3] pointed out that it was not uncommon to find a patient wearing a 5.00 D difference in the corrections for the two eyes who had no measurable hyperphoria with a Maddox rod during downward gaze through the lenses. As already indicated, an additional problem which may occur in anisometropia is that of asthenopia due to the image size differences between the two eyes (i.e., induced aniseikonia). However, Ellerbrock[4] recommended that prismatic corrections should be provided for anisometropic patients who are uncomfortable at the reading position before any consideration is given to aniseikonia and its correction.

Duke-Elder[5] suggested that, as a general rule, a differential vertical prismatic effect at the reading level is not apt to cause symptoms of asthenopia unless it amounts to 1 prism diopter or more. The practitioner should keep in mind the possibility of a differential prismatic effect at the reading level for any patient having *1 diopter or more* of anisometropia in the vertical meridian, since for a reading level of 10 mm below the distance optical centers each diopter of anisometropia will cause 1 prism diopter of differential vertical prismatic effect. Note that a comparison of the *spherical powers* of the two lenses is not sufficient as, in the above example, since a cylindrical correction for only one eye (or a stronger cylinder for one eye than for the other) can greatly increase the difference in the powers of the two lenses in the vertical meridian.

Any one of eight procedures may be used to compensate for differential vertical prismatic effects at the reading level:

1. Lowering the distance optical centers
2. Prescribing single-vision lenses for reading only
3. Prescribing dissimilar bifocal segments for the two eyes
4. Prescribing compensated bifocal segments
5. Prescribing prism segments
6. Prescribing a "slab-off" lens
7. Prescribing a Fresnel Press-on prism
8. Prescribing contact lenses

1. *Lowering the Distance Optical Centers.* For the patient who is not yet presbyopic, a possible method of alleviating a vertical imbalance at the reading level is that of lowering the optical centers of the lenses. For example, if the optical centers are dropped 3 mm relative to their normal positions, the amount of vertical imbalance would be reduced by a factor of 0.3 multiplied by the amount of anisometropia in the vertical meridian. Of course, the amount of vertical imbalance not present at the reading level will be present at the positions of the original optical centers (but opposite in direction). For this reason, this method is apt to prove unsatisfactory and should be used cautiously if at all.

2. *Prescribing Single-Vision Lenses for Reading Only.* Some pre-presbyopic patients may be willing to accept the idea of wearing two pairs of glasses, one for distance vision and one for reading. The distance lenses are centered in the usual manner, with the optical centers positioned in front of the pupils for distance vision. For the near lenses, the optical centers are positioned lower in the lenses, usually from 5 to 10 mm lower than their placement for distance vision. The patient should be observed while reading, and the points where the lines of sight cross the lenses marked with a grease pencil or a felt pen. The most effective way to do this is to use a small mirror, held next to the reading material, while the examiner sits at a table opposite the patient.

For *presbyopic* patients, two pairs of glasses can be prescribed, a pair of bifocals for general wear and a pair of single-vision glasses with the optical centers lowered for reading. The bifocals are worn for distance vision and for short periods of near work, but for prolonged periods of near work the single-vision glasses are worn.

3. *Prescribing Dissimilar Bifocal Segments.* Segments of different size or construction, with their centers at different levels, may be used to compensate for an induced vertical imbalance at the reading level. The amount of compensation is equal to the vertical separation of the segment centers multiplied by the power of the add. As an extreme example, assume that an Ultex A segment (having its center 19 mm below the segment top) is worn on one eye and an Executive segment (having its center at the top of the segment) is worn on the other, with the segment tops at the same level for both eyes (see Figure 10-4). If the power of the bifocal addition is +2.00 D, this arrangement would compensate for an induced vertical prismatic effect of

$$1.9(+2.00) = 3.8 \text{ prism diopters.}$$

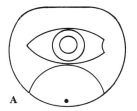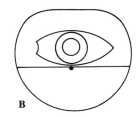

**FIGURE 10-4.** Prescribing dissimilar bifocal segments: An Ultex A bifocal is worn on one eye and an Executive bifocal on the other eye.

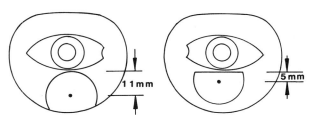

**FIGURE 10-5.** Prescribing dissimilar bifocal segments: A round fused bifocal is worn on one eye and a straight-top fused bifocal is worn on the other eye.

If a 22-mm round fused bifocal (having its segment center located 11 mm below the segment top) is worn on one eye and a straight-top fused bifocal (having its segment center 5 mm below the segment top) on the other (Figure 10-5), for a +2.00 D add the amount of compensation would be

$$0.6(+2.00) = 1.2 \text{ prism diopters.}$$

In order to minimize the difference in the appearance of the segments, it is desirable for the two segments to be as nearly similar in shape as possible. For example, one may use two round fused bifocals with different diameters, or an Ultex A and an Ultex B bifocal. Judicious selection of the two segments allows a wide range of vertical imbalances to be compensated. Table 10-1, listing conventional bifocal lenses according to the distance from the segment top to the segment center, is useful for this purpose. The segment with its center at the *greatest* distance from the segment top is always prescribed for the eye requiring the *most plus or least minus power* in the vertical meridian. Although the use of dissimilar segments is a straightforward, inexpensive, and readily available method of solving the problem, it is likely that many patients will resist the idea of wearing differently shaped segments for the two eyes.

4. *Prescribing Compensated Bifocal Segments.* A more cosmetically acceptable method of compensating for differential prismatic effects is to make use of compensated ribbon segments. The 14-mm-deep

TABLE 10-1
**Conventional Bifocal Lenses Listed According to the Distance from the Segment Top to the Segment Center**

| Bifocal | Segment Top to Segment Center (mm) |
|---|---|
| Executive-style one piece | 0.0 |
| B&L Panoptik (22 x 25) | 3.5 |
| B&L Panoptik (24 x 16.5) | 4.0 |
| Vision-Ease C (22 x 16) | 4.0 |
| Univis F (22 x 16) | 4.0 |
| Kurova CT (22 x 16) | 4.0 |
| AO Tillyer (20 x 14.4) | 4.5 |
| AO Sovereign (22 x 15.5) | 4.5 |
| AO Panoptik (24 x 16.4) | 4.5 |
| Univis B (22 x 9) | 4.5 |
| Kurova B (22 x 9) | 4.5 |
| Vision-Ease B (22 x 9) | 4.5 |
| Most straight-top fused | 5.0 |
| Tillyer Masterpiece S | 5.5 |
| AO Sovereign (25 x 17.5) | 5.5 |
| Univis R (22 x 14) | 7.0 |
| Vision-Ease R (22 x 14) | 7.0 |
| Tillyer D (20 round) | 10.0 |
| Kryptok (22 round) | 11.0 |
| Other 22 round fused | 11.0 |
| Ultex B (22 round) | 11.0 |
| Hydray CX B (22 round) | 11.0 |
| Ultex A (38 x 18) | 19.0 |
| Other 38 x 18 one piece | 19.0 |
| Ultex AL (38 x 33) | 19.0 |
| RH Hydray (40 x 20) | 20.0 |

*R-Compensated* segment has been designed for this purpose. Although the normal position of the optical center of the R segment is at the center of the segment (7 mm below the segment top), it can be specified anywhere from 4 to 10 mm below the segment top. With the segment centers in the extreme positions of 4 and 10 mm below the segment tops, one segment is "upside down" with respect to the other, as shown in Figure 10-6. For a +2.00 D add, the

resulting prismatic compensation would be

$$0.6(+2.00) = 1.2 \text{ prism diopters.}$$

An R-Compensated segment is produced by regrinding the front surface of a regular R-segment lens blank. The resurfacing process moves the segment center from its original position to a higher or lower position. To move the segment center to a lower position, more of the upper edge of the segment is ground away than the lower edge. To move the center higher, more of the lower segment edge is removed than the upper edge.

As a result of this resurfacing process, it is obvious that the two segments are *not* identical in shape; however, not enough difference exists to be cosmetically unacceptable. The use of R-Compensated segments therefore constitutes a subtle application of the concept of dissimilar segments. This method is more expensive than other methods of using dissimilar segments, and is limited to an amount of prism equal to 0.6 multiplied by the power of the add, but the extra cost is often justified by the improved appearance.

As stated above, the segment with its center at the *greatest* distance from the segment top is always prescribed for the eye requiring the *most plus or least minus power* in the vertical meridian.

5. *Prescribing Prism Segments.* A few bifocal manufacturers make available segments incorporating vertical prism power. These lenses are expensive, excessively thick, and are "factory orders," involving a delay of several months. Their use is not recommended, since other methods (particularly the use of R-Compensated segments and the provision of a "slab-off" lens, discussed below) are effective, less expensive, and do not require a factory order.

6. *Prescribing a "Slab-Off" Lens.* A "slab-off" lens is made by a procedure known as *bi-centric grinding.* As shown in Figure 10-7A, the front surface is finished in the usual manner, after which a dummy lens is cemented onto the front surface (Figure 10-7B). The front surface is then reground, using the same tool originally used for that surface, but ground in such a way that the dummy is ground away in the upper portion of the lens but remains attached to the lower portion. The back surface is then finished with the remaining dummy considered as an integral part of the blank (Figure 10-7C). The thickness of the blank would then be equal at the top and bottom unless the prescription calls for prism in the distance portion. When the lens is finished, the remaining dummy on the lower portion is then removed (Fig-

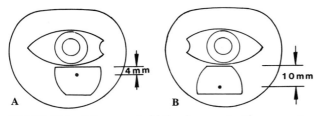

FIGURE 10-6. **R-Compensated bifocal segments: The segment optical center is 4 mm below the segment top for one eye and 10 mm below the segment top for the other eye.**

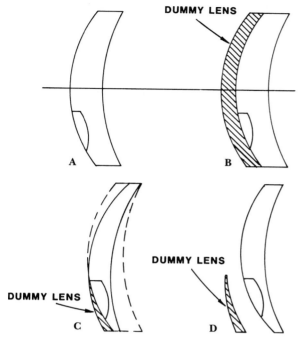

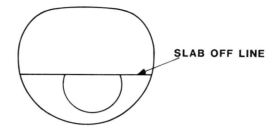

FIGURE 10-8. **A straight-top fused bifocal with slab-off prism, resulting in a horizontal line across the lens.**

FIGURE 10-7. **Bi-centric grinding, producing a "slab-off" lens. This procedure produces a horizontal line across the lens.**

ure 10-7D). The dummy is a base-down prism, so this procedure results in the removal of base-down prism in the lower portion of the lens, or in the addition of *base-up* prism in the lower portion of the lens.

The center of curvature of the front surface of the lens, in the lower portion, is displaced upward, with the result that the front surface has *two* centers of curvature, one for the upper portion and one for the lower portion, even though the upper and lower portions have the same curvature. This process results in a unique optical axis for each of the two portions of the lens.

Thus, bi-centric grinding results in the removal of base-down prism (whether used on a minus lens or a plus lens), without changing the refractive power below the slab-off line. It is therefore used for the lens that induces the lesser amount of base-up (or the greater amount of base-down) prism: This is always the lens having the *less plus or more minus power* in the vertical meridian. For example, in the case of the prescription

OD −1.00 DS
OS −4.00 DS

bi-centric grinding would be used for the *left* lens.

Bi-centric grinding results in a horizontal line all the way across the lens (Figure 10-8), providing a

very acceptable cosmetic result when the line coincides with the segment top of a straight-top bifocal. It may be done on either the front or the back surface of the lens. *Front surface* bi-centric grinding is used for single-vision glass lenses, straight-top fused bifocals, and one-piece glass bifocals (such as the Ultex) having the bifocal on the back surface. *Back surface* bi-centric grinding is used for one-piece straight-top glass Executive-style bifocals and trifocals and for plastic single-vision lenses, bifocals, and trifocals.

Recently, Younger Optics and Aire-O-Lite have introduced plastic single-vision and bifocal lenses with a slab-off molded or cast with *base-down* prism in the segment area, rather than individually generating a base-up prismatic effect by bi-centric grinding. These lenses are known as *reverse slab-off* lenses. Because the compensating prism is base down rather than base up, the slab-off lens is prescribed for the eye requiring the *more plus* or *less minus* power in the vertical meridian. These lenses can be maintained in the laboratory's inventory as semifinished lenses, making possible much faster delivery than for the individually produced slab-off lenses.

Slab-off lenses are not usually produced for less than 1.25 prism diopters of compensation, because in such cases a sharp, horizontal, straight line is difficult to produce. But when more than 1.25 prism diopters of compensation is required, the use of a slab-off lens is the method of choice. The slab-off line is relatively inconspicuous and in straight-top fused bifocals the slabbed-off area extends beyond the segment to the distance portion of the lens.

Although bi-centric grinding is more expensive than the use of R-Compensated segments, it has the advantage that a larger amount of prism can be provided. The amount of prism possible is limited only by the thickness of the lens, whereas for the R-Compensated bifocal the amount of prism possible (as already noted) is limited to 0.6 times the power of the add. Bi-centric grinding can be done by any prescription laboratory, so does not require a factory order.

In the case of a finished lens the amount of slab-off may be verified, using a lensometer, by comparing the vertical prismatic effects of the two lenses at the reading level. The amount of the slab-off is the difference between the calculated amount of vertical imbalance and the amount found by the lensometer at the reading level. Peters[6] has suggested a much simpler method of determining the amount of slab-off. His method, illustrated in Figure 10-9, employs a standard lens gauge (with separations of the fixed legs of 20.8 mm). The lens gauge is first used to measure the front surface power *above* the slab-off line (Figure 10-9A). The second reading is then taken with the three points of the lens gauge arranged vertically and with the middle pin directly on the slab-off line (Figure 10-9B). The difference between the initial reading and the second reading (in diopters of refracting power) is the amount of slab-off, in prism diopters.

7. *Prescribing a Fresnel Prism.* A Fresnel prism, as described in Chapter 5, is a plastic "press-on" prism, based on the Fresnel principle, that can be placed on

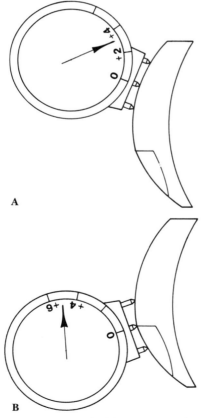

**A**

**B**

FIGURE 10-9. **Using a standard lens gauge to determine the amount of slab-off prism.**

an ordinary spectacle lens. A Fresnel prism can be placed on the back of the segment area of a bifocal lens, as a temporary measure prior to ordering a more expensive correction with compensated segments or bi-centric grinding. Disadvantages of Fresnel prisms are that they give the lens a "striated" appearance and lower the wearer's visual acuity by about one line of letters on the Snellen chart.

8. *Prescribing Contact Lenses.* For an anisometrope who is motivated to wear them, the prescribing of contact lenses can be an effective method of avoiding differential vertical prismatic effects at the reading level. They may be considered as the treatment method of choice for some pre-presbyopic patients, since well-fitted single-vision contact lenses remain centered with respect to the pupils when the eyes move downward for reading.

Although bifocal contact lenses have achieved only limited acceptance by both practitioners and wearers, it is likely that the reading level for a wearer of these lenses is considerably higher than that for a wearer of bifocal spectacle lenses. This should result in a much smaller differential prismatic effect than that induced by the wearing of bifocal spectacles. The very popular "monovision" method of fitting presbyopes with contact lenses (fitting a distance lens for one eye and a reading lens for the other) would also be effective in eliminating differential prismatic effects at the reading level. Finally, differential prismatic effects also can be avoided for a presbyope if he or she is willing to wear contact lenses for general wear, supplemented by spectacle lenses (centered for the reading level) for reading.

## Should the Prismatic Effect be Fully Compensated?

For an assumed reading level at a given distance below the distance optical centers, the vertical imbalance may be precisely calculated. The amount of calculated imbalance, the amount of close work, and the degree of compensation as determined by dissociated and associated phorias will assist in determining whether any correction for vertical imbalance should be prescribed. If a *full correction* is applied for a given reading level, this correction, being a constant, will be an *undercorrection* for any level lower than the assumed reading level or an *overcorrection* for any level higher than the assumed reading level. Since the assumed reading level is largely an estimate, it is usually best to provide an *undercorrection*, no matter which method of compensation is used, since the patient may read at a higher level than assumed.

## ANISEIKONIA

### 10.4. Introduction

It has been known for many years that the correction of anisometropia by spectacles can introduce unequal image sizes for the two eyes, with consequent disturbances in binocular vision. Donders[7] was one of the early writers who was aware of the dimensional changes in the retinal images introduced by the correction of anisometropia.

The researches initiated by Ames, and implemented by his associates at the Dartmouth Eye Institute,[8] established the basic principles underlying the measurement of image size differences and their equalization in clinical practice. Lancaster,[9] in 1938, coined the term "aniseikonia," deriving it from Greek word roots, literally meaning "not equal images."

Aniseikonia is defined as the relative differences in the sizes and/or shapes of the *ocular* images of the two eyes. The term ocular image includes not only the *retinal* image formed by the dioptrics of the eye (together with any correcting lens) but also the modification of the retinal image by the distribution of the nerve endings in the retina and their representation in the visual cortex. Since the absolute sizes of the ocular images, as just defined, are incapable of measurement, we measure the relative difference in their sizes, and state this difference in terms of the *percentage magnification* of afocal magnifying lenses (or "size" lenses) required to equalize the two images.

### 10.5. Etiology of Aniseikonia

As a result of the lateral separation of the two eyes of 60 to 70 mm, differences in the sizes and shapes of the retinal images are routinely experienced in everyday life. It is this separation of the eyes, providing two different views of all surfaces and contours, which forms the basis for normal binocular stereopsis and space perception. The resulting differences in retinal image size and shape are more pronounced when viewing a nearby object that is displaced to one side, requiring asymmetrical convergence of the eyes (see Figure 10-10). As a result of this asymmetrical convergence, the retinal image of the nearer, more converged, eye is *larger* than that of the farther or less converged eye. These image differences are com-

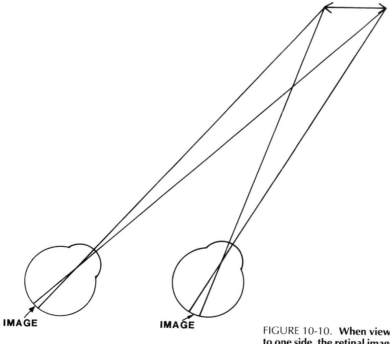

**IMAGE**

**LEFT EYE**

**IMAGE**

**RIGHT EYE**

FIGURE 10-10. **When viewing a nearby object that is displaced to one side, the retinal image of the nearer, more converged, eye is larger than that of the less converged eye.**

pensated psychologically and cause no symptoms, but provide clues which assist in spatial localization of objects.

The term *anomalous aniseikonia* is used to include all forms of aniseikonia other than the "normal" aniseikonia just described. The etiology of anomalous aniseikonia can be classified as either *anatomical* or *optical*. Anatomical causes for aniseikonia include such factors as the degree of separation (density) of the retinal receptors, and the functional organization of the terminal neural visual pathways in the cortex. For example, one could argue that if the neural elements were more widely separated in the retina of one eye than in that of the other, the perceived image would be smaller because fewer retinal elements would be stimulated. Ogle[10] reported that in about one-third of the cases of aniseikonia the measured image differences were seemingly unrelated to the dioptric characteristics of the anisometropia present, indicating that the aniseikonia could have been anatomical in origin.

*Optical* aniseikonia has two origins: *inherent* and *induced*. Inherent aniseikonia depends solely upon the dioptric system of the eye itself, whereas induced aniseikonia is due to the magnification properties of the lenses worn for the correction of a refractive error, more specifically for the correction of either anisometropia or astigmatism.

The types of image size differences occurring in aniseikonia may be classified as follows.

*Symmetrical Differences*

1. Overall, in which the ocular image of one eye is increased or decreased in size (equally in all me-

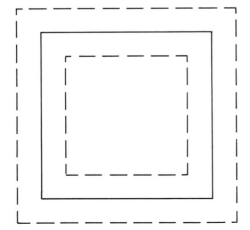

FIGURE 10-11.  **Overall aniseikonia. The ocular image of one eye is increased or decreased in size (equally in all meridians).**

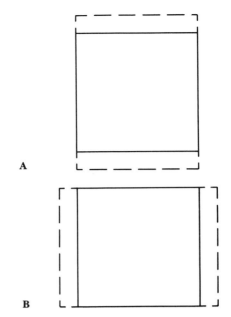

FIGURE 10-12.  **Meridional aniseikonia. The ocular image of one eye is increased or decreased symmetrically in one meridian: A, in the vertical meridian; B, in the horizontal meridian.**

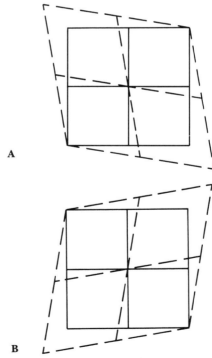

FIGURE 10-13.  **Meridional aniseikonia. The ocular image of one eye is increased or decreased symmetrically in an oblique meridian.**

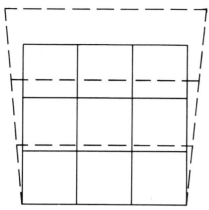

FIGURE 10-14. **Asymmetrical aniseikonia. A progressive increase or decrease in image size in one meridian, as would be produced by a flat prism.**

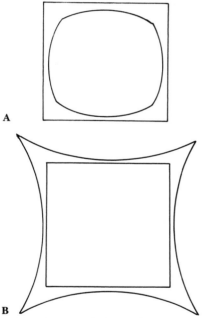

FIGURE 10-15. **Asymmetrical aniseikonia. A progressive increase or decrease in image size in all directions from the visual axis, as illustrated (A) by barrel distortion and (B) by pincushion distortion.**

ridians) as compared to the other eye. This is shown in Figure 10-11.

2. Meridional, in which the ocular image of one eye is increased or decreased symmetrically in one meridian, compared to that of the other eye. The meridian of magnification may be vertical (Figure 10-12A), horizontal (Figure 10-12B), or oblique (Figures 10-13A and B).

*Asymmetrical Differences*

1. A progressive increase or decrease in the image size in one meridian as would be produced by an ordinary flat prism (Figure 10-14).
2. A progressive increase or decrease in image size in all directions from the visual axis, as illustrated by barrel or pincushion distortion (Figure 10-15). Although barrel (Figure 10-15A) and pincushion (Figure 10-15B) distortion aren't usually thought of as manifestations of aniseikonia, they nevertheless are embraced by its definition.

## 10.6. Significance of Aniseikonia

The correction of ametropia and oculomotor imbalance provides optimum visual acuity and comfort for most patients. However, some patients fail to obtain relief even though they see clearly and even though ocular or systemic abnormalities have been ruled out by a routine visual examination. After careful correction of the ametropia and alleviation of oculomotor imbalance by appropriate lenses, prisms, or training, the occurrence of any unexplained visual symptoms should suggest the possibility of aniseikonia.

Aniseikonia has its primary effects on fusion and spatial organization. To produce a single mental image from a single object, the images formed by the two retinas must undergo sensory fusion, or unification. The greater the degree of similarity between the two images, the easier is the process of fusion. A significant difference in the sizes or shapes of the two images impedes the fusional process and compromises the binocular status.

The perception of space is served by two mechanisms. The first of these, binocular in nature, is an innately determined mechanism derived from the slight disparity of the retinal images as a result of the lateral separation of the two eyes, and gives rise to normal binocular stereoscopic perception. The second mechanism, uniocular in nature, consists mostly of learned clues such as geometrical perspective, motion parallax, overlay, light and shadow, and the expected sizes and distances of known objects. Our orientation in space is a result of the constant interrelation and summation of the uniocular and binocular factors.

Normal stereoscopic vision is based on the normal disparity of the retinal images: When significant amounts of disparity are present, the result is anomalous spatial perception with apparent changes in one's environment.

Aniseikonia, with an abnormal size or shape relationship between the two ocular images, causes anomalous binocular spatial localization with rotation and distortion of binocular stereoscopic perception. The effects which image size differences evoke are so predictable and so characteristic that they are utilized to measure aniseikonia by means of an instrument called a *space eikonometer*.

The distortion of stereoscopic space perception affects the integration and efficiency of the space orientation process. The symptoms encountered by patients with aniseikonia most likely result from efforts to retain former patterns of perception or to adapt to the new visual condition.

### Symptoms of Aniseikonia

The symptoms of aniseikonia mimic those reported by patients having uncorrected errors of refraction and/or oculomotor imbalance. They are different, however, in that the symptoms persist (or even have their onset) when lenses have been prescribed to correct the ametropia and/or oculomotor imbalance. Patients complain not so much about distortion of space but about how their eyes "feel." Ocular symptoms include asthenopia in and around the eyes, usually associated with concentrated use of the eyes. Other symptoms include headaches, photophobia, giddiness, and nervousness.

An important diagnostic test for aniseikonia as the cause of asthenopia is monocular occlusion: If the symptoms persist under monocular occlusion, aniseikonia should be strongly suspected as their underlying cause.

## 10.7. The Spectacle Magnification Formula

The size of the retinal image, formed by the optical system of the eye, can be discussed in terms of either *spectacle magnification* or *relative spectacle magnification*. Spectacle magnification concerns the change in the retinal image size of a single eye brought about by a correcting lens (either a spectacle lens or a contact lens), whereas relative spectacle magnification concerns the size of the retinal image in corrected ametropia as compared to that of the emmetropic eye.

The ratio of the retinal image size with the correcting lens compared to that without the lens, or spectacle magnification, may be expressed mathematically as

$$SM = \frac{\text{retinal image size in the corrected eye}}{\text{retinal image size in the uncorrected eye}}.$$

$$(10.1)$$

Spectacle magnification provides the factor by which the initial size of the retinal image is magnified or minified. *Percentage* spectacle magnification may be expressed as

$$\%SM = (SM - 1)(100),  \quad (10.2)$$

where SM refers to the spectacle magnification expressed as a ratio as in the previous equation. Hence, spectacle magnifications of 1.05 and 0.94, respectively, indicate a magnification of 5% and a minification of 6%.

Spectacle magnification is a type of angular magnification that is always present in ophthalmic correcting devices. The simplest way to approach the spectacle magnification of a lens is to imagine the lens as being separated into two components. A lens intended for distance vision can be conceived as being composed of an *afocal* (telescopic) component and a *power* component, as shown in Figure 10-16. The afocal component has the same thickness as the composite (real) lens, while the power component has zero thickness.

The front surface of the afocal component is the same as that of the composite lens, but since the afocal component has zero back vertex power, its back surface can be computed in the following manner, using the classic back vertex power formula:

$$F_V = \frac{F_{1A}}{1 - \dfrac{t}{n} F_1} + F_{2A},$$

where $F_V$ = the back vertex power of the afocal component, $F_{1A}$ = the front surface power of the afocal component, and $F_{2A}$ = the back surface power of the afocal component. Since the unit is afocal, $F_V = 0$, and therefore

$$F_{2A} = -\frac{F_{1A}}{1 - \dfrac{t}{n} F_{1A}}.  \quad (10.3)$$

Considering now the power component, the front surface curvature of this component matches the curvature of the back surface of the afocal component, and hence its power is numerically equal to the power of the back surface of the afocal component, but *opposite in sign*. If we let $F_{1P}$ equal the front surface power of the power component,

$$F_{1P} = +\frac{F_{1A}}{1 - \dfrac{t}{n} F_{1A}}.  \quad (10.4)$$

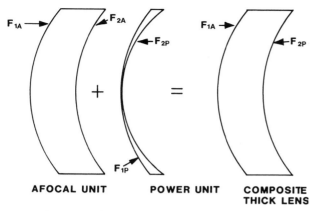

**AFOCAL UNIT**    **POWER UNIT**    **COMPOSITE THICK LENS**

FIGURE 10-16. **A thick lens intended for distance vision can be thought of as being composed of an afocal (telescopic) component and a power component of zero thickness.**

The back surface of the power component is also the back surface of the composite lens. Since the power component is infinitely thin, its power is simply equal to the sum of its two surface powers. If we let $F_P$ equal the back vertex power of the power component, and if we let $F_{1P}$ and $F_{2P}$ equal, respectively, the front and back surface powers of the power component,

$$F_P = F_{1P} + F_{2P}$$

or

$$F_P = \frac{F_{1A}}{1 - \dfrac{t}{n} F_{1A}} + F_{2P}. \qquad (10.5)$$

This is identical with the back vertex power of the composite lens, and can be measured by means of a lensometer.

It should be understood that the afocal component, taken by itself, has no effect on the clarity of vision when a distant object is seen through the lens (since it does not change the vergence of the incident rays), but it does bring about a change in the apparent size of the object and is therefore said to have *magnifying power*, which can be expressed by computing its angular magnification.

In Figure 10-17, an incident ray of light, MP, is directed toward the center of curvature, $C_1$, of the front surface. This ray is not refracted at the front surface, since it is normal to it, but is refracted at the second surface, being deviated in the direction PQ. If E is located in the focal plane of the front surface, a ray of light from E through $C_2$ will pass through the back surface unrefracted but will be refracted at the front surface in the direction OS, parallel to MP. If E also lies in the primary focal plane of the back surface, the ray PQ must be parallel to OE and hence the parallel incident rays (SO and MP) emerge from the

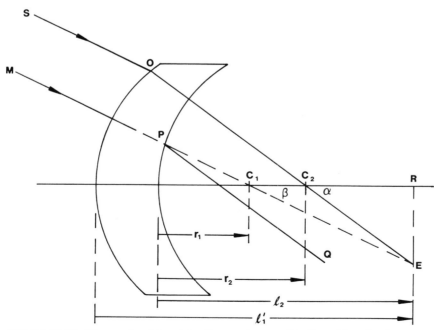

FIGURE 10-17. **Diagram for determining the angular magnification of the afocal component.**

lens as a parallel bundle. This is in accordance with the principle that the afocal component has zero back vertex power.

The angular magnification is defined as the ratio of the slope of the emerging rays to the slope of the incident rays with respect to the optic axis of the lens, and can be expressed as follows:

$$AM_A = \frac{\alpha}{\beta} = \frac{\dfrac{RE}{C_2R}}{\dfrac{RE}{C_1R}} = \frac{C_1R}{C_2R} = \frac{l_1' - r_1}{l_2 - r_2},$$

but since

$$l_1' = \frac{n}{F_1}, \quad \text{and} \quad r_1 = \frac{n-1}{F_1},$$

$$l_1' - r_1 = \frac{n - n + 1}{F_1} = \frac{1}{F_1}.$$

And since

$$l_2 = -\frac{n}{F_2}$$

then

$$l_2 = \frac{\dfrac{n}{F_1}}{1 - \dfrac{t}{n}F_1},$$

$$r_2 = \frac{1 - n}{F_2}$$

and

$$r_2 = \frac{1 - n}{-\dfrac{F_1}{1 - \dfrac{t}{n}F_1}},$$

$$r_2 = \frac{\dfrac{n-1}{F_1}}{1 - \dfrac{t}{n}F_1},$$

$$l_2 - r_2 = \frac{\dfrac{n - n + 1}{F_1}}{1 - \dfrac{t}{n}F_1} = \frac{\dfrac{1 - \dfrac{t}{n}F_1}{F_1}}{F_1}$$

$$AM_A = \frac{l_1' - r_1}{l_2 - r_2} = \frac{\dfrac{1}{F_1}}{\dfrac{1 - \dfrac{t}{n}F_1}{F_1}}$$

$$AM_A = \left(\frac{1}{1 - \dfrac{t}{n}F_1}\right). \tag{10.6}$$

The principles involved in computing the angular magnification of the power unit are illustrated in Figure 10-18. The ray TC from an object Q at infinity passes through the optical center, C, of the lens. The ray SP from Q, being parallel to TC, is refracted at P and passes through the center of the entrance pupil, E, and crosses the ray TC at the point Q' which lies in the secondary focal plane of the lens. Without the lens, Q would be seen in the direction OR, but with the lens it is seen in the direction OP. The angular magnification of the power unit is derived as follows:

$$AM_P = \frac{\alpha}{\beta} = \frac{f'}{f' - h} = \frac{1}{1 - \dfrac{h}{f'}},$$

but since

$$\frac{1}{f'} = F_P,$$

$$AM_P = \left(\frac{1}{1 - hF_P}\right). \tag{10.7}$$

The angular magnification (spectacle magnification) of the composite lens is equal to the product of the angular magnifications of the afocal unit (also called the *shape factor* $M_S$) and the power unit (also called the *power factor*, $M_P$).

$$AM = SM = (M_S)(M_P)$$

$$SM = \left(\frac{1}{1 - \dfrac{t}{n}F_1}\right)\left(\frac{1}{1 - hF_V}\right)$$

where $F_1$ = the power ($F_{1A}$) of the front surface of the lens, $F_V$ = the back vertex power ($F_P$) of the composite lens, $t$ = the lens thickness, $n$ = the index of refraction of the lens material, and $h$ = the distance from the back vertex of the lens to the entrance pupil of the eye (it is normally assumed that the distance from the front surface of the cornea to the entrance pupil of the eye is equal to 3 mm).

Inspection of the shape factor shows that spectacle magnification increases with an increase in front sur-

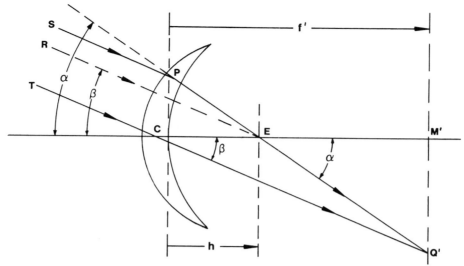

FIGURE 10-18. **Diagram for determining the angular magnification of the power component.**

face power or with an increase in lens thickness, and decreases with an increase in the index of refraction of the lens material. Inspection of the power factor indicates that spectacle magnification varies directly with back vertex power, the retinal image size increasing with increasing plus lens power and decreasing with increasing minus lens power.

It is evident that any spectacle lens, by virtue of its power, will bring about a change in retinal image size: As a plus lens is moved closer to the eye, decreasing the value of $h$, there is a *decrease* in retinal image size, whereas when a minus lens is moved closer to the eye there is an *increase* in image size.

It can be seen from the spectacle magnification formula that it is possible to construct a series of lenses having the same back vertex power but with various spectacle magnification values, depending upon the thickness, the power of the front surface, and the position of the lens relative to the entrance pupil of the eye. However, if the lens should have *zero power*, only the afocal unit is used, and its distance from the entrance pupil of the eye does not affect its spectacle magnification. If one or both surfaces are *toric* rather than spherical, the spectacle magnification must be computed separately for each of the principal meridians.

## 10.8. Clinical Application of the Spectacle Magnification Formula

The use of the spectacle magnification formula will be illustrated by the following examples.

### EXAMPLE 1

A patient who has previously worn no correction is found to require $-2.00$ D spheres for both eyes. What will be the spectacle magnification if the front surface power is $+6.00$ D, the center thickness is 2.2 mm, the index of refraction is 1.523, and the vertex distance is 12 mm?

If we assume that the distance from the front surface of the cornea to the entrance pupil is 3 mm, the value of $h$ (for a vertex distance of 12 mm) will be 15 mm. Therefore,

$$SM = \left( \frac{1}{1 - \dfrac{0.0022(6)}{1.523}} \right) \left( \frac{1}{1 - 0.015(-2)} \right)$$

$$= (1.009)(0.971) = 0.980.$$

$$\%SM = (SM - 1)(100)$$
$$= (0.980 - 1)(100)$$
$$= -2.0\%.$$

The above example shows why even a small amount of minus power can cause a patient to notice that "everything looks smaller" with the new correction.

### EXAMPLE 2

A patient who has previously worn no correction is found to require $+2.00$ D spheres for both eyes. What will be the spectacle magnification if the front surface power is $+8.00$ D, the center thickness is 2.5 mm, the index of refraction is 1.523, and the vertex distance is 12 mm?

$$SM = \left( \frac{1}{1 - \dfrac{0.0025(8)}{1.523}} \right) \left( \frac{1}{1 - 0.015(+2)} \right)$$

$$= (1.013)(1.031) = 1.044.$$

$$\%SM = (SM - 1)(100)$$

$$= (1.044 - 1)(100)$$

$$= +4.4\%.$$

**Important Note:** It is clear from the above examples and from examination of Eq. (10.8) that spectacle magnification is always greater than unity for a plus lens, and always less than unity for a minus lens. This is true not only for a spectacle lens but also for a contact lens. Although the *amount* of magnification will be different for a contact lens than for a spectacle lens because of the much shorter distance between the lens and the entrance pupil of the eye, the *direction* of the magnification change will be identical for corrections of equal sign: That is, spectacle magnification is greater than unity for either a spectacle lens or a contact lens of plus power, and is less than unity for either a spectacle lens or a contact lens of minus power.

If a correcting lens were to be infinitely thin and placed at the entrance pupil of the eye, the shape factor and the power factor would each be equal to 1.0, with the result that spectacle magnification would be *unity* whatever the refractive error. This condition is most nearly approximated by either a contact lens or an intraocular lens.

The computation of spectacle magnification is useful in predicting the effect on the image size (for a single eye) in any of the following circumstances:

1. A change in a patient's prescription, for any reason
2. A change in the vertex distance of a spectacle lens
3. The change from spectacles to contact lenses or vice versa
4. A change in the thickness of a lens
5. A change in the "bend," or "form," of a lens

In addition, an understanding of the principles of spectacle magnification makes it possible to manipulate the variables in such a way as to equalize the sizes of the retinal images of the two eyes, thus minimizing induced aniseikonia. This will be discussed further in Section 10.21.

## 10.9. Comparison of Spectacle and Contact Lens Magnification

If we assume both lenses to be infinitely thin, the retinal image formed by either a spectacle lens or a contact lens is directly proportional to the size of the image formed at the far point of the eye. Therefore, the retinal image size is *directly proportional* to the secondary focal length of the correcting lens, or *inversely proportional* to the power of the lens. That is,

$$\frac{\text{retinal image size with a contact lens}}{\text{retinal image size with a spectacle lens}} = \frac{\dfrac{1}{F_C}}{\dfrac{1}{F_S}} = \frac{F_S}{F_C}$$

or

$$\frac{\text{retinal image size with a spectacle lens}}{\text{retinal image size with a contact lens}} = \frac{\dfrac{1}{F_S}}{\dfrac{1}{F_C}} = \frac{F_C}{F_S}.$$

### EXAMPLE 1

A patient wears +14.00 D spectacles at a vertex distance of 12 mm. If the patient is fitted with contact lenses of appropriate refracting power, fitted "on K" (i.e., so that the radius of curvature of the back surface of each lens is equal to the apical radius of the corresponding cornea in the flattest meridian), what will be (a) the required contact lens power and (b) the percentage change in "spectacle magnification" when the contact lenses are worn rather than glasses?

(a) The required contact lens power will be

$$\frac{1}{\dfrac{1}{14} - 0.012} = \frac{1}{0.0714 - 0.012} = +16.83 \text{ D.}$$

(b) The change in spectacle magnification will be

$$\frac{F_S}{F_C} = \frac{+14.00}{+16.83} = 0.832.$$

The percentage spectacle magnification change will be

$$\%SM \text{ change} = (0.832 - 1)100 = -16.8\%.$$

### EXAMPLE 2

A patient wears a +16.83 D contact lens, fitted "on K." If the patient is fitted with appropriate spectacle lenses worn at a vertex distance of 12 mm, what will be (a) the required power of the spectacle lens and (b) the percentage change in spectacle magnification?

Upon casual inspection it may seem that the answer is a +16.8% increase in spectacle magnification. However, this is not the case:

(a) The required spectacle lens power will be

$$\frac{1}{\dfrac{1}{16.83} + 0.012} = \frac{1}{0.5941 + 0.012} = +14.00 \text{ D.}$$

(b) The change in spectacle magnification will be

$$\frac{F_c}{F_g} = \frac{+16.83}{+14.00} = 1.202.$$

The percentage spectacle magnification change will be

%SM change = (1.202 − 1)100 = +20.2%.

### EXAMPLE 3

A patient wears −10.00 D contact lenses fitted "on K." If the patient is fitted with appropriate spectacle lenses worn at a vertex distance of 12 mm, what will be (a) the required power of the spectacle lenses and (b) the percentage change in spectacle magnification?

(a) The required spectacle lens power will be

$$\frac{1}{\dfrac{1}{10} - 0.012} = \frac{1}{0.10 - 0.012} = -11.36 \text{ D.}$$

(b) The change in spectacle magnification will be

$$\frac{F_C}{F_S} = \frac{-10.00}{-11.36} = 0.88.$$

The percentage spectacle magnification change will be

%SM change = (0.88 − 1)100 = −12.0%.

Therefore, when going from contact lenses to glasses, for this myopic patient the retinal image size decreased by 12%.

### EXAMPLE 4

A patient wears spectacle lenses having a power of −11.36 D at a vertex distance of 12 mm. If this patient is fitted with appropriate contact lenses fitted "on K," what will be (a) the required power of the contact lenses and (b) the percentage increase in spectacle magnification?

(a) The required spectacle lens power will be

$$\frac{1}{\dfrac{1}{-11.36} - 0.012} = \frac{1}{-0.0880 - 0.012} = -10.00 \text{ D.}$$

(b) The change in spectacle magnification will be

$$\frac{F_S}{F_C} = \frac{-11.36}{-10.00} = 1.136.$$

The percentage spectacle magnification change will be

%SM change = (1.136 − 1)100 = +13.6%.

Therefore, in going from spectacle lenses to contact lenses, the retinal image size increased 13.6% for this patient.

**Important Note:** In none of the four examples was the actual spectacle magnification calculated, for either spectacles or contact lenses. What was calculated was the *percentage change* in spectacle magnification when going from one form of correction to the other.

## 10.10. Spectacle Magnification in Astigmatism

When a lens incorporates a correction for astigmatism, the spectacle magnification will differ in the two principal meridians. If the lenses compensate for the patient's refractive error, the retinal image will be distorted, being larger in one principal meridian than the other. The retinal image will be larger in the direction of the *axis of the minus cylinder* which corrects the astigmatism.

If we assume that the vertex distance for a spectacle lens is 12 mm and the distance from the anterior corneal surface to the entrance pupil of the eye is 3 mm, calculation of the power factor,

$$\frac{1}{1 - hF_V},$$

for each of the principal meridians will indicate that the percentage difference between the two meridians will be about 1.5% per diopter of astigmatism for a spectacle lens, but only about 0.3% per diopter of astigmatism for contact lenses. Although the value of the shape factor may alter these amounts slightly, these differences in spectacle magnification can be substantial. This could be a source of perceptual difficulties that some patients experience when changing from spectacles to contact lenses or vice versa.

Although contact lenses tend to be more difficult to fit on highly astigmatic eyes than on eyes having mainly spherical ametropia, if an astigmatic patient does not object to being fitted with contact lenses, their use should be seriously considered since the difference in image size for the two principal meridians will be decreased by a factor of 5. In addition, in many cases of astigmatism the use of contact lenses will markedly reduce the amount of induced aniseikonia as compared to that induced by spectacle lenses. This will be discussed further in Section 10.14.

## 10.11. Relative Spectacle Magnification

To compare the retinal image size of one eye with that of another eye, *relative spectacle magnification* may be used. Relative spectacle magnification is defined as the ratio of the retinal image size (for an object at infinity) of the corrected ametropic eye in question to that of the standard emmetropic eye. It can be expressed by the formula

$$\text{RSM} = \frac{\text{image size for a corrected ametropic eye}}{\text{image size for a standard emmetropic eye}}.$$
(10.9)

When a patient's anisometropia is corrected by spectacles or by contact lenses, the concept of relative spectacle magnification may be used to compare the sizes of the clear retinal images of each of the two eyes to the size of the retinal image of the standard emmetropic eye. The ratio of the actual retinal image sizes for the two eyes can then be expressed as the ratio of the relative spectacle magnification of one eye to that of the other:

$$\text{magnification ratio} = \frac{\text{RSM}_R}{\text{RSM}_L}.$$
(10.10)

In order for such a comparison to be made, it must be known if the patient's anisometropia is due primarily to *refractive* or *axial* differences in the two eyes. Clinical indications of the presence of refractive ametropia include (a) significant differences in the keratometric readings for the two eyes; (b) the development of cataracts, more advanced in one eye than in the other; and (c) aphakia. In the absence of A-scan ultrasonographic equipment for measuring the axial length of the eye, the main indication for axial ametropia is the presence of structural changes at the posterior pole of the eye as viewed with the ophthalmoscope due to axial elongation.

As pointed out in the discussion in Chapter 5, it was found by Sorsby et al.[11] that for small amounts of ametropia the axial length of the eye is usually within the range for the emmetropic eye; but for eyes having ametropia of ±4.00 D or more, the axial length is almost always outside the range for the emmetropic eye.

If the source of the ametropia of each eye is known, the following relationship can be developed. The size of the retinal image of a corrected ametropic eye is directly proportional to the equivalent focal length of the system or inversely proportional to the equivalent power of the system. Therefore,

$$\text{RSM} = \frac{f'_E}{f'_{ST}},$$
(10.11)

where $f'_E$ is the secondary equivalent focal length of the lens/eye system and $f'_{ST}$ is the secondary equivalent focal length of a standard emmetropic eye. Consequently,

$$\text{RSM} = \frac{F_{ST}}{F_E},$$
(10.12)

where $F_{ST}$ is the equivalent power of the standard emmetropic eye and $F_E$ is the equivalent power of the system composed of the correcting spectacle lens, $F_{SP}$, and the refracting power of the ametropic eye, $F_A$. Therefore, if the distance from the secondary principal point of the spectacle lens to the primary principal point of the eye is denoted by $d$, it follows that the equivalent power of the lens/eye system is

$$F_E = F_{SP} + F_A - dF_{SP}F_A$$
(10.13)

and, substituting for $F_E$ in Eq. (10.12),

$$\text{RSM} = \frac{F_{ST}}{F_{SP} + F_A - dF_{SP}F_A}.$$
(10.14)

## 10.12. Relative Spectacle Magnification in Axial Ametropia

If the source of the ametropia is purely axial, then the power of the ametropic eye, $F_A$, is equal to the power of the standard emmetropic eye, $F_{ST}$, or: $F_A = F_{ST}$. Substituting $F_{ST}$ for $F_A$ in Eq. (10.14),

$$\text{RSM}_{(axial\ ametropia)} = \frac{F_{ST}}{F_{SP} + F_{ST} - dF_{SP}F_{ST}}.$$
(10.15)

If the spectacles are placed at the anterior focal point of the eye,

$$d = \frac{1}{F_{ST}},$$

and

$$\text{RSM}_{axial\ ametropia} = \frac{F_{ST}}{F_{SP} + F_{ST} - \dfrac{1}{F_{ST}}F_{SP}F_{ST}}$$

$$= \frac{F_{ST}}{F_{ST}} = 1.00.$$
(10.16)

This equation mathematically illustrates *Knapp's law*, which states that for an axially ametropic eye, if the correcting lens is placed so that its secondary principal point coincides with the anterior focal point

of the eye, the size of the retinal image will be the same as if it were the standard emmetropic eye.

If it is assumed that the refracting power of the standard emmetropic eye is $+58.50$ D, the distance from the first primary principal point of the eye to the anterior focal point of the eye is 17.1 mm. If the distance from the cornea to the first principal plane (or, with negligible error, the center of the entrance pupil) of the eye is considered to be 3 mm, the anterior focal point of the axially ametropic eye can be considered as 14 mm from the cornea.

It should be recalled that, for the purely axially ametropic eye, the optical system of the eye is of *normal* focal length (i.e., that of the standard emmetropic eye), with the result that the primary focal length is the same for the axially ametropic eye as for the emmetropic eye. Reference to Figure 10-19 illustrates Knapp's law diagrammatically, showing that, if the correcting lens is placed so that its secondary principal point coincides with the primary focal point of the eye, the resulting retinal image size is the same as that of the standard emmetropic schematic eye.

The student should understand that four conditions are necessary in order for Knapp's law to apply:

1. The ametropia must be *purely axial.*
2. The correcting lens must be located so that its secondary principal point coincides with the primary focal point of the eye. The position of the principal planes moves as the "bend" of a lens is increased—moving toward the surface with the greatest curvature. The secondary principal point of a plus meniscus lens is located a short distance anterior to the front surface of the lens and the secondary principal point of a minus meniscus lens is located a short distance behind the back surface. Therefore the back vertex of a plus meniscus lens would have to be located somewhat *closer* than 14 mm from the cornea and a minus meniscus lens would have to be placed *farther* than 14 mm from the cornea.

3. The refracting power of the eye must be equal to that of the standard emmetropic eye.
4. The shape factor of the correcting lens must be unity (in reality the shape factor for both plus and minus lenses is *greater than* 1 if the front surface is convex).

An additional factor to be considered is that Knapp's law applies only to *induced* aniseikonia. As stated earlier, ametropic eyes may suffer *anatomical* aniseikonia, due to the differences in receptor densities for the two eyes (this would be particularly expected in axial ametropia, the density of the receptors being spread out in a large axially ametropic eye and more closely packed in a small, hyperopic eye). This condition, in addition to the four conditions listed above, makes the clinical application of Knapp's law a somewhat inexact procedure.

In the case of axial ametropia corrected by a *contact lens*, the power of a contact lens, fitted "on K," can be considered as equal to the ocular refraction, $F_O$. Consequently, substituting $F_O$ for $F_{SP}$ in Eq. (10.15), we obtain

$$\text{RSM} = \frac{F_{ST}}{F_O + F_{ST} - dF_OF_{ST}}, \qquad (10.17)$$

but since $d$ may be considered as negligible for a contact lens,

$$\text{RSM} = \frac{F_{ST}}{F_O + F_{ST}} \qquad (10.18)$$

when a contact lens is used for the correction of axial ametropia.

It is apparent that if contact lenses are to correct an anisometropic patient whose anisometropia is axial in origin, the relative spectacle magnification will *not* be unity, and differences in retinal image sizes for the two eyes will result.

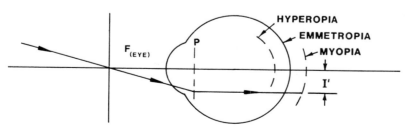

FIGURE 10-19. **Diagram illustrating Knapp's law: If the correcting lens for an axially ametropic eye is placed in the primary focal plane of the eye, the resulting image size is the same as that for the emmetropic schematic eye.**

## 10.13. Relative Spectacle Magnification in Refractive Ametropia

If the source of the ametropia is assumed to be refractive, the power of the ametropic eye, $F_A$, is *not* equal to the power of the standard emmetropic eye, $F_{ST}$, but the *axial length* of the eye is equal to that of the standard emmetropic eye. If the spectacle correction, $F_{SP}$, is known, the effective power of the spectacle correction at the primary principal point of the eye (the ocular refraction, $F_O$) added to the power of the ametropic eye, $F_A$, must be equal to the power of the standard emmetropic eye, $F_{ST}$. Therefore,

$$F_{ST} = \frac{F_{SP}}{1 - dF_{SP}} + F_A,$$

or

$$F_{ST} = \frac{F_{SP} + F_A - dF_{SP}F_A}{1 - dF_{SP}}.$$

Since the general equation (10.14) is

$$RSM = \frac{F_{ST}}{F_{SP} + F_A - dF_{SP}F_A},$$

and substituting for $F_{ST}$,

$$RSM = \frac{\dfrac{F_{SP} + F_A - dF_{SP}F_A}{1 - dF_{SP}}}{F_{SP} + F_A - dF_{SP}F_A},$$

or

$$RSM = \frac{1}{1 - dF_{SP}} \qquad (10.19)$$

for refractive ametropia.

**Important Note:** It is obvious that the relative spectacle magnification for ametropia that is *refractive* in origin is essentially equal to the *spectacle magnification*, if the lens is considered to be a thin lens. That is, in refractive ametropia,

$$SM = RSM.$$

Therefore, if spectacle lenses are to be used to correct an eye whose ametropia is refractive in origin, the relative spectacle magnification will *not* be unity: The correcting lens is likely to bring about a marked change in image size as compared to the situation in axial ametropia in which a "normal" retinal image size may result. More specifically, a plus lens will create a magnification greater than unity while a minus lens will create a magnification less than unity.

Inspection of Eq. (10.19) indicates that, for refractive ametropia, relative spectacle magnification increases with increasing values of $d$, and will be at a minimum when $d$ is at a minimum. Therefore, if a *contact lens* rather than a spectacle lens is fitted, the vertex distance will be zero and the value of $d$ will be at a minimum. If the distance $d$ is considered to be negligible (i.e., if it is considered that the distance from the corneal surface to the principal plane of the eye is equal to zero), the relative spectacle magnification approaches unity. Therefore, in a situation in which the practitioner wishes to minimize the amount of relative spectacle magnification for an eye that is known to have *refractive* ametropia, the prescribing of a contact lens should be considered.

## 10.14. Relative Spectacle Magnification in Astigmatism

When a patient has a significant amount of astigmatism, the *cornea* is almost always the source of the astigmatism, with the lens being responsible for astigmatism only on rare occasions. However, in either case astigmatism is obviously a form of *refractive* ametropia. It follows, then, that the magnification difference between the two principal meridians of an astigmatic eye will be a maximum if glasses are worn and a minimum if contact lenses are worn. As pointed out in Section 10.10, when astigmatism is corrected by a minus cylinder in the form of a spectacle lens, the retinal image size will be greatest in the axis meridian of the correcting minus cylinder and least in the power meridian.

For a single eye, the fact that the shape of the retinal image with a correcting lens is not a true representation of the shape of the object may not constitute a problem for the wearer. Our concern here, however, is whether or not the retinal image shapes *differ for the two eyes*. For example, if a patient has 2.00 D of with-the-rule astigmatism in each eye, corrected with plus cylinders, axis 90°, each lens will have its greatest magnification in the horizontal meridian but the shape and orientation of the retinal images will be the same for both eyes (Figure 10-20A). However, if the patient has 2.00 D of oblique astigmatism with symmetrically placed plus cylinder axes (e.g., 45° and 135° for the right and left eyes, respectively), the shapes of the two retinal images will be the same, but the orientation will be decidedly different (Figure 10-20B), the greatest magnification in this case being at 135° for the right eye and 45° for the left. As a result of these meridional magnification

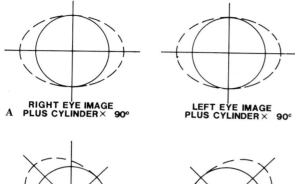

FIGURE 10-20. **(A) With 2.00 D of astigmatism in each eye corrected with plus cylinders with their axes at 90 degrees, the shape and orientation of the retinal images will be the same for both eyes. (B) with 2.00 D of symmetrical astigmatism corrected with plus cylinders with axes at 45 and 135 degrees, the shapes of the retinal images will be the same for the two eyes but the orientation will differ.**

differences, a significant amount of aniseikonia will be induced.

As noted in Section 10.10, the magnification difference for the two principal meridians will be approximately 1.5% per diopter of astigmatism for a spectacle lens, but only about 0.3% per diopter for a contact lens. Therefore, when a significant amount of astigmatism is present (particularly oblique astigmatism with symmetrically placed axes, or in with-the-rule astigmatism in one eye and against-the-rule in the other), a much smaller amount of meridional aniseikonia will be induced if contact lenses are worn rather than glasses.

Many highly astigmatic patients who have worn spectacles for a number of years will have no complaints whatsoever that could be traced to induced meridional aniseikonia: They have adapted to the situation very well. If such a patient is switched to contact lenses, he or she may initially have difficulty adapting to the lenses due to the decrease in the accustomed amount of induced aniseikonia. This, however, is not sufficient reason *not* to switch the patient to contact lenses, but he or she should be warned of the possible adaptation problems.

On the other hand, if a spectacle-wearing high astigmat has difficulty adapting to spectacles, he or she should be advised that the adaptation problem

may be minimized if contact lenses rather than spectacles are worn. In any case, since contact lenses are considered by most people as mainly a *cosmetic* form of visual correction—something one can wear to avoid wearing glasses—an astigmatic patient (like any other patient) should not be urged to wear contact lenses unless some degree of motivation for their wear is present.

## 10.15. The Dilemma of Relative Spectacle Magnification

Although relative spectacle magnification is a rigorous mathematical concept, making use of this concept in routine optometric care often presents the practitioner with an insolvable dilemma. Relative spectacle magnification can be determined accurately only if both the equivalent power and the axial length of the eye are known. Unfortunately, the equivalent power of the eye is difficult to determine; and A-scan ultrasonography, although a highly developed and accurate technology for measurement of the axial length of the eye, is not available in most practitioners' offices. Nevertheless, relative spectacle magnification is a useful concept in the presence of anisometropia.

**The Question:** Is the anisometropia axial or refractive?

**The Answer:** Examine the keratometer findings!

If the keratometer findings are essentially equal for the two eyes, it is a fair assumption that the equivalent power (due to the corneal power, anterior chamber depth, and lens power) is the same for the two eyes and therefore the anisometropia is *axial* in origin. On the other hand, if the keratometer findings manifest the same dioptric difference as the magnitude of the anisometropia, it is likely that the axial lengths are equal and that the anisometropia is *refractive* in origin.

The use of the keratometer findings in determining whether anisometropia is axial or refractive will be illustrated by the following examples.

### EXAMPLE 1
Given the following clinical findings, is the patient's anisometropia due to axial or refractive ametropia?

| | |
|---|---|
| Keratometer: | OD 44.50 D |
| | OS 44.50 D |
| Refraction: | OD −2.50 DS |
| | OS plano |

Since the keratometer findings are the same for both eyes, the fact that the right eye is myopic while the left eye is emmetropic can best be explained by the right eye being longer than the left eye. The ametropia is therefore axial in origin, and the relative spectacle magnification will approach unity when spectacle lenses are worn (and differ significantly from unity if contact lenses are worn).

**EXAMPLE 2**

Given the following clinical findings, is the patient's anisometropia due to axial or refractive ametropia?

| | |
|---|---|
| Keratometer: | OD 44.50 D |
| | OS 42.00 D |
| Refraction: | OD −2.50 DS |
| | OS plano |

Since the patient's anisometropia can be accounted for entirely by the difference in the keratometer findings, the anisometropia is due to the fact that the right eye has *refractive* myopia. In this situation, the relative spectacle magnification will approach unity when contact lenses are worn and will differ significantly from unity if spectacles are worn.

At the University of Houston, the keratometer findings and axial lengths (using A-scan ultrasonography) of both eyes of a number of anisometropic subjects have been measured, and it has been found that for amounts of anisometropia as low as 1.50 to 2.00 D the anisometropia can almost always be accounted for by a difference in the *axial lengths* of the eyes. Following is a typical example:

| | |
|---|---|
| Keratometer: | OD 42.50 D |
| | OS 42.75 D |
| Refraction: | OD +2.75 D |
| | OS +4.25 D |
| Axial length: | OD 23.6 mm |
| | OS 23.1 mm |

In the absence of a significant difference in the keratometer findings, the anisometropia can be assumed to be due to the fact that the more hyperopic eye is 0.5 mm *shorter* than the less hyperopic eye. (Since the accuracy of our ultrasonic measurements has been found to be on the order of 0.2 mm, when five readings are taken, the 0.5-mm difference far exceeds the error of measurement.)

If these preliminary results are corroborated by those of a larger study now in progress, they will tend to add credence to the value of keratometer findings in predicting whether anisometropia is due to axial or refractive ametropia.

## 10.16. Clinical Considerations in Anisometropia and Astigmatism

On the basis of the foregoing discussion, the clinical management of anisometropia and astigmatism, insofar as the question of induced aniseikonia is concerned, may be summarized as follows.

*Axial Ametropia*

In uncorrected anisometropia due to axial ametropia, the equivalent powers (corneal power and lens power) are the same for the two eyes but the retinal image sizes are different due to the axial length difference. If spectacle lenses are worn, the retinal image size difference will be offset by the spectacle magnification produced by the correcting lenses, resulting (as stated by Knapp's law) in the relative spectacle magnification for each eye being essentially unity and in little or no aniseikonia being induced.

If, on the other hand, axial ametropia is corrected by contact lenses, the spectacle magnification is much less than that for spectacle lenses with the result that the relative spectacle magnification for each eye will differ significantly from unity and a significant amount of aniseikonia will be induced.

In summary, if the ametropia is axial in origin, the prescribing of spectacle lenses is the treatment of choice if one wishes to minimize induced aniseikonia.

*Refractive Ametropia*

In uncorrected anisometropia due to refractive ametropia, the axial lengths of the two eyes are considered to be equal with the result that the retinal image size for each eye is essentially equal to that of the standard emmetropic eye. If spectacle lenses are worn, the spectacle magnification produced by the lenses will bring about an inequality in the sizes of the two retinal images with the result that relative spectacle magnification for each eye will differ from unity and a significant amount of aniseikonia will be induced.

But if contact lenses are worn rather than spectacles, the lenses will bring about very little spectacle magnification with the result that relative spectacle magnification will differ little from unity, and a minimum amount of aniseikonia will be induced.

In summary, if the ametropia is refractive in origin, the prescribing of contact lenses is the treatment of choice if one wishes to minimize induced aniseikonia. A comparison of image sizes in ametropia to

TABLE 10-2
**Retinal Image Sizes in Ametropia as Compared to Emmetropia**

| Ametropia | Uncorrected | Spectacle Correction | Contact Lens Correction |
|---|---|---|---|
| Axial myopia | Larger than emmetropia | Equal to emmetropia | Larger than emmetropia |
| Axial hyperopia | Smaller than emmetropia | Equal to emmetropia | Smaller than emmetropia |
| Refractive myopia | Equal to emmetropia | Smaller than emmetropia | Equal to emmetropia |
| Refractive hyperopia | Equal to emmetropia | Larger than emmetropia | Equal to emmetropia |

those in emmetropia is given in Table 10-2; the indication of the presence or absence of aniseikonia in uncorrected and corrected anisometropia is given in Table 10-3.

*Astigmatism*

Since astigmatism is a form of *refractive* ametropia, the wearing of contact lenses rather than glasses will result in a minimum of induced meridional aniseikonia for any patient who has high astigmatism with the axes positioned symmetrically for the two eyes.

TABLE 10-3
**Presence or Absence of Aniseikonia in Uncorrected and Corrected Anisometropia**

| Ametropia | Uncorrected | Spectacle Correction | Contact Lens Correction |
|---|---|---|---|
| Axial anisometropia | Present | Absent | Present |
| Refractive anisometropia | Absent | Present | Present |

# PRESCRIBING TO ELIMINATE OR MINIMIZE INDUCED ANISEIKONIA

When discussing the topic of prescribing to eliminate or minimize induced aniseikonia, one must take into consideration not only the optical principles of aniseikonia itself (described earlier in this chapter) but also the clinical indications of the presence of aniseikonia, methods of measuring and estimating image size differences, methods of avoiding induced aniseikonia with ordinary spectacle lenses, and, finally, the design of "eikonic" lenses—lenses that produce magnification in such a way as to avoid or minimize induced aniseikonia.

Our intention in writing this textbook has been to concentrate on the optical principles and practical applications of lenses used for the correction of visual anomalies, with a minimum of attention given to patient care procedures as such. However, insofar as aniseikonia is concerned, it is difficult (if not impossible) to separate the strictly *optical* aspects of the management of aniseikonia from the *patient care* aspects.

We have decided therefore to include in our discussion all aspects of the management of aniseikonia in what we hope the student and practitioner will find to be a useful treatment of the subject.

## 10.17. Indications of the Presence of Aniseikonia

The most important clues to the presence of aniseikonia are the following:

1. The presence of a significant amount of anisometropia or astigmatism.
2. The presence of an impaired binocular vision status with reduced fusional amplitudes and/or reduced quality of fusion.
3. Complaints of symptoms, usually asthenopia and headache, following the use of the eyes, occur-

ring *after* proper correction of any refractive error and/or oculomotor imbalance.

4. The report of more comfortable and satisfactory vision when using only one eye, especially for prolonged visual tasks. Since aniseikonia is a binocular anomaly, *monocular occlusion* is a useful diagnostic procedure: If occlusion brings relief of symptoms, the symptoms may possibly arise from aniseikonia. On the other hand, monocular vision is more disturbing to some people than stressful binocular vision, so if monocular occlusion fails to bring relief, aniseikonia cannot absolutely be eliminated as a source of the problem.

5. The presence of symptoms having their onset coincident with the replacement of a broken lens or with the prescription of new glasses. In such a case the symptoms may be due to a change in base curve or thickness or a change from a negative to a positive toric form (or vice versa) or, in the case of new glasses, to a change in lens powers or vertex distance. The use of a lens gauge and thickness calipers is essential for monitoring lens design.

6. A report of space distortion such as slanting floors, tilted walls, or the ground appearing to be too close or too far away.

Many patients experience symptoms of distortion and discomfort when glasses are first worn or when a significant change has been made in the prescription. Most patients "get used to" these symptoms within about two weeks. However, this "adaptation" does not mean that the aniseikonia is no longer present, but only that the individual has undergone a conditioning process whereby he or she responds to the familiar empirical clues for space localization and learns to ignore the stereoscopic clues. One would not expect physiological compensation for the aniseikonia to occur, because neither the innate retinal correspondence nor the disparity of the ocular images has changed.[12] This conclusion is reinforced by the fact that when a patient who has adapted to such space distortions is asked to view a *leaf room* (having a visual environment devoid of empirical clues for the perception of space, as shown in Figure 10-21) the space distortions will instantly reappear.[13]

The important consideration is whether or not the patient can make this form of "adaptation" without undue discomfort and without compromising visual efficiency. Many patients make this adaptation without consequence, although others can make the adaptation only with severe asthenopia or loss of visual efficiency. One of us has made a long-standing promise to students to deliver a "tolerance level indicating device" which, when connected to the patient, will separate those who can adapt to their new lenses from those who can't. The list of orders is lengthy!

## 10.18. Measurement of Image Size Differences

A clinical instrument designed to measure the differences in the magnification of the retinal images of the two eyes is known as an *eikonometer*. In 1951, American Optical introduced what they referred to as the Office Model of the Space Eikonometer (Figure 10-22). The design of this instrument was based on that of Ogle's space eikonometer, a research instrument which was large enough to fill a small room. The space eikonometer makes it possible to measure image size differences on the basis of their effects on binocular space perception. For many years American Optical made available to practitioners a "translation" service, which interpreted Space Eikonometer findings and designed a pair of lenses that corrected the aniseikonia and provided the appropriate refractive correction. These lenses were first known as *iseikonic* lenses, but the shortened version, *eikonic*, seems now to be preferred. A few years ago American Optical discontinued the manufacture of the Space Eikonometer and no longer provides the translation service.

The Keystone View Company has developed a set of stereoscopic cards which are reproductions of the Space Eikonometer target, used to estimate the

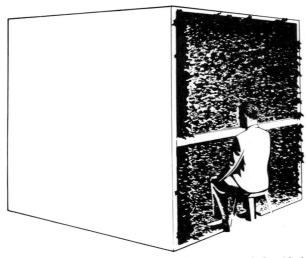

FIGURE 10-21. **A leaf room: The visual environment is devoid of empirical clues for the perception of space. (From R. E. Bannon,** *Clinical Manual on Aniseikonia*, **American Optical Corp., Buffalo, N.Y., 1954, Figure 28.)**

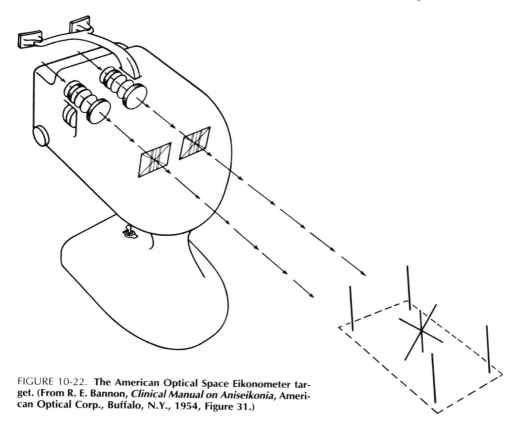

FIGURE 10-22. **The American Optical Space Eikonometer target. (From R. E. Bannon, *Clinical Manual on Aniseikonia*, American Optical Corp., Buffalo, N.Y., 1954, Figure 31.)**

amount of aniseikonia present. The cards are designed to be used with the Keystone Orthoscope, a stereoscope with "minimum-distortion" lenses. Whereas the American Optical Space Eikonometer uses adjustable afocal magnification units by means of which the magnification can be changed optically to correct the manifested aniseikonia, the Keystone series of cards offers a number of graded stereoscopic targets with built-in aniseikonic errors. The stereogram that appears normal to the patient (that is, demonstrates no distortion of binocularly perceived space) indicates the approximate amount of aniseikonic error.

The nature of the test of aniseikonia with both the A.O. Space Eikonometer and the Keystone Orthoscope requires that the patient have single binocular vision, normal retinal correspondence, and at least 20/60 visual acuity in each eye. A patient without these visual characteristics cannot be measured for aniseikonia with either of these instruments. Bannon[14] has stated that the patient's intelligence level should be equivalent to that of a normal 6-year-old in order to understand the questions, make the necessary observations, and give unequivocal responses.

## Screening Devices

Because of these limitations and because of the interest in simplification and reduction in cost, many clinicians have been interested in designing "screening" devices for detecting and prescribing for the approximate correction of aniseikonia. However, an instrument designed for screening is of no value unless it is as sensitive and as accurate as an instrument designed for the measurement of aniseikonia.

The following brief review illustrates some of these screening devices. Elvin[15] used the Turville Infinity Balance Test and size lenses to measure the aniseikonia in the vertical meridian of an anisometrope (Figure 10-23), and arrived at a finding of 2.4% overall aniseikonia, the correction of which provided clear, comfortable vision. Brecher[16] used a Maddox rod placed in front of one eye and two separated light sources, arranged in the meridian of investigation. The patient saw two spots of light and two streaks: The difference between the spot and streak separation constituted the amount of aniseikonia; and afocal size lenses were used before the eye without the Maddox rod, in order to measure the size difference.

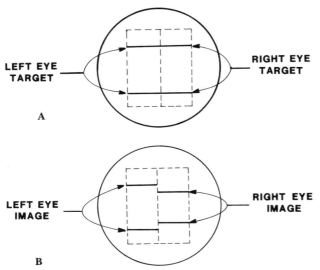

FIGURE 10-23. **The Turville Infinity Balance Test used to measure aniseikonia in the vertical meridian: A, no aniseikonia is present; B, aniseikonia is present in the vertical meridian. (From F. T. Elvin,** *Amer. J. Optom. Arch. Amer. Acad. Optom.,* **Vol. 26, pp. 78–82, 1949.)**

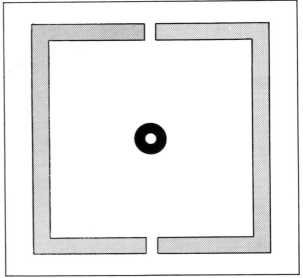

FIGURE 10-24. **Zeiss Polatest target for measurement of aniseikonia. From** *Illus. for The Polatest,* **1966 (Carl Zeiss, Inc., Thornwood, New York).**

Brecher, Winters, and Townsend[17] proposed the measurement of aniseikonia by alternating between the images of the two eyes slowly enough to prevent binocular fusion and still fast enough for a subjective comparison of image size.

Baldwin[18] described an instrument called the Zeiss Polatest, designed for binocular refraction, which made use of a target (Figure 10-24) intended to measure the amount of aniseikonia in the vertical meridian by estimating the vernier displacement in terms of target line thickness. Grolman,[19] in 1971, patented a series of Vectographic Nearpoint Cards designed to facilitate binocular refraction at the near point. One of these cards, No. 4 (Figure 10-25), is intended for use in screening the differences in overall and meridional image size. A determination of the magnitude of an indicated meridional or overall aniseikonia could be made by eliciting from the patient an estimate of the vernier displacement in terms of target line thickness.

Most of the screening devices lack reliability and can detect only gross differences in image size. If an individual has normal binocular vision, image size differences greater than 5% are rarely encountered.[20] Image size differences exceeding that amount result in loss of binocularity and are not apt to create asthenopia or other problems. Most patients struggling with aniseikonia have *small* differences in image size, and these small differences

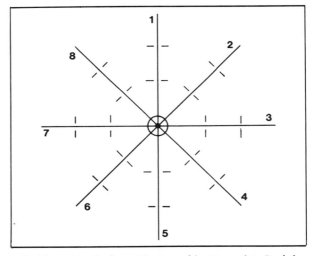

FIGURE 10-25. **Grolman Vectographic Nearpoint Card for screening differences in overall image size. For each radial line, the vernier line on one side is seen by the right eye and that on the other side is seen by the left eye. From** *Introduction for Vectograph Nearpoint Cards* **(Reichert Opthalmic Instruments, Buffalo, New York).**

would be missed by gross screening methods. For example, Bannon[21] reported that most of the image size differences in clinical cases were between 1.0 and 2.0%—differences too small to be detected with most screening instruments (the scale on the A.O. Space

Eikonometer extends only to an image size difference of ±4%).

A correction for aniseikonia based on accurate measurement by the Space Eikonometer can be made with the greatest of confidence, whereas a close approximation of the amount of aniseikonia can be obtained with the stereograms designed for use with the Keystone Orthoscope. Although the Space Eikonometer is no longer manufactured, its use may be available upon referral to another practitioner or to an optometric institution.

## 10.19. Estimating the Amount of Aniseikonia

If neither the Space Eikonometer nor the Keystone instrumentation is available, the practitioner may reasonably estimate the amount of aniseikonia present and make use of some method of determining the necessary correction. Linksz and Bannon[22] state that when it is not possible to make eikonometer measurements for a patient who has worn an anisometropic correction but is uncomfortable, then a correction for the *estimated* aniseikonia is better than doing nothing at all, and that there is a place for intelligent guessing about aniseikonia.

The same authors proceed to state that at least a nodding familiarity with aniseikonia and its effects will enable the practitioner to avoid or minimize the effects of aniseikonia when prescribing and designing the patient's glasses.

Anisometropia and aniseikonia are often concomitant, although not highly correlated.[23] The problem, as stated earlier, is that of differentiating axial from refractive anisometropia. If this determination can be made, it follows that, if anisometropia is axial, ordinary spectacle lenses constitute the most suitable form of refractive correction, whereas if the anisometropia is refractive, contact lenses are indicated for maximal visual efficiency and comfort.

As stated in Section 10.15, a comparison of the keratometric and refractive findings may provide some insight into the origin of the anisometropia. Although spherical anisometropia may be refractive, axial, or both, it may be incorrect to assume that, in a given case, anisometropia is *entirely* axial or refractive. However, clinically significant degrees (over 2.00 D) of anisometropia are typically axial in nature.[24] In addition, an easy clinical test can assist the practitioner in deciding whether a patient's anisometropia is axial or refractive in origin: While the spherical correction is worn (without the minus cylinder correction, if required), the eyes are dissociated by a vertical prism and a simple direct comparison of the size of the two ocular images is made. If the images are perceived to be equal in size, axial anisometropia is assumed. On the other hand, if the images appear to be of unequal size, refractive anisometropia is assumed.

All astigmatism, as stated earlier, is considered to be *refractive* in origin, and routine refractive changes occurring in adults are also usually refractive in origin. After years of refractive stability, the factors most likely to change are the dioptric elements of the eye rather than the length. For example, the increase in manifest hyperopia in early presbyopia is a common phenomenon, and although this usually occurs binocularly it occasionally occurs mainly in one eye. Also, it is a rather common experience that sudden changes in refraction can occur due to changes in the index of refraction of the crystalline lens due to progressive nuclear sclerosis. Frequently, the change occurs more rapidly in one eye than the other, and the correction of such refractive changes may induce aniseikonia.

In the case of the refractive type of ametropia, Ogle[25] estimated that the percentage change in magnification of the retinal image in the corrected eye as compared to that in the normal eye would be 1.5% per diopter of correction in the case of corneal astigmatism, and 2% for a refractive error arising from the crystalline lens.

Linksz and Bannon,[22] on the other hand, have suggested that if the image size difference is associated with anisometropia of refractive origin we can expect it to be about 1.5% per diopter of anisometropia, but since ametropia may be partly axial, an estimate of 1.0% per diopter is more realistic. The eye having the least hyperopia or the more myopia would be expected to have the smaller retinal image, and therefore would be the one to require magnification. If the patient has astigmatism, the estimate must be made separately for each principal meridian.

In the same article, Linksz and Bannon noted that there are cases on record in which a spherical ametropia of several diopters has been fully corrected by ordinary lenses, and eikonometer measurements have shown no significant aniseikonia. These would, of course, be cases of *axial* ametropia, and their findings are consistent with those of Sorsby et al.,[11] who found that ametropia of ±4.00 D is usually due to an abnormally long or short axial length. Linksz and Bannon commented that most of these cases have been myopes, and that clinical experience shows that there are many cases of spherical ametropia in which this result is *not* obtained, indicating that ametropia may not always be due to axial factors.

Ryan[26] has published tables of relative spectacle magnification based on *power* magnification only (ignoring the shape factor in the spectacle magnification formula) which show that, for a 12-mm vertex distance, relative spectacle magnification changes at the rate of 1.4% per diopter for purely refractive ametropia and only 0.25% per diopter for purely axial ametropia. He made the point that if the lens is worn closer to the eye than the primary focal plane (closer than a vertex distance of 12 mm) the relative spectacle magnifications of the axial and refractive components vary *in opposite directions* and therefore tend to compensate for one another if present in the right proportions.

## 10.20. Avoiding or Minimizing Induced Aniseikonia

A number of procedures have been recommended for prescribing in such a way as to avoid or to minimize induced aniseikonia for patients who have anisometropia or astigmatism.

### Anisometropia

In anisometropia, a partial refractive correction for the more ametropic eye has long been advocated. Donders[7] recommended the prescribing of reading lenses of equal power for presbyopic patients having anisometropia, if they have not previously worn a full correction for the anisometropia and have not experienced discomfort. Such a procedure has the advantage of not inducing aniseikonia and its associated asthenopic symptoms, but on the other hand it fails to provide optimum visual acuity for each eye and fails to provide normal binocular vision.

As for the younger patient, the best thing may be to prescribe a correction that provides maximum visual acuity in each eye and the best possible binocular vision. A partial correction that reduces vision and impairs binocularity may cause failure of the vision requirements set forth by schools, industry, military services, civil services, and driver licensing bureaus. However, full correction may produce aniseikonia with annoying symptomatology, although children tend to adapt to distortions more readily than adults. Once again it should be recalled that large amounts of ametropia tend to be at least partially *axial* in origin, and in such cases a full correction in spectacle lenses may induce little or no aniseikonia.

### Astigmatism

Rayner[27] has stated that the spectacle-wearing patient with intact binocular vision most likely to have aniseikonia would be one with *cylindrical* anisometropia of 1.00 to 1.50 D, with a difference in the axes. Because the powers (and possibly the base curves) vary in the principal meridians of a cylindrical or toric lens for an astigmatic eye, unequal magnification effects occur in the various meridians of the eye. This meridional magnification produces monocular distortion and rotation (tilting) of all lines toward the meridian of greatest magnification (except those meridians parallel to the principal meridians). The amount of tilting of vertical lines is known as the *declination error*. The maximum tilting of vertical lines occurs when the principal meridians of a correcting lens are 45° and 135°.

Monocular distortion and tilting are not, by themselves, a problem, as their effects are small. Ogle and Madigan[28] have shown that when the axes are at 45° and 135°, providing a maximum amount of distortion, the resulting tilt of vertical lines is only 0.43° per diopter of astigmatism. When binocular vision is present, a significant clinical problem occurs. These declination errors, when existing between the images of the two eyes, are important sources of induced aniseikonia because they produce a unique type of false spatial localization for objects seen stereoscopically. A startling effect is generated, in that objects appear to be *inclined*, with the top either away from or toward the observer (see Figure 10-26). For example, if a +1.00 D cylinder axis 135° is worn for the right eye and a +1.00 D cylinder axis 45° is worn for the left eye, for a vertical pole 3 m away the declination error is only 0.43° for each eye, seen monocularly—an amount hardly detectable. However, in *binocular* vision, the vertical pole will now appear to be tilted toward the subject *more than 30°*.[28]

As mentioned in Section 10.17, most patients "adapt" to anomalous spatial localization by learning to supress stereoscopic clues, depending instead upon empirical clues acquired through experience, such as perspective, known size and shape, knowledge of location, and perpendicularity. Some patients make this adaptation easily while others adapt with difficulty or not at all, and are extremely sensitive to anomalous spatial localization.

In order to reduce anomalous spatial orientation, one can minimize the monocular distortion which produces it. There are two things that can be done for spectacle lenses to reduce monocular distortion with no sacrifice in visual acuity.

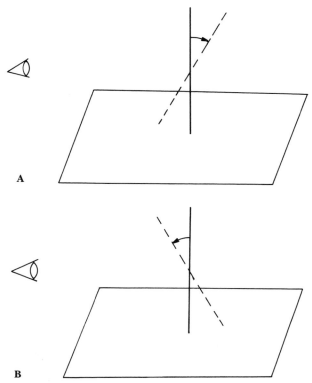

FIGURE 10-26. **Inclination of objects away from and toward an observer due to declination error (tilt of vertical lines) arising from correcting cylinders with oblique axes.**

1. Specify in the prescription *negative toric* lens construction. With a spherical front surface, the shape factor of the spectacle magnification formula is constant for the two principal meridians, and hence does not contribute to the meridional size difference.

2. Use as short a vertex distance as possible. Since astigmatism is *refractive* in origin, meridional size differences due to the power factor are reduced as the vertex distance is made shorter. Of course, if we reduce the vertex distance to zero, with contact lenses, the monocular distortion arising from astigmatism is essentially eliminated.

Monocular distortion can be further reduced by manipulation of the correction, although this will result in a failure to fully correct the refractive error and therefore will reduce the sharpness of the retinal image. Such manipulations include the following:

1. Reduction of the power of the cylinder, but at the correct axis. Since the amount of meridional magnification is a function of the power of the cylinder, reduction of the power decreases meridional magnification. This will leave some astigmatism uncorrected, which will of course cause some blur. Tentatively, when this is done the spherical power should be altered accordingly, to maintain the same spherical equivalent, but the final sphere power should be one that subjectively provides the best visual acuity.

2. If the cylinder axes are oblique, the axes can be rotated toward the 180° or 90° position, depending upon which is closer. The rationale underlying this manipulation is that distortion is usually better tolerated when the axis is either horizontal or vertical. The direction of the meridional magnification is determined primarily by the axis position of the correcting cylinder, whether or not it is the correct axis.

If the axis of a cylinder is rotated to make the direction of meridional magnification more tolerable, the cylinder should be reduced in power in order to minimize the amount of residual astigmatism. The Jackson cross cylinder, used at the new axis location for the refinement of cylinder power, will automatically give the optimum cylinder power to result in a minimum of residual astigmatism for a unique axis error.[29] The optimum power for the new axis position may also be determined by rotating any two-meridian astigmatic chart (such as the "rotating T") in alignment with the principal meridians for the new axis setting, and balancing the sharpness of the two lines by manipulating cylinder power.

Patients who wear glasses only intermittently for distance vision may not adapt to space distortions during short periods of wear. In such cases, the astigmatic correction may be altered as previously discussed or may even be eliminated altogether.

Some presbyopic patients with slight or moderate amounts of astigmatism will refuse to wear any glasses for distance vision but need assistance for near work. Again, it is probably best to alter the astigmatic correction or even to eliminate it, because adaptation to space distortions may not readily occur.

## 10.21. Prescribing and Lens Design

When it has been concluded that a patient's symptomatology is due to the presence of induced aniseikonia, the following factors, suggested by Bannon,[30] may be considered when deciding whether or not to prescribe an aniseikonic correction:

1. The age of the patient
2. The nature of the previous correction
3. The patient's occupation and hobbies
4. The patient's temperament
5. The patient's interest in the appearance of the glasses
6. The cost of the correction
7. The nature of the patient's complaint, and the practitioner's assessment of the likelihood of partial or total relief

As shown by Eq. (10.8), developed in Section 10.7, spectacle magnification depends upon the combination of the shape factor ($M_S$) and the power factor ($M_P$):

$$SM = (M_S)(M_P)$$

$$= \left( \frac{1}{1 - \frac{t}{n} F_1} \right) \left( \frac{1}{1 - hF_V} \right),$$

the shape factor, $M_S$, being due to the front surface curvature, $F_1$, and the thickness, $t$, of the lens; and the power factor, $M_P$, being due to the power, $F_V$, of the lens and the distance, $h$, from the lens to the center of the entrance pupil of the eye.

Although the refractive power of the lens, $F_V$, participates in the power factor, the powers of the patient's lenses obviously cannot be changed if the ametropia is to be properly corrected. Therefore, the following factors are the only ones that can be altered without significantly affecting the refractive power of the lenses:

1. The front surface power of the lens
2. The center thickness of the lens
3. The distance of the lens from the center of the entrance pupil

The spectacle magnification formula is an *exact* formula, and a rather awkward one, but by using approximations the formula can be simplified with no significant degree of error. Thus, if $t/n$ is small, as it usually is, the spectacle magnification formula can be expanded to

$$M \cong 1 + \frac{t}{n} F_1 + hF_V.$$

Since

$$\%M = 100(M - 1),$$

$$\%M \cong 100 \frac{t}{n} F_1 + 100hF_V,$$

with $t$ and $h$ expressed in meters.

If $n = 1.50$ and $t$ and $h$ are expressed in millimeters,

$$\%M = \frac{tF_1}{15} + \frac{hF}{10}. \qquad (10.20)$$

This formula gives a first approximation of the effect of changes in the front surface power, thickness, and vertex distance on spectacle magnification. An examination of this approximate formula indicates the approximate magnification effects of varying each of these parameters, which may be stated by the following "rules of thumb":

1. If all other factors are held constant, for every diopter increase in front surface power, the percentage magnification is increased (for both plus and minus lenses) by *1/15* of the center thickness. The percentage change in shape magnification per diopter of change in front surface power is given by

$$\Delta\%M_S = \frac{\Delta F_1(t)}{15}. \qquad (10.21)$$

2. If all other factors are held constant, for every millimeter increase in center thickness the percentage magnification is increased (for both plus and minus lenses) by *1/15* of the front surface power. The percentage change in shape magnification per diopter of change in thickness is given by

$$\Delta\%M_S = \frac{\Delta t(F_1)}{15}. \qquad (10.22)$$

3. If all other factors are held constant, for every millimeter the vertex distance is changed, the percentage magnification is changed by *1/10* of the refractive power. Magnification *increases* as a plus lens is moved away from the eye and a minus lens is moved toward the eye; and magnification *decreases* as a plus lens moves toward the eye and a minus lens moves away from the eye. The percentage change in the power magnification, per millimeter of vertex distance change, is given by the relationship

$$\Delta\%M_P = \frac{\Delta h(F_V)}{10}.$$

It must be noted that these "rules of thumb" do not work independently of each other. For example, if the front surface curvature is steepened and the back vertex power remains constant, the back surface curvature must be steepened essentially the same amount in order to maintain the specified back vertex power, with the result that the back vertex of the lens will be *farther* from the eye. For plus lenses this causes an additional increase in magnification. Increasing the front surface curvature for a minus lens also produces an increase in magnification, but this is combined with a decrease in magnification due to a consequent increase in the vertex distance.

Linksz and Bannon[22] have pointed out that the decrease in magnification brought about by increased vertex distance of a minus lens begins to cancel out the magnification due to the increased curvature as the power of the lens approaches −2.00 D. They suggest that since strong minus lenses have thick edges, it is possible to change the position of the bevel and thus bring the lens closer to the eye, hence preserving the magnification achieved by the increase in front surface power. Linksz and Bannon have developed tables (reproduced here as Tables 10-4 and 10-5) which illustrate the approximate magnification changes associated with changes in front surface curvature and vertex distance (eyewire distance) for lenses of various powers.

Table 10-4 may be used in the following manner. Suppose that the practitioner determines (by measurement or by estimation) that a patient's right lens, having a back vertex power of +2.00 D and a center thickness of 2.7 mm, requires an increase in magnification of 1%. The table shows that, for this lens, an increase in magnification of 1.1% will result if the front surface power is increased by 4.00 D. It should be pointed out that this table takes into consideration the change in *vertex distance* which results when the front surface power is changed. For example, the table shows that, for a −3.00 D lens, an increase in front surface power of 2.00 D actually causes a *minification* of 0.2%, due to the increase in vertex distance which accompanies the steepening of the front surface.

Magnification changes associated with changes in vertex distance are shown in Table 10-5. In this table, a positive sign indicates an increase in vertex distance and a negative sign indicates a decrease in vertex distance. As mentioned previously, an increase in the vertex distance results in an increase in magnification for a plus lens and a decrease in magnification for a minus lens. On the other hand, a decrease in the vertex distance produces a decrease in magnification for a plus lens and an increase in magnification for a minus lens.

For a minus lens the practitioner may decide not to alter the front surface power but to bring about the desired magnification by changing *only* the vertex distance. For example, if one wishes to bring about a 1.2% increase in the magnification of a −6.00 D lens, this table shows that this may be accomplished by moving the lens toward the eye (by moving the bevel forward) 2 mm.

It is obvious from Table 10-5 that much greater magnification changes can be obtained with plus lenses than with minus lenses. For this reason, the

TABLE 10-4

**Approximate Magnification Changes Associated with Changes in Front Base Curves for Various Lens Powers—Percentage**

| Vertex Power of Lens (D) | Average Center Thickness (mm) | Change in Front Base Curve | | | | | |
|---|---|---|---|---|---|---|---|
| | | −4.00 D | −2.00 D | +2.00 D | +4.00 D | +6.00 D | +8.00 D |
| +8.00 | 7.0 | −3.5 | −1.7 | +1.7 | +3.5 | +5.2 | +6.9 |
| +7.00 | 6.2 | −3.0 | −1.5 | +1.5 | +3.0 | +4.5 | +6.0 |
| +6.00 | 5.4 | −2.6 | −1.3 | +1.3 | +2.6 | +3.9 | +5.2 |
| +5.00 | 4.6 | −2.2 | −1.1 | +1.1 | +2.2 | +3.3 | +4.4 |
| +4.00 | 3.9 | −1.8 | −0.9 | +0.9 | +1.8 | +2.8 | +3.7 |
| +3.00 | 3.2 | −1.5 | −0.7 | +0.7 | +1.5 | +2.2 | +2.9 |
| +2.00 | 2.7 | −1.1 | −0.6 | +0.6 | +1.1 | +1.7 | +2.2 |
| +1.00 | 2.2 | −0.8 | −0.4 | +0.4 | +0.8 | +1.2 | +1.6 |
| 0.00 | 1.8 | −0.5 | −0.2 | +0.2 | +0.5 | +0.7 | +1.0 |
| −1.00 | 1.4 | −0.2 | −0.1 | +0.1 | +0.2 | +0.3 | +0.3 |
| −2.00 | 1.0 | +0.1 | +0.1 | −0.1 | −0.1 | −0.2 | −0.3 |
| −3.00 | 0.9 | +0.4 | +0.2 | −0.2 | −0.4 | −0.5 | −0.7 |
| −4.00 | 0.9 | +0.6 | +0.3 | −0.3 | −0.6 | −0.8 | −1.1 |
| −5.00 | 0.9 | +0.8 | +0.4 | −0.4 | −0.8 | −1.1 | −1.5 |
| −6.00 | 0.9 | +1.0 | +0.5 | −0.5 | −1.0 | −1.4 | −1.9 |
| −7.00 | 0.9 | +1.2 | +0.6 | −0.6 | −1.2 | −1.7 | −2.3 |
| −8.00 | 0.9 | +1.4 | +0.7 | −0.7 | −1.4 | −2.0 | −2.7 |

Source: From A. Linksz and R.E. Bannon, *International Ophthalmologic Clinics*, Vol. 5, No. 2, pp. 515–534, Little, Brown, Boston, 1965.

TABLE 10-5
**Approximate Magnification Changes Associated with Changes in Eyewire Distance for Various Lens Powers— Percentage**

| Vertex Power of Lens (D) | Change in Eyewire Distance Lens Moved toward the Eye | | | |
|---|---|---|---|---|
| | −1 mm | −2 mm | −3 mm | −4 mm |
| +10 | −1.0 | −2.0 | −3.0 | −4.0 |
| + 8 | −0.8 | −1.6 | −2.4 | −3.2 |
| + 6 | −0.6 | −1.2 | −1.8 | −2.4 |
| + 4 | −0.4 | −0.8 | −1.2 | −1.6 |
| + 2 | −0.2 | −0.4 | −0.6 | −0.8 |
| − 2 | +0.2 | +0.4 | +0.6 | +0.8 |
| − 4 | +0.4 | +0.8 | +1.2 | +1.6 |
| − 6 | +0.6 | +1.2 | +1.8 | +2.4 |
| − 8 | +0.8 | +1.6 | +2.4 | +3.2 |
| −10 | +1.0 | +2.0 | +3.0 | +4.0 |

| | Lens Moved away From Eye | | | |
|---|---|---|---|---|
| | +1 mm | +2 mm | +3 mm | +4 mm |
| +10 | +1.0 | +2.0 | +3.0 | +4.0 |
| + 8 | +0.8 | +1.6 | +2.4 | +3.2 |
| + 6 | +0.6 | +1.2 | +1.8 | +2.4 |
| + 4 | +0.4 | +0.8 | +1.2 | +1.6 |
| + 2 | +0.2 | +0.4 | +0.6 | +0.8 |
| − 2 | −0.2 | −0.4 | −0.6 | −0.8 |
| − 4 | −0.4 | −0.8 | −1.2 | −1.6 |
| − 6 | −0.6 | −1.2 | −1.8 | −2.4 |
| − 8 | −0.8 | −1.6 | −2.4 | −3.2 |
| −10 | −1.0 | −2.0 | −3.0 | −4.0 |

Source: From A. Linksz and R.E. Bannon, *International Ophthalmologic Clinics*, Vol. 5, No. 2, pp. 515–534, Little, Brown, Boston, 1965.

design of eikonic lenses should always begin with the lens having the *most plus or least minus power*, the shape magnification of this lens being reduced to a minimum.

Although magnification can be increased by increasing lens thickness, any increase in thickness must be modest in order to keep from making the lenses unattractive or excessively heavy. However, when attempting to increase magnification of a minus lens, if thickness is increased while maintaining a constant front bevel (such as a Hide-a-Bevel), a substantial increase in magnification can be obtained. Assuming a constant front bevel, the change in magnification by a change in thickness would be given by the following formula[31]:

$$\Delta\%M = \Delta t \left( \frac{F_1}{15} - \frac{F_V}{10} \right). \qquad (10.23)$$

In cases in which there is a correction for astigmatism, the power in each of the two principal meridians must be considered separately.

Berens and Bannon[32] recommend that the induced aniseikonia should be only *partially* corrected. They compare this to the partial correction of heterophoria by prisms. An approximate correction provides not only therapeutic value but diagnostic information as well. If an approximate correction provides partial relief from the symptoms, it may be worthwhile to have a thorough eikonometric examination to determine a more exact correction.

## 10.22. The Use of Eikonic Lenses in Fit-Over Form

Before making up the correction in final form, some practitioners have found it useful to conduct a confirmation test using a fit-over afocal magnifying lens with the patient's glasses, the fit-over lens being placed before the lens with the smaller predicted image size. If relief is reported after wearing the fit-over lens for a few days, one may feel confident in prescribing the correction. Some practitioners find it helpful to *reverse* the afocal fit-over lens, putting the lens before the opposite eye, to increase rather than correct the image size difference: If symptoms are intensified by the reversal, this tends to confirm the diagnosis.

Since afocal telescopic lenses change the magnification of the retinal image without changing the power of the refractive correction, they are frequently called *size lenses*. Since the back vertex power is zero, the magnification of an afocal lens is given by the shape factor alone,

$$M_S = \frac{1}{1 - \dfrac{t}{n} F_1}. \qquad (10.24)$$

Since

$$F_V = \frac{F_1}{1 - \dfrac{t}{n}} + F_2,$$

$$F_1 = \frac{F_V - F_2}{1 + \dfrac{t}{n}(F_V - F_2)},$$

and, substituting for $F_1$, in Eq. (10.24),

$$M_S = \frac{1}{1 - \dfrac{t}{n}\left( \dfrac{F_V - F_2}{1 + \dfrac{t}{n}(F_V - F_2)} \right)}$$

## TABLE 10-6

**Plano-Powered Lenses with Desired Percentage Magnification: Required Center Thickness *(t)* in Millimeters and Inside 1.53 Curve ($D_2$) for a Selected 1.53 Front Curve ($D_1$)**

| % Mag. | 1.53 $D_1$ | 2.00 | 2.50 | 3.00 | 3.50 | 4.00 | 4.50 | 5.00 | 5.50 | 6.00 | 6.50 | 7.00 | 7.50 | 8.00 | 8.50 |
|---|---|---|---|---|---|---|---|---|---|---|---|---|---|---|---|
| 0.50 | $t$ | 3.84 | 3.07 | 3.56 | 2.19 | 1.92 | 1.71 | 1.54 | 1.40 | 1.28 | 1.18 | 1.10 | 1.02 | 0.96 | 0.90 |
|  | $D_2$ | 2.01 | 2.51 | 3.02 | 3.52 | 4.02 | 4.52 | 5.03 | 5.53 | 6.03 | 6.53 | 7.04 | 7.54 | 8.04 | 8.54 |
| 0.75 | $t$ | 5.74 | 4.60 | 3.83 | 3.28 | 2.87 | 2.55 | 2.30 | 2.09 | 1.91 | 1.77 | 1.64 | 1.53 | 1.44 | 1.35 |
|  | $D_2$ | 2.02 | 2.52 | 3.02 | 3.53 | 4.03 | 4.53 | 5.04 | 5.54 | 6.05 | 6.55 | 7.05 | 7.56 | 8.06 | 8.56 |
| 1.00 | $t$ | 7.64 | 6.11 | 5.09 | 4.37 | 3.82 | 3.39 | 3.06 | 2.78 | 2.55 | 2.35 | 2.18 | 2.04 | 1.91 | 1.80 |
|  | $D_2$ | 2.02 | 2.53 | 3.03 | 3.54 | 4.04 | 4.55 | 5.05 | 5.56 | 6.06 | 6.57 | 7.07 | 7.58 | 8.08 | 8.59 |
| 1.25 | $t$ | 9.52 | 7.62 | 6.35 | 5.44 | 4.76 | 4.23 | 3.81 | 3.46 | 3.17 | 2.93 | 2.72 | 2.54 | 2.38 | 2.24 |
|  | $D_2$ | 2.03 | 2.53 | 3.04 | 3.54 | 4.05 | 4.56 | 5.06 | 5.57 | 6.08 | 6.58 | 7.09 | 7.59 | 8.10 | 8.61 |
| 1.50 | $t$ | 11.4 | 9.12 | 7.60 | 6.52 | 5.70 | 5.07 | 4.56 | 4.15 | 3.80 | 3.51 | 3.26 | 3.04 | 2.85 | 2.68 |
|  | $D_2$ | 2.03 | 2.54 | 3.05 | 3.55 | 4.06 | 4.57 | 5.08 | 5.58 | 6.09 | 6.60 | 7.11 | 7.61 | 8.12 | 8.63 |
| 1.75 | $t$ | 13.3 | 10.6 | 8.85 | 7.58 | 6.63 | 5.90 | 5.31 | 4.83 | 4.42 | 4.08 | 3.79 | 3.54 | 3.32 | 3.12 |
|  | $D_2$ | 2.04 | 2.54 | 3.05 | 3.56 | 4.07 | 4.58 | 5.09 | 5.60 | 6.11 | 6.61 | 7.12 | 7.63 | 8.14 | 8.65 |
| 2.00 | $t$ | 15.1 | 12.1 | 10.1 | 8.64 | 7.56 | 6.72 | 6.05 | 5.50 | 5.04 | 4.65 | 4.32 | 4.03 | 3.78 | 3.56 |
|  | $D_2$ | 2.04 | 2.55 | 3.06 | 3.57 | 4.08 | 4.59 | 5.10 | 5.61 | 6.12 | 6.63 | 7.14 | 7.65 | 8.16 | 8.67 |
| 2.25 | $t$ | 17.0 | 13.6 | 11.3 | 9.70 | 8.49 | 7.55 | 6.79 | 6.17 | 5.66 | 5.22 | 4.85 | 4.53 | 4.24 | 3.99 |
|  | $D_2$ | 2.05 | 2.56 | 3.07 | 3.58 | 4.09 | 4.60 | 5.11 | 5.62 | 6.14 | 6.65 | 7.16 | 7.67 | 8.18 | 8.69 |
| 2.50 | $t$ | 18.8 | 15.1 | 12.5 | 10.8 | 9.41 | 8.36 | 7.53 | 6.84 | 6.27 | 5.79 | 5.38 | 5.02 | 4.70 | 4.43 |
|  | $D_2$ | 2.05 | 2.56 | 3.08 | 3.59 | 4.10 | 4.61 | 5.13 | 5.64 | 6.15 | 6.66 | 7.18 | 7.69 | 8.20 | 8.71 |
| 2.75 | $t$ | 20.7 | 16.5 | 13.8 | 11.8 | 10.3 | 9.18 | 8.26 | 7.51 | 6.88 | 6.35 | 5.90 | 5.51 | 5.16 | 4.86 |
|  | $D_2$ | 2.06 | 2.57 | 3.08 | 3.60 | 4.11 | 4.62 | 5.14 | 5.65 | 6.17 | 6.68 | 7.19 | 7.71 | 8.22 | 8.73 |
| 3.00 | $t$ | 22.5 | 18.0 | 15.0 | 12.8 | 11.2 | 9.99 | 8.99 | 8.17 | 7.49 | 6.91 | 6.42 | 5.99 | 5.62 | 5.29 |
|  | $D_2$ | 2.06 | 2.58 | 3.09 | 3.61 | 4.12 | 4.64 | 5.15 | 5.67 | 6.18 | 6.70 | 7.21 | 7.73 | 8.24 | 8.76 |
| 3.25 | $t$ | 24.3 | 19.4 | 16.2 | 13.9 | 12.1 | 10.8 | 9.71 | 8.83 | 8.09 | 7.47 | 6.94 | 6.48 | 6.07 | 5.71 |
|  | $D_2$ | 2.07 | 2.58 | 3.10 | 3.61 | 4.13 | 4.65 | 5.16 | 5.68 | 6.20 | 6.71 | 7.23 | 7.74 | 8.26 | 8.78 |
| 3.50 | $t$ | 26.1 | 20.9 | 17.4 | 14.9 | 13.1 | 11.6 | 10.4 | 9.49 | 8.70 | 8.03 | 7.45 | 6.96 | 6.52 | 6.14 |
|  | $D_2$ | 2.07 | 2.59 | 3.11 | 3.62 | 4.14 | 4.66 | 5.18 | 5.69 | 6.21 | 6.73 | 7.25 | 7.76 | 8.28 | 8.80 |
| 3.75 | $t$ | 27.9 | 22.3 | 18.6 | 15.9 | 13.9 | 12.4 | 11.2 | 10.1 | 9.30 | 8.58 | 7.97 | 7.44 | 6.97 | 6.56 |
|  | $D_2$ | 2.08 | 2.59 | 3.11 | 3.63 | 4.15 | 4.67 | 5.19 | 5.71 | 6.23 | 6.74 | 7.26 | 7.78 | 8.30 | 8.82 |
| 4.00 | $t$ | 29.7 | 23.7 | 20.8 | 17.0 | 14.8 | 13.2 | 11.9 | 10.8 | 9.89 | 9.13 | 8.48 | 7.91 | 7.42 | 6.98 |
|  | $D_2$ | 2.08 | 2.60 | 3.12 | 3.64 | 4.16 | 4.68 | 5.20 | 5.72 | 6.24 | 6.76 | 7.28 | 7.80 | 8.32 | 8.84 |
| 4.25 | $t$ | 31.5 | 25.2 | 21.0 | 18.0 | 15.7 | 14.0 | 12.6 | 11.4 | 10.5 | 9.68 | 8.99 | 8.39 | 7.86 | 7.40 |
|  | $D_2$ | 2.09 | 2.61 | 3.13 | 3.65 | 4.17 | 4.69 | 5.21 | 5.73 | 6.26 | 6.78 | 7.30 | 7.82 | 8.34 | 8.86 |
| 4.50 | $t$ | 33.2 | 26.6 | 22.2 | 19.0 | 16.6 | 14.8 | 13.3 | 12.1 | 11.1 | 10.2 | 9.49 | 8.86 | 8.31 | 7.82 |
|  | $D_2$ | 2.09 | 2.61 | 3.14 | 3.66 | 4.18 | 4.70 | 5.23 | 5.75 | 6.27 | 6.79 | 7.32 | 7.84 | 8.36 | 8.88 |
| 4.75 | $t$ | 35.0 | 28.0 | 23.3 | 20.0 | 17.5 | 15.6 | 14.0 | 12.7 | 11.7 | 10.8 | 10.0 | 9.33 | 8.75 | 8.23 |
|  | $D_2$ | 2.10 | 2.62 | 3.14 | 3.67 | 4.19 | 4.71 | 5.24 | 5.76 | 6.29 | 6.81 | 7.33 | 7.86 | 8.38 | 8.90 |
| 5.00 | $t$ | 36.7 | 29.4 | 24.5 | 21.0 | 18.4 | 16.3 | 14.7 | 13.4 | 12.3 | 11.3 | 10.5 | 9.80 | 9.18 | 8.64 |
|  | $D_2$ | 2.10 | 2.63 | 3.15 | 3.68 | 4.20 | 4.73 | 5.25 | 5.78 | 6.30 | 6.83 | 7.35 | 7.88 | 8.40 | 8.93 |

Source: From A. Rayner, *Amer. J. Optom. Arch. Amer. Acad. Optom.*, Vol. 43, pp. 617–632, 1966.

# TABLE 10-7

**Plano-Powered Lenses with Desired Percentage Magnification: Required Center Thickness *(t)* in Millimeters and Inside 1.53 Curve ($D_2$) for a Selected 1.53 Front Curve ($D_1$).**

| % Mag. | | 1.53 $D_1$ 9.00 | 9.50 | 10.00 | 10.50 | 11.00 | 11.50 | 12.00 | 12.50 | 13.00 | 13.50 | 14.00 | 14.50 |
|--------|------|------|------|-------|-------|-------|-------|-------|-------|-------|-------|-------|-------|
| 0.50 | $t$ | 0.85 | 0.81 | 0.77 | 0.73 | 0.70 | 0.67 | 0.64 | 0.61 | 0.59 | 0.57 | 0.55 | 0.53 |
|      | $D_2$ | 9.05 | 9.55 | 10.05 | 10.55 | 11.06 | 11.56 | 12.06 | 12.56 | 13.07 | 13.57 | 14.07 | 14.57 |
| 0.75 | $t$ | 1.28 | 1.21 | 1.15 | 1.09 | 1.04 | 1.00 | 0.96 | 0.92 | 0.88 | 0.85 | 0.82 | 0.79 |
|      | $D_2$ | 9.07 | 9.57 | 10.08 | 10.58 | 11.08 | 11.59 | 12.09 | 12.59 | 13.10 | 13.60 | 14.11 | 14.61 |
| 1.00 | $t$ | 1.70 | 1.61 | 1.53 | 1.45 | 1.39 | 1.33 | 1.27 | 1.22 | 1.18 | 1.13 | 1.09 | 1.05 |
|      | $D_2$ | 9.09 | 9.60 | 10.10 | 10.61 | 11.11 | 11.62 | 12.12 | 12.63 | 13.13 | 13.64 | 14.14 | 14.65 |
| 1.25 | $t$ | 2.12 | 2.01 | 1.90 | 1.81 | 1.73 | 1.66 | 1.59 | 1.52 | 1.47 | 1.41 | 1.36 | 1.31 |
|      | $D_2$ | 9.11 | 9.62 | 10.13 | 10.63 | 11.14 | 11.64 | 12.15 | 12.66 | 13.16 | 13.67 | 14.18 | 14.68 |
| 1.50 | $t$ | 2.53 | 2.40 | 2.28 | 2.17 | 2.07 | 1.98 | 1.90 | 1.82 | 1.75 | 1.69 | 1.63 | 1.57 |
|      | $D_2$ | 9.14 | 9.64 | 10.15 | 10.66 | 11.17 | 11.67 | 12.18 | 12.69 | 13.20 | 13.70 | 14.21 | 14.72 |
| 1.75 | $t$ | 2.95 | 2.79 | 2.65 | 2.53 | 2.41 | 2.31 | 2.21 | 2.12 | 2.04 | 1.97 | 1.90 | 1.83 |
|      | $D_2$ | 9.16 | 9.67 | 10.18 | 10.68 | 11.19 | 11.70 | 12.21 | 12.72 | 13.23 | 13.74 | 14.25 | 14.75 |
| 2.00 | $t$ | 3.36 | 3.18 | 3.03 | 2.88 | 2.75 | 2.63 | 2.52 | 2.42 | 2.33 | 2.24 | 2.16 | 2.09 |
|      | $D_2$ | 9.18 | 9.69 | 10.20 | 10.71 | 11.22 | 11.73 | 12.24 | 12.75 | 13.26 | 13.77 | 14.28 | 14.79 |
| 2.25 | $t$ | 3.77 | 3.57 | 3.40 | 3.23 | 3.09 | 2.95 | 2.83 | 2.72 | 2.61 | 2.52 | 2.43 | 2.34 |
|      | $D_2$ | 9.20 | 9.71 | 10.23 | 10.74 | 11.25 | 11.76 | 12.27 | 12.78 | 13.29 | 13.80 | 14.32 | 14.83 |
| 2.50 | $t$ | 4.18 | 3.96 | 3.76 | 3.58 | 3.42 | 3.27 | 3.14 | 3.01 | 2.89 | 2.79 | 2.69 | 2.60 |
|      | $D_2$ | 9.23 | 9.74 | 10.25 | 10.76 | 11.28 | 11.79 | 12.30 | 12.81 | 13.33 | 13.84 | 14.35 | 14.86 |
| 2.75 | $t$ | 4.59 | 4.35 | 4.13 | 3.93 | 3.75 | 3.59 | 3.44 | 3.30 | 3.18 | 3.06 | 2.95 | 2.85 |
|      | $D_2$ | 9.25 | 9.76 | 10.28 | 10.79 | 11.30 | 11.82 | 12.33 | 12.84 | 13.36 | 13.87 | 14.39 | 14.90 |
| 3.00 | $t$ | 4.99 | 4.73 | 4.49 | 4.28 | 4.09 | 3.91 | 3.75 | 3.60 | 3.46 | 3.33 | 3.21 | 3.10 |
|      | $D_2$ | 9.27 | 9.79 | 10.30 | 10.82 | 11.33 | 11.85 | 12.36 | 12.88 | 13.39 | 13.91 | 14.42 | 14.94 |
| 3.25 | $t$ | 5.40 | 5.11 | 4.96 | 4.63 | 4.42 | 4.22 | 4.05 | 3.89 | 3.74 | 3.60 | 3.47 | 3.35 |
|      | $D_2$ | 9.29 | 9.81 | 10.33 | 10.84 | 11.36 | 11.87 | 12.39 | 12.91 | 13.42 | 13.94 | 14.46 | 14.97 |
| 3.50 | $t$ | 5.80 | 5.49 | 5.22 | 4.97 | 4.74 | 4.54 | 4.35 | 4.17 | 4.01 | 3.87 | 3.73 | 3.60 |
|      | $D_2$ | 9.32 | 9.83 | 10.35 | 10.87 | 11.39 | 11.90 | 12.42 | 12.94 | 13.46 | 13.97 | 14.49 | 15.01 |
| 3.75 | $t$ | 6.20 | 5.87 | 5.58 | 5.31 | 5.07 | 4.85 | 4.65 | 4.46 | 4.29 | 4.13 | 3.98 | 3.85 |
|      | $D_2$ | 9.34 | 9.86 | 10.38 | 10.89 | 11.41 | 11.93 | 12.45 | 12.97 | 13.49 | 14.01 | 14.53 | 15.04 |
| 4.00 | $t$ | 6.59 | 6.25 | 5.93 | 5.65 | 5.40 | 5.16 | 4.95 | 4.75 | 4.57 | 4.40 | 4.24 | 4.09 |
|      | $D_2$ | 9.36 | 9.88 | 10.40 | 10.92 | 11.44 | 11.96 | 12.48 | 13.00 | 13.52 | 14.04 | 14.56 | 15.08 |
| 4.25 | $t$ | 6.99 | 6.62 | 6.29 | 5.99 | 5.72 | 5.47 | 5.24 | 5.03 | 4.84 | 4.66 | 4.49 | 4.34 |
|      | $D_2$ | 9.38 | 9.90 | 10.43 | 10.95 | 11.47 | 11.99 | 12.51 | 13.03 | 13.55 | 14.07 | 14.60 | 15.12 |
| 4.50 | $t$ | 7.38 | 6.99 | 6.64 | 6.33 | 6.04 | 5.78 | 5.54 | 5.32 | 5.11 | 4.92 | 4.75 | 4.58 |
|      | $D_2$ | 9.41 | 9.93 | 10.45 | 10.97 | 11.50 | 12.02 | 12.54 | 13.06 | 13.59 | 14.11 | 14.63 | 15.15 |
| 4.75 | $t$ | 7.77 | 7.37 | 7.00 | 6.66 | 6.36 | 6.08 | 5.83 | 5.60 | 5.38 | 5.18 | 5.00 | 4.83 |
|      | $D_2$ | 9.43 | 9.95 | 10.48 | 11.00 | 11.52 | 12.05 | 12.57 | 13.09 | 13.62 | 14.14 | 14.67 | 15.19 |
| 5.00 | $t$ | 8.16 | 7.73 | 7.35 | 7.00 | 6.68 | 6.39 | 6.12 | 5.88 | 5.65 | 5.44 | 5.25 | 5.07 |
|      | $D_2$ | 9.45 | 9.98 | 10.50 | 11.03 | 11.55 | 12.08 | 12.60 | 13.13 | 13.65 | 14.18 | 14.70 | 15.23 |

Source: From A. Rayner, *Amer. J. Optom. Arch. Amer. Acad. Optom.*, Vol. 43, pp. 617–632, 1966.

$$= \cfrac{1}{\cfrac{1 + \dfrac{t}{n}(F_V - F_2) - \dfrac{t}{n}(F_V - F_2)}{1 + \dfrac{t}{n}(F_V - F_2)}}$$

$$= \cfrac{1 + \dfrac{t}{n}(F_V - F_2)}{1 + \dfrac{t}{n}(F_V - F_2) - \dfrac{t}{n}(F_V - F_2)}$$

$$= 1 + \dfrac{t}{n}(F_V - F_2),$$

and since for an afocal lens, $F_V = 0$,

$$M_S = 1 - \frac{t}{n}F_2. \qquad (10.25)$$

One may easily design a set of afocal lenses that will serve as a trial eikonic lens ("size lens") set. Once the thickness has been selected, either Eq. (10.24) or (10.25) may be used to design the lenses. For overall magnifiers, spherical curves would of course be used. Meridional magnifying lenses, having maximum magnification in one meridian and no magnification at all in the opposite meridian, are bi-toric lenses, made by grinding toric surfaces both on the front and on the back of the lens, with the axes precisely aligned.

Rayner[27] has provided the lens specifications for afocal lenses with various magnifications. Tables 10-6 and 10-7 may be used to select the front surface power, thickness, and back surface power that will provide the desired magnification. The most useful areas of each of these tables fall between the heavy lines which exclude either very thin lenses or very thick lenses (more than 8 mm thick). For example, a 3.00% afocal magnifier could be manufactured with a front surface power of +6.00 D, a center thickness of 7.49 mm, and a back surface power of −6.18 D, or with a front surface power of +12.00 D, a center thickness of 3.75 mm and a back surface power of −12.36 D.

## 10.23. Frame Selection for Eikonic Lenses

The correct fitting and adjustment of the frame are critical for an eikonic correction. Since these lenses are thicker and therefore heavier than most other lenses, the eye size should be kept small, both to reduce weight and to allow fitting the lens at as short a vertex distance as possible. A plastic frame will conceal the thick edge better than will a metal frame or a rimless mounting. An eikonic lens tends to have highly curved surfaces and a correspondingly curved bevel, and therefore may be difficult to mount without warping the frame: A *symmetrical* lens shape is therefore recommended, that is, a shape having a small difference between the "A" and "B" measurements.

The patient should be seen often during the first few weeks after the correction is dispensed, in order to check the adjustment and to provide reassurance and additional instruction if needed.

## 10.24. Aniseikonia: Clinical Considerations in Anisometropia and Astigmatism

In summary, in cases of significant anisometropia or astigmatism:

1. The practitioner can predict the amount of image size difference that will result for the two eyes, using one of the rules of thumb given above and remembering that while high refractive errors are most likely to be axial, moderate or low refractive errors are more likely to be due to axial-refractive combinations.

2. Once the prediction has been made, tables such as those prepared by Linksz and Bannon and by Rayner may be used to determine the lens parameters necessary to minimize the induced aniseikonia.

3. It is normally preferable to design lenses that will only *partially* correct for the induced aniseikonia.

## References

1. Fischer, L.G. Analytical Aids for Determining Induced Vertical Prism Imbalance in Anisometropia. *Amer. J. Optom. Physiol. Opt.*, Vol. 53, pp. 249−258, 1976.

2. Ellerbrock, V.J., and Fry, G.A. Effects Induced by Anisometropia Corrections, *Amer. J. Optom. Arch. Amer. Acad. Optom.*, Vol. 19, pp. 444−459, 1942.

3. Cusick, P., and Hawn, H. Prism Compensation in Cases of Anisometropia. *Arch. Ophthal.*, Vol. 25, pp. 651−654, 1941.

4. Ellerbrock, V.J. A Clinical Evaluation of Compensation for Vertical Prismatic Imbalances. *Amer. J. Optom. Arch. Amer. Acad. Optom.*, Vol. 25, pp. 309−325, 1948.

5. Duke-Elder, S. *Ophthalmic Optics and Refraction*, Vol. 5. Henry Kimpton, London, 1970.

6. Peters, H.B. Measurement of a "Slab-Off" Ophthalmic Lens with a Lens Gauge. *Amer. J. Optom. Arch. Amer. Acad. Optom.*, Vol. 26, pp. 16–18, 1949.

7. Donders, F.C. *On the Anomalies of Accommodation and Refraction of the Eye.* New Sydenham Society, London, 1864. (Reproduced by the Hatton Press, London, 1952.)

8. Burian, H.M. History of Dartmouth Eye Institute. *Arch. Ophthal.*, Vol. 40, pp. 163–175, 1948.

9. Lancaster, W. Aniseikonia. *Arch. Ophthal.*, Vol. 20, pp. 907–912, 1938.

10. Ogle, K.N. Some Aspects of the Eye as an Image Forming Mechanism. *J. Opt. Soc. Amer.*, Vol. 33, pp. 506–512, 1943.

11. Sorsby, A., Benjamin, D., Davey, J.B., Sheridan, M., and Tanner, J.M. *Emmetropia and Its Aberrations.* Her Majesty's Stationery Office, London, 1957.

12. Bannon, R.E. *Clinical Manual on Aniseikonia*, p. 95. American Optical Corp., Buffalo, N.Y., 1954.

13. Ogle, K.N. *Researches in Binocular Vision*, p. 282. W. B. Saunders, Philadelphia, 1950.

14. Bannon, R.E. Space Eikonometry in Aniseikonia. *Amer. J. Optom. Arch. Amer. Acad. Optom.*, Vol. 30, pp. 86–93, 1953.

15. Elvin, F.T. Aniseikonia—A Simplified Method of Measurement and a Case Report. *Amer. J. Optom. Arch. Amer. Acad. Optom.*, Vol. 26, pp. 78–82, 1949.

16. Brecher, G.A. A New Method of Measuring Aniseikonia. *Amer. J. Ophthal.*, Vol. 34, pp. 1016–1021, 1951.

17. Brecher, G.A., Winters, D.M., and Townsend, C.A. Image Alternation for Aniseikonia Determination. *Amer. J. Ophthal.*, Vol. 45, pp. 253–258, 1958.

18. Baldwin, W.R. Binocular Testing and Distance Correction with the Berlin Polatest. *J. Amer. Optom. Assoc.*, Vol. 34, pp. 115–125, 1962.

19. Grolman, B. Vectographic Near Point Cards (Instructions). American Optical Corp., Buffalo, N.Y., 1971.

20. Ogle, K.N. *Researches in Binocular Vision*, p. 300. W. B. Saunders, Philadelphia, 1950.

21. Bannon, R.E. Aniseikonia and Binocular Vision. *Amer. J. Optom. Arch. Amer. Acad. Optom.*, Vol. 26, pp. 240–250, 1949.

22. Linksz, A., and Bannon, R.E. Aniseikonia and Refractive Problems. *International Ophthalmologic Clinics*, Vol. 5, No. 2, pp. 515–534. Little, Brown and Co., Boston, 1965.

23. Carleton, E.H., and Madigan, L.F. Relationships between Aniseikonia and Ametropia. *Arch. Ophthal.*, Vol. 18, pp. 237–247, 1937.

24. Sorsby, A., Leary, G.A., and Richards, M.J. The Optical Components in Anisometropia. *Vis. Res.*, Vol. 2, pp. 43–51, 1962.

25. Ogle, K.N. *Researches in Binocular Vision*, p. 264. W. B. Saunders, Philadelphia, 1950.

26. Ryan, V.I. Predicting Aniseikonia in Anisometropia. *Amer. J. Optom. Physiol. Opt.*, Vol. 52, pp. 96–105, 1975.

27. Rayner, A.W. Aniseikonia and Magnification in Ophthalmic Lenses. *Amer. J. Optom. Arch. Amer. Acad. Optom.*, Vol. 43, pp. 617–632, 1966.

28. Ogle, K.N., and Madigan, L.F. Astigmatism at Oblique Axes and Binocular Stereopsis Spatial Localization. *Arch. Ophthal.*, Vol. 33, pp. 116–127, 1945.

29. Guyton, D.L. Prescribing Cylinders: The Problem of Distortion. *Survey Ophthal.*, Vol. 22, pp. 117–188, 1977.

30. Bannon, R.E. *Clinical Manual on Aniseikonia*, p. 27. American Optical Corp., Buffalo, N.Y., 1954.

31. Brown, R.H., and Enoch, J.M. Combined Rules of Thumb in Aniseikonia Prescriptions. *Amer. J. Ophthal.*, Vol. 69, pp. 118–126, 1970.

32. Berens, C., and Bannon, R.E. Aniseikonia: A Present Appraisal and Some Practical Considerations. *Arch. Ophthal.*, Vol. 70, pp. 181–188, 1963.

## Questions

1. Given:

   OD −4.00 DS −2.00 DC x 180
   OS −3.00 DS −0.50 DC x 180
   +2.00 D add; D-25 bifocal

   The reading center is 10 mm below the distance center. What is the slab-off prism required to eliminate the vertical imbalance at the reading center?

2. A +5.00 DS add +2.00 D, 22-mm-wide, 16-mm-high straight top bifocal is made with the optical center of the distance lens and the top of the segment at the geometric center of the finished lens. If the lens is made with a 3 prism diopter slab-off, what is the total prismatic power 10 mm below the geometric center?

3. Given:

   OD −1.00 DS −0.75 DC x 180
   OS +1.00 DS −0.75 DC x 180
   +2.50 D add
   Segs 18 mm high
   Distance optical center located at geometrical center (GC)
   Reading level 10 mm below distance centers
   Lens size 46 x 40 mm

   If you decide to reduce the vertical imbalance at the reading level by using R-Compensated segs, what are the R segs that will come closest to leaving the patient with a residual vertical imbalance of 1 prism diopter?

4. A patient required the following lens correction:

   OD −0.75 DS −2.00 DC x 180
   OS +0.75 DS −1.00 DC x 180
   +2.00 D add
   Segs 20 mm high
   Distance optical center located at GC
   Reading level 10 mm below distance centers
   Lens size 50 x 44 mm

   You decide to prescribe different segment types for the eyes to reduce the vertical imbalance at the reading level. Which of the following pairs of segment types will come closest to eliminating the vertical imbalance *without overcorrection* at the reading level?

    (a) OD: Kryptok; OS: Ultex A
    (b) OD: Ultex A; OS: Executive
    (c) OD: 16 x 22 mm wide flat top; OS: Ultex A
    (d) OD: Kryptok; OS: Executive
    (e) OD: Executive; OS: Kryptok

5. Given:
    OD $-1.25$ DS $-1.75$ DC x 180
    OS $+1.50$ DS $-0.75$ DC x 180
    Refractive index = 1.50
    Distance optical center located at GC
    Reading level 8 mm below distance centers
    Lens size 46 x 40 mm

A slab-off prescription is used to totally correct the vertical imbalance at the reading level. If the slab-off line is 18 mm high, what is the difference in thickness of the upper and lower edges of the slabbed-off lens?

6. A lens clock is used to measure the amount of slab-off prism on a flat-top 22-mm bifocal lens with a correction of $+0.50$ DS $-1.00$ DC x 180, $+2.00$ D add and 5 prism diopters of slab-off. The base curve of the lens is $+6.50$ D. When the legs of the lens clock properly straddle the line perpendicularly, what is the clock dial reading?

7. Given:
    OD $-1.00$ DS
    OS $-4.00$ DS

The distance center is located at the GC. The reading level is 10 mm below the distance centers; the lens size is 46 x 40; the slab-off line is 16 mm high. If the left lens is slabbed off 3 prism diopters, where is the optical center of the slabbed-off portion located?

8. Given the following prescription:
    OD $+2.00$ DS $-0.75$ DC x 180
    OS $-0.50$ DS $-0.75$ DC x 180
    Reading level 10 mm below distance optical centers

The patient views a near object through these lenses. You measure a vertical phoria of 1.5 prism diopters right hyperphoria in this position. If you remove the lenses and repeat the measurement, what vertical phoria would you expect to find?

9. Given the following prescription:
    OD $-0.50$ DS $-0.75$ DC x 180
    PD $+2.00$ DS $-0.75$ DC x 180
    Reading level 10 mm below distance optical centers

The patient views a near object through these lenses. You measure a vertical imbalance of 2.5 prism diopters left hyperphoria in this position. If you remove the lenses and repeat the measurement, what is the expected vertical phoria?

10. A $+3.00$ DS $+1.00$ DC x 45 lens of ophthalmic crown glass with an index of refraction of 1.52 is made with minus cylinder construction, a center thickness of 3.0 mm, and a base curve of $+8.00$ D. What is the difference in magnification in the two principal meridians if the lens is fitted 18.0 mm from the entrance pupil?

11. A $-4.00$ D lens with an index of refraction of 1.50 is worn 15 mm from the entrance pupil. If the thickness is 6 mm, what must be the back surface power to produce a total spectacle magnification of 4%?

12. A $+8.00$ D lens with a front surface power of $+12.00$ D has an axial thickness of 4.5 mm. The lens is made of glass with a refractive index of 1.50 and the lens rests 15 mm from the entrance pupil. We wish to increase the total spectacle magnification of this lens to 20% (1.20) by changing the thickness. If the back surface power is held constant, what must be the thickness of the lens?

13. In order to produce a pair of plano lenses, $n = 1.50$, for a case of aniseikonia, the right lens is made on a base curve of $+6.00$ D (3.0 mm in thickness). If the left lens (3.00 mm in thickness) is to have a 1% greater overall magnification than the right lens, what would be the base curve of the left lens?

14. An aphakic patient who has been wearing contact lenses of the appropriate power decides to switch to spectacle lenses. A spectacle refraction done at a vertex distance of 12 mm is found to be $+14.00$ D. In switching from contact lenses to spectacles what is the *percentage change* in image size?

15. A patient who has been wearing contact lenses of $-10.00$ D decides to switch to spectacle lenses worn at a vertex distance of 12 mm. If we assume both lenses to be thin, what is the *percentage change* in retinal image size in switching to spectacle lenses from contact lenses?

16. If a lens with a back vertex power of $-10.00$ D is fit 3.00 mm closer to the face, how much would the magnification approximately change?

17. If a lens with a thickness of 3 mm is made with an increase in front surface curvature of 5.00 D, what is the approximate change in magnification?

**18.** If a lens with a front surface power of $+10.00$ D is made 3.00 mm thicker, how much is the magnification changed approximately?

**19.** A plus toric lens with a power of $-3.00$ DS $-2.00$ DC x 90 and a refractive index of 1.523 is worn 15 mm in front of the entrance pupil. The lens has a center thickness of 3.0 mm and the front surface power in the vertical meridian is $+4.00$ D. We wish to have the same spectacle magnification in the horizontal meridian as in the vertical meridian. What is the surface power required in the horizontal meridian of the front surface?

# CHAPTER ELEVEN

# Lenses for High Refractive Errors

Patients who have high refractive errors must contend with a number of problems that do not exist for the majority of spectacle wearers. The most obvious of these problems are the greater thickness and weight of the lenses. Perhaps less obvious is the fact that lenses of high refractive power are subject to aberrations, magnification properties, and visual field limitations that can be extremely annoying to the wearer.

The problem of lens weight tends to be greater for wearers of high plus lenses, whereas the problem of lens thickness tends to be greater for wearers of high minus lenses, because of the thick edges of these lenses. Magnification and visual field problems, on the other hand, exist to a much greater degree for wearers of high plus lenses, because of their increased magnification and the decreased extent of the visual field, than for wearers of high minus lenses.

If we consider those patients having refractive errors in excess of ±10.00 D, the largest group of patients in this category consists of aphakics—those individuals who have had one or both crystalline lenses removed, due to cataracts, and therefore require very strong plus lenses. High hyperopia (apart from aphakia) is relatively uncommon, very few people requiring lenses stronger than +4.00 or +5.00 D. As for myopia, although very few myopes are in the range of 10.00 D or greater, a considerable number of myopes require lenses stronger than −4.00 or −5.00 D.

Fortunately, the practitioner has at his disposal many methods of reducing lens weight and thickness, and controlling aberrations as well as magnification.

# LENSES FOR APHAKIA AND HIGH HYPEROPIA

*Aphakia* ("without lens"), a term proposed by Donders,[1] is used to indicate the absence of the crystalline lens from the dioptric system of the eye. The lens is usually removed by surgery; but in some cases aphakia results from trauma, the lens being lost due to perforation of the globe or displacement from the pupil (subluxation). Rarely, the lens may be congenitally subluxated or even congenitally absent.

## 11.1. Management of the Cataract Patient

As life expectancy has increased in recent years, the proportion of geriatric patients seen in optometric offices has correspondingly increased. One of the most common causes of acquired visual loss in older patients is cataract; and if an individual lives long enough, some form of cataract will invariably develop. Not surprisingly, when even a small amount of visual loss has been experienced, a patient may become apprehensive about the possibility of going blind.

Being told that he or she has a cataract is an emotional experience for many patients, raising imaginary as well as realistic expectations. The timing of this information is a delicate matter, and either of two approaches may be used. One approach is to notify the patient as soon as any sign of cataract development is discovered; the other (and usually the more desirable) approach is to mention the word "cataract" only when the opacity becomes a problem to the patient. Using this approach, the optometrist limits the use of the word "cataract" to lens opacities which produce uncorrectable visual loss. In a sense, the practitioner defines the word cataract in the way that appears to serve the patient the best.

If the word cataract were used as a label for *any* lens opacity, then it would have to be applied to any of the large variety of congenital and developmental lenticular opacities that are routinely seen in eyes of patients of all ages having no reduction in vision. Although these opacities could technically be termed cataracts, most practitioners note their presence on the record form, possibly telling the patient or the parent about the presence of a "small opacity," but not using the term cataract.

Senescent cataracts usually develop slowly, but their course is often unpredictable both in rate and in severity. Regardless of the rate of development, the time will arrive when the patient should be informed of the cataract's presence. Imparting this information involves a certain amount of counseling, including education concerning the nature of the cataractous process, and requires tact and sensitivity to individual differences.

Once senile lens changes have been found to be present, the practitioner should carefully monitor the visual loss and its effect, if any, on all the daily activities of the patient. Cataracts tend to affect *contrast sensitivity* more than they affect visual acuity,[2] with the result that a cataract patient's complaints may be related directly to the loss of contrast sensitivity. Although contrast sensitivity testing is still considered largely as a research tool, clinical methods of testing for this function, for example, by means of the Arden plates,[3] are available.

At some point the optometrist may decide to refer the patient to an ophthalmologist for consultation. It is a good idea for the patient to be seen by the ophthalmic surgeon long before the lens is completely opaque, so that ophthalmoscopic examination of the fundus will be possible. Otherwise, a successful cataract operation may result in little or no improvement in vision due to the presence of macular disease, such as age-related macular degeneration.

However, in those cases in which the patient is not seen by the optometrist or ophthalmologist until the lens is completely opaque, examination by laser interferometry can provide evidence of the integrity of the macula. The laser interferometer is mounted on a slit lamp, and the laser beam is sufficiently small and intense so that it will pass through what appears to be an opaque lens. Interference fringes produce line gratings on the retina, and by varying the frequency and orientation of the line gratings, a subjective measurement of visual acuity, independent of refractive error, is possible. If the patient demonstrates good acuity with laser interferometry, good acuity is expected to result after surgery. Unfortunately, laser interferometry will not penetrate a very dense cataract, with the result that the patient cannot see the red laser light. In such cases visual acuity following surgery cannot be predicted.[4] Other techniques of assessing the integrity of the retina include tests of color perception, tests of entoptic phenomena, two-point discrimination tests, and visually evoked potential (VEP).

The indications for cataract surgery are variable, depending not only on the state of the cataract and the degree of macular function but also on the patient's age, general health, visual acuity, occupation, and leisure-time needs. The characteristics of the cataract, such as the stage of development, location, and size, are important considerations. A hypermature cataract should usually be removed, even if unilateral, to prevent phacolytic glaucoma or phacoanaphylaxis.[5] In some cataracts, even in the early stages, the lens swells sufficiently to block the anterior chamber angle, putting the patient at risk for an attack of angle closure glaucoma. Therefore, a patient whose lens is swollen to the extent of interfering with the anterior chamber angle (as indicated by gonioscopy or by estimation of the angle width by means of the slit lamp) should be referred for possible surgery even though visual acuity is still relatively good.

Whereas it was formerly customary to perform cataract surgery at some arbitrary level of visual acuity, other criteria are also of importance. What is important in making the decision for cataract surgery is the patient's functional disability induced by the cataract. With bilateral cataracts, when a patient can no longer read or move about safely, a recommendation for surgery is usually accepted.

As a general rule, a patient who has bilateral cataracts will accept a recommendation for surgery when the corrected visual acuity with the better eye has reduced to approximately 20/50. The reason for this is that the reading of ordinary newsprint requires from 20/40 to 20/50 vision. However, a patient who does very little reading or other critical work may not require surgery until the corrected acuity in the better eye reaches 20/60 or worse, while a relatively young, active patient may have great difficulty with visual acuity in the 20/30 to 20/40 range. Some types of senile cataracts (particularly nuclear and posterior

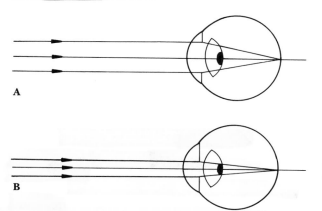

FIGURE 11-1. **When the pupil is reduced in size due to bright light, it may limit the bundle of rays available to the retina to the opaque portion of the lens, particularly in posterior subcapsular cataract: (A), large pupil; (B), small pupil, in bright light.**

subcapsular) can cause a severe visual handicap for distance vision, particularly on a sunny day, even though the patient can achieve reasonably good acuity in the practitioner's refracting room. There are two reasons for this: (1) Scattering of light by the opacities in the lens is a greater problem outdoors, in bright light, than it is indoors, and (2) as shown in Figure 11-1, the reduced pupil size in bright light tends to limit the bundle of rays available to the retina to the opaque portion of the lens (particularly in the case of a posterior subcapsular cataract, which is located very close to the nodal point of the eye).

The optometrist's responsibility does not end with the election to have surgery: He or she should be familiar with the current methods of cataract surgery and acquainted with the skills and professional philosophies of accessible surgeons. Candid discussion with the surgeon should establish joint responsibility for the management of the patient during the postsurgical period.

Prior to surgery, the practitioner will want to counsel the patient concerning the comparative advantages and disadvantages of the methods of visual correction to be used after surgery. This counseling session has the additional benefit of allowing the patient to understand the differences between normal and aphakic vision. In addition to correction with spectacle lenses, other options are contact lenses (either "hard" or "soft," and either daily wear or extended wear) and intraocular lenses. If an intraocular lens is to be used, the decision should be made prior to surgery, since the procedure is much less apt to result in complications if the lens is implanted at the time of the cataract surgery (a primary implant) rather than as a separate operation at a later time (a secondary implant). It has been estimated that intraocular lenses are now used in the majority of cataract operations.

Most patients expect some guidance from the optometrist in the selection of the type of aphakic correction that should be used, and the optometrist is in a better position to advise the patient if he or she has some understanding of the attitudes and capabilities of the patient. The practitioner should have an appreciation of the patient's physical abilities, visual requirements, tolerance of health aids, emotional stability, and ability and willingness to follow instructions. Of course, after presentation of all the facts, the final decision rests with the patient.

In bilateral cataract, surgery is normally done first on the eye with the more advanced cataract: In order to prevent both eyes being susceptible to complications at the same time, surgery on the second eye is done at a later date. When the cataract is *unilateral,* the decision whether to recommend surgery is more difficult. Unless the eye is at risk for angle closure glaucoma, a unilateral cataract is usually left alone until the visual acuity in the affected eye falls to the point that binocular vision is adversely affected. If good binocular vision is essential to the patient's normal activities, a unilateral cataract may be operated on at any stage, but the patient will be able to achieve binocular vision only by the use of a contact lens or an intraocular lens. On the other hand, if a unilateral cataract is operated on and spectacles, rather than a contact lens or an intraocular lens, are used, binocular vision will be impossible so the operated eye is usually not corrected. This allows suppression of vision for the operated eye.

The rapidly increasing popularity of intraocular lenses has been fostered, to some extent, by a belief on the part of many that aphakic spectacle lenses are thick and heavy, and difficult to adapt to because of the resulting aberrations and high magnifications. Although adaptation to aphakic lenses remains a problem (as indicated in the following sections), the student should understand that improvements in the design of aphakic lenses have occurred at a rapid pace during the past several years—during the same period of time in which both intraocular lenses and extended-wear contact lenses have become increasingly popular.

Unless an intraocular lens has been implanted, the patient is given a pair of single-vision spectacles immediately after surgery, for temporary wear. The power of these lenses is usually about +11.00 D. After a period of about 6 weeks, the eyes normally will have healed and the corneal curvature will have stabilized sufficiently so that permanent aphakic spectacle lenses can be ordered.

## 11.2. Optical Consequences of Cataract Surgery

There are many optical consequences of cataract surgery, even when no postoperative complications occur, which can affect the patient's visual performance. The cornea may be distorted, inducing against-the-rule or oblique astigmatism. The size, shape, and mobility of the pupil may be affected, inducing aberrations and altering the amount and distribution of light reaching the retina as well as the depth of focus of the eye.

The absence of the lens in the dioptric system of the eye is responsible for some significant changes in visual function. Since the lens is responsible for ac-

commodation, all accommodation is abolished and the patient is essentially an absolute presbyope (however, apparent accommodation has been claimed to take place in an aphakic patient[6]).

The removal of the crystalline lens also deprives the eye of about one-quarter of its refractive power, usually resulting in very high hypermetropia. Gullstrand's "exact" schematic eye has a refracting power of +58.64 D and an anterior focal length of 17.05 mm, whereas Gullstrand's schematic aphakic eye has a refracting power of +43.05 D (due to the cornea only) and an anterior focal length of 23.22 mm. The optical system of the aphakic eye is greatly simplified, essentially being reduced to a single spherical refracting surface (the cornea) separating air from the aqueous and vitreous, which may be regarded as a single medium having a uniform index of refraction.

Additional optical problems exist when aphakia is corrected with spectacle lenses. These will be discussed in Section 11.4.

## 11.3. Predicting the Power of an Aphakic Lens

Formulas have been proposed to predict the power of the spectacle lens required postoperatively in aphakia. Laurance[7] estimated that the power of the crystalline lens in the spectacle plane averages about +11.00 D, and suggested that the power of a spectacle lens required to correct an aphakic eye could be calculated by the use of the formula

$$F_{aphakic} = +11.00 \text{ D} + \tfrac{1}{2}F_{pre\text{-}aphakic}. \quad (11.1)$$

Thus, for an eye having 2.00 D of hyperopia before surgery, the required aphakic lens would be approximately +12.00 D, while for a pre-aphakic eye having 4.00 D of myopia the required aphakic lens would be approximately +9.00 D. Emmetropia would be achieved after surgery if the patient were myopic by 22.00 D. Sanders, Retzlaff, and Kraff[8] described a formula for predicting aphakic refraction based on linear regression analysis. Their formula is

$$F_{aphakic} = 80.4 - 1.65 L - 0.7 K, \quad (11.2)$$

where $L$ is the axial length of the eye, in millimeters, determined by ultrasonography, and $K$ is the preoperative keratometer finding, in diopters.

This formula correctly predicted the postoperative spectacle correction to within 1.0 D in 82% of cases (98 of 120 eyes), and within 2.0 D in 99% of cases (119 of 120 eyes).

## 11.4. Problems with Aphakic Spectacles

Of the three types of optical devices currently available for the correction of aphakia, spectacle lenses produce the most disturbing visual problems for the wearer: Unlike a contact lens or an intraocular lens, a spectacle lens must be mounted at an appreciable distance from the eye, and does not move when the eye rotates. Fortunately, most of the problems due to aphakic spectacles can be minimized, if not completely solved, by the careful selection of aphakic lenses and frames.

Inherent optical defects of convex lenses which go unnoticed in lenses of low power become so noticeable in high plus lenses as to cause a number of problems for the wearer. These problems include the following:

1. Increased retinal image size
2. Decreased field of view
3. Presence of a ring scotoma
4. Increase in ocular rotations
5. Increased lens aberrations
6. Motion of objects in the field of view
7. Appearance of the wearer
8. Demands on convergence

Each of these problems, and their solutions (to the extent that solutions are available), will be discussed.

### Increased Retinal Image Size

As discussed in Section 11.2, Gullstrand's schematic emmetropic eye has a refracting power of +58.64 D and an anterior focal length of 17.05 mm, while the aphakic schematic eye has a refracting power of +43.05 D and an anterior focal length of 23.23 mm. If a correcting lens is placed at the anterior focal point of the eye, the size of the retinal image is directly proportional to the anterior focal length and inversely proportional to the refracting power of the eye. That is,

$$\frac{\text{retinal image size in aphakia}}{\text{retinal image size in emmetropia}} = \frac{23.23}{17.05} = \frac{58.64}{43.05}$$

$$= 1.36,$$

or an increase in retinal image size of 36%.

In practice, the lens is placed much closer to the eye than 23.23 mm, and the closer distance will decrease the size of the retinal image; but even then the

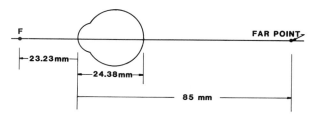

FIGURE 11-2. **Diagram for determining the power of the correcting lens for an aphakic eye, based on Gullstrand's schematic eye.**

retinal image size will be appreciably larger than it would be in the emmetropic phakic eye.

To find the power of the correcting lens required at the anterior focal point of an aphakic eye, the far point of the eye must be located. The length of Gullstrand's schematic eye is 24.38 mm. Therefore, with a refractive power of 43.05 D, the point conjugate to the retina, or the far point of the eye, may be found as follows (see Figure 11-2):

$$F = \frac{n'}{l'} - \frac{n}{l},$$

where $F$ = the refracting power of the eye, $n'$ = the index of refraction of the aphakic eye (1.336), $l'$ = the axial length of the eye (24.38 mm), $n$ = the index of refraction of air (1.0), and $l$ = the distance from the far point to the corneal vertex. Hence,

$$43.05 = \frac{1.336}{0.02438} - \frac{1}{l}, \qquad 43.05 = 54.80 - \frac{1}{l},$$

and

$$l = \frac{1}{11.75} = 0.085 \text{ m} = 85.0 \text{ mm}.$$

Therefore, the point conjugate to the retina is a virtual object point located 85.0 mm behind the cornea. The distance from this point to the anterior focal point of the eye is

$$85.0 + 23.23 = 108.23 \text{ mm}.$$

The lens needed at the anterior focal point, to correct this aphakic eye, has a power of

$$\frac{1}{0.10823} = +9.24 \text{ D}.$$

If the correcting lens is placed at a vertex distance of 12 mm rather than at the anterior focal point of the eye, the required power would then be

$$\frac{1}{0.085 - 0.012} = \frac{1}{0.097} = +10.31 \text{ D}.$$

The equivalent power of a thin lens having a power of $+10.31$ D and a $+43.05$ D cornea separated by a distance of 12 mm is equal to

$$F_e = F_1 + F_2 - \frac{t}{n}F_1F_2$$

$$= +10.31 + 43.05 - 0.012(10.31)(43.05)$$

$$= +48.03 \text{ D}.$$

The retinal image size for this aphakic eye would, therefore, be equal to

$$\frac{58.64}{48.03} = 1.22,$$

or 1.22 times larger than the retinal image of the schematic emmetropic eye, an increase of 22%.

Up to this point, the correcting spectacle lens was considered to be infinitely thin. It is obvious that a lens of approximately $+10.00$ D, having a diameter of 44 mm or more, is not a thin lens. It is also clear that even when the back surface power of the lens is relatively low (about $-3.00$ D), the front surface power must be rather high in a lens of this power. Since the formula for the shape factor is

$$\frac{1}{1 - \frac{t}{n}F_1},$$

it is apparent that both the large thickness value and the steeply curved front surface tend to increase the magnification. For a lens that is 6 mm thick, with a front surface power of $+12.00$ D, the shape factor would be approximately

$$\frac{1}{1 - \frac{0.006}{1.5}(+12)} = \frac{1}{1 - 0.048} = 1.05.$$

Given the magnitude of the shape factor for high plus lenses, it cannot be ignored. The total spectacle magnification of the $+10.31$ D lens with a front surface power of $+12.00$ D and a center thickness of 6.0 mm, placed at a vertex distance of 12 mm, is therefore equal to

$$SM_T = (1.22)(1.05) = 1.28, \text{ or } 28\%.$$

In bilateral aphakia, if both eyes somewhat resem-

ble Gullstrand's emmetropic eye, the retinal image sizes for the two eyes may be essentially the same, but each would be magnified about 28% as compared to the retinal image of the emmetropic eye. This greatly increased magnification means that the aphakic patient must adapt to new size–distance relationships in his or her daily environment: Not only do familiar objects appear to be much larger, they also appear to be much closer. The unilateral aphakic patient will have a more serious problem, since the magnification differences between the two eyes will make binocular vision impossible with spectacle lenses. This matter will be discussed further in Section 11.8. Ironically, the same magnification that causes so many problems when aphakia is corrected by a spectacle lens sometimes enables the aphakic patient to achieve central visual acuity that exceeds the best visual acuity obtained prior to surgery. This single positive outcome of the greatly increased magnification may permit the prescribing of a weaker reading addition than that for the phakic presbyope of the same age, therefore extending the range of vision through

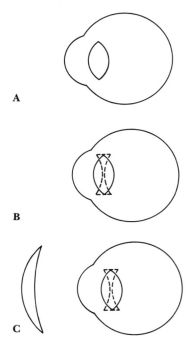

FIGURE 11-3. **The magnification effects of an aphakic eye: A, a normal emmetropic eye, prior to removal of the crystalline lens; B, the optical result of the removal of the lens may be represented as an infinitely thin minus lens inserted at the center of the crystalline lens; C, the combination of a convex spectacle lens and the hypothetical concave lens forms a Galilean telescopic system for an otherwise normal eye. (Based on an explanation by K. N. Ogle,** *Researches in Binocular Vision,* **Saunders, 1950, p. 258, by permission of Mayo Foundation.)**

the bifocal segment and reducing the convergence demand.

Ogle[9] has offered a unique explanation of the magnification effects in an aphakic eye. Figure 11-3A represents a normal emmetropic eye in which the crystalline lens has not been removed. The optical result of the removal can be represented as though an infinitely thin minus lens had been inserted at the center of the crystalline lens (Figure 11-3B). The resulting hyperopia is then corrected by a convex spectacle lens placed before the eye at a specified distance from the hypothetical minus lens (Figure 11-3C). The combination of the convex spectacle lens and the hypothetical concave lens forms an afocal Galilean telescopic system for an otherwise normal eye.

The angular magnification of a Galilean telescope, as developed in Section 12.2, is

$$M = -\frac{f_1}{f_2} = -\frac{F_2}{F_1}, \qquad (11.3)$$

where $F_1$ is the power of the objective lens (the correcting lens, in this case) and $F_2$ is the power of the eyepiece (the hypothetical minus lens).

For an afocal Galilean telescope, the equivalent power is

$$F_e = F_1 + F_2 - \frac{t}{n}F_1F_2 = 0,$$

or

$$-F_1 = F_2 - \frac{t}{n}F_1F_2$$

$$-F_1 = F_2\left(1 - \frac{t}{n}F_1\right).$$

Substituting for $F_1$ in Eq. (11.3),

$$M = \frac{F_2}{F_2\left(1 - \dfrac{t}{n}F_1\right)} = \frac{1}{\left(1 - \dfrac{t}{n}F_1\right)}.$$

But since $n = 1$,

$$M = \frac{1}{1 - tF_1},$$

where $t$ is the distance between the correcting lens and the hypothetical thin minus lens at the center of the crystalline lens.

If the average distance from the correcting lens to the center of the crystalline lens is 17.5 mm, and if the

power of the correcting lens is +10.50 D, then

$$M = \frac{1}{1 - 0.0175(10.50)}$$

$$= \frac{1}{1 - 0.184} = \frac{1}{0.816}$$

$$= 1.23,$$

which is essentially the same magnification as that obtained by using the spectacle magnification formula.

*Decreased Field of View*

The base-to-the-center prismatic effect of a high plus lens (see Figure 11-4) reduces the size of the field of view through the lens, just as the base-to-the-edge prismatic effect increases the size of the field of view through a minus lens.

In discussing the field of view of a spectacle lens, we can think in terms of either (1) the peripheral field of view or (2) the macular field of view, or field of fixation. The peripheral field of view is the field of view for the steadily fixating eye, subtended at the entrance pupil (see Figure 11-5), and is given by the equation

$$\tan \phi = y(E - F), \qquad (11.4)$$

where $\phi$ = one-half of the angular field of view, $y$ = one-half the lens aperture, in meters, $E$ = the vergence of light at the entrance pupil of the eye, and $F$ = the power of the correcting lens.

On the other hand, macular field of view is the field of view for the moving eye, subtended at the center of rotation of the eye (Figure 11-6), and is given by the equation

$$\tan \theta = y(S - F) \qquad (11.5)$$

where $\theta$ = one-half of the angular field of view, $y$ =

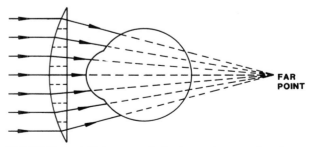

FIGURE 11-4. **A high-powered plus lens acts as a prism, base toward the center, whose power gradually increases toward the periphery.**

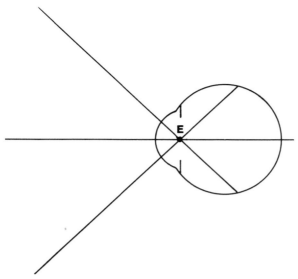

FIGURE 11-5. **The peripheral field of view for a steadily fixating eye, subtended at the center of the entrance pupil.**

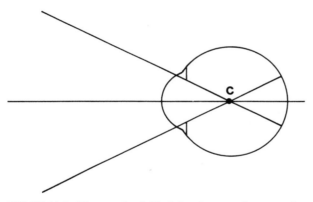

FIGURE 11-6. **The macular field of view for a moving eye, subtended at the center of rotation of the eye.**

one-half the lens aperture, in meters, $S$ = the vergence of light at the center of rotation of the eye, and $F$ = the power of the correcting lens.

The relationship between the field of view and the field of fixation can be demonstrated by the use of an example.

### EXAMPLE

Consider an aphakic patient who requires a +12.50 D spectacle lens having a horizontal width of 50 mm, fitted at a distance of 13 mm from the entrance pupil of the eye, and 25 mm from the center of rotation of the eye. What are the peripheral and macular fields of view?

Peripheral field of view:

$$\tan \phi = 0.025(76.92 - 12.50) = 1.6106$$
$$\phi = 58.165°$$
$$2\phi = 116.33°.$$

Macular field of view:

$$\tan \theta = 0.025(40.00 - 12.50) = 0.6875$$
$$\theta = 34.51°$$
$$2\theta = 69.02°.$$

These fields of view may be compared to those for the same eye in the pre-aphakic condition by assuming that the refractive error before surgery was +2.00 D. By substituting +2.00 for +12.50 D in each of the above equations, the results for the pre-aphakic condition would be:

Peripheral field of view:

$$\tan \phi = 0.025(76.92 - 2.00) = 1.873$$
$$\phi = 61.90°$$
$$2\phi = 123.80°.$$

Macular field of view:

$$\tan \theta = 0.025(40.00 - 2.00) = 0.950$$
$$\theta = 43.53°$$
$$2\theta = 87.06°.$$

A comparison of these results with those for the +12.50 D lens shows that the aphakic patient suffers more from loss of macular field of view than from loss of peripheral field of view: The loss of peripheral field is only about 7 degrees, whereas the loss of macular field is 17 degrees. This loss of both peripheral and macular field of view when wearing a high plus lens is responsible for the *ring scotoma* (to be discussed below).

Inspection of Eq. (11.5) shows that the macular field of view can be made larger by increasing the size of the lens (increasing the value of $y$) or by fitting the lens closer to the eye (decreasing the stop distance and therefore the value of $S$).

### Presence of a Ring Scotoma

The presence of a ring scotoma, seen by a wearer of an aphakic lens, may be understood with the help of Figure 11-7. As described previously, the base-toward-the-center prismatic effect of a strong plus lens causes an angular "gap" in object space, all the way around the lens, where the patient sees nothing either through the lens or beyond the lens unless a head movement is made. Beyond this scotomatous area the wearer has uncorrected vision, while central to the scotoma he has, of course, corrected vision.

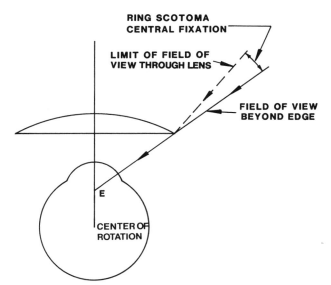

FIGURE 11-7. **Ring scotoma occurring in central fixation (subtended at the center of the entrance pupil) for an aphakic eye.**

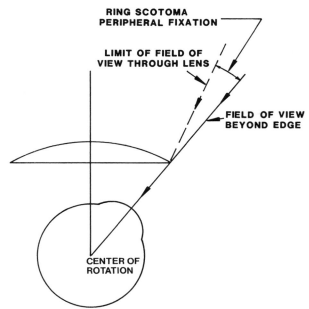

FIGURE 11-8. **Ring scotoma occurring in peripheral fixation (subtended at the center of rotation) for an aphakic eye. Note that the scotoma has moved inward, as compared to the scotoma for central fixation shown in Figure 11-7.**

The angular size of the gap is equal to the prismatic effect present at the edge of the lens. For example, if an aphakic patient wears a +12.00 D lens, 44 mm wide, centered in front of the center of the pupil, the prismatic effect at the edge of the lens will be

$$2.2(12) = 26.4 \text{ prism diopters},$$

and the angular extent of the scotoma is approximately 15 degrees.

As the eye rotates, the scotoma moves, hence the term "roving ring scotoma." As shown in Figure 11-8, the scotoma moves in the direction *opposite* to the direction in which the eyes are moved; as the eye rotates outward, the scotoma moves centrally (inward), and vice versa. The term "jack-in-the-box phenomenon" has been used to refer to the fact that an object will jump in and out of the field of view as it moves out of and into the ring scotoma. This event can also occur with a stationary object that can be seen in indirect vision at the periphery of the object field, but the object vanishes into the roving ring scotoma when attempts are made to fixate it directly by rotating the eye. The "jack-in-the-box phenomenon" also occurs when the eyes are held stationary: An object moving across the field of vision may disappear into the ring scotoma and then appear on the other side of the scotoma, apparently popping out of nowhere.

The roving ring scotoma causes little difficulty at a normal reading distance of 40 cm (where it is only 15 degrees wide, extending from 35 to 50 degrees), and

at a distance of 20 feet or beyond (as when driving a car), the central field is sufficiently wide to allow adequate vision with the result that the scotoma presents no problem. But at *intermediate distances*, especially between about 2 and 10 feet, the ring scotoma often creates a difficult problem. Thus, when the wearer is in an ordinary room, he or she experiences a relatively small area of clear vision enclosed by a wedge of blindness. Moreover, when the eyes move, this wedge-shaped ring of blindness moves also, so that objects (as well as people) suddenly appear and disappear. As the eyes move peripherally to fixate an object of attention, the ring scotoma moves in the opposite direction, so the wearer does not immediately see the face of another person in the room, clumsily bumps into furniture when moving about, knocks over ashtrays and drinking glasses, stumbles on uneven ground, or misses a step on a flight of stairs with disastrous results. The problems experienced by the spectacle-wearing aphakic are accurately (and amusingly) described in an article by an anonymous aphakic physician.[10]

The following factors have an effect on the size and the position of the ring scotoma:

1. *Lens Power.* The stronger the lens, the larger the ring scotoma.

2. *Vertex Distance.* The farther the lens from the eye, the smaller the ring scotoma; but, unfortunately, moving the lens farther out makes the ring scotoma move centrally. Therefore, the closer the vertex distance the less bothersome the ring scotoma will be.
3. *Lens Size.* As lens size increases, the ring scotoma increases in size, but it also moves toward the periphery. Increasing the lens size therefore has the same general effect as moving the lens closer to the eye.
4. *Pupil Size.* The smaller the pupil, the larger the ring scotoma.
5. *Lens Thickness.* As the lens thickness increases, so does the magnification and the size of the ring scotoma.
6. *Base Curve.* With increasing steepness of the base curve, both the magnification and the size of the ring scotoma increase.

### Increase in Ocular Rotations

When a spectacle lens is worn, the angle the eye must turn in changing fixation from one object point to another increases or decreases in comparison with the angle that would be required without the lens.

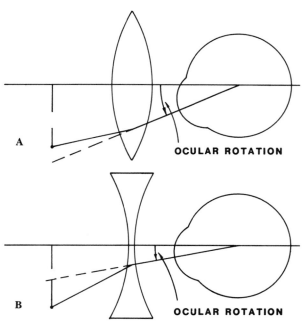

**FIGURE 11-9. Effect of a spectacle lens on the ocular rotation required to fixate a peripheral object: A, increased for a plus lens; B, decreased for a minus lens.**

The excursions of the eye are increased for plus lenses (Figure 11-9A) and decreased for minus lenses (Figure 11-9B). The increase or decrease in ocular rotation that occurs with corrective lenses is due to the presence of *prismatic effects*, the amount of prismatic deviation depending on the power of the lens and the angular separation of the two alternately fixated points.

In plus lenses of the power required for the correction of aphakia, the increase in the amplitude of ocular excursions is significant. Numerically, the increase in ocular rotation exceeds the magnification of the retinal image: This is because the reference point for angular magnification is the entrance pupil of the eye, whereas the reference point for ocular excursions is the center of rotation of the eye.

### Increased Lens Aberrations

As discussed in Chapter 6, any ophthalmic lens is subject to aberrations which degrade the optical image, but for plus lenses in the powers needed for the correction of aphakia, these aberrations can be quite severe. It will be recalled that these aberrations include:

1. Spherical aberration
2. Coma
3. Oblique astigmatism (or marginal astigmatism)
4. Curvature of image (power error, or mean oblique error)
5. Distortion
6. Chromatic aberration (both axial and transverse)

Some aberrations degrade the image more than others. Spherical aberration and coma occur only for large aperture systems, and are ignored by lens designers. In any event, aphakic patients, most of whom have small pupils, will be affected very little by these aberrations. The off-axis aberrations of oblique astigmatism and curvature of image (mean oblique error) are significant errors that can affect the patient's visual acuity and contrast sensitivity. They are important aberrations, and are amenable to correction.

As discussed in Chapter 6, in order to eliminate oblique astigmatism, one may plot a family of *Tscherning ellipses.* For each ellipse, the front or back surface power is plotted as a function of the total refracting power of the lens. The front or back surface required for a particular lens power can be determined by visual inspection of such an ellipse. Similar ellipses can be constructed for zero power error (zero mean oblique error) in which the circle of least confusion of the image formed by the lens falls on the eye's far-point sphere, and ultimately on the retina.

The limits of these ellipses are approximately −23.00 D and +8.00 D, depending on a number of factors including (a) whether the lens is designed for zero oblique astigmatism or for zero mean oblique error, (b) the refractive index of the lens, (c) the object distance, and (d) the stop distance (the distance from the lens to the center of rotation of the eye). Outside these limits, neither oblique astigmatism nor mean oblique error can be eliminated if spherical surfaces are used. However, they can be minimized by the judicious selection of spherical surfaces.

For plus lenses above +8.00 D, either oblique astigmatism or power error can be eliminated by making at least one of the surfaces aspheric, but both cannot be eliminated with the same choice of curves. Thus, if oblique astigmatism is eliminated, substantial power error exists; and if the power error is eliminated, substantial oblique astigmatism is present. Hence, the lens designer must consider these two aberrations together, attempting to reach a balance between them. The same problem exists when using spherical surfaces for lens powers falling within the range of Tscherning's ellipse.

As discussed in Section 6.20, *aspheric surfaces* are rotationally symmetrical surfaces whose cross sections are noncircular. It should be noted that *toric* surfaces form circular cross sections in their principal meridians, and are therefore not considered as aspheric. Most aspheric surfaces used for ophthalmic lens design are *conicoid* surfaces, obtained by rotating a conic section (ellipse, parabola, or hyperbola) about an axis of symmetry of the conic. The *circle* is considered to be a special case of the ellipse.

Smith and Bailey[11] have presented the following mathematical formulation of the conic section:

$$C[Y + (1 + Q)x^2] - 2x = 0$$

where x is the axis of revolution, C is the curvature of the surface at the origin, or vertex (0,0), and Q is a quantity that specifies the type of conic, its shape, and degree of asphericity. The type of conicoid can be determined from the value of Q, as follows:

| | |
|---|---|
| $Q < -1$ | Hyperboloid |
| $Q = -1$ | Paraboloid |
| $-1 < Q < 0$ | Ellipsoid with the x-axis as the major axis |
| $Q = 0$ | Sphere |
| $Q > 0$ | Ellipsoid with the y-axis as the major axis. |

For ophthalmic lenses, it is convenient to designate the *x-axis* as the optical axis and C as the curvature in the paraxial region.

Currently there are two different approaches to the design of aspheric surfaces for ophthalmic lenses. One approach is to manufacture the surface as a *continuous* aspheric surface, as done with the American Optical *Fulvue* lens. The curvature of such a surface reduces continuously from the center toward the edge. This curvature reduction results in a drop in power toward the edge, as shown in Figure 11-10.

In the other approach, the lens surface is made up of a series of different zones arranged in an annular pattern about the center, with each zone being essentially spherical but with progressively decreasing surface power, the farther the zone is from the center. In order to eliminate any obvious dividing lines on the surface, the tangents to the curves of adjacent zones are arranged to coincide to the boundary between adjacent zones, and the junctions between adjacent zones are smoothed by polishing with flexible pads. Such a surface is known as a *zonal* aspheric. The Armorlite *Multi-Drop* lens, previously known as the Welsh *Four-Drop* lens, is an example of this design. Zonal aspherics can be ground and finished with tools used for working spherical surfaces. The surfaces of the glass molds used for casting the front surfaces of CR-39 aspheric lenses are formed by a series of spherical generator cuts.

The continuous aspheric surface can be analyzed by conventional third-order aberrational theory, if a mathematical description of the surface is available. It is difficult to analyze the zonal aspheric surface unless the zonal surface powers and the outer and inner zonal diameters are known and the radial surface powers of the zones are constant.

Smith and Atchison[12] have shown that for a lens having an aspheric front surface in the form of a conicoid, the third-order solution for *oblique astigmatism* is a cubic equation. Therefore, for any front

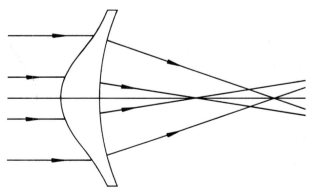

FIGURE 11-10. **The reduced curvature toward the periphery of an aspheric lens results in a drop in refracting power.**

surface asphericity and equivalent lens power, there are one, two, or three solutions for the back surface power for which oblique astigmatism is essentially zero. The results for the elimination of third-order *oblique error* (power error) show the same trends as those for the elimination of third-order oblique astigmatism. However, in both cases, the back surface power is often too steep for the solutions to be of practical value.

Distortion, a term frequently misused in describing undesirable optical effects, results from changing magnification across the field of view of a lens. With distortion, the shape and/or the size of an object is not faithfully reproduced. For an aphakic patient wearing high plus lenses, distortion creates the familiar *pincushion* effect, causing straight lines to look curved; and the apparent location of an object changes as the observer's eye turns to a different portion of the lens. This happens because of magnification changes. Distortion causes some of the "navigational" difficulties so vividly described in the literature by aphakic patients. Control of distortion requires steeply curved lenses, which would be unsightly and difficult to manufacture. Fortunately, design factors that minimize oblique astigmatism and power error (mean oblique error) also decrease distortion. There is some evidence that newly corrected aphakic patients can adapt quickly to some lens distortion.[13]

Axial and transverse chromatic aberration cannot be eliminated when only a single refractive medium is used. Transverse chromatic aberration is the more bothersome of the two, and occurs because the image locations (due to axial chromatic aberration) and magnifications differ for different wavelengths. It affects primarily the tangential meridian of oblique astigmatism. Therefore, by controlling the tangential error, the effect of transverse chromatic aberration can be attenuated.

Hence, the aberrations of oblique astigmatism and curvature of image are the most important, and have been the main targets of the lens designer. Secondary attention has been directed toward the correction of distortion, which is still a significant aberration in current aspheric lens designs (although considerably less so than in a lens having spherical curves on both the front and back surfaces).[14]

### Motion of Objects in the Field of View

When a wearer of high plus lenses holds the eyes steady and moves the head to look at any object not in direct view, a marked reversed motion of the field of view (an "against" motion) is experienced.

This motion is often referred to as "swim." It occurs when the field from which the visual axes are moving away is seen in indirect vision, through progressively peripheral portions of the lens with the prismatic base in the direction of the head movement (see Figure 11-11).

"Swim" can be avoided by moving the eyes from one fixation point to another while the head is held stationary. However, this action may produce distortion effects. The best procedure is to turn the head slowly, so that the head and eyes move simultaneously.

### Appearance

Among the strong objections to cataract lenses has been the matter of appearance. The appearance of an aphakic spectacle wearer may convey two impressions. The first impression is that of the apparent enlargement of the eyes and their unusual appearance behind bulbous, thick lenses. The second impression is the general awkwardness of the patient's bearing and movements. Both of these impressions originate from the magnification and distortion effects of the lenses. The high plus lens required for the aphakic eye causes a large amount of angular magnification, no matter in which direction one looks through the lens. The magnification and distortion of the image experienced by the patient alter the behavior in the environment to the extent that the patient's bearing seems awkward to others.

Aphakic lenses of *lenticular* construction present a particularly bothersome appearance. Such a lens provides the patient's prescription only in the central portion of the lens, as shown in Figure 11-12; the peripheral portion is thin and light in weight and is not optically useful, but serves as a carrier for the central portion. These lenses provide good optical performance in the central portion, which is usually 30 to 40 mm in diameter, and are lighter in weight than full-field lenses. However, they have a poor cosmetic appearance. The magnification is confined to the central portion of the lens, and produces a typical "bull's eye" or "fried egg" appearance. This identifies the wearer as an individual who has a very serious eye defect.

### Demands on Convergence

If the distance centers of aphakic lenses are placed in front of the centers of the wearer's pupils, convergence of the visual axes toward a near fixation point

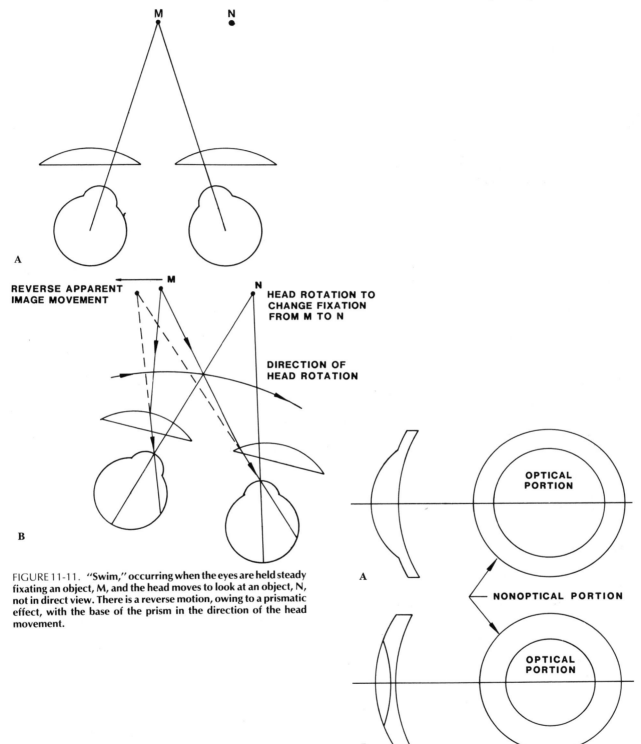

FIGURE 11-11. "Swim," occurring when the eyes are held steady fixating an object, M, and the head moves to look at an object, N, not in direct view. There is a reverse motion, owing to a prismatic effect, with the base of the prism in the direction of the head movement.

FIGURE 11-12. **Lenticular aphakic lenses: A, one piece (used for plastic lenses); B, fused (used for glass lenses).**

creates a *base-out* prismatic effect. Because of this base-out effect, the eyes must converge, when reading, considerably more than would be the case for a wearer of lenses of low power. This can place a strain on positive fusional vergence and may lead to discomfort.

This effect can be reduced if the distance centers are placed closer together than the patient's distance PD. This produces a slight base-in effect at distance, but relieves the load on positive fusional vergence at near. This practice is particularly recommended when the patient is exophoric at distance.

## 11.5. Parameters of Aphakic Lenses

In addition to the unique optical phenomena and effects that occur with aphakic lenses of high power, there are lens specifications (including bend and asphericity) and wearing parameters of spectacles that influence the dioptric value and the visual efficacy of the prescription. These factors can alter the prescription by significant amounts and demand the same meticulous attention to detail as do the cataract surgery, the refractive technique, and the lens design. These critical lens specifications and fitting parameters include vertex distance, pantoscopic tilt, centration, lens weight and thickness, and adjustment of the spectacles.

### Vertex Distance

Since most aphakic lenses are of very high plus power, even a small variation in the position of the lens before an aphakic eye produces a substantial change in the optical effect. In order to correct the ametropia properly, aphakic lenses should be fitted at the vertex distance that was used during the refraction. If the vertex distance must be different because of the structure of the patient's face or because of the style of the frame, compensation in the power must be made, using the effective power formula (see Section 3.6).

The reader will recall that if a plus lens is moved toward the eye, the secondary focal length must be shortened by the distance the lens has been moved, thus making the lens stronger. Of course, if the plus lens is moved farther from the eye, the secondary focal length must be lengthened by the distance moved, thus making the lens weaker. For example, if a +12.00 D lens is moved 3 mm closer to the eye, the new focal length of the lens must be

$$\frac{1}{12} - 0.003 = 0.083 - 0.003 = 0.080 \text{ m},$$

and the power of the lens is

$$\frac{1}{0.08} = +12.50 \text{ D}.$$

For a plus lens, an identical amount of movement closer to the eye causes a greater refractive change than when the lens is moved farther from the eye.

Not only does the effective power of the lens change when the position of the lens is changed, but the size of the retinal image changes also. With a plus lens, the image is made smaller as the vertex distance is decreased, even though the power of the lens must be made stronger. We have shown in Section 11.4 that, in an aphakic eye, the size of the retinal image is 36% larger than that in the normal eye if the correction (+9.34 D) is placed at the anterior focal point of the aphakic eye. However, if the lens is moved to 12 mm from the cornea, the power of the lens must now be +10.31 D, but the size of the retinal image is now only 22% larger than that for the normal emmetropic phakic eye.

### Pantoscopic Tilt

All lenses perform best and provide the widest field of view with minimum aberrations when the optic axis of the correcting lens passes through the center of rotation of the eye. All lens designers assume that lenses will be worn so that this occurs. This means that when the visual axis of the eye passes through the optical center of the lens, the visual axis and the optic axis of the lens will coincide.

Usually, spectacle lenses are tilted around a horizontal axis to improve their appearance on the wearer's face. The amount of tilt is typically 8 or 10°, as discussed in Section 2.19 and illustrated in Figure 2-27; and the optical center should be lowered 0.5 mm for every degree of pantoscopic tilt in order for the optic axis to pass through the center of rotation of the eye. However, in aphakic lenses we must *reverse* this procedure. In order to minimize thickness and weight, the optical center of the lens should fall at the geometrical center; and the optic axis should then be made to pass through the center of rotation by altering the vertical position of the frame on the face and by adjusting the pantoscopic tilt. The reversed procedure provides minimum weight and thickness of the lens in addition to proper optical performance.

If the optic axis of the lens does not pass through the center of rotation, then the visual axis will be oriented obliquely to the optic axis when looking through the optical center, with the result that the effective power of the prescription will be changed. With lenses of low power, the induced changes are very small, but with aphakic lenses of high power, the induced changes are significant.

Because of the fact that the correct prescription is found when the lenses are verified by means of a lensometer, the induced changes are often overlooked as a source of trouble. Formulas for calculating the induced changes when the optic axis of the lens fails to pass through the center of rotation of the eye are given in Section 2.20.

### Positions of Optical Centers

Special care must be taken to determine the distance and near PD accurately. Small errors in PD measurements produce significant effects when working with aphakic lenses. The monocular PD should be taken, preferably by a method of corneal reflection, to determine the positions of the visual axes. The visual axis does not usually pass through the center of the pupil, even in phakic eyes, and this may be especially true when, as a result of cataract surgery, the pupil is irregular or deformed.

### Weight and Thickness

Lenses for the correction of aphakia are thick and heavy. Weight can be reduced by using CR-39 plastic, but because of the lower index of refraction, thickness may increase slightly. Thickness can be reduced by using high-index glass, but the weight remains about the same and the chromatic aberration almost doubles. The use of lenticular aphakic lenses reduces both thickness and weight, but these lenses are not acceptable to some patients because of their unattractive appearance. Lens thickness and weight can be minimized by using frames with symmetrical (and small) eye sizes, placing the optical center at the geometrical center of the lens, and ordering the lenses with essentially knife-edge thickness.

### Spectacle Adjustments

It is often difficult to maintain aphakic spectacles in the proper position. Being heavy, they tend to slip down on the nose. Once the spectacles are removed from the face, the patient has difficulty seeing them, and, in groping for them, they are often knocked to the floor. They rarely break, but are easily bent out of adjustment.

Small changes in position and alignment introduce large errors which greatly reduce the visual efficiency of an already precarious optical system. Frequent office visits should be scheduled, so that the frames can be realigned and readjusted in order to provide the patient with optimum vision and comfort.

## 11.6. Aphakic Lens Design

The design of spectacle lenses for aphakia has seen a number of improvements during the past decade. These design improvements have occurred both as a result of the availability of new lens materials and manufacturing processes and as a result of a better understanding on the part of lens designers of the optimum performance requirements for lenses of high plus power. It is ironic that these important advances in aphakic spectacle lens designs have occurred in tandem with the rapidly increasing popularity of extended-wear contact lenses and intraocular lenses for the correction of aphakia. Advances in aphakic lens design can be considered in terms of the achievement of the following design goals:

1. Reducing lens weight
2. Reducing lens thickness
3. Reducing spectacle magnification
4. Increasing the field of view
5. Minimizing lens aberrations
6. Choice of optimum multifocal style and position
7. Protection from "glare" and ultraviolet radiation

Each of these design goals, and methods used to achieve them, will be discussed.

### Reducing Lens Weight

This design goal is undoubtedly the easiest one to achieve, simply by using plastic rather than glass for the manufacture of aphakic lenses. Since CR-39 material has a specific gravity of 1.32 compared to 2.54 for crown glass, the use of this material provides a saving of lens weight close to 50%. Although newer plastic materials (such as polycarbonate, having a higher index of refraction and a lower specific gravity than CR-39) are available, almost all currently available aphakic lenses are made of CR-39 material. An additional method of reducing lens weight involves the restriction of frames to small eye sizes, making possible the use of lenses of less weight.

### Reducing Lens Thickness

Although lens thickness is to some extent a cosmetic problem, a reduction in lens thickness not only reduces lens weight but, as already noted, reduces the magnification brought about by the lens.

#### LENTICULAR-APHAKIC LENSES

As described in Section 11.5, an early method of reducing lens thickness was the use of a "lenticular" form, the optical zone of the lens having a diameter of 30 to 40 mm, being surrounded by a nonoptical "carrier" lens. For many years, glass aphakic lenses were made in lenticular form, and some plastic lenses are also made in this form. The disadvantage of a lenticular lens is that it gives the wearer's eyes a "bull's eye" or "fried egg" appearance, calling attention to the visual anomaly.

#### FULL-DIAMETER ASPHERIC LENSES

Most of the full-diameter plastic lenses for aphakia are currently made with an aspheric front surface. This construction not only reduces the effects of aberrations but also reduces lens thickness and weight. In order to keep lens thickness to a minimum, frames with large eye sizes must be avoided. Davis and Torgersen[15] have pointed out that more thickness and weight reduction occurs by decreasing the eye size as little as 2 mm than by making a spherical lens aspheric and retaining the same size.

### Reducing Spectacle Magnification

As shown by inspection of the spectacle magnification formula [equation (10.8)], magnification can be reduced by reducing lens thickness or by flattening the front curve of the lens. Atchinson and Smith[16] found that "flatback" aphakic lenses having aspheric front surfaces will reduce spectacle magnification and look more attractive compared to lenses with steeper forms. However, the amount of asphericity (power drop) seems insufficient to markedly improve peripheral visual performance. Reducing vertex distance will also reduce spectacle magnification,

and, as already noted, will move the ring scotoma farther out, making it less bothersome.

Benton and Welsh[17] have recommended what they refer to as "minimal effective diameter" (MED) fitting for aphakic lenses. This simply means that the frame size (eye size plus DBL) is chosen such that the frame PD is equal to the patient's distance PD, so that no decentration is required and the lenses can be ground with knife-thin edges. As shown in Figure 11-13, if an aphakic patient is fitted with a modern large-size frame requiring several millimeters of inward decentration, the center thickness of the lenses will have to be considerably greater than if no decentration were necessary. In the event that the frame PD cannot be made to exactly match the patient's distance PD, it is usually best to select a frame PD that is slightly *narrower* (rather than wider) than the patient's PD, and not to decenter the lenses: This will result in the creation of base-in prism, which will help to reduce the expected exophoria at near.

### Increasing the Field of View

The field of view can be increased by fitting the lens at as close a vertex distance as possible. Increasing the lens size also helps (but of course adds thickness, weight, and magnification). As already mentioned, reducing vertex distance and increasing lens size both increase the size of the ring scotoma, but this is of little concern since the scotoma is moved farther out into the periphery. The field of view will also be increased by the use of an aspheric front surface, which reduces the power in the periphery of the lens.

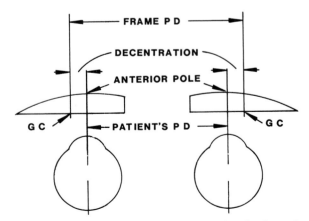

FIGURE 11-13. **If an aphakic patient is fitted with a large-size frame requiring inward decentration, the thickness of each lens (both at the center and at the nasal edge) will be greater than if the lenses were not decentered. (GC = geometrical center.)**

*Minimizing Lens Aberrations*

It should be recalled that when discussing the aberrations of oblique astigmatism, curvature of image, and distortion, we are concerned with the aberrations experienced when the eye rotates (using macular vision) to look through a peripheral point in the lens—not with the peripheral field of view of the stationary eye. Varying the form (or bend) of the lens and the use of an aspheric front surface are the designer's principal tools in controlling these aberrations.

As indicated in the discussion of spectacle magnification in Chapter 10, the goals of reducing spectacle magnification and reducing lens aberrations are at odds with one another when it comes to manipulating the form (i.e., the base curve selection) of the lens. Although a base curve in the neighborhood of −20.00 D would be effective in minimizing distortion, both the magnification and the appearance of such a lens would be unacceptable.

The aberrations encountered upon peripheral gaze through lenses with spherical surfaces are large—as large as a few diopters of error at 30 degrees of obliquity, as noted by Whitney.[18] The tangential errors are the largest, with the additive effect of transverse chromatic aberration. A steeper back surface curve (−3.50 D) has the advantage of reducing distortion, lens reflections, and sensitivity to differences in vertex distance. Shallower back surface curves (for example, −1.75 D) reduce both lens thickness and spectacle magnification, and help compensate for the power increase as frames slip forward.

The use of an aspheric surface does much more than help control peripheral aberrations: By allowing greater control over the lens periphery an aspheric surface can provide a greater visual field, reduce lens weight and thickness, reduce magnification, reduce the size of the ring scotoma, and improve the appearance of the lens.

The degree of asphericity (power drop) should be varied with lens power. Ideally, a different aspheric surface should be used for each increment of spherical power. Good lens design requires a wide variety of aspheric lens blanks; and the number of aspheric front curves available is one index of the quality of an aphakic lens series.

**Important Note:** The term "base curve" has taken on a slightly different meaning for aspheric aphakic lenses than the one used for ordinary spherical lenses. Traditionally, the term base curve refers to the standardized curvature for a series of lens powers. The same concept applies to aspheric lens design, but the degree of asphericity varies on the basis of base curve. Hence, in aspheric lens design the number of base curves not only indicates the number of different central curvatures available but indicates the number of degrees of asphericity as well.

It is unfortunate that some lens manufacturers have implied that the performance of an aphakic lens depends *solely* upon the amount of asphericity, or power drop. A large number of factors must be taken into consideration when judging the quality of the image produced by a strong plus lens: These include errors of sphere (curvature of image), cylinder (oblique astigmatism), and distortion for direct foveal vision and for specified angles of gaze, as well as distortion for indirect peripheral vision.

The choice of aphakic lenses should be based on (*a*) accurate technical information, with *numerical* detail, and (*b*) the *number* of base curves, and instructions on how they should be used. Table 11-1 lists some of the current full-field aspheric bifocal aphakic lenses, with the number of base curves and the segment type. All of the lenses are CR-39 plastic, having an index of refraction of 1.498 to 1.499, a nu value of 57.8, and a specific gravity of 1.32.

*Choice of Multifocal Style and Position*

Most plastic aspheric full-diameter multifocal aphakic lenses are made up in the form of either round or

TABLE 11-1
**Currently Available CR-39 Full-Field Aspheric Aphakic Bifocal Lenses**

| Manufacturer | Lens | Base Curves | Segment Type |
|---|---|---|---|
| American Optical | FulVue | 13 | 22 round |
| Armorlite | Multiple Drop[a] | 3 | 22 round 22 straight top |
| Signet | Hyperaspheric | 4 | 22 round 22 straight top |
| Silor | Super Modular | 5 | 22 round |
| Sola | Hi-Drop | 4 | 20 round 22 straight top |
| Frieder | Double Drop | 2 | 22 round 25 straight top |

[a]Formerly the Welsh Four Drop.

straight-top bifocals (note that although the segment looks very much like a fused segment, it is actually a one-piece segment). It is important to realize that although the front surface of the lens is aspheric, the segment surface is *spherical*. This has important implications for the shape of the segment.

ROUND SEGMENTS

The "round" segments used in aphakic lenses are usually not perfectly round, because the segment shape is modified by the intersection of the spherical segment surface with the aspheric surface of the distance portion of the lens. The result is a pleasant oval shape which widens more rapidly than a circular segment as the eyes are lowered.

The round segment has several desirable attributes in aphakic lenses. First, there is a reduction in the total prismatic displacement at the reading level, because the base-down effect of the segment balances some of the base-up prismatic effect due to the distance portion of the lens. An aspheric front surface has rotational symmetry, having a center (pole) of its own. The inside surface should be centered directly behind the pole of the front surface, so that the optical center will fall at the pole of the aspheric surface. With round-top segments, the bifocal inset may be "customized" by rotating the lens about the center of the aspheric curve, thus moving the segment center laterally to the desired position without changing the segment height significantly. In this manner, the correct bifocal inset can be obtained without altering the distance center and the optics of the aspheric surface.

STRAIGHT-TOP SEGMENTS

The straight-top bifocal segment is relatively wide at the top, has very little image jump as the line of sight enters the segment, and has practically no differential displacement occurring at the reading level. However, unlike round segments, straight-top segments are manufactured with *predetermined* segment inset, relative to the center of the aspheric surface. If the prescription is fabricated in such a way as to alter that inset, the distance optical center will not coincide with the center of the aspheric surface, and the patient will not get the full benefit of the aspheric design.

Due to the increasing amount of oblique astigmatism induced as the visual axis moves downward through the lens, bifocal segments for aphakic lenses should be set as high as possible. This applies both to round and to straight-top segments. It is usually a good idea to make the addition as weak as possible, to lessen the demand for convergence and to increase the "arm's length" range through the segment. For vision at closer than the usual reading distance, clear vision may be obtained by pushing the glasses down (only a few millimeters) on the nose.

SEGMENT INSET

In order for the reading fields for the two eyes to coincide, aphakic lenses may require a larger amount of inset than one would ordinarily provide. American Optical[19] has provided a table, reproduced here as Table 11-2, showing the amount of inset required for the two reading fields to coincide, based on the patient's distance PD, the distance prescription, and the reading distance.

Since an aphakic patient is unable to accommodate, when wearing aphakic bifocals the patient is limited to the ranges of clear vision provided by the depth of focus surrounding each of the two focal powers. Consequently, it would seem that the aphakic patient would be an ideal candidate for a trifocal. On the contrary, trifocals are almost useless in aphakia, and are not recommended: The presence of the intermediate segment requires an excessive downward movement to reach the near segment, with the result that the prismatic effects and aberrations at the reading level preclude useful vision. An exception to this general rule is a patient who was myopic before surgery, and requires a distance correction no greater than +8.00 D.

Some older, senile aphakics, who would be in danger of stumbling if bifocals were used, may be fitted with single-vision lenses for distance vision, and a separate pair of glasses in the form of bifocals or single-vision glasses for reading. The patient should

TABLE 11-2

**Amount of Segment Inset for Each Lens Required for the Two Reading Fields to Coincide, for a Lens Having a −3.00 D Back Surface and a Stop Distance of 27 mm**

| *Lens Power* | | *+10.00 D* | | *+12.00 D* | | *+14.00 D* | | *+16.00 D* | |
|---|---|---|---|---|---|---|---|---|---|
| *Reading Distance (cm)* | | *33* | *40* | *33* | *40* | *33* | *40* | *33* | *40* |
| Distance | 56 | 3.5 | 2.9 | 3.9 | 3.3 | 4.4 | 3.7 | 5.0 | 4.2 |
| PD | 60 | 3.7 | 3.1 | 4.2 | 3.5 | 4.7 | 4.0 | 5.3 | 4.5 |
| (mm) | 64 | 4.0 | 3.4 | 4.4 | 3.8 | 5.0 | 4.2 | 5.7 | 4.8 |
| | 68 | 4.2 | 3.6 | 4.7 | 4.0 | 5.3 | 4.5 | 6.0 | 5.1 |
| | 72 | 4.5 | 3.8 | 5.0 | 4.2 | 5.6 | 4.8 | 6.5 | 5.4 |

Source: From *Optical, Cosmetic and Mechanical Properties of Ophthalmic Lenses*, Lens Information Kit, Table of Seg Insets, p. 4.201, American Optical Co., 1968.

be instructed that, when the reading glasses are not available, reading may be done for short periods of time by pushing the distance glasses downward on the nose.

### Protection from "Glare" and Ultraviolet Radiation

In answers to a questionnaire presented to thirteen spectacle-wearing aphakics by Atchinson,[20] *glare* was ranked as the second worst perceptual disturbance (ranking just behind the reduced visual field). When asked in the same questionnaire to identify the factors most important in spectacle lens design, glare again ranked second (again, the reduced visual field ranked first). These findings are understandable, because as the eye ages the crystalline lens becomes increasingly yellow, resulting in a significant decrease in the transmission of visible light. After removal of the lens, the eye therefore is more sensitive to visible radiation simply because more of it now reaches the retina.

In addition to being a filter for visible light, the crystalline lens contains pigments which act as an ultraviolet filter, preventing much of this radiation from reaching the retina. Removal of the crystalline lens subjects the retina to potential ultraviolet damage for two reasons: (1) In the absence of the lens, the cornea will concentrate radiant energy in such a way that the retina receives an increased amount of radiation per unit area; (2) the coefficient of absorption by the retina is greater for ultraviolet radiation than for visible light. Ham, Muller, and Sliney[21] have shown that near ultraviolet and short wavelengths of the visible spectrum can produce retinal damage; and Lerman[22] has suggested that cystoid macular edema, which has a high incidence in aphakia, may be due to increased ultraviolet radiation reaching the retina.

Although there is disagreement as to the cutoff of ultraviolet radiation that is desirable, it would seem that a lens that filters out most of the ultraviolet below 380 nm and progressively transmits wavelengths between 380 and 500 nm would eliminate most of the near ultraviolet and visible wavelengths that are potentially harmful.

In addition to the ultraviolet-absorbing filter, an aphakic patient may wish to have one of the light tints for general wear. Such a tint has the further advantage of improving the appearance of the lenses. If the patient complains of discomfort when outdoors, a separate pair of prescription sunglasses, containing an ultraviolet absorber, should be provided. Ultraviolet-absorbing lenses are discussed in Chapter 7.

## 11.7. Frames for Aphakic Lenses

A sturdy, lightweight, easily adjustable frame is required for aphakic lenses. Usually the best frame to select is either a metal frame or a combination plastic and metal frame. Frames with thick eyewires should be avoided, as such an eyewire may increase the width of the ring scotoma. Adjustable pads are a necessity, and for most patients jumbo pads are recommended. The pads should be adjusted so that the lenses fit as close to the eyes as possible (consistent with the wearer's facial features).

In order to meet the criterion of "minimal effective diameter," the patient should be encouraged not only to select a frame with a relatively small eye size, but also to select one having a relatively symmetrical lens shape with no pronounced upsweep and no sharp angles. A useful guide for symmetry is that the effective diameter (ED) of the frame should be no more than 2 mm greater than the marked eye size (the *A* dimension of the box system). A frame so selected means that there will be little variation in thickness of the lens edge, and if the lens edge is made as thin as possible the lens will have optimum appearance and lightness of weight. In order to avoid decentration, the frame PD should match the patient's PD.

For a patient with a narrow PD, in order to keep from making the eye size smaller than necessary, the bridge (DBL) should be kept as narrow as physically possible, thus allowing the lenses to be as large as possible for the required frame PD. For a patient who has a wide PD, the eye size can be kept relatively small by making the DBL larger than usual.

If a patient insists on a stylish frame, lens thickness can be kept to a minimum only by using lenticular lenses. The practitioner or dispenser should keep on hand an assortment of attractive sample frames that can be used with aphakic lenses.

## 11.8. The Unilateral Aphakic

The principal problem associated with the correction of unilateral aphakia by spectacles is the difference in the sizes of the retinal images of the two eyes. A second difficulty consists of the large differential prismatic effects encountered in the peripheral portions of the lenses.

As shown in Section 11.4, the spectacle magnification brought about by a typical aphakic spectacle lens is in the neighborhood of 30%. For the *bilateral* aphakic, this amount of spectacle magnification can be tolerated, after a period of adaptation, because it is essentially the same for both eyes; but for the *unilateral* aphakic, the magnification difference for the two eyes is so great that binocular vision is impossible.

Suppose, for example, that the aphakic eye of a unilateral aphakic patient required the lens described in the example in Section 11.4, having a back vertex power of +13.00 D, a front surface power of +16.00 D, a center thickness of 8.0 mm, an index of refraction of 1.498, and a vertex distance of 10 mm, bringing about a spectacle magnification of 31.5%. Suppose also that the patient's unoperated eye required a lens having a back vertex power of +2.00 D, a front surface power of +8.00 D, a center thickness of 2.0 mm, an index of refraction of 1.498, and a vertex distance of 10 mm. The spectacle magnification brought about by this lens would be

$$SM = \left(\frac{1}{1 - \dfrac{t}{n}F_1}\right)\left(\frac{1}{1 - hF_v}\right)$$

$$= \left(\frac{1}{1 - \dfrac{0.002(8)}{1.498}}\right)\left(\frac{1}{1 - 0.013(2)}\right)$$

$$= (1.011)(1.0267) = 1.038$$

$$\%SM = +3.8\%$$

The difference between the spectacle magnifications for the two lenses would therefore be 31.5 − 3.8, or 27.7%. Manipulation of the base curve, thickness, etc., of either lens would bring about only an insignificant decrease in this difference.

This substantial amount of aniseikonia, together with the differential peripheral prismatic effects for the two eyes that occur when a unilateral aphakic wears spectacle lenses, are responsible for disturbances in sensory and motor fusion and in stereoscopic spatial projection, making binocular vision impossible.

The large magnification difference for the spectacle-wearing unilateral aphakic is manifested by a difference in the *rate of movement* across the visual field of the images of objects seen by the two eyes. This confusing movement of diplopic images is apparently what makes the magnification difference intolerable to the patient. The practitioner, however, has three options available: He or she may not correct the operated eye, may prescribe a "balance" lens or an "occluder" lens for the unoperated eye, or may prescribe a *contact lens* for the aphakic eye.

Most patients tend to resist wearing no correcting lens before the operated eye or wearing a balance lens or an occluder lens before the unoperated eye. However, if the visual acuity of the corrected unoperated eye is still reasonably good (in the neighborhood of 20/40), the problems of adjusting to an aphakic lens are such that the patient may be much happier to see with the unoperated eye than with the operated eye. Inasmuch as unilateral aphakia is in many cases only temporary, with an operation scheduled for the second eye in the near future, the patient may have to suffer the inconvenience of being monocular for only a relatively short period of time.

In those cases in which the monocular situation is to be relatively permanent, and when an intraocular lens has been ruled out, there's no doubt that the best method of achieving binocular vision is to fit a contact lens on the aphakic eye. In Section 11.4 it was found that a +10.31 D spectacle lens was needed at a vertex distance of 12 mm to correct an eye that was the optical equivalent of Gullstrand's schematic eye prior to surgery. If a contact lens were used to correct such an eye, its power would be

$$\frac{1}{0.097 - 0.012} = \frac{1}{0.085} = +11.76 \text{ D}.$$

The resulting power of this aphakic schematic eye, corrected with a contact lens, would be the sum of the power of the contact lens and the power of the cornea, thus,

$$+11.76 \text{ D} + 43.05 \text{ D} = +54.81 \text{ D}.$$

Since the size of the retinal image is inversely proportional to the refracting power of the eye,

$$\frac{\text{retinal image size corrected by a contact lens}}{\text{retinal image size in emmetropia}} = \frac{+58.64}{+54.81} = 1.07.$$

Hence, there is a retinal image size difference of 7% between the Gullstrand schematic aphakic eye corrected with a contact lens and the Gullstrand emmetropic schematic eye. Ogle[23] has stated that individuals are rarely found who have binocular vision in the presence of aniseikonia greater than 5 or 6%. This figure would indicate that, in cases of unilateral aphakia corrected by a contact lens, the amount of residual aniseikonia would make normal binocular vision difficult, if not impossible. Yet, many unilateral aphakic patients corrected with contact lenses appear to have comfortable vision.

The actual amount of residual aniseikonia present in unilateral aphakia corrected with contact lenses depends upon the following factors:

1. The refractive error present before the onset of the cataract in the aphakic eye together with the origin of the ametropia (axial or refractive). This is the most important factor.
2. The refractive error in the phakic (unoperated) eye, as well as its origin (axial or refractive) and whether it is corrected by a spectacle lens or a contact lens.
3. Any alteration of the power of the cornea due to the surgery.
4. The individual variations in the dimensions of the primary refracting components of the eye in question, which often differ greatly from those of the standard schematic eye (making calculations on the basis of the schematic eye misleading).
5. The shape factor of the spectacle magnification formula. This is often considered to be equal to unity, the lens (either spectacle or contact) being considered as a thin lens. However, the finite thickness of an actual lens, together with the curvatures of the surfaces of the lens, cause the shape factor to vary considerably from unity and from one lens to another.

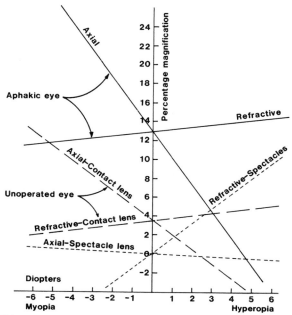

FIGURE 11-14. **Residual aniseikonia based on pre-cataract data. (From K. N. Ogle, H. M. Burian, and R. E. Bannon, *Arch. Ophthal.*, Vol. 59, pp. 639–652, Copyright 1958, American Medical Association.)**

Each of the above factors obviously may have an effect on the amount of residual aniseikonia present when unilateral aphakia is corrected by contact lenses. The ametropia that was present before the development of the cataract, as well as whether it was axial or refractive, is especially important (although often neglected) in determining the amount of residual aniseikonia in the correction of unilateral aphakia. Ogle, Burian, and Bannon[24] plotted the residual aniseikonia based on pre-cataract data, as shown in Figure 11-14, and noted the following:

1. If the ametropia prior to the onset of the cataract was *axial* in origin and equal in amount for the two eyes:

(*a*) In high myopia (5 to 6 D), correction of unilateral aphakia with contact lenses for both eyes would result in a residual aniseikonia of 15%, whereas the use of a contact lens for the aphakic eye and a spectacle lens for the unoperated eye would result in an increase of residual aniseikonia to 26%.

(*b*) In high hyperopia (again, 5 to 6 D), the correction of unilateral aphakia with contact lenses for both eyes would leave a residual aniseikonia of only 2%, whereas the use of a contact lens for the aphakic eye and a spectacle lens for the unoperated eye would reduce the residual aniseikonia to approximately zero.

2. If the ametropia prior to the onset of the cataract was *refractive* in origin, and of the same magnitude for each eye, the amount of residual aniseikonia if both eyes are fitted with contact lenses will be about 10% for both the high myope and the high hyperope. However, if a contact lens is worn on the aphakic eye and a spectacle lens is worn on the unoperated eye, the amount of residual aniseikonia for the high myope would increase to about 24%, but would decrease to only about 5% for the high hyperope.

3. If both eyes were emmetropic before the cataract onset, and if a contact lens is worn on the aphakic eye and no lens is worn on the unoperated eye, the estimated aniseikonia would be about 13%. If contact lenses are worn on both eyes, the residual aniseikonia would be about 10%.

The predicted amounts of aniseikonia for each of these examples, with contact lenses worn on both eyes or with a contact lens on the aphakic eye and a spectacle lens on the unoperated eye, are summarized in Table 11-3.

These examples show that for the high myope prior to cataract onset (whether axial or refractive) residual aniseikonia can be minimized by fitting contact lenses for both eyes. On the other hand, for the high hyperope prior to cataract onset (whether axial

TABLE 11-3
**Predicted Amounts of Aniseikonia, Depending upon the Refractive State of the Aphakic Eye prior to Cataract Onset, Based on Data Plotted in Figure 11-14**

| Pre-cataract Refraction | Origin | Lens for Aphakic Eye | Lens for Phakic Eye | Residual Aniseikonia (%) |
|---|---|---|---|---|
| −5 to −6 D | Axial | Contact | Contact | 15 |
| | | Contact | Spectacle | 26 |
| −5 to −6 D | Refractive | Contact | Contact | 10 |
| | | Contact | Spectacle | 24 |
| +5 to +6 D | Axial | Contact | Contact | 2 |
| | | Contact | Spectacle | 0 |
| +5 to +6 D | Refractive | Contact | Contact | 10 |
| | | Contact | Spectacle | 5 |
| Emmetropia | — | Contact | Contact | 10 |
| | | Contact | No lens | 13 |

Source: From K.N. Ogle, H.M. Burian, and R.E. Bannon, *Arch. Ophthal.*, Vol. 59, p. 644, 1958.

TABLE 11-4
**Parameters for the Design of a Reverse Galilean Telescope for the Aphakic Eye in Unilateral Aphakia, Using an Overcorrection for the Contact Lens, F(+), as the Ocular, and a Minus Spectacle Lens, F(−), as the Objective**

| Aphakic Minification (%) | Vertex Distance (mm) | F(+) (D) | F(−) (D) | Phakic Magnification (%) |
|---|---|---|---|---|
| 12.3 | 15.4 | 8.00 | 9.12 | 14.1 |
| 11.2 | 13.9 | 8.00 | 9.00 | 12.5 |
| 10.0 | 11.1 | 9.00 | 10.00 | 11.1 |
| 9.4 | 15.8 | 6.00 | 6.62 | 10.4 |
| 8.6 | 10.7 | 8.00 | 8.75 | 9.4 |
| 8.0 | 13.9 | 5.75 | 6.25 | 8.6 |
| 7.4 | 11.9 | 6.25 | 6.75 | 8.0 |
| 6.7 | 9.5 | 7.00 | 7.50 | 7.1 |
| 6.0 | 10.6 | 5.75 | 6.12 | 6.5 |
| 5.6 | 9.1 | 6.25 | 6.62 | 6.0 |

Source: From J.M. Enoch, *Amer. J. Optom. Arch. Amer. Acad. Optom.*, Vol. 45, p. 236, 1968.

or refractive) residual aniseikonia can be minimized by fitting a contact lens on the aphakic eye and a spectacle lens on the unoperated (phakic) eye.

Enoch[25] has suggested that binocular vision may be restored in unilateral aphakia by fitting the aphakic eye with a reverse Galilean telescope, bringing about a *minification* of the retinal image. The plus power necessary for the eyepiece, or ocular, of the telescope is added to the contact lens prescription for the aphakic eye, while the minus power for the telescope objective is supplied in the form of a spectacle lens. He estimated the residual aniseikonia to be in the 6 to 8% range, and provided a table (selected values of which are reproduced here as Table 11-4) to indicate the amount of minification available for various vertex distances. If a vertical imbalance occurs at the reading level, Enoch suggested the use of a slab-off prism in the spectacle lens.

Whether the aphakic eye is fitted with a contact lens or a spectacle lens, a bifocal addition or a single-vision lens for reading must be supplied for that eye. Since most unilateral aphakics are absolute presbyopes, the same bifocal addition or reading lens can usually be supplied for both eyes. If the patient is still able to accommodate with the unoperated eye, an add can be given only for the aphakic eye, or unequal adds can be prescribed.

## 11.9. Determining the Final Aphakic Prescription

Contrary to the usual method of refraction, the frame for an aphakic patient should be selected *before* the refraction is performed. The sample frame should be of the correct eye size and DBL, and should be adjusted to the patient's face so that the vertex distance can be determined. An instrument called the Distometer has been designed for this purpose. Since the optical center of the lens should be placed at the geometrical center, the distance from the geometrical center to the level of the center of the pupil will determine the amount of pantoscopic tilt to ensure that the optic axis of the lens will pass through the center of rotation of the eye. If the pantoscopic tilt exceeds about 8°, the frame may have to be positioned lower on the face than it would ordinarily be fitted, by adjusting the nose pads, in order to meet the criterion that the optical center should be lowered 1 mm for each 2° of pantoscopic tilt.

There are two methods of refraction that should lead to accurate results. One method is to use the refractor for a careful refraction (sphere, cylinder

FIGURE 11-15. **Trial lens clip, which attaches to the frame for verifying the lens power needed for the frame the patient will be wearing.**

power, cylinder axis, as usual) and then to refine that result with a trial frame at the correct vertex distance and pantoscopic tilt. After the trial frame refraction, the trial frame with the correcting lenses is carefully placed in a lensometer to determine the final prescription in back vertex power.

The second method involves a technique known as "over-refraction." American Optical, Signet Optical, Armorlite, and Frieder all make over-refraction kits which consist of several frames suitable for aphakic corrections, containing the manufacturer's aphakic lenses in assorted powers and frame PDs. Also included in the kit is a trial lens clip (Figure 11-15) which attaches to the frame for verifying the power needed with the frame in which the patient's lenses will be ultimately mounted, correctly adjusted for vertex distance and pantoscopic tilt. By utilizing a frame having fitting characteristics similar to the one the patient will be wearing, vertex distance and pantoscopic tilt during refraction will be similar to those during actual wear of the glasses. By using lenses of design and power similar to those which will be dispensed, the optics employed during the final refraction and the final spectacle design will be comparable.

When the finished glasses are received from the laboratory, the frame must be fitted very precisely on the patient's face. The distance optical centers should be "spotted," and the frame should be adjusted so that the optical center of each lens is located 1 mm below the center of the pupil (with the patient's head held erect and the eyes in the primary position) for each 2 degrees of pantoscopic tilt. The reflection method of alignment, as described in Section 9.17, should be used to verify that the optic axis passes through the center of rotation of the eye.

## 11.10. The High Hyperope

In general, high hyperopes tend to have the same problems with their glasses as do aphakics: excessive lens weight and thickness, magnified appearance of the wearer's eyes, increased retinal image size, lens aberrations, and decreased field of view. However, very few hyperopes require lenses anywhere near the normal range for aphakic powers—from about +10.00 to +14.00 D. The refractive error distribution curve for unselected subjects shows *very few* hyperopes of more than 5 or 6 D.

With the wide use of plastic lenses, the high hyperope need no longer put up with intolerably *heavy* lenses. *Thickness*, as with aphakic lenses, can be minimized by selecting a frame size that will result in little or no decentration so that the lenses can have knife-thin edges. The use of an aspheric front surface, also, will help to minimize lens thickness.

Magnification of the retinal image is not as much of a problem for the high hyperope as it is for the aphakic, not only because the lens power is usually much lower than that for an aphakic eye, but also because the hyperope normally has become accustomed to the magnification over a period of years whereas the aphakic is suddenly presented with extremely high magnification. Nevertheless, as in the case of aphakic lenses, it is possible to minimize both the magnified appearance of the wearer's eyes and the retinal image magnification and to maximize the field of view by placing the lenses as close to the eyes as possible and by choosing base curves that are as shallow as possible, compatible with good optical performance.

When viewed from the front, the surfaces of a plus meniscus lens act as a pair of reflecting surfaces which form images that have a tendency to hide the wearer's eyes from observers, and also to annoy the wearer. The intensity of these images may be greatly reduced by the use of an antireflective coating. In addition, the number of reflected images potentially visible to an observer may be reduced by reducing the object field of view. The size of the object field of view for a spherical surface is directly proportional to the size of the reflecting surface and inversely proportional to its radius of curvature. Thus, the size of the object field of view may be reduced by selecting as small an eye size as possible that is acceptable to the patient and making the base curve as shallow as possible in terms of optical performance. Of course, a small eye size has the advantage of reducing the lens thickness and weight, and a shallow base curve has the advantage of reducing magnification effects.

# LENSES FOR HIGH MYOPIA

## 11.11. Problems Caused by High Minus Lenses

LENS WEIGHT AND THICKNESS

The problems caused by the lenses required for the correction of high myopia are quite different from those caused by lenses for the correction of aphakia or hyperopia. Lens weight is somewhat less of a problem for the myope than for the hyperope or the aphakic, and is usually not a problem at all if plastic lenses are used. Lens thickness, however, is more of a problem with the myope because thick *edges* are much more obvious to the observer than are thick *centers*.

MINIFICATION

High minus lenses obviously bring about a minification of the wearer's eyes, to the observer, as well as minifying the retinal image. Both the apparent size of the wearer's eyes and the retinal image size can be increased by fitting the lenses as close to the wearer's eyes as possible. Many myopes have solved this problem for themselves by wearing *contact lenses* rather than glasses. For the high myope, minification of the wearer's eyes is probably at least as great a cosmetic deterrent as are edge thickness and edge reflections. Contact lenses not only allow the eyes to appear normal in size, but bring about an increase in retinal image size that is readily apparent to the high myope.

FIELD OF VIEW

Minus lenses actually *increase* both the peripheral and macular fields of view, as compared to those for an unpowered lens. The increased macular field of view means that rather than having a roving ring scotoma as the aphakic has, the myope might be expected to experience a ring-shaped area of overlapping *double-vision*—an area that involves relatively clear vision through the lens and blurred vision outside the lens, as shown in Figure 11-16. However, the "ring of diplopia" is so far in the periphery that it is seldom, if ever, noticed.

With the exception of the minification caused by the lenses, the problems of wearing high minus lenses have to do almost entirely with the *appearance* of the lenses. Problems having to do with the appearance of high minus lenses include not only their thick edges and the minified appearance of the wearer's eyes, but also the *rings* that an observer can see, caused by reflections from the lens edges.

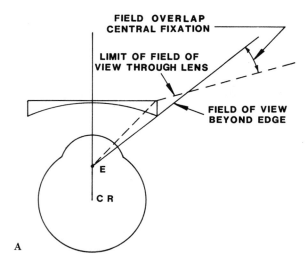

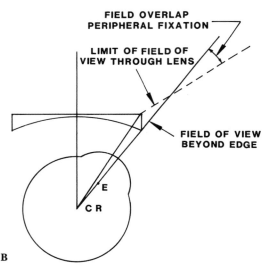

FIGURE 11-16. **The ring-shaped area of diplopia, or field overlap, which may be seen (but is seldom noticed) by the wearer of a high minus lens: A, for central fixation; B, for peripheral fixation. CR = center of rotation.**

## 11.12. Minimizing Edge Thickness

The relationship between the edge thickness and center thickness of a lens is discussed in Sections 3.12 to 3.14, and formulas relating power to center and edge thickness of a lens are given in Eq. (3.13), for a spherical lens, and in Eq. (3.14), for a sphero-cylindrical lens. In Eq. (3.13),

$$t_C - t_P = \frac{F_A h^2}{2(n-1)},$$

where $t_C$ is the center thickness, $t_P$ is the edge thickness, $F_A$ is the approximate power of the lens, $h$ is one-half the diameter (chord length) of the lens, and $n$ is the index of refraction of the lens material.

For a lens of a given power, edge thickness can be minimized (1) by keeping the lens size as small as possible and by using a frame whose PD is equal to the patient's distance PD so that no decentration is required (analogous to the "minimal effective diameter" method of fitting aphakic lenses) and (2) by using a material having as high an index of refraction as possible. However, the current popularity of large lens sizes makes it difficult for the practitioner or dispenser to interest most patients in keeping lens size to a minimum; and the popularity of CR-39 plastic lenses, although decreasing lens weight, has the consequent disadvantage of increasing the edge thickness.

However, it will be recalled that *polycarbonate* material is not only lighter in weight than ophthalmic crown glass, but also has a higher index of refraction, with the result that a lens made of this material will be both lighter and thinner than one made of glass. Table 11-5 lists the indices of refraction, specific gravities, and nu values of CR-39 and polycarbonate materials, as compared to those of ophthalmic crown glass.

The edge thickness that can be expected for lenses made of ophthalmic crown glass, as compared to CR-39 and polycarbonate materials, will be shown by the use of an example.

### EXAMPLE

Given the prescription −8.00 D sphere, to be mounted in a frame having an eye size of 50 mm, the lens having a center thickness of 2.0 mm. If the patient's PD is equal to the frame PD, making

no decentration necessary, what will be the edge thickness at the thickest points of the lens edge (assumed to be in the 180 meridian) for:

(a) an ophthalmic crown glass lens ($n = 1.523$)
(b) a CR-39 lens ($n = 1.498$)
(c) a polycarbonate lens ($n = 1.586$)

(a) For the ophthalmic crown glass lens:

$$t_C - t_P = \frac{-8(0.025)^2}{2(1.523 - 1)} = \frac{-0.005}{1.046} = -0.0048 \text{ m.}$$

$$t_P = 0.0048 + 0.002 = 0.0068 \text{ m} = 6.8 \text{ mm.}$$

(b) For the CR-39 lens:

$$t_C - t_P = \frac{-0.005}{2(1.498 - 1)} = \frac{-0.005}{0.996} = -0.0050 \text{ m.}$$

$$t_P = 0.0050 + 0.002 = 0.0070 \text{ m} = 7.0 \text{ mm.}$$

(c) For the polycarbonate lens:

$$t_C - t_P = \frac{-0.005}{2(1.586 - 1)} = \frac{-0.005}{1.172} = -0.00427 \text{ m.}$$

$$t_P = 0.00427 + 0.002 = 0.00627 \text{ m}$$
$$= 6.27 \text{ mm.}$$

Whereas the edge thickness difference of only 0.2 mm between CR-39 and glass may be considered negligible, the difference of over 0.7 mm between CR-39 and polycarbonate is approximately 10% of the edge thickness and is sufficient to make a noticeable difference.

An additional alternative is the use of one of the available *high-index* glass materials. Currently available high-index glasses, together with their respective indices of refraction, specific gravities, and nu values, are listed in Table 11-6.

The most commonly used high-index glasses are *High-Lite*, made by Schott, and similar materials made by Chance-Pilkerton and by Hoya. Most have an index of refraction of 1.70 and a nu value that is substantially lower than that of ophthalmic crown glass, with the result that the chromatic dispersion is much greater than that for ophthalmic crown glass.

Using the above example, the edge thickness for a high-index (1.70) glass lens would be:

$$t_C - t_P = \frac{-0.005}{2(1.70 - 1)} = \frac{-0.005}{1.400} = -0.0036 \text{ m.}$$

$$t_P = 0.0036 + 0.002 = 0.0056 \text{ m} = 5.6 \text{ mm.}$$

The resultant saving in edge thickness, as compared to ophthalmic crown glass, is 1.2 mm, or close to 20%.

TABLE 11-5
**Characteristics of Currently Available Plastic Materials, Compared to Ophthalmic Crown Glass**

| Name | Index of Refraction | Specific Gravity | Nu Value |
|---|---|---|---|
| Ophthalmic crown | 1.523 | 2.54 | 58.9 |
| CR-39 | 1.498 | 1.32 | 58 |
| Polycarbonate | 1.586 | 1.20 | 30 |

TABLE 11-6
**Characteristics of Currently Available High-Index Glasses, Compared to Ophthalmic Crown Glass**

| Manufacturer | Name | Index of Refraction | Specific Gravity | Nu Value |
|---|---|---|---|---|
| All manufacturers | Ophthalmic crown | 1.523 | 2.54 | 58.9 |
| Schott | 1.60 crown | 1.60 | 2.62 | 40.7 |
| Chance Pilkington | Slimline 640 | 1.60 | 2.58 | 41 |
| Schott | High-Lite | 1.70 | 2.99 | 31 |
| Chance Pilkington | Slimline 730 | 1.70 | 2.99 | 31 |
| Chance Pilkington | Slimline 750 | 1.70 | 3.38 | 51 |
| Hoya | LHI | 1.70 | 2.99 | 40.2 |
| Chance Pilkington | Slimline 825 | 1.805 | 3.35 | 25 |
| Hoya | THI | 1.806 | 4.56 | 40.7 |

If one of the recently introduced 1.80 index glasses were used for this lens, an even greater saving in edge thickness would result:

$$t_C - t_P = \frac{-0.005}{2(1.80 - 1)} = \frac{-0.005}{1.60} = -0.0031 \text{ m.}$$

$$t_P = 0.0031 + 0.002 = 0.0056 \text{ m} = 5.1 \text{ mm.}$$

The charts shown in Figure 11-17 indicate the expected edge thickness for minus lenses at various distances from the optical center, for lenses made of ophthalmic crown glass (Figure 11-17A); 1.70 index glass (Figure 11-17B); 1.80 index glass (Figure 11-17C); CR-39 plastic (Figure 11-17D); and polycarbonate plastic (Figure 11-17E). Although these charts are designed to be used for minus lenses having a center thickness of 2.0 mm, they can be used for predicting the *center* thickness of plus lenses: For example, if it is assumed that the edge thickness of a plus lens should be 1 mm, the predicted center thickness of a given plus lens would be equal to the edge thickness (as found in the chart) minus 1 mm.

In view of the higher specific gravity of high-index glass, the fact that this material produces thinner lenses does not necessarily mean that the lenses will weigh less. Tucker[26] has compared the weights of CR-39 plastic, ophthalmic crown glass, and high-index glass (1.70) for a 50-mm lens size and for a minimum edge or center thickness of 2 mm. Her results are shown plotted in Figure 11-18. In this diagram, it can be seen that the curves for ophthalmic crown glass and high-index glass cross at back vertex powers of approximately ±8.00 D. Below 8.00 D, lenses made of high-index glass are heavier than lenses made of ophthalmic crown glass, but above 8.00 D they are lighter; so for lenses having a diameter of 50 mm, if the goal is to reduce *both* thickness and weight, these lenses should be used only for powers in excess of ±8.00 D.

Note, however, that the saving in weight is relatively small, amounting to only 1 or 2 grams for lenses having powers of ±12.00 D. On the other hand, for lenses of relatively low power, the *excess* weight of high-index lenses amounts to only 1 or 2 grams. It should be understood that the cross-over values for these curves will vary both with lens size and with the criterion for minimum edge or center thickness.

All high-index glass lenses should have an antireflective coating. Since the intensity of the reflections increases with the index according to Fresnel's equation, the reflections from an uncoated high-index lens will be quite noticeable. As discussed in Section 7.27, the ideal antireflective coating will have an index of refraction that is equal to the square root of the index of refraction of the glass to be coated. Magnesium fluoride has an index of refraction of 1.38, which is considerably higher than the square root of the index of ophthalmic crown (which is 1.234). However, since the square root of the index of glass having an index of 1.70 is 1.304, an antireflective coating of magnesium fluoride will function more efficiently for a high-index glass than for ophthalmic crown glass.

### 11.13. Minimizing Edge Reflections

As discussed in Sections 7.24 through 7.30, there are reflections from the surfaces and edges of an opthalmic lens that are not noticed by the wearer but may be annoying or distracting to an observer. These reflections tend to enhance the "glassiness" of the lenses and to obscure the eyes of the wearer, thus serving as an impediment to communication and presenting a problem particularly for an individual who must regularly appear in public.

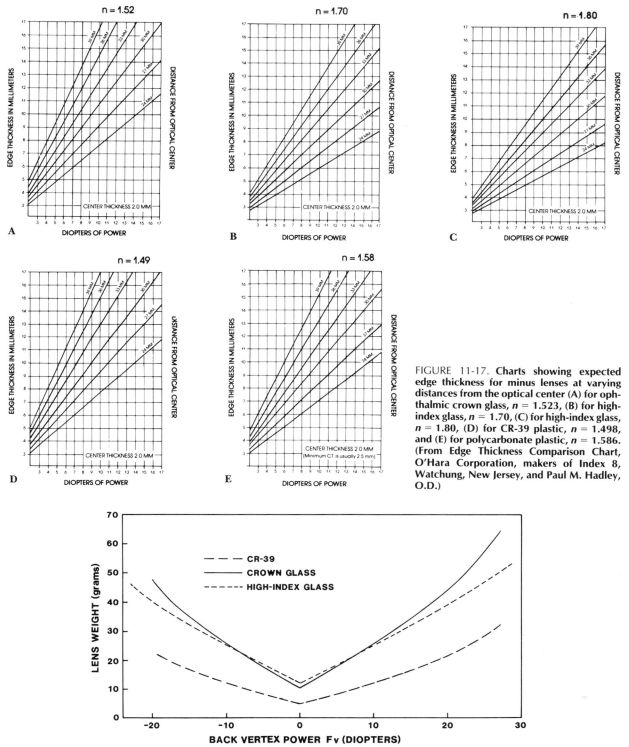

FIGURE 11-17. Charts showing expected edge thickness for minus lenses at varying distances from the optical center (A) for ophthalmic crown glass, $n = 1.523$, (B) for high-index glass, $n = 1.70$, (C) for high-index glass, $n = 1.80$, (D) for CR-39 plastic, $n = 1.498$, and (E) for polycarbonate plastic, $n = 1.586$. (From Edge Thickness Comparison Chart, O'Hara Corporation, makers of Index 8, Watchung, New Jersey, and Paul M. Hadley, O.D.)

FIGURE 11-18. Comparison of the weights of lenses made of CR-39 plastic, ophthalmic crown glass, and high-index glass for a 50-mm lens with a minimum edge or center thickness of 2 mm. (From T. Tucker, *Ophthal. Optician*, Vol. 16, pp. 795–804, 1976.)

In addition to the reflections formed by the surfaces of the lenses, the edges of the lenses are responsible for the *multiple rings* seen just inside the edges of high minus lenses. These rings (described in Section 7.31 and illustrated in Figure 7-46) are images of the edge of the lens, and are caused by internal reflections which begin at the lens edge and reflect their way toward the center of the lens until they strike the surface at an angle which permits refraction by the lens and into the eye of the observer. This process is then repeated, with the result that multiple rings are seen. These rings are more pronounced for oblique angles of view than for straight-ahead viewing.

Although the use of an *antireflective coating* is helpful in attenuating the intensity of the rings on straight-ahead viewing, for rings seen upon oblique viewing the most effective method of attenuation is the use of one of the available forms of *edge treatment*. Methods of edge treatment include:

1. *Edge Coating.* This helps to reduce the granular appearance of the edge.
2. *Edge Painting.* Using a neutral gray color, painting the edge tends to make the rings less noticeable.
3. *Using a Semitransparent Edge.* For a glass lens, a semitransparent edge can be produced by using an edging wheel that results in a fine, semitransparent grain of the bevelled edge rather than the usual coarse white grain.
4. *Buffing the Edge.* For a plastic lens, buffing produces the same effect as a semitransparent edge on a glass lens.
5. *Tinting the Lens.* Another effective procedure for a plastic lens is that of tinting (dyeing) the lens after it has been edged. Even a relatively light tint is effective in reducing the multiple ring effect.
6. *Hide-a-Bevel Technique.* As shown in Figure 7-47, this procedure results in a flat edge with a small, narrow bevel, protruding from the edge. Due to the angle between the flat edge and the lens surface, the reflected rays that are responsible for the multiple-ring effect are not apt to enter an observer's eye.

The use of a combination of these methods can be very effective: For example, for a plastic lens one may use the Hide-a-Bevel technique, also buffing the edge and tinting the lens.

## 11.14. Lenticular Lenses

In 1933, Obrig[27] introduced a lenticular lens for the correction of myopia, called the *Myo-Disc*. This lens was made by grinding a small concave disk on the back surface of a plano lens, as shown in Figure 11-19A. This lens was remarkably thin and light in weight. Other lenticular lenses are to a great extent similar to the original Myo-Disc lens, having a powered portion approximately 30 mm in diameter, surrounded by a "carrier" lens having either plano or low plus power as shown in Figure 11-19B. As with plus lenticular lenses, these lenses have the advantage of decreased lens thickness and weight, but the disadvantage of causing a "bull's eye" appearance, thus calling attention to the fact that the individual is wearing very strong lenses.

Most practitioners would consider the use of a minus lenticular lens only for a patient having *very* high myopia, for example, in excess of 11.00 or 12.00 D. Even for lenses in this power range, the practitioner should consider all of the options short of using a lenticular lens, as discussed in the preceding paragraphs. Keeping lens size to a minimum and making use of the methods described in minimizing edge thickness (including the use of high-index materials) and minimizing lens reflections will in many cases provide a more cosmetically acceptable pair of

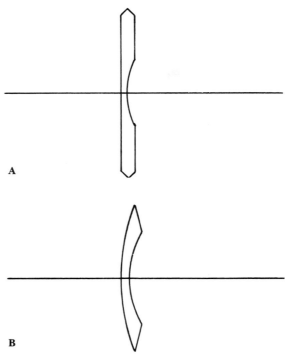

A

B

FIGURE 11-19. **Lenticular lenses for myopia: A, Myo-Disc lens, made by grinding a small concave disk on the back surface of a flat plano lens; B, lens having a powered portion 30 mm in diameter and a plus-powered carrier.**

glasses than could be provided by the use of minus lenticular lenses. Also, the use of contact lenses should obviously be considered, particularly in view of the fact that, as shown by Gumpelmeyer,[28] very high myopes tend to have significantly better visual acuity with contact lenses than with glasses, even beyond that predicted by the greatly increased spectacle magnification.

The saving in edge thickness due to the use of a minus lenticular lens for a very high myope may be demonstrated by the use of an example.

### EXAMPLE

For a patient requiring a lens having a power of −15.00 D sphere, using a lens having a center thickness of 2.0, made of CR-39 material having an index of refraction of 1.498, and using a frame requiring no decentration: (*a*) What would be the edge thickness at the thickest point if a 50-mm-wide full-diameter lens were to be used? (*b*) What would be the thickness at the thickest point of the transition between the optical portion and the carrier if a lenticular lens having a 30-mm-wide optical zone were to be used?

(*a*) For the full-diameter lens:

$$t_C - t_P = \frac{-15(0.025)^2}{2(1.498 - 1)} = -0.0094 \text{ m.}$$

$$t_P = 0.0094 + 0.002 = 0.0114 \text{ m}$$
$$= 11.4 \text{ mm.}$$

(*b*) For the minus lenticular lens:

$$t_C - t_P = \frac{-15(0.015)^2}{2(1.498 - 1)} = -0.0034 \text{ m.}$$

$$t_P = 0.0034 + 0.002 = 0.0054 \text{ m}$$
$$= 5.4 \text{ mm.}$$

This example shows that the edge thickness at the thickest point of a full-diameter lens would be more than twice the transition thickness for a minus lenticular lens. If the lenticular lens were to have a *convex* carrier, the edge thickness would be considerably less than the 5.4-mm transition thickness. For those very high minus prescriptions for which full-diameter lenses would result in unacceptable edge thicknesses, the "bull's eye" appearance of minus lenticular lenses may be minimized by using lightly tinted lenses having an antireflective coating.

## 11.15. Fresnel Press-on Lenses

Fresnel Press-on lenses are made in much the same manner as Fresnel Press-on prisms, described in Sec-

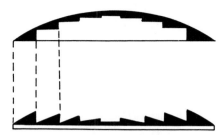

FIGURE 11-20. **A Fresnel Press-on lens.**

tion 4.19: However, the grooves in the material form a concentric pattern on the surface rather than forming parallel lines (see Figure 11-20). Successive concentric grooves cut away from the center have slightly increased apical angles which provide the increased deviating power from the center to the edge. The angle of each groove can be set independently, with the result that any spherical or aspheric surface can be generated. The generation of an aspheric surface is neither more expensive nor more difficult than the generation of spherical surfaces.

Fresnel lenses, like Fresnel prisms, provide a thin and lightweight alternative to conventional glass or plastic lenses. Fresnel lenses (in either plus or minus powers) offer unique advantages in many circumstances, including (*a*) the correction of very high refractive errors; (*b*) temporary use as a trial or diagnostic lens; (*c*) additions with high power or eccentric placement of segments; and (*d*) for use when a lens power or construction would not be available or would result in exorbitant costs or long delays.

The grooves cut in the lenses produce scattering of light which may result in a slight reduction in visual acuity. The most important factor in determining resolution through the lenses is the *width* of the grooves: The smaller the groove width, the more precise the focus and (theoretically) the better the resolution. However, as groove width gets smaller, diffraction begins to reduce resolution: Therefore, the groove width chosen must represent a balance between precise focus and image deterioration from diffraction. The resolution is dependent also upon the orientation of the object.[29] Components of an object which are parallel to the orientation of the grooves have the poorest resolution, while components at right angles to the grooves undergo little deterioration.

The advantages of Fresnel lenses of thinness, light weight, large aperture, unlimited placement, and in-office application must be weighed against the disadvantages of reduced contrast, a slight reduction in visual acuity, and the visibility of the concentric grooves to observers.

Fresnel lenses in plus powers are available from +0.50 D to +20.00 D, while in minus lenses they are available from −1.00 D to −14.00 D; precut straight-top (D-25) segments are available from +0.50 to +8.00 D. Cylinder powers are not available in Fresnel Press-on lenses: If a patient's prescription contains a cylindrical component, the cylinder must be placed on the carrier lens.

There appears to be no end to the potential applications of Fresnel lenses: It is only limited by the practitioner's imagination.

## References

1. Donders, F.C. *On the Anomalies of Accommodation and Refraction of the Eye*, p. 309. The New Sydenham Society, London, 1864.
2. Hess, R., and Woo, G. Vision through Cataracts. *Invest. Ophthal. Vis. Sci.*, Vol. 17, pp. 428–431, 1978.
3. Arden, G. Measuring Contrast Sensitivity with Gratings: A New Technique for Early Diagnosis of Retinal and Neurological Disease. *J. Amer. Optom. Assoc.*, Vol. 50, pp. 35–39, 1979.
4. Enoch, J., Bedell, H., and Kaufman, H. Interferometric Visual Acuity Testing in Anterior Segment Disease. *Arch. Ophthal.*, Vol. 97, pp. 1916–1919, 1979.
5. Frayer, W. C., and Scheie, H. G. Cataract Surgery. In T.D. Duane, ed., *Clinical Ophthalmology*, Vol. 5, Ch. 7, p. 1. Harper and Row, New York, 1976.
6. Bettman, J.W. Apparent Accommodation in Aphakic Eyes. *Amer. J. Ophthal.*, Vol. 33, pp. 921–923, 1950.
7. Laurance, L. *Visual Optics and Sight Testing*, p. 117. The School of Optics, London, 1920.
8. Sanders, D.R., Retzlaff, J., and Kraff, M.C. Comparison of Empirically Derived and Theoretical Aphakic Refraction Formulas. *Arch. Ophthal.*, Vol. 101, pp. 965–967, 1983.
9. Ogle, K.N. *Optics: An Introduction for Ophthalmologists*, 2nd ed., p. 202. Charles C Thomas, Springfield, 1968.
10. Anonymous. The Adjustment to Aphakia. *Amer. J. Ophthal.*, Vol. 35, p. 118, 1952.
11. Smith G. and Bailey I.L., Aspheric Spectacle Lenses—Design and Performance, Part One: Introduction, The Optician, Vol. 181, pp. 21–26, 1981.
12. Smith G. and Atchison D.A., Effect of Conicoid Asphericity on the Tscherning Ellipses of Ophthalmic Spectacle Lenses. *J. Opt. Soc. Am.*, Vol. 73, pp. 441–445, 1983.
13. Fowler, C.W. Measurement of Distortion in Single Vision Ophthalmic Lenses. In *Transactions of the First International Congress*, Vol. 1, British College of Ophthalmic Opticians, London, pp. 148–157, 1984.
14. Fowler, C.W. Aspheric Spectacle Lens Design for Aphakia. *Amer. J. Optom. Physiol. Opt.*, Vol. 61, pp. 737–740, 1984.
15. Davis, J.K., and Torgersen, D.L. The Properties of Lenses for Correction of Aphakia. *J. Amer. Optom. Assoc.*, Vol. 54, pp. 685–693, 1983.
16. Atchinson, D.A., and Smith, G. Assessment of Aphakic Spectacle Lenses: High Power Flatback Lenses for the Correction of Aphakia. *Austral. J. Optom.*, Vol. 63, pp. 258–263, 1980.
17. Benton, C.D., and Welsh, R.C. *Spectacles for Aphakia*, p. 39, Charles C Thomas, Springfield, 1966.
18. Whitney, D.B. What Aspheric Design Can and Cannot Do for the Aphakic Patient. *Optom. Wkly.*, pp. 45–52, June 17, 1976.
19. *Optical, Cosmetic and Mechanical Properties of Ophthalmic Lenses*. Lens Information Kit, p. 4.201. Lens Development Dept., American Optical Co., Southridge, MA, March 1968.
20. Atchison, D.A. Problems of Spectacle Wearing Aphakics. *Austral. J. Optom.*, Vol. 66, pp. 216–218, 1983.
21. Ham, W.T., Jr., Muller, H.A., and Sliney, D.H. Retinal Sensitivity to Damage from Short Wavelength Light. *Nature*, Vol. 260, pp. 153–155, 1975.
22. Lerman, S.L. *Radiant Energy and the Eye*, p. 176, Macmillan, New York, 1980.
23. Ogle, K.N. *Researches in Binocular Vision*, p. 260. W.B. Saunders, Philadelphia, 1950.
24. Ogle, K.N., Burian, H.M., and Bannon, R.E. On the Correction of Unilateral Aphakia with Contact Lenses. *Arch. Ophthal.*, Vol. 59, pp. 639–652, 1958.
25. Enoch, J.M. A Spectacle-Contact Lens Combination Used as a Reverse Galilean Telescope in Unilateral Aphakia. *Amer. J. Optom. Arch. Amer. Acad. Optom.*, Vol. 45, pp. 231–240, 1968.
26. Tucker, T. The Use of High Index Glass. *Ophthal. Optician*, Vol. 16, pp. 795–804, 1976.
27. Obrig, T.E. *Modern Ophthalmic Lenses and Optical Glass*, p. 106. Chilton Co., New York, 1935.
28. Gumpelmeyer, T.F. Special Considerations in the Fitting of Corneal Lenses in High Myopia. *Amer. J. Optom. Arch. Amer. Acad. Optom.*, Vol. 47, pp. 879–886, 1970.
29. Kapash, R.T., and Barkan, E. Fresnel Optics and Human Visual Performance. Presented at the Annual Meeting of the American Academy of Optometry, Toronto, Canada, December 1971.

## Questions

1. According to Laurance, a pre-aphakic eye having 4.00 D of hyperopia would need how much spectacle correction after surgery?

2. According to Sanders et al., an eye that is 24.0 mm long with a pre-aphakic K reading of 44.00 D would require how much spectacle correction after cataract surgery?

3. Where is the far point located for an aphakic eye that has a refractive power of 42.00 D, an axial length of 24.0 mm, and a refractive index of 1.336?

4. If a spectacle lens is to be worn at a distance of 11 mm from the cornea of the patient in Problem 3, what must be its power?

5. What is the angular size of the "ring scotoma" of a $+12.50$ DS aphakic prescription that is made as a 50-mm round centered lens?

6. If a $+12.00$ D aphakic spectacle wearer with a vertex distance of 12 mm is fitted with contact lenses, what must be the power of the contact lenses?

7. What is the difference in the edge thickness between a $-10.00$ D, 50-mm round centered glass lens made of ophthalmic crown glass of refractive index 1.523 ($n = 1.523$) and one made with an index of refraction of 1.80 if both lenses have a center thickness of 1.5 mm?

8. Why are nuclear and subcapsular cataracts more troublesome outdoors than indoors?

9. Identify the factors affecting the size and position of the ring scotoma and describe the effect of each factor.

10. Discuss how oblique astigmatism and the curvature of the field are addressed in the design of lenses above $+8.00$ D.

11. Why is it best to place the optical center of aphakic lenses at the geometrical center of the lens?

12. Discuss the advantages and disadvantages of trifocals for aphakic spectacle wearers.

13. Why is it important to provide an ultraviolet filter for an aphakic patient?

14. Why should all high-index glasses have an anti-reflective coating?

15. List methods that can be used to reduce reflections from the edge of high minus lenses.

16. If the optical center is placed at the geometrical center of the lens, how can the optic axis of the lens be made to pass through the center of rotation of the eye?

17. Describe the design of aphakic lenses that minimize weight and thickness.

18. Discuss the advantages and disadvantages of lenticular aphakic lenses.

19. What is the meaning of "base curve" when applied to an aphakic lens with an aspheric surface?

# CHAPTER TWELVE

# Optical Principles of Lenses for Low Vision

"Low vision," or "partial sight," has been defined by Mehr and Freid[1] as "reduced central acuity or visual field loss which even with the best optical correction provided by regular lenses still results in visual impairment from the performance standpoint." Most definitions of low vision do not include a visual acuity "cutoff": However, reading ordinary newsprint requires vision in the neighborhood of 6/12 to 6/15 (20/40 to 20/50), so a person may be assumed to have low vision if regular spectacle lenses fail to provide vision of this magnitude.

Optical aids designed for patients having low vision differ from those for other patients in that the main function of these aids is to bring about magnification, with or without changing the vergence of incident light.

A large number of methods of providing magnification are available. For distance vision, telescopic lens systems are used, in the form of professionally prescribed headborne telescopic systems, nonprescription headborne telescopic systems, and hand-held telescopes. For near vision, a telescope with a "reading cap" may be used, but more commonly a high plus lens, either in the form of a "high add" or reading glasses, is prescribed. A large number of additional aids are available for near vision, including hand magnifiers, stand magnifiers, projection magnifiers, and even closed-circuit television.

## 12.1. Methods of Providing Magnification

Since a patient with low vision, by definition, cannot obtain satisfactory visual acuity in spite of the correction of any existing refractive error, any method of improving visual acuity requires the *enlarging of the retinal image* so that the image of a fixated object (such as a letter on a Snellen chart) will stimulate a larger number of photoreceptors. This, of course, means that some form of *magnification* must be provided.

The extent to which magnification actually improves visual acuity depends, to a great extent, upon the *cause* of the low vision. For example, if only a tiny central retinal area is available for vision (as in advanced cases of retinitis pigmentosa), enlarging the retinal image may serve only to stimulate nonfunctioning photoreceptors and not improve visual acuity or may even make it worse. Inasmuch as a discussion of the possible causes of low vision is beyond the scope of a textbook on the subject of optics, the authors will assume that (as is usually the case) any magnification of the retinal image will be accompanied by a proportional increase in visual acuity. For example, if the best visual acuity for a given eye when tested at a distance of 3 m (10 feet) is 3/30, or 10/100, a visual aid providing an angular magnification of 2x would be expected to result in a visual acuity of 3/15, or 10/50. Note that since visual acuity testing for low-vision patients is routinely done at a distance of 3 m (10 feet), this testing distance will be used for all visual acuity examples.

The intent of magnification, then, is to increase the size of the retinal image of the object of regard. Although the retinal image size may be specified in terms of the angle subtended at any reference point within the eye, a satisfactory point is the center of the eye's exit pupil. Since the centers of the entrance and exit pupils are conjugate points, any alteration of the angular subtense of the object at the center of the entrance pupil produces a proportional change in angular size of the retinal image measured from the center of the exit pupil. The use of the centers of the entrance/exit pupils as reference points has the addi-

tional advantage that all equations relating to them are valid for both in-focus and out-of-focus retinal images. Hence, if the angular subtense of the object for the eye's optical system is increased, there will also be an increase in the angular size of the retinal image.

In order to provide magnification for patients having low vision, the methods used increase the angular subtense of the object for the eye's optical system. The methods commonly used can be placed in four basic categories·

1. Relative distance magnification
2. Relative size magnification
3. Projection magnification
4. Angular magnification

When two or more of these methods are used in combination, the *total magnification* is defined as the *product* of each of the individual magnifications.

The discussion in this chapter will concentrate on methods of providing *angular magnification*, but a brief description of each of the first three methods follows.

*Relative distance magnification* is that magnification resulting from decreasing the distance of the object from the eye. In order to specify the amount of magnification, it is necessary to make use of an initial "standard" or "reference" distance, usually either 40 cm or 25 cm. As shown in Figure 12-1, an object ($h$ millimeters high) is at a distance $q$ millimeters from the entrance pupil of the eye and subtends an angle, $\alpha$, at the entrance pupil. If the object is moved so that it is at a distance $q'$ from the entrance pupil, it now subtends an angle of $\alpha'$.

The relative distance magnification, $M_D$, is found by the relationship

$$M_D = \frac{\tan \alpha'}{\tan \alpha} = \frac{h/q'}{h/q} = \frac{q}{q'}. \quad (12.1)$$

For example, if an object is moved from the reference distance of 40 cm to 10 cm, the relative distance magnification is

$$M_D = \frac{40}{10} = 4\times.$$

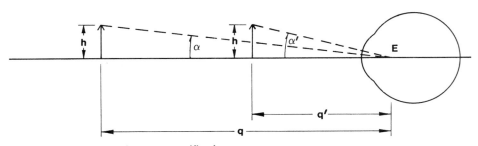

FIGURE 12-1. **Relative distance magnification.**

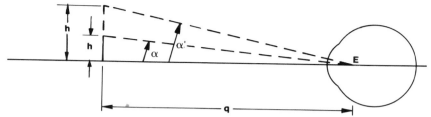

FIGURE 12-2. **Relative size magnification.**

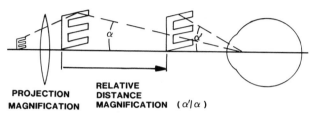

PROJECTION MAGNIFICATION    RELATIVE DISTANCE MAGNIFICATION ($\alpha'/\alpha$)

FIGURE 12-3. **Projection magnification used in combination with relative distance magnification.**

The use of short fixation distances requires large amounts of accommodation: For example, a 10-cm working distance requires 10 D of accommodation. If the patient's accommodation is insufficient, a clear retinal image can be obtained by the use of a plus lens of sufficient power. In this example, the use of a +10.00 D lens would require no accommodation at all.

*Relative size magnification* is that magnification obtained by increasing the size of the object at its original position, as shown in Figure 12-2. Large-print books, magazines, and newspapers are examples of this form of magnification. In order to quantify relative size magnification, the same fixation distance must be used for both object sizes. For example, if at 40 cm an object is 0.5 mm high but is increased to 2.0 mm high, the relative size magnification is

$$M_S = \frac{2.0}{0.5} = 4\times.$$

*Projection magnification* is that magnification resulting from the formation of an enlarged image, on a screen, of an opaque or transparent object. The projection system may be either optical or electronic, an example of the latter being a closed-circuit television system. Projection magnification makes possible a high level of magnification at a viewing distance that is convenient for the user, and has the advantage that it may be used in combination with relative distance magnification; that is, a lens bringing about relative distance magnification may be used to view the projected image, as shown in Figure 12-3. With such a combination, the aberrations accompanying the use of very strong optical aids are greatly reduced. Projection magnification can be considered as a special form of relative size magnification.

*Angular magnification* of an optical system (see Figure 12-4) is specified by the ratio of the angle subtended by the image of an object formed by the system to the angle subtended by the object when viewed directly, both angles being subtended at the center of the entrance pupil of the eye, or

$$AM = \frac{\text{angle subtended by the image of an object}}{\text{angle subtended by the object when viewed directly}}.$$

Thus, angular magnification compares the apparent increase in the size of the object seen through the optical system to the size of the object seen without the optical system in place.

Although relative size magnification and relative distance magnification can be specified in terms of angular subtense, the term angular magnification is usually reserved for the magnification produced by the optical system itself.

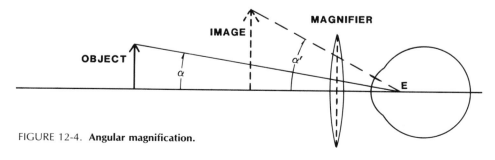

FIGURE 12-4. **Angular magnification.**

# _____OPTICAL AIDS FOR DISTANCE VISION_____

## 12.2. Afocal Telescopes

An afocal telescope is defined as an optical system that provides angular magnification without bringing about a change in vergence. Both of the classical afocal telescope designs—the Galilean and the Keplerian telescopes—are used as aids for low vision.

In their simplest construction afocal telescopes contain two optical elements, the objective and the eyepiece. In all telescopes designed to provide angular magnification the objective lens is positive in power and is placed toward the object to be viewed. The eyepiece, placed close to the eye of the observer, is much stronger in power than the objective and may be either a positive or negative lens.

With all afocal telescopes the basic optical principle is that the secondary focal plane of the objective lens coincides with the primary focal plane of the eyepiece lens. Parallel rays incident upon the objective lens form an image at the secondary focal plane, and this image becomes the object for the eyepiece lens. Since the object is located at the primary focal plane of the eyepiece, the rays will emerge from the system parallel, but will now form a greater angle with the optic axis than that formed by the incident rays.

The _Galilean telescope_, shown in Figure 12-5, uses a positive-powered objective lens (O) and a negative-powered eyepiece (E). The objective lens forms a _virtual_ image of height $h$, with the chief ray through the optical center of the objective lens forming an angle $\alpha$ with the optical axis. The ray passing through the optical center of the eyepiece helps form the tip of the virtual image and emerges undeviated from

the eyepiece lens. This ray forms an angle $\alpha'$ with the optic axis.

Thus parallel rays enter the objective lens at angle $\alpha$ to the optic axis, and a parallel bundle emerges from the eyepiece lens at angle $\alpha'$ from the optic axis. The ratio $\alpha'/\alpha$ compares the angular size of the image to the angular size of the object. In Figure 12-5, $F_o$ = power of objective lens; $F_e$ = power of eyepiece lens; $f_o'$ = secondary focal length of objective lens; $f_e$ = primary focal length of eyepiece lens; thus

$$M = \frac{\alpha'}{\alpha},$$

or, more exactly,

$$M = \frac{\tan \alpha'}{\tan \alpha} = \frac{h/f_e}{h/f_o'} = \frac{f_o'}{f_e},$$

or

$$M = -\frac{F_e}{F_o}. \qquad (12.2)$$

The separation, $d$, of the objective lens and the eyepiece can be determined as follows:

$$d = f_o' - f_e \qquad (12.3)$$

or

$$d = \frac{1}{F_o} + \frac{1}{F_e},$$

and

$$F_o = \frac{F_e}{dF_e - 1} \qquad (12.4)$$

$$F_e = \frac{F_o}{dF_o - 1} \qquad (12.5)$$

If the value of $F_o$ in Eq. (12.4) is substituted in Eq. (12.2),

$$M = \frac{-F_e}{\dfrac{F_e}{dF_e - 1}} = -(dF_e - 1),$$

or

$$M = 1 - dF_e. \qquad (12.6)$$

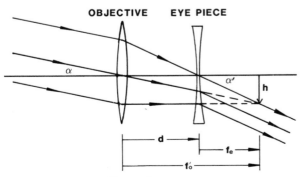

**OBJECTIVE    EYE PIECE**

FIGURE 12-5. **The Galilean telescope.**

If the value of $F_e$ in Eq. (12.5) is substituted in Eq. (12.2),

$$M = \frac{F_o}{\frac{dF_o - 1}{-F_o}} = \frac{1}{-(dF_o - 1)},$$

or

$$M = \frac{1}{1 - dF_o}. \qquad (12.7)$$

Galilean telescopes can be shown to have the following characteristics:

1. Since for a Galilean telescope $F_e$ is always a negative lens and $F_o$ is always a positive lens, in the formula

$$M = \frac{-F_e}{F_o},$$

the magnification will have a positive sign, which indicates that the image formed will be *erect*.

2. The length, $d$, of the telescope is the secondary focal length of the objective lens minus the primary focal length of the eyepiece. As will be seen later, the Galilean telescope will be *shorter* than the Keplerian telescope if both have the same magnification. For constant magnification, increasing the power of both the objective and the eyepiece produces a shorter telescopic unit.

### EXAMPLE

An afocal Galilean telescope has a $+10.00$ D objective lens and a $-25.00$ D eyepiece lens. What is the magnifying power of this telescope?

$$M = \frac{-F_e}{F_o} = -\frac{-25.00}{10.00} = 2.5 \times.$$

What will be the separation of the two lenses?

$$d = f_o' - f_e$$
$$= +0.10 - 0.04 = 0.06 \text{ m} = 6 \text{ cm}.$$

The *Keplerian telescope*, shown in Figure 12-6, has a positive-powered objective lens and a stronger positive-powered eyepiece lens. A parallel bundle of rays incident upon the objective lens at an angle $\alpha$ to the optic axis forms a real image of height $h$ at the secondary focal plane of the objective lens. Since the primary focal plane of the eyepiece lens coincides with the secondary focal plane of the objective lens, the rays emerge parallel from the eyepiece lens at an angle $\alpha'$ to the optic axis. The magnification brought about by the telescope is

$$M = \frac{\alpha'}{\alpha} = \frac{\tan \alpha'}{\tan \alpha}$$
$$= \frac{h/f_e}{h/f_o'} = \frac{f_o'}{f_e} = \frac{-F_e}{F_o}.$$

The separation, $d$, of the objective lens and the eyepiece lens of the Keplerian telescope is given by the relationship

$$d = f_o' - f_e.$$

Since the secondary focal length of the objective lens is positive and the primary focal length of the

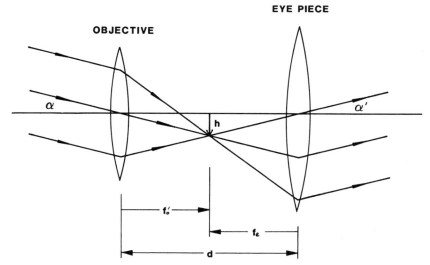

FIGURE 12-6. **The Keplerian telescope.**

eyepiece lens is negative, the separation is the *sum* of these two focal lengths. Hence, as mentioned earlier, a Keplerian telescope will always be *longer* than a Galilean telescope of the same magnifying power.

Since $F_o$ and $F_e$ are both positive in power, in the formula

$$M = \frac{-F_e}{F_o},$$

the magnification will be negative, indicating that the image is *inverted*. Therefore, all Keplerian telescopes used as low-vision aids must incorporate an image-erecting lens or prism system and are referred to as *terrestrial* telescopes. On the other hand, when a sense of orientation is not important, an image-erecting system is not incorporated into the Keplerian telescope, and such a telescope is described as an *astronomical* telescope.

The basic formulas developed for Galilean telescopes, given below, are valid also for Keplerian telescopes. However, the magnification will be positive in sign for the Galilean telescope (image erect) and negative (image inverted) for the Keplerian telescope, unless the telescope is fitted with an erecting system.

$$M = \frac{-F_e}{F_o}$$

$$d = f_o' - f_e$$

$$M = \frac{1}{1 - dF_o}$$

$$= 1 - dF_e.$$

**EXAMPLE**

An afocal Keplerian telescope has a +10 D objective lens and a +30.00 D eyepiece lens. What is the magnification?

$$M = \frac{-F_e}{F_o} = \frac{-30}{10} = -3\times.$$

What will be the separation of the two lenses?

$$d = f_o' - f_e$$
$$= 0.10 - (-0.033) = 0.133 \text{ m} = 13.3 \text{ cm}.$$

Galilean and Keplerian afocal telescopic systems may be compared in the following respects:

1. A Galilean telescope is shorter than a Keplerian telescope of similar magnification.
2. The image formed by a Galilean telescope is erect while that formed by a Keplerian telescope is inverted, requiring an erecting system when used as a low-vision aid.

3. The design of a Galilean telescope is usually simpler than that of a Keplerian telescope.
4. Keplerian telescopes are moderately heavy compared to Galilean telescopes.
5. The image quality and brightness across the field of view are usually better in Keplerian telescopes than Galilean telescopes.
6. The magnification available for telescopes of Keplerian design (up to 10×) is higher than that available for those of Galilean design (usually less than 4×).
7. Keplerian telescopes usually have larger fields of view than Galilean telescopes. The size and the position of the *exit pupil* of the telescope are important factors in determining the field of view. In the Galilean telescope the exit pupil is virtual and usually falls inside the telescope, between the objective and the eyepiece, at some distance in front of the entrance pupil of the eye. However, in the Keplerian telescope the exit pupil is *real* and is located a short distance behind the eyepiece, close to the eye's entrance pupil. The position of the exit pupil of the telescope in relation to the eye's entrance pupil increases the field of view of the Keplerian telescope as compared to the Galilean telescope.

In discussing afocal telescopes it is assumed that they are used with emmetropic eyes or those that are made artifically emmetropic, with the distance correction in place. The distance correction can, however, be incorporated into the eyepiece of the telescope.

Some telescopes are adjustable, and with such telescopes it is possible to correct for *spherical* ametropia (for distance vision) by varying the distance between the objective lens and the eyepiece. If the wearer of a telescope is hyperopic with respect to a distant object the separation of the lenses is *increased* as compared to the afocal setting, whereas if the wearer is myopic the separation is *decreased*. This applies to both the Galilean and Keplerian designs.

The tube length required to correct spherical ametropia is illustrated by the following:

**EXAMPLE 1**

What tube length is required for an uncorrected 5.00 D myope to have clear distance vision through a 4× Galilean telescope having a +5.00 D objective lens?

$$M = +4 = \frac{-F_e}{F_o} = \frac{-F_e}{+5}$$

$$F_e = -20.00 \text{ D}.$$

$$d = f_o' - f_e$$
$$= 0.20 - 0.05 = 0.15 \text{ m}$$
$$= 15 \text{ cm.}$$

Since the telescope is to be used by an uncorrected 5.00 D myope, the vergence leaving the eyepiece must be $-5.00$ D. The eyepiece lens must therefore be moved a distance, $Z$, from where the vergence upon it is $+15.00$ D rather than $+20.00$ D:

$$\frac{+20.00}{1 - Z(+20.00)} = +15.00 \text{ D}$$

$$Z = -\frac{5}{300} = -0.0167 \text{ m} = -1.67 \text{ cm.}$$

Therefore, the value of $d$, for the 5.00 D myopic eye ($d_{\text{ametropia}}$), as compared to its value for the emmetropic eye ($d_{\text{emetropia}}$), is given by

$$d_{\text{ametropia}} = d_{\text{emmetropia}} + Z$$
$$= 15 + (-1.67)$$
$$= 13.33 \text{ cm.}$$

### EXAMPLE 2
What tube length is required for an uncorrected 5.00 D myope to have clear distance vision through a 10× Keplerian telescope having a +5.00 D objective lens?

$$M = -10 = \frac{-F_e}{F_o} = \frac{-F_e}{+5}$$

$$F_e = +50.00 \text{ D.}$$

$$d = f_o' - f_e$$
$$= 0.20 - (-0.02) = 0.22 \text{ m}$$
$$= 22 \text{ cm.}$$

Since the telescope is to be used by an uncorrected 5.00 D myope, the vergence leaving the system must be $-5.00$ D. The eyepiece lens must therefore be moved a distance of $Z$, from where the vergence upon it is $-55.00$ D rather than $-50.00$ D:

$$\frac{-50.00}{1 - Z(-50.00)} = -55.00 \text{ D}$$

$$Z = -\frac{5}{2750} = -0.0018 \text{ m} = -0.18 \text{ cm.}$$

Therefore, the value of $d$, for the 5.00 D myopic eye ($d_{\text{ametropia}}$), as compared to its value for the emmetropic eye ($d_{\text{emmetropia}}$), is given by

$$d_{\text{ametropia}} = d_{\text{emmetropia}} + Z$$
$$= 22 + (-0.18)$$
$$= 21.82 \text{ cm.}$$

It should be noted that, when using the adjustable feature to correct ametropia for distant objects, the effective magnification is changed. For a myope, the angular magnification is reduced with a Galilean telescope and increased with a Keplerian telescope, whereas for a hyperope, the angular magnification is increased with a Galilean telescope and decreased with a Keplerian telescope. Smith[2] has developed an expression which provides the effective magnification:

$$M' = M + \frac{d(M)(A)}{M - 1},$$

where $M'$ = the new effective magnification, $M$ = the original magnification, $d$ = the original separation, and $A$ = the refractive correction (having a minus sign for a myopic correction and a plus sign for a hyperopic correction).

### EXAMPLE 1
A 2× afocal Galilean telescope has an objective lens of +25.00 D and an eyepiece lens of $-50.00$ D. If the telescope is adjusted to provide a correction for a 5.00 D myope for distance vision, what will be the effective magnification?

$$d = f_o' - f_e$$
$$= 0.04 - 0.02$$
$$= 0.02 \text{ m} = 2 \text{ cm.}$$

$$M' = M + \frac{d(M)(A)}{M - 1}$$
$$= 2 + \frac{(0.02)(2)(-5)}{2 - 1}$$
$$= 2 + (-0.2)$$
$$= 1.8×.$$

### EXAMPLE 2
If a 2× afocal Keplerian telescope with an objective lens of +25.00 D and a +50.00 D eyepiece lens is adjusted for a 5.00 D myope for distance vision, what is the new effective magnification?

$$d = f_o - f_e$$
$$= 0.04 + 0.02$$
$$= 0.06 \text{ m} = 6 \text{ cm.}$$

$$M' = M + \frac{d(M)(A)}{M - 1}$$
$$= -2 + \frac{(0.06)(-2)(-5)}{-2 - 1} = -2 + \frac{0.6}{-3}$$
$$= -2 + (-0.2)$$
$$= -2.2×.$$

## 12.3. Headborne Telescopic Systems

Until recently the great majority of afocal telescopes used as low-vision aids were of Galilean design. Keplerian telescopes, sometimes referred to as "expanded-field" telescopes, have, however, seen much greater use in recent years. As discussed by Bailey,[3] although these telescopes have the advantage of having a wider field of view than Galilean telescopes, they have the disadvantage of being longer and heavier, and are about three times more costly.

Headborne telescopes, mounted in spectacle frames, are available from many manufacturers as *full-diameter* telescopes, the telescope occupying the entire aperture of the spectacle lens; and available from Designs for Vision in what they call a *Bioptic* telescope, a small-diameter telescopic unit mounted in the upper part of a carrier lens with the patient's prescription provided in the carrier lens. Telescopes of Galilean design are available as full-diameter telescopes in magnifying powers ranging from 1.3× to 2.2× and as Bioptic telescopes in powers ranging from 2.2× to 4×; Keplerian "expanded-field" telescopes are available in powers ranging from 2× to 8×. Bailey[3] has commented that in his experience the most useful of the Galilean Bioptic telescopes has been the 2.2×, while Robert Browning, of the University of Houston Low Vision Clinic, (personal communication) has found that the most useful Keplerian telescopes are the 3× and the 4×.

As already stated, the practitioner can usually expect an improvement in visual acuity proportional to the angular magnification provided by the aid. For example, if without the telescopic aid a patient's corrected visual acuity is 3/60 (10/200), a 2× afocal telescope should improve visual acuity to 3/30 (10/100) and a 4× afocal telescope should improve visual acuity to 3/15 (10/50).

### Problems Associated with Telescopes for Distance Vision

It should be understood that a telescopic system brings about angular magnification at the expense of making drastic changes in the sizes of perceived objects and in the extent of the visual field. A 2× tele-scope not only makes an object appear to be twice as *large*, but twice as *near*, and also brings about an apparent increase in the rate of motion of moving objects, in addition to greatly reducing the visual field and providing extreme peripheral aberrations. The perceptual problems encountered by a patient wearing a telescopic lens are similar to those experienced by a spectacle-wearing aphakic patient, but are greatly intensified.

Because of these perceptual problems, full-diameter telescopic spectacles are indicated mainly for relatively sedentary activities rather than for walking and other forms of locomotion. However, a Bioptic telescope can be worn for locomotion (and even for driving an automobile) by some patients: With such a telescope the carrier lens, which provides no magnification, is used for routine locomotion, and the head is tilted downward slightly when a magnified view through the telescope is desired.

## 12.4. Nonprescription Telescopes

A large number of manufacturers make available nonprescription headborne telescopic systems, often categorized as "sport-scopes," intended only for occasional use. These aids may also be used by normally sighted individuals, in much the same way as prism binoculars are used, and therefore are available at a fraction of the cost of telescopic prescription spectacles.

Other forms of inexpensive telescopic systems include monocular clip-on telescopic units (designed to be clipped on to ordinary spectacles) and hand-held monocular telescopes. These aids are intended mainly for momentary use, such as identifying a street sign or the route number of a bus, and are available in magnifying powers as high as 6× to 10×. Genensky,[1] a researcher for Rand Corporation who is partially sighted, recommends the use of ordinary prism binoculars for occasional use for distance vision, on the basis that they can easily be worn around the neck, are always available when needed, and are more convenient to use than a monocular telescope even though only one eye may be used with the binoculars.

# OPTICAL AIDS FOR NEAR VISION

## 12.5. Microscopic Lenses

One of the simplest low-vision aids for near vision is the use of high-powered addition, in the form of either bifocals or single-vision lenses for near vision only. The term "microscopic," often used to describe such an aid, does not imply a compound lens system as used in a laboratory microscope.

Although microscopic lenses, reading magnifiers, and loupes are often grouped together, a distinction should be made between them. All are basically plus-powered systems that allow the viewing of an object at a short distance with the object located inside the optical system's focal length. *Microscopic lenses* are usually mounted in a conventional frame at a very short vertex distance. A *reading magnifier*, on the other hand, is used at a relatively great distance from the eye: As a result, the diameter of the lens is large in order to provide an adequate field of view. Increasing the size also increases the weight of the lens, with the result that a reading magnifier is normally hand-held or mounted on a stand. *Loupes*, like spectacle lenses, are usually worn on the head with an appropriate device or attached to the spectacle frame. However, the vertex distance is considerably longer than ordinarily used for spectacles, and hence the required distance falls between that used for spectacles and that used for reading magnifiers.

In order to be an effective low-vision aid, the power of the addition of a microscopic lens must be extremely high. Semifinished bifocals kept in stock by ophthalmic laboratories seldom have adds exceeding +4.00 D, and even on special order additions available for fused bifocals are usually restricted to about +4.50 D because of the increased counter-sink depth and the resulting increase in lens thickness. However, on special order, one-piece bifocals may be obtained with additions as high as +20.00 D.

Wild[5] has pointed out that if the segment is placed on the front surface of a high-powered one-piece bifocal, the magnification is increased compared to that obtained by placing it on the back surface. This also has the advantage that if a cylindrical correction is needed, it is placed on the back surface.

As already stated, high additions may be ordered also in the form of single-vision lenses for near only. However, beyond a total power of +10.00 D (in either single-vision lenses or bifocals) ordinary corrected-curve ophthalmic lenses are subject to excessive aberrations, so the use of lenses designed specifically for low-vision patients should be considered.

Many manufacturers have microscopic lenses available, both in bifocal and in single-vision form. For example, Designs for Vision makes a bifocal microscope in magnifying powers from 2× through 10× (+4.00 through +40.00 D), with the distance prescription ground into the carrier lens, and also makes full-diameter microscopes in magnifying powers ranging from 2× to 20×.

American Optical has available glass microscopic lenses, in bifocal form, in magnifying powers of from 2× to 8× (+8.00 to +32.00 D) and plastic aspheric single-vision microscopic lenses in magnifying powers from 6× to 12×. American Optical also markets plastic microscopic lenses in half-eye frames, in refracting powers from +4.00 to +12.00 D.

### Dioptric Power Notation for Microscopic Lenses

For simplicity in calculation, low-vision lenses are often considered to be single *thin* lenses. For a thin lens, no distinction is made between back vertex power, front vertex power, and equivalent power. In reality, most low-vision magnifiers are *thick* lenses, and the appropriate power notation for such lenses is *equivalent power*. It will be recalled that equivalent power is the negative reciprocal of the distance between the primary principal point and the primary focal point, or the reciprocal of the distance between the secondary principal point and the secondary focal point. In computing the relative sizes of object and image for any thick lens system, object and image distances must be measured from the principal planes. To compare sizes of retinal images accurately for thick low-vision aids, the equivalent powers of the aids must be known. Manufacturers' specifications normally list the equivalent power of low-vision devices.

A disadvantage of using equivalent power (as opposed to back or front vertex power) is that the practitioner cannot easily identify the location of the primary or secondary principal plane of a lens or optical system. However, Bailey[6] has described a method of determining equivalent power that can be used in the practitioner's office.

The general formula for angular magnification can be developed from Figure 12-7, in which an object, Y, is placed inside the primary focal point of the lens at a distance *l* from the lens. The lens rests at

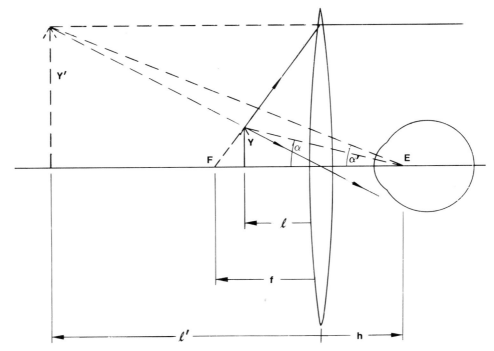

FIGURE 12-7. **Angular magnification of a microscopic lens when the object is placed inside the primary focal plane.**

a distance $h$ from the center of the entrance pupil of the eye. The object and the image subtend angles $\alpha$ and $\alpha'$, respectively, at the center of the entrance pupil. A formula for the angular magnification may be derived as follows:

$$M = \frac{\tan \alpha'}{\tan \alpha},$$

where

$$\tan \alpha' = \frac{y'}{l' - h} \quad \text{and} \quad \tan \alpha = \frac{y}{l - h}.$$

Substituting,

$$M = \left(\frac{\dfrac{y'}{l' - h}}{\dfrac{y}{l - h}}\right) = \left(\frac{y'}{l' - h}\right)\left(\frac{l - h}{y}\right)$$

$$= \left(\frac{y'}{y}\right)\left(\frac{l - h}{l' - h}\right).$$

Since

$$\frac{y'}{y} = \frac{l'}{l},$$

$$M = \left(\frac{l'}{l}\right)\left(\frac{l - h}{l' - h}\right).$$

Since

$$\frac{1}{l'} = \frac{1}{l} + F,$$

$$l' = \left(\frac{1}{\dfrac{1}{l} + F}\right) = \left(\frac{1}{\dfrac{1 + Fl}{l}}\right).$$

Therefore,

$$M = \left(\frac{l}{1 + Fl}\right)\left(\frac{1}{l}\right)\left(\frac{l - h}{\dfrac{l}{1 + Fl} - h}\right)$$

$$= \left(\frac{1}{1 + Fl}\right)\left(\frac{l - h}{\dfrac{l - h(1 + Fl)}{1 + Fl}}\right),$$

and

$$M = \frac{l - h}{l - h(1 + Fl)}. \tag{12.8}$$

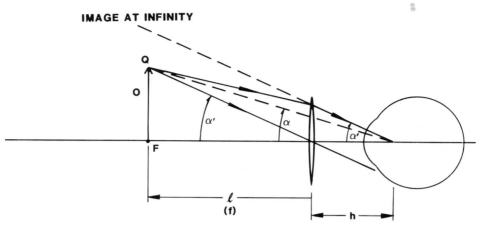

FIGURE 12-8. **Angular magnification of a microscopic lens when the object is placed at the primary focal plane.**

In this derivation, the distance $l$ is less than the distance $f$, and the object lies inside the primary focal plane of the lens. Maximum magnification occurs when $l = f$, that is, when the object is located at the primary focal plane of the lens and the image is located at infinity. Therefore, when the object is located at the primary focal plane (see Figure 12-8), Eq. (12.8) reduces initially to

$$M = \frac{l - h}{l},$$

and by substituting $-(1/F)$ for $l$,

$$M = \frac{-\dfrac{1}{F} - h}{-\dfrac{1}{F}} = \frac{\dfrac{1}{F} + h}{\dfrac{1}{F}}$$

$$M = \frac{\dfrac{1 + hF}{F}}{\dfrac{1}{F}} = 1 + hF. \tag{12.9}$$

This formula gives the amount of *angular magnification* for an object in the primary focal plane of the lens, that is, the ratio of the angular subtense of the object seen through the lens to the angular subtense seen without the lens, with the object at a constant distance from the eye. The *total magnification* resulting from a microscopic lens is the product of the *angular magnification* and the *relative distance magnification*. Although microscopic lenses provide some angular magnification, the amount is small because

of the closeness of the lens to the eye. Most of the increase in retinal image size originates from relative distance magnification. In situations requiring large amounts of relative distance magnification, the eye is unable to maintain sustained accommodation for such a close distance (or in the case of an absolute presbyope is unable to accommodate at all). Therefore, the primary function of a high-powered plus lens is to allow the patient to see near objects clearly with little or no accommodation. For example, if a +8.00 D lens is placed 15 mm from the eye, the angular magnification for an object in the primary focal plane is

$$\begin{aligned} M &= 1 + hF \\ &= 1 + 0.015(8) \\ &= 1 + 0.12 \\ &= 1.12\times, \end{aligned}$$

and the percentage magnification is equal to

$$100(M - 1) = 100(1.12 - 1) = 12\%.$$

However, if the reading card was originally held at a distance of 40 cm from the eye, the distance of the card from the eye when viewed through the lens is now

$$\begin{aligned} 1/8 \text{ m} + 0.015 \text{ m} &= 0.125 \text{ m} + 0.015 \text{ m} \\ &= 0.140 \text{ m} = 14 \text{ cm.} \end{aligned}$$

Therefore, the *relative distance magnification* is

$$40/14 = 2.857\times.$$

*Total magnification,* also called *effective magnification,* is the product of the angular magnification and the relative distance magnification:

$$1.12(2.857) = 3.2\times.$$

**IMAGE AT INFINITY**

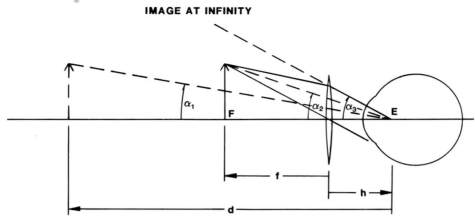

FIGURE 12-9. **Total magnification, or effective magnification: The product of the relative distance magnification and the angular magnification.**

If $d$ (in Figure 12-9) is the reference viewing distance and the quantity $(f - h)$ is the distance of the object from the eye with the lens before the eye, then the angular magnification is equal to

$$\frac{\tan \alpha_3}{\tan \alpha_2} = 1 + hF$$

and the relative distance magnification is

$$\frac{d}{f - h},$$

and the total magnification is

$$M = \text{(relative distance magnification)}$$
$$\times \text{(angular magnification)}$$

$$= \frac{d}{f - h}(1 + hF)$$

$$= \frac{dF}{fF - hF}(1 + hF)$$

$$= \frac{dF}{\dfrac{f}{-f} - hF}(1 + hF)$$

$$= \frac{dF}{-1 - hF}(1 + hF)$$

$$= \frac{-dF}{1 + hF}(1 + hF)$$

$$M = -dF. \qquad (12.10)$$

This formula emphasizes the need for establishing a standard reference viewing distance. Although arbitrary, historically this distance has long been taken as 25 cm, a distance known as the "least distance of distinct vision." Hence formula (12.10) is often stated as

$$M = -(-0.025)F,$$

or

$$M = \frac{F}{4}, \qquad (12.11)$$

and magnification so stated has been known as *effective* magnification. Thus, a +4.00 D magnifier designed to be used when viewing an object at 25 cm has an effective magnifying power of 1×, while a +8.00 D magnifier designed to view an object at 12.5 cm has an effective magnifying power of 2×, and so forth.

Most optometrists would prefer to use 40 cm as the standard reference distance, since they are accustomed to testing their patients' near vision at this distance.

Magnifiers are sometimes described in terms of *conventional magnification*, which is defined as the ratio of the angular subtense of the image formed at the standard reference distance from the eye by the optical system to the angular subtense of the object placed at the standard reference distance *without* the optical system in place (see Figure 12-10). It is assumed that the ametropia is corrected and that the same amount of accommodation is required with and without the optical system.

From Figure 12-10 it is apparent that

$$M = \frac{O'}{O} = \frac{l'}{l} = \frac{L}{L'}$$

$$= \frac{L' - F}{L'}$$

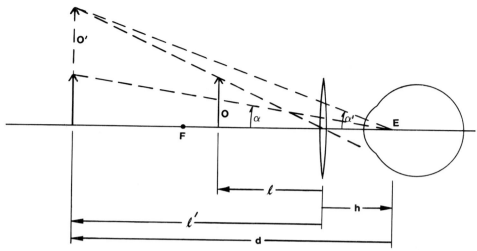

FIGURE 12-10. **Conventional magnification: the ratio of the angular subtense of the image formed by the optical system at the standard reference distance, *d*, from the eye, to the angular subtense of the object placed at the standard reference distance *without* the optical system in place.**

$$= \frac{L'}{L'} - \frac{F}{L'}$$

$$= 1 - \frac{F}{L'} = 1 - l'F,$$

and since $l' = h + d$,

$$M = 1 - (h + d)F.$$

If the magnifying lens is held close to the eye so that the distance $h$ can be ignored and if the standard reference distance, $d$, is equal to $-0.25$ m, then the conventional magnification can be stated as

$$M = 1 - (-0.25)F = 1 + \frac{F}{4}. \quad (12.12)$$

When commercially available microscopic low-vision aids are rated only by their magnification, it may be impossible to know the underlying assumptions used in making such ratings. It is more reliable, therefore, to simply specify the *equivalent power* rather than the magnifying power. Specification of magnification in terms of equivalent power has the advantage that it doesn't presuppose a standard reference distance, and knowing the equivalent power of a given microscopic lens the practitioner can then easily determine the magnification (and the expected visual acuity) that it will provide. For example, if a presbyopic patient's best corrected near acuity with a +2.50 D add using a reduced Snellen chart at a distance of 40 cm is 20/100, then a microscopic lens having an equivalent power of +10.00 D would en-

able the patient to read at 10 cm, providing a magnification of 4×, with an expected visual acuity of 20/25.

Although the use of equivalent power is the most desirable method of specifying the magnifying power of microscopic lenses, most manufacturers still specify magnification on the basis of the standard reference distance of 25 cm. Table 12-1 shows the magnifying power (based on the 25-cm distance) of microscopic lenses of various equivalent powers, with corresponding reading distances for absolute presbyopes.

### Problems Associated with Microscopic Lenses

The following problems are associated with the use of microscopic reading aids:

TABLE 12-1
**Magnifying Power of Microscopic Lenses on the Basis of a Standard Reading Distance of 25 cm, Together with the Necessary Reading Distance for an Absolute Presbyope**

| Equivalent Power | Magnifying Power | Reading Distance |
|---|---|---|
| +4.00 D | 1× | 25 cm |
| +6.00 D | 1.5× | 16.7 cm |
| +8.00 D | 2× | 12.5 cm |
| +10.00 D | 2.5× | 10 cm |
| +12.00 D | 3× | 8.3 cm |
| +16.00 D | 4× | 6.25 cm |
| +20.00 D | 5× | 5 cm |

1. *Aberrations.* Lenses of the very high powers required for microscopic aids often produce severe aberrations unless they are carefully designed. Aberrations may, however, be controlled by using aspheric surfaces or multiple elements, adding considerably to the expense.

2. *Illumination.* When the reading material is held very close to the eyes it is difficult to provide adequate illumination on the reading material: A large amount of ambient light is blocked by the patient's head.

3. *Centration of Lenses.* Often a patient who requires a microscopic aid at near has such a severe visual loss in each eye that an aid is prescribed only for the better eye. For example, if the patient's reduced Snellen acuity at 40 cm is 20/100 with the right eye but less than 20/200 for the left eye, the practitioner may decide to prescribe the aid only for the right eye. On the other hand, if acuity is approximately equal for the two eyes, the patient will very likely be able to make use of binocular vision if an aid is prescribed for each eye. In such a case, the practitioner must know where to place the *centers* of the microscopic aids for the two eyes. Since reading will be done at a very close distance, obviously more inward decentration will be required than the "2 mm in" normally used for bifocal lenses.

If it is assumed that the stop distance (the distance between the back vertex of the lens and the center of rotation of the eye) is 27 cm, inspection of Figure 12-11 (note the similar triangles ABC and A'B'C) shows that the required amount of decentration can be determined by the use of the following formula:

$$\text{total decentration} = \frac{27(\text{distance PD})}{(\text{reading distance}) + 27}.$$
$$(12.13)$$

For example, for a reading distance of 10 cm and an interpupillary distance of 60 mm, the total decentration would be

$$\frac{27(60)}{100 + 27} = 12.8 \text{ mm, rounding to 13 mm,}$$

with the result that each lens should be decentered inward 6.5 mm.

Note that Eq. (12.13) does not require the power of the microscope, only the reading distance at which the microscope will be used. Table 12-2 shows the required amount of decentration for typical reading distances, assuming a distance PD of 64 mm and a stop distance of 27 mm.

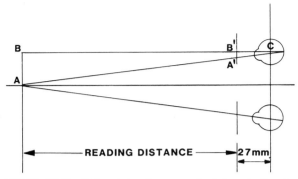

FIGURE 12-11. **Determining the amount of decentration for microscopic lenses.**

TABLE 12-2
**Decentration Required for Microscopic Lenses, for a Distance PD of 64 mm and a Stop Distance of 27 mm**

| Reading Distance | Decentration (Total) | Decentration (Each Lens) |
|---|---|---|
| 40 cm | 4 mm | 2 mm |
| 33 cm | 5 mm | 2.5 mm |
| 25 cm | 6 mm | 3 mm |
| 20 cm | 8 mm | 4 mm |
| 17 cm | 9 mm | 4.5 mm |
| 12.5 cm | 11 mm | 5.5 mm |
| 10 cm | 14 mm | 7 mm |

In discussing decentration of microscopic lenses, Bailey[7] made the point that when a reading distance closer than 10 cm is necessary, binocularity becomes impractical so only a lens for the better eye should be prescribed. He suggested that when lenses are to be prescribed for both eyes, the amount of decentration may be determined by the following rule of thumb: For each diopter of working distance (i.e., add for near), give 1.5 mm of total decentration (both eyes), and if the distance PD is greater than 65 mm, add 1 mm of decentration.

4. *Maintenance of the Proper Viewing Distance.* Although a microscopic lens can greatly improve the near acuity of a low-vision patient, many patients have difficulty adapting to the very short reading distance that is necessary. For example, a microscopic lens with an equivalent power of +20.00 D requires a reading distance of 5 cm. Holding the reading material at the correct viewing distance and keeping the facial plane parallel to the page is often a most difficult task. Since the depth of focus of a strong plus lens is very small, movement of the reading material only a slight distance from the primary

focal plane of the lens provides large changes in vergence of the light emerging from the lens. If the reading material falls within the focal point, the emergent light will be divergent and the retinal image will be blurred unless accommodation is used. Therefore, for presbyopic or aphakic patients the reading material must be held in the primary focal plane of the lens in order for the print to be seen clearly. Many elderly patients cannot hold reading material steadily at the required distance for long periods of time.

## 12.6. Telescopic Lenses for Near Vision

Even if an individual has a high amplitude of accommodation, it is virtually impossible to accommodate for near objects through an afocal telescope. This is because of what is called the "vergence amplification effect": When diverging rays are incident upon a telescope, the emerging rays have considerably more divergence than the incident rays (as shown in Figure 12-12). Because of this effect, if an attempt is made to accommodate for a near object through an afocal telescope, the amount of accommodation necessary will be much greater than indicated by the object distance.

An approximate equation for the vergence amplification effect has been proposed by Bailey[8]:

$$L_2' = M^2 L_1, \qquad (12.14)$$

where $L_2' =$ the emerging vergence, $M =$ the magnifying power of the telescope, and $L_1 =$ the incident vergence.

For a $2\times$ telescope used at a distance of 40 cm, the vergence amplification effect would be

$$2^2(-2.50) = 4(-2.50) = -10.00 \text{ D},$$

and therefore the amount of accommodation required at that distance would be 10 D. For a $4\times$ telescope used at the same distance, the vergence amplification effect would be

$$4^2(-2.50) = 16(-2.50) = -40.00 \text{ D}.$$

The problem of the vergence amplification effect can be solved by placing a plus lens of appropriate power either in front of or behind the telescope. It should be understood that when a plus lens is placed *in front of* a telescope to be used for near vision, parallel rays of light will enter the telescope with the result that the vergence amplification effect will not occur; whereas if a plus lens is placed *behind* the telescope, the vergence amplification effect will have

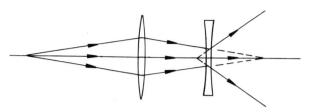

FIGURE 12-12. **The vergence amplification effect: When diverging rays are incident upon a telescope, the emerging rays have a greater vergence than the incident rays.**

taken place with the result that a lens of much higher plus power will have to be used.

Therefore, a telescopic system to be used for near vision is usually fitted with what is called a "reading cap" on the front of the telescope. The reading cap is easily removable, and can be carried by the patient and worn when needed for near vision. The reading cap must have the power determined by the desired reading distance; that is, for a reading distance of 40 cm, the reading cap must have a power of +2.50 D, or for a distance of 20 cm, it must have a power of +5.00 D. The combination of an afocal telescope and a reading cap is known as a *telemicroscope*.

For an individual whose ametropia is corrected, the working distance for a telemicroscope is equal to the front focal length of the reading cap lens, which is also the front focal lengh of the telemicroscope itself (since the telescopic unit is afocal). If a lens of front vertex power $F_N$ is placed before the objective of an afocal telescope, the resulting telemicroscope will have a front vertex power $F_N$, with a front focal length of $f_N$ (or $1/F_N$). The front focal length of the reading cap determines the working distance of the lens and is inversely proportional to the relative distance magnification brought about by the reading cap.

As previously discussed, relative distance magnification is based on a standard reference distance for which magnification is considered to be unity; in order to avoid confusion as to whether the reference distance is 25 or 40 cm, it is more appropriate to define or compare low-vision devices in terms of their equivalent power.

The total magnification of a telemicroscope is the product of the angular magnification of the afocal telescope and the relative distance magnification of the reading cap. This total magnification is equal to the relative distance magnification of a single plus lens, and is directly proportional to the equivalent power. Hence, different telemicroscopes of the same equivalent power will produce the same magnification for the wearer, although the working distances will be different.

The equivalent power of a telemicroscope can be found by

$$F_E = F_1 + F_2 - dF_1F_2,$$

where $F_E$ = the equivalent power, $F_1$ = the sum of powers of the reading cap and the objective lens, $F_2$ = the power of the eyepiece lens, and $d$ = the separation of the objective and eyepiece lenses.

### EXAMPLE

An afocal Galilean telescope has the following specifications:

$$F_1 = +10.00 \text{ D}$$

$$F_2 = -20.00 \text{ D}$$

$$d = 5 \text{ cm}$$

What is the magnifying power of the telescope?

$$M = \frac{-F_2}{F_1} = -\frac{(-20)}{10} = 2\times.$$

If a +4.00 D reading cap is placed on the objective lens, the equivalent power can be found by the use of the formula

$$F_E = F_1 + F_2 - dF_1F_2,$$

$$F_E = +14 + (-20) - 0.05(14)(-20)$$

$$= -6 + 14$$

$$= +8.00 \text{ D}.$$

An alternative method can be developed: The front vertex power of the telemicroscope is equal to the front vertex power of the reading cap, and thus the front vertex power of the telemicroscope is

$$F_N = \frac{F_2}{1 - dF_2} + F_1,$$

where $F_N$ = the front vertex power of the telemicroscope, $F_1$ = the sum of the powers of the reading cap and the objective lens, $F_2$ = the power of the eyepiece lens, and $d$ = the separation of the objective and eyepiece lenses.

$$F_N = \frac{F_2}{1 - dF_2} + \frac{F_1(1 - dF_2)}{1 - dF_2}$$

$$= \frac{F_1 + F_2 - dF_1F_2}{1 - dF_2}$$

$$= \frac{F_E}{1 - dF_2}. \tag{12.15}$$

In Section 12.2 it was found that the magnification of an afocal telescope can be expressed as

$$M = 1 - dF_2.$$

Thus

$$F_N = \frac{F_E}{M},$$

or

$$F_E = F_N M. \tag{12.16}$$

Equation (12.16) shows that the equivalent power of a telemicroscope is equal to the *power of the reading cap times the magnifying power of the afocal telescope.*

In the example worked previously, the equivalent power resulting by placing a +4.00 D reading cap on a 2× Galilean telescope was +8.00 D. In the equation

$$F_E = F_N M,$$

$$F_E = +4(2) = +8.00 \text{ D}.$$

In each example, if the standard reference distance were 25 cm instead of 40 cm, the magnification would be

$$M = dF = 0.25(8) = 2\times.$$

This simple relationship—that the equivalent power of a telemicroscope is equal to the power of the reading cap times the angular magnification of the afocal telescope—is useful in selecting the combination of reading cap and afocal telescope that will provide the required magnification. For example, each of the following telemicroscopes provides an equivalent power of 24 D, and all provide the same magnification: a 4× telescope with a +6.00 D reading cap (Figure 12-13A); a 6× telescope with a +4.00 D reading cap (Figure 12-13B); and an 8× telescope with a +3.00 D reading cap (Figure 12-13C). What varies, of course, is the *working distance*, being 16.67 cm for telemicroscope A, 25 cm for telemicroscope B, and 33.3 cm for telemicroscope C. Each of these telemicroscopes would provide the same magnification as a +24.00 D *microscopic* lens, which would have a working distance of 4.17 cm (Figure 12-13D).

When compared to the use of a microscopic lens, or "high add," a telescope with a reading cap has the following advantages and disadvantages:

1. By virtue of the power of the afocal telescope, the patient can obtain *greater magnification* at the same working distance as that for a microscopic lens. For example, if only a +10.00 D microscopic lens were worn for a working distance of 10 cm, the angular magnification would be 2.5× (assuming a reference distance of 25 cm), but with the addition of the 2× afocal telescope, the equivalent power of the resulting telemicroscope, according to Eq. (12.16), would be 10(2), or +20.00 D, and the total magnification at that reading distance would be +20/4, or 5×.

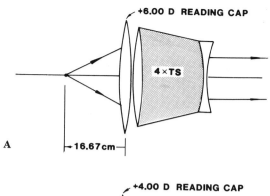

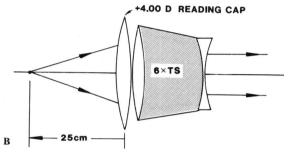

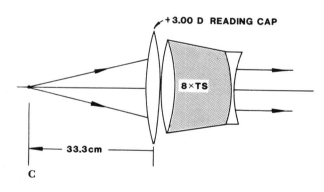

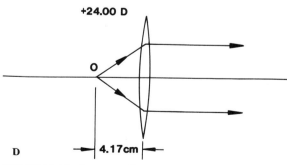

FIGURE 12-13. **Four magnifiers with the same equivalent power (the same magnification) but varying working distances: (A) 4× telescope with a +6.00 D reading cap and a reading distance of 16.67 cm; (B) 6× telescope with a +4.00 D reading cap and a reading distance of 25.00 cm; (C) 8× telescope with a +3.00 D reading cap and a reading distance of 33.30 cm; (D) +24.00 D microscopic lens having a working distance of 4.17 cm.**

2. Conversely, the magnifying power provided by the afocal telescope means that, for a given amount of magnification, a patient wearing a telescope with a reading cap can read at a *greater distance* than if only a microscopic lens were worn. Again using the above example, a patient wearing a 2× afocal telescope and a +10.00 D reading cap would be able to read at a distance of 10 cm with a total magnification of 5×; but if only a 5× (or +20.00 D) microscopic lens were worn, the patient would have to read at a distance of 5 cm, with the same amount of magnification.

3. An afocal lens with a reading cap has the disadvantage of a *smaller field of view* than would be present if only a microscopic lens were worn.

All manufacturers of telescopic spectacles have reading caps available in various powers. Designs for Vision makes a low-vision aid called a *Trioptic* consisting of an afocal telescope (a Bioptic) with a reading cap, mounted in the upper part of a carrier lens, and a microscopic lens mounted in the lower part of the carrier lens.

## 12.7. Hand Magnifiers

A large number of hand magnifiers are available, at relatively low cost. Many patients who have a low-vision problem will have already visited the local drug or variety store and purchased one or more of these aids. They are particularly useful for short-term visual tasks such as looking at a telephone number, but can also be used for prolonged reading.

Although the hand magnifier is a relatively simple device, the optical principles involved in its use are complex because the magnification cannot be represented by a unique number. The magnification depends not only on the equivalent power but also on how the magnifier is used: Both the distance from the lens to the eye and the distance from the object to the lens affect the amount of magnification. Each of these distances can be altered independently over a wide range, according to the wishes of the user.

Normally the magnification of a hand magnifier is specified by the manufacturer. The stated magnification is usually based on the formula

$$M = \frac{F_E}{4}$$

The use of this formula assumes that the standard reference distance for unaided vision is 25 cm and that the object is located at the primary focal point of the lens. In this situation, light from the object will emerge parallel from the lens with the result, as

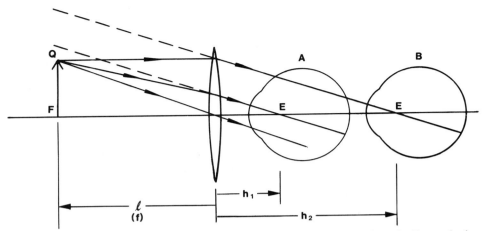

FIGURE 12-14.  **With the object in the primary focal plane of a hand magnifier, the retinal image size is constant and independent of the distance between the magnifier and the eye.**

shown in Figure 12-14, that parallel light will be incident upon the eye regardless of the distance from the eye to the lens.

The fact that parallel light is incident upon the eye regardless of the distance from the eye to the lens is responsible for an important property of a hand magnifier: When an object is located in the primary focal plane of the magnifier, the magnification (and therefore the retinal image size) is constant *regardless of the distance between the magnifier and the eye.*

As shown in Figure 12-14, the retinal image size is determined by the angle between the optic axis and the ray from the object point, Q, subtended at the entrance pupil. If the eye is moved from its original position A to position B, farther from the lens, the ray that passes through the entrance pupil will create a retinal image that is equal in linear size to that in the original position. This occurs because of the fact that all rays leaving the lens are parallel, with the result that any ray passing through the center of the entrance pupil crosses the optic axis at the same angle.

It should be understood, however, that even though the retinal image size is constant for all magnifier-to-eye distances, the field of view decreases as the distance is increased.

Another explanation may be helpful in accounting for the constancy of the retinal image size. If the object is kept in the focal plane of the magnifier and if the object and the magnifier are moved, in unison, away from the eye, the *angular magnification* will increase. However, the *relative distance magnification* decreases, and the product of the angular magnification and the relative distance magnification is a constant.

If the object is positioned in the primary focal plane of the magnifier, no accommodation need be exerted if the viewer's ametropia is properly corrected. The equivalent power of the system is therefore simply the equivalent power of the magnifier.

When the object is positioned at a distance *closer* to the magnifier than the primary focal length, the image will fall behind the magnifier, closer than infinity, as shown in Figure 12-15. In order to see the

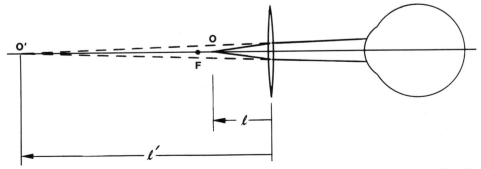

FIGURE 12-15.  **When an object is positioned closer to a hand magnifier than its primary focal length, the image will be located behind the magnifier, closer than infinity.**

image clearly, an emmetropic (or corrected ametropic) observer must resort either to accommodation or to the use of a reading addition. Note that in this situation the retinal image size is *no longer constant* for all viewing distances. In this event the equivalent power of the system is equal to

$$F_E = F_M + F_A - dF_M F_A,$$

where $F_E$ = the equivalent power, $F_M$ = the power of the magnifier, $F_A$ = the amount of accommodation or power of add, and $d$ = the distance from the magnifier to the spectacle plane.

When the distance from the magnifier to the spectacle plane is equal to the focal length of the magnifier,

$$F_M = \frac{1}{d},$$

and

$$F_E = F_M + F_A - d\left(\frac{1}{d}\right)F_A,$$

and therefore

$$F_E = F_M,$$

with the result that the equivalent power of the optical system is equal *solely* to the power of the magnifier, being independent of the amount of accommodation or of the power of the reading addition. The image of the system will be seen clearly provided that the object is placed so that the image falls at the distance for which the eye is accommodated or at the focal distance of the reading addition.

When the distance from the magnifier to the spectacle plane is *less* than the focal length of the magnifier, the resultant equivalent power will be *greater* than the power of the magnifier, reaching a maximum value when the magnifier is placed in the spectacle plane (in which case $d = 0$). In this case, the equivalent power is equal to the sum of the power of the magnifier and the power of the reading addition. On the other hand, when the distance from the magnifier to the spectacle plane is *greater* than the focal length of the magnifier, the resultant equivalent power will be *less* than the power of the magnifier. The greater the separation, the less will be the resultant equivalent power. These principles will be illustrated by the following examples:

### EXAMPLE 1
A simple magnifier having a refracting power of +20.00 D is used by a presbyope who wears a +2.50 D add, the magnifier being held 5 cm in front of the spectacle lens ($d$ = 5 cm). What is the equivalent power of the magnifier/reading-add system?

$$F_E = F_M + F_A - dF_M F_A,$$
$$= +20.00 + 2.50 - 0.05(+20.00)(+2.50)$$
$$= +22.50 - 2.50$$
$$= +20.00 \text{ D}.$$

### EXAMPLE 2
Using the same magnifier and the same reading add, the magnifier is held in contact with the spectacle lens ($d$ = 0). What is the equivalent power of the combination?

$$F_E = F_M + F_A - dF_M F_A,$$
$$= +20.00 + 2.50 - 0(+20.00)(+2.50)$$
$$= +22.50 + 0$$
$$= +22.50 \text{ D}.$$

### EXAMPLE 3
Again using the same magnifier and the same reading add, the magnifier is moved to a position 10 cm in front of the spectacle plane. What is the equivalent power of the combination?

$$F_E = F_M + F_A - dF_M F_A,$$
$$= +20.00 + 2.50 - 0.1(+20.00)(+2.50)$$
$$= +22.50 - 5.00$$
$$= +17.50 \text{ D}.$$

To summarize these results, if a magnifier is placed at its focal distance from the spectacle plane, its equivalent power will always be equal to the power of the magnifier (whether or not a bifocal is used), as shown in Example 1. When a bifocal is used with the hand magnifier, if the magnifier is held at a distance in front of the spectacle plane shorter than its focal length the equivalent power will be greater than that of the magnifier alone (Example 2), whereas if the magnifier is held at a distance in front of the spectacle plane greater than its focal length the equivalent power will be less than that of the magnifier alone (Example 3).

For objects located at arm's length (i.e., beyond the distance for which a bifocal addition is intended) magnification is maximum when the object is placed in the primary focal plane of the magnifier and the distance spectacles are used.

Since the magnification of a hand magnifier depends upon how it is used, a hand magnifier should be specified not in terms of a fixed magnifying power but by its *equivalent power*.

As described above (Example 2) it is possible for a low-vision patient to use a hand magnifier by placing it in contact with his or her lenses, using it as a "high add," and holding the reading material at a suitably close distance. Faye[9] has made the point that some patients who resist wearing a high add in the form of a spectacle lens are willing to use a hand magnifier, and that for these patients the use of a hand magnifier held against the spectacle correction serves as a simple and inexpensive method for the patient to gain experience in using a high addition. After practicing with this combination, some patients actually request to be fitted with a high addition in spectacle form.

When an object is located in the primary focal plane of a hand magnifier, the *linear field of view* is given by the following expression[10]:

$$w = \frac{y}{F_E(d)},$$

where $w$ = the linear width of the field of view, $y$ = the lens diameter in meters, $F_E$ = the equivalent power of the magnifier, and $d$ = the distance of the magnifier to the spectacle plane in meters.

For example, if a 50-mm round +20.00 D hand magnifier is held 10 cm from the eye, and the object is located at the primary focal point of the magnifier, then the field of view of the magnifier is

$$w = \frac{0.05}{20(0.1)}$$

$$= \frac{0.05}{2}$$

$$= 0.025 \text{ m, or } 25 \text{ mm.}$$

If the object is not located in the primary focal plane of the magnifier, the field of view behaves in the following manner:

1. If the distance from the eye to the magnifier is *greater* than the focal length of the magnifier, the linear field of view will *increase* slightly.
2. If the distance from the eye to the magnifier is *less* than the focal length of the magnifier, the linear field of view will *decrease* slightly.

## 12.8. Stand Magnifiers

Although a few focusable stand magnifiers are available, most stand magnifiers are of the *fixed-focus* type, having a fixed object-to-lens distance. The magnifier support is placed directly upon the reading material, with the result that both the object distance and the

image distance are constant. If the plane of the reading material coincides with the primary focal plane of the magnifier, the image will be located at infinity. Usually, the reading material is located just inside the focal plane of the magnifier with the result that a virtual, erect image is formed behind the magnifier, between the plane of the reading material and infinity. In order for the image to be seen clearly, an emmetrope or a corrected ametrope would have to accommodate or wear a suitable reading addition.

When a stand magnifier is prescribed for a low-vision patient, the practitioner should be able to locate the image plane in order to instruct the patient concerning the proper viewing distance and to estimate the amount of accommodation or the power of the reading addition required to see the image clearly. The location of the image plane can be found by neutralizing the emerging divergence with a plus lens, causing the image to be formed at infinity (see Figure 12-16).

Since a stand magnifier is intended to be used so that the virtual image is located at a standard reading distance, it is possible to calculate the *equivalent power* of the magnifier, which permits such a magnifier to be compared to magnification systems (such as high add) used close to the eye.

The equivalent power may be found by combining the refractive power of the magnifier and the

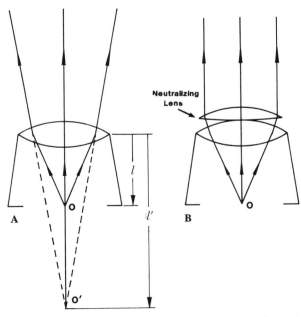

FIGURE 12-16. **(A) Location of the image plane of a stand magnifier. (B) The power of the neutralizing lens which renders the emergent light parallel indicates the position of the image plane,** *O'.*

amount of accommodation in play (or the power of the reading addition), using the formula used for a hand magnifier,

$$F_E = F_M + F_A - dF_M F_A,$$

where $F_E$ = the equivalent power, $F_M$ = the power of the magnifier, $F_A$ = the amount of accommodation or power of add, and $d$ = the distance from the magnifier to the spectacle plane.

The amount of accommodation or power of the reading addition is

$$F_A = \frac{1}{d - l'}, \qquad (12.17)$$

where $l'$ is the negative image distance behind the stand. In the general equivalent power equation,

$$F_E = F_M + F_A - dF_M F_A,$$

or

$$F_E = F_A[F_M/F_A + 1 - dF_M].$$

Substituting the expression for $F_A$ from Eq. (12.17) within the brackets,

$$F_E = F_A[F_M(d - l') + 1 - dF_M]$$
$$= F_A[dF_M - F_M l' + 1 - dF_M]$$
$$= F_A(1 - F_M l')$$
$$= F_A \left(1 - \frac{F_M}{L'}\right)$$
$$F_E = F_A \left(\frac{L' - F_M}{L'}\right). \qquad (12.18)$$

The expression

$$\left(\frac{L' - F_M}{L'}\right) \qquad (12.19)$$

has been referred to by Bailey[11] as the MULTACC factor ($M_T$).

Therefore

$$F_E = F_A(M_T),$$

where

$$M_T = \left(\frac{L' - F_M}{L'}\right),$$

and since, in the simple lens formula, $L' - F = L$,

$$M_T = \frac{L}{L'} = \frac{l'}{l}.$$

Therefore, the MULTACC factor is an expression for lateral or transverse magnification.

Bailey recommends that all fixed-focus stand magnifiers be labeled in terms of refracting power, $F_M$, the image position, $l'$, and the MULTACC factor, $M_T$, therefore allowing easy calculation of the equivalent power. A stand magnifier will create different equivalent powers depending upon the separation of the eye and the magnifier, offering different resolution abilities because of the fact that resolution ability is directly proportional to equivalent power.

In order to illustrate how the equivalent power changes with the manner in which a stand magnifier is used, consider a stand magnifier having a refracting power, $F_M$, of +20.00 D, and an image distance, $l'$, of −20 cm (and therefore an image vergence, $L'$, of −5.00 D).

$$M_T = \frac{L' - F}{L'} = \frac{-5 - 20}{-5} = \frac{-25}{-5} = 5.$$

### EXAMPLE 1
If the magnifier is located in the spectacle plane and if the distance between the spectacle plane and the image is 20 cm, what is the equivalent power?

$$F_A = \frac{1}{0.2} = +5.00 \text{ D}.$$

Thus

$$F_E = F_A(M_T) = 5(5) = +25.00 \text{ D}.$$

### EXAMPLE 2
If the magnifier is located 20 cm from the spectacle plane and the distance between the spectacle plane and the image is 40 cm (20 cm + 20 cm), what is the equivalent power?

$$F_A = \frac{1}{0.4} = +2.50 \text{ D}.$$

Thus

$$F_E = F_A(M_T) = 2.50(5) = +12.50 \text{ D}.$$

Hence changing the distance from the spectacle plane to the magnifier from zero to 20 cm changes the accommodation (or power of the required reading addition) from 5.00 to 2.50 D and the equivalent power from +25.00 to +12.50 D. Since resolution ability is directly proportional to equivalent power, this change will reduce the observer's resolution ability by one-half.

### EXAMPLE 3
If the stand magnifier is used as described in Example 2, by a patient who can barely read 25-

point print at 40 cm while wearing a +2.50 D add, what size print can the patient resolve?

With the stand magnifier specified in Example 2, the equivalent power of the combination of the +20.00 D stand magnifier and the +2.50 D addition is

$$F_E = F_A(M_T) = 2.50(5) = +12.50 \text{ D}.$$

The +12.50 D system should improve resolution by a factor of 5 as compared to that obtained with the +2.50 D add, so that the patient can now be expected to resolve 5-point print.

## 12.9. The Paperweight Magnifier

The "paperweight magnifier" is a very popular reading aid. It is a thick, plano-convex lens that is held in contact with the reading material and is moved along the page as the person reads; it is therefore a modification of the stand magnifier, having a magnifier-to-print distance of zero.

Although the magnifying power of a paperweight magnifier is relatively low, it has the advantage of very good light-gathering properties and is very easy to use, particularly for an older person with unsteady hands. When given a choice of a large number of commercially available aids, many low-vision patients will choose such a magnifier.

The magnifying power of a paperweight magnifier can be determined by finding the image position and applying the linear magnification formula for a single spherical refracting surface, thus,

$$M = \frac{h'}{h} = \frac{nl'}{n'l} = \frac{L}{l'}.$$

### EXAMPLE 1

A paperweight magnifier has a spherical surface with a radius of curvature of 5 cm, a thickness of 3 cm, and an index of refraction of 1.53. (a) What is the position of its image? (b) What is its magnification?

(a) Using the single spherical refracting surface formula (see Figure 12-17),

$$\frac{n'}{l'} - \frac{n}{l} = \frac{n' - n}{r}$$

$$\frac{1}{l'} - \frac{1.53}{-0.03} = \frac{1 - 1.53}{-0.05}$$

$$\frac{1}{l'} + 51 = +10.6$$

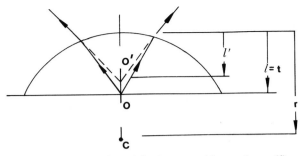

FIGURE 12-17. **Location of the image position and magnification for a paperweight magnifier (see Example 1 in text).**

$$\frac{1}{l'} = -40.4$$

$$l' = -0.0248 \text{ m} = -2.48 \text{ cm}.$$

(b) The magnification can be determined by use of the expression

$$M = \frac{nl'}{n'l}$$

$$= \frac{1.53(-0.0248)}{1(-0.03)} = \frac{0.0379}{0.03}$$

$$M = 1.26\times.$$

An interesting form of the paperweight magnifier occurs when the radius of curvature is equal to the center thickness, that is, when the magnifier is in the form of a *hemisphere*, as shown by the following example.

### EXAMPLE 2

A paperweight magnifier has a spherical surface with a radius of curvature of 5 cm, a thickness of 5 cm, and an index of refraction of 1.53. (a) What is the position of its image? (b) What is its magnification?

(a) Referring to Figure 12-18A and using the formula for a spherical refracting surface,

$$\frac{n'}{l'} - \frac{n}{l} = \frac{n' - n}{r}$$

$$\frac{1}{l'} - \frac{1.53}{-0.05} = \frac{1 - 1.53}{-0.05}$$

$$\frac{1}{l'} + 30.6 = +10.6$$

$$\frac{1}{l'} = -20$$

$$l' = -0.05 \text{ m} = -5 \text{ cm}.$$

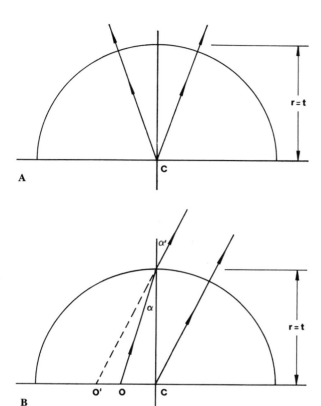

**FIGURE 12-18. Location of the image position and magnification for a paperweight magnifier whose thickness is equal to its radius of curvature, being hemispherical in shape (see Example 2 in text).**

The image of this paperweight magnifier is therefore located *in the plane of the object.*

(*b*) Referring to Figure 12-18B, the magnification will be

$$M = \frac{nl'}{n'l}$$

$$= \frac{1.53(-0.05)}{1(-0.05)} = \frac{0.0379}{0.05}$$

$$M = 1.53\times.$$

Hence, for a paperweight magnifier in the form of a hemisphere, the *magnification is equal to the index of refraction.*

## References

1. Mehr, E.B., and Freid, A.M. *Low Vision Care.* Professional Press, Chicago, 1975.

2. Smith, G. Magnification of Afocal Telescopes When Used Focally. *Austr. J. Optom.*, Vol. 64, No. 5, pp. 202–205, 1964.

3. Bailey, I.L. New "Expanded Field" Bioptic Systems. *Optom. Monthly*, Vol. 69, pp. 895–898, 1978.

4. Genensky, S.M. Binoculars: A Long-Ignored Aid for the Partially Sighted. *Amer. J. Optom. Physiol. Opt.*, Vol. 51, pp. 648–673, 1974.

5. Wild, B.W. A Low Vision Aid Design for Reading. *Optom. Weekly*, Vol. 59, pp. 36–37, Apr. 11, 1968.

6. Bailey, I.L. Traditional Methods for Measuring the Magnification of Telescopes. *Optom. Monthly*, Vol. 70, pp. 128–131, Feb. 1979.

7. Bailey, I.L. Centering High Addition Spectacle Lenses. *Optom. Monthly*, Vol. 69, pp. 981–984, Oct. 1978.

8. Bailey, I.L. Telescopes: Their Use in Low Vision. *Optom. Monthly*, Vol. 69, pp. 634–638, 1978.

9. Faye, E. *Clinical Low Vision.* Little, Brown, Boston, 1976.

10. Johnston, A.W. Technical Note: The Relationship between the Magnification and Field of View for Simple Magnifiers. *Austr. J. Optom.*, Vol. 65, No. 2, pp. 74–77, 1982.

11. Bailey, I.L. The Use of Fixed Focus Stand Magnifiers. *Optom. Monthly.* Vol. 72, pp. 37–39, 1981.

## Questions

1. A patient with a best corrected acuity of 10/80 must read 20/40 on the job. What approximate magnification will be needed?

2. If an object is moved from 40 to 15 cm, what is the relative distance magnification?

3. An afocal Keplerian telescope has an objective lens of +5.00 D and an eyepiece lens of +12.50 D.
   (a) What is the magnifiying power for a distant object?
   (b) What is the length of the telescope?

4. An afocal 4× Galilean telescope has a telescopic length of 7.5 cm. What is the power of the objective lens and the eyepiece lens?

5. An afocal 4× Galilean telescope has an objective lens of +5.00 D. What telescopic length is required for an uncorrected 5.00 D hyperope to have clear distance vision?

6. If a 4× afocal Keplerian telescope with an objective lens of +10.00 D and a +40.00 D eyepiece lens is adjusted to allow an uncorrected 5.00 D hyperope to see clearly at distance, what is the new effective magnification?

7. A microscopic lens has an equivalent power of +20.00 D. If the lens is used 15 mm from the eye, what is the angular magnification for an object placed at its primary focal point?

8. If the object was originally placed at a distance of 40 cm from the spectacle plane, what is the total magnification of the lens described in Problem 7?

9. If a 2× afocal telescope is used for an object at 50 cm without a reading cap, what is the approximate amount of accommodation required to see the object clearly?

10. If a 4× afocal telescope is fitted with a +4.00 D reading cap, what is the equivalent power of the combination?

11. What would be the working distance of the tele-microscope described in Problem 10?

12. What is the total magnification relative to a reference distance of 25 cm of a 2× afocal Galilean telescope with a +10.00 D cap?

13. A printer needs 4× magnification in order to perform his duties. However, the working distance must be 25 cm. What combination of afocal telescope and reading cap will provide the desired magnification?

14. A +12.00 D hand magnifier is used in conjunction with a +3.00 D spectacle addition at a distance of 20 cm from the spectacle plane. What is the equivalent power of the magnifier/reading-add system?

15. What is the linear field of view for a 60-mm round +15.00 D hand magnifier held 10 cm from the eye, with the object at the primary focal plane of the magnifier?

16. A +25.00 D stand magnifier is 20 cm from the spectacle plane. The virtual image formed by the stand magnifier is located 20 cm from the stand magnifier. What is the equivalent power of the stand magnifier/reading add when used by a presbyopic patient wearing a +2.50 D add?

17. What is the magnification of a plastic paperweight magnifier with an index of refraction of 1.49, a curved surface with a radius of curvature of 5 cm, and a thickness of 4 cm?

18. What is the magnification of a glass hemispheric paperweight magnifier with an index of refraction of 1.60 and a curved surface whose radius of curvature is 6 cm?

# CHAPTER THIRTEEN

# Optics of
# Contact Lenses

Contact lenses differ from spectacle lenses in many respects. Some of these differences are obvious, while others are not so obvious.

*The Contact Lens as a Thick Lens*. Without a doubt the least obvious difference is that while a spectacle lens can often be considered as a *thin lens*, a contact lens must always be considered as a *thick lens*. Although a contact lens is much thinner than a spectacle lens, the fact that it is so steeply curved means that the use of the "approximate power" formula will result in serious errors.

*Effective Power.* One of the more obvious ways in which contact lenses differ from glasses is that contact lenses *touch the eyes*. One result of the proximity of the contact lens to the eye is the fact that the *effective power* of a correcting lens changes as it is brought toward the eye. This means that for a myope a contact lens must be *weaker* than a spectacle lens for the same eye, whereas for a hyperope a contact lens must be *stronger* than a spectacle lens for the same eye.

*The Contact-Lens/Eye System*. The fact that contact lenses touch the eyes means that a contact lens (particularly if it is a "hard" lens) can alter the refractive state of the eye. Conversely, the eye can alter the refractive power of the contact lens (particularly if it is a "soft" contact lens). These effects are of particular importance when we are dealing with *astigmatism*: A spherical hard contact lens tends to mask (or to eliminate) corneal astigmatism, while a spherical soft lens tends to conform to the toricity of the cornea and thus has little or no effect on astigmatism. If astigmatism is to be corrected with soft contact lenses, *toric* lenses must be used.

*Magnification Effects*. Still another effect of the proximity of a contact lens to the eye is the change in *magnification* that occurs when one changes from glasses to contact lenses. Since the power factor of the spectacle

415

magnification formula includes the distance $h$ (the distance from the correcting lens to the center of the entrance pupil of the eye), it follows that moving a correcting lens closer to the eye brings about a change in magnification. For a myope, a contact lens provides a *larger* retinal image than a spectacle lens, whereas for a hyperope a contact lens provides a *smaller* retinal image. Thus, a high myope who switches from glasses to contact lenses will often be delighted by the fact that "everything looks bigger," whereas a hyperope (particularly an aphakic) may be relieved to find that objects appear to be more nearly their "normal" sizes with contact lenses than with glasses.

***Changes in Accommodative Demand***. A myope who wears spectacles has the advantage that the minus lenses reduce the amount of accommodative demand, as compared to that for an emmetropic eye; but if contact lenses are worn rather than glasses, the myope loses this advantage. On the other hand, when a hyperope wears spectacles, the plus lenses increase the accommodative demand as compared to that for the emmetropic eye, with the result that the hyperope will need to accommodate less when switching to contact lenses. This change in accommodative demand has an important implication for the correction of *presbyopia*: It often happens that a myope who is on the verge of becoming presbyopic will need an add for near work *earlier* if contact lenses are worn rather than glasses, whereas a hyperope in the same situation who switches from glasses to contact lenses may not need an add for near as soon as if glasses were worn.

***Changes in Accommodative Convergence***. The increased accommodative demand while wearing contact lenses means that the myope will use *more* accommodative convergence while wearing contact lenses than while wearing glasses, whereas the hyperope will use *less*. Therefore, an esophoric myope will have to use more *negative fusional vergence* while wearing contact lenses than while wearing glasses, and an exophoric hyperope will have to use more *positive fusional vergence* while wearing contact lenses than while wearing glasses.

These effects on the demand for accommodation and fusional vergence are negligible for small refractive errors but may be of importance for larger refractive errors, particularly if the AC/A ratio (the ratio between accommodative convergence and accommodation) is also large.

***Changes in Prismatic Effects***. The fact that correctly fitted contact lenses remain *centered* on the eyes, in spite of convergence or divergence of the visual axes, is responsible for an additional change in fusional vergence demand. If a pair of spectacle lenses are centered for the wearer's distance PD—the usual situation—the lenses induce prismatic power when worn for near vision: Minus lenses induce *base-in* prism, and plus lenses induce *base-out* prism. Therefore, an exophoric myope, when switching from glasses to contact lenses, will be at a disadvantage because of the lack of base-in prismatic effect for near work, whereas an esophoric hyperope switching to contact lenses will be at a similar disadvantage due to the lack of base-out prismatic effect for close work.

Fortunately, the effects on fusional vergence demand due to the lack of decentration tend to cancel those effects due to changes in accommodative demand. However, for some patients one effect can outweigh the other to the extent that it causes an adaptation problem.

The absence of induced prismatic effects due to contact lenses provides a decided advantage for an anisometrope wearing contact lenses. When an anisometrope wears spectacles, a vertical prismatic imbalance is induced at the reading level; but if contact lenses are worn, no vertical imbalance is present if the contact lenses remain centered on the eyes during downward gaze.

*Aberrations and Field of View.* The fact that well-fitted contact lenses are centered on the eyes, and remain so when the eyes are moved, is responsible for two additional optical differences between contact lenses and glasses. These are the differences in the *aberrations* experienced by the wearer and in the wearer's *field of view.*

It will be recalled that the aberrations of most importance in the design of spectacle lenses are oblique astigmatism, curvature of image, and distortion.

Oblique astigmatism and curvature of image occur only when the wearer rotates his eyes to look through the periphery of the lens—this occurs when spectacles are worn, but a contact lens wearer is unable to fixate an object through the periphery of the lens. Distortion, on the other hand, occurs partially as a result of the distance between the aperture of the lens and the aperture (pupil) of the eye. Since this distance is at a minimum when a contact lens is worn, distortion, too, is at a minimum.

Most contact lens wearers are pleased to find that they have a larger *field of view* when wearing contact lenses than when wearing glasses. Although it can be shown that the *peripheral field of view* (of the stationary eye) may be no larger when wearing a contact lens than when wearing a spectacle lens—in many cases it is actually *smaller*—a contact lens wearer has the advantage that there is no limit to the *macular field of view* (for the moving eye), since in the absence of a spectacle rim the macular field is limited only by the wearer's field of fixation (i.e., his ability to move his eye in order to maintain macular vision).

## BASIC CONTACT LENS OPTICS

### 13.1. The Contact Lens as a Thick Lens

Any lens may be considered as a "thin lens" if its radii of curvature are large compared with its thickness. For such a lens, the *approximate power formula* may be used:

$$F_a = F_1 + F_2.$$

However, the surfaces of a contact lens are so highly curved that the sagitta of the lens cannot be considered to be small in terms of the chord length, so in spite of its obvious thinness a contact lens must be considered a thick lens. The necessity for treating a contact lens as a thick lens will be shown by an example.

## EXAMPLE

Given a contact lens having both the front and back radii of curvature ($r_1$ and $r_2$) equal to 7.5 mm, and a center thickness ($t$) equal to 0.20 mm, made of a material having an index of refraction of 1.490. Find the approximate power and the back vertex power of this lens.

**Approximate Power.** The values of $F_1$ and $F_2$ are

$$F_1 = \frac{1.490 - 1}{0.0075} = +65.33 \text{ D.}$$

$$F_2 = \frac{1 - 1.490}{0.0075} = -65.33 \text{ D.}$$

Therefore, the approximate power of this lens is

$$F_1 + F_2 = +65.33D + (-65.33D) = 0.00 \text{ D.}$$

**Back Vertex Power.** Using the back vertex power formula,

$$F_V = \frac{F_1}{1 - \dfrac{t}{n} F_1} + F_2$$

$$= \frac{+65.33}{1 - \dfrac{0.0002}{1.490}(65.33)} + (-65.33)$$

$$= +65.91 - 65.33 = +0.58 \text{ D.}$$

This example shows that the use of the approximate power formula, rather than the back vertex power formula, would result in an error of more than 0.50 D. For a thicker lens (such as an aphakic lens) the error would be much greater.

Since contact lenses, like spectacle lenses, are specified in terms of back vertex power, the power of a contact lens is measured in the same way the power of a spectacle lens is measured: A lensometer is used, and the lens is mounted so that the concave side of the lens rests against the lensometer aperture. However, due to the steepness of the contact lens, the lensometer should be fitted with a smaller "stop" than ordinarily used for a spectacle lens (see Figure 13-1). Most modern lensometers are supplied with such a stop.

### The Base Curve of a Contact Lens

The concave side of a contact lens is designated as the *base curve* of the lens. The base curve radius is selected mainly to meet the criteria for a comfortable and physiologically acceptable fit on the wearer's eye,

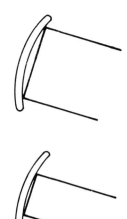

FIGURE 13-1. **The use of a small lensometer stop for measuring the back vertex power of a contact lens.**

including good centration and adequate movement. However (particularly in the case of a hard contact lens), the base curve radius also has important *optical* significance.

In the fitting of hard contact lenses, the base curve radius is usually selected so as to be equal (or nearly so) to the radius of curvature of the front surface of the cornea, as measured by the keratometer. This method of fitting a contact lens is known as the "parallel fit" or "on K" method. In a laboratory order for a contact lens, the practitioner specifies the base curve radius, $r_2$, the back vertex power, $F_v$, and the center thickness, $t$; and on the basis of this information the laboratory selects the necessary front curve radius, $r_1$.

When the finished lens is received from the laboratory, the practitioner verifies the base curve radius by means of an instrument known as a *radiuscope*. As described in Section 13.18 and illustrated in Figures 13-35A and B, the radiuscope measures the actual (physical) distance between the back surface of the lens and its center of curvature.

## EXAMPLE

A practitioner orders a contact lens having a base curve radius of 8.0 mm, a back vertex power of −3.00 D, and a center thickness of 0.20 mm. If the lens is to be made from a material having an index of refraction of 1.490, what will be the power ($F_1$) and the radius of curvature ($r_1$) of the front curve?

$$F_2 = \frac{1 - 1.490}{0.008} = -61.25 \text{ D.}$$

In order to find $F_1$, we begin with the back vertex formula,

$$F_V = \frac{F_1}{1 - \frac{t}{n}F_1} + F_2$$

and solve for $F_1$.

$$F_V - F_2 = \frac{F_1}{\left(1 - \frac{t}{n}F_1\right)}$$

$$(F_V - F_2)\left(1 - \frac{t}{n}F_1\right) = F_1$$

Expanding the left-hand side of the equation and rearranging, we have

$$F_V - F_2 = F_1\left(1 + F_V\frac{t}{n} - F_2\frac{t}{n}\right).$$

Therefore

$$F_1 = \frac{F_V - F_2}{1 + \frac{t}{n}(F_V - F_2)}$$

$$= \frac{-3.00 - (-61.25)}{1 + \frac{0.0002}{1.490}[-3.00 - (-61.25)]}$$

$$= +57.80 \text{ D.}$$

$$r_1 = \frac{1.490 - 1}{57.80} = 8.48 \text{ mm.}$$

## 13.2. The Effective Power of a Contact Lens

As discussed in Section 3.6, the term "effective power" is used to indicate the change in lens power required when a lens is moved from one position to another in front of the eye. For example, a plus lens is "more effective" (has more effectivity) when it is moved farther from the eye, whereas a minus lens is "more effective" when moved closer to the eye. In equation (3.16) in which the letters A and B refer to the original and altered positions of the lens,

$$F_B = \frac{F_A}{1 - dF_A},$$

the letter $d$, indicating the distance the lens has been moved, is given a plus sign if the lens is moved toward the eye and a minus sign if the lens is moved away from the eye.

### EXAMPLE 1
A 4.00 D myope, refracted at a vertex distance of 12 mm, is going to be fitted with contact lenses (see Figure 13-2). What will be the required power of the contact lenses (assuming that the contact lens does not change the refractive power of the eye; that is, its back surface is parallel to the air/tear interface)?

$$F_B = \frac{-4.00}{1 - 0.012(-4.00)} = -3.82 \text{ D.}$$

### EXAMPLE 2
An aphakic is refracted at a distance of 15 mm and is found to require a +13.00 D lens (see Figure 13-3). What will be the required power of a contact lens for this eye (assuming again that the contact lens does not change the refraction of the eye)?

$$F_B = \frac{+13.00}{1 - 0.015(13.00)} = +16.15 \text{ D.}$$

These examples show that for a myopic eye the required power of a contact lens will be *less* than the power of a spectacle lens, while for a hyperopic eye the required power of a contact lens will be *greater* than that for a spectacle lens. It can easily be demonstrated that, for refractive errors of less than ±3.00 D, the difference in the power of a contact lens and a

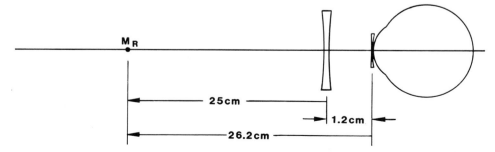

FIGURE 13-2. **Effective power of a contact lens for a 4.00 D myope, refracted at a vertex distance of 12 mm.**

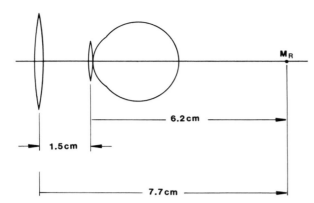

FIGURE 13-3. **Effective power of a contact lens for an aphakic requiring a +13.00 D lens when refracted at a vertex distance of 15 mm.**

spectacle lens will be on the order of no more than 0.12 D, and can be considered to be negligible. As for the aphakic eye, the second example shows that the difference between the power of a spectacle lens and a contact lens for the same eye is on the order of 3.00 D.

## 13.3. Calibration of the Keratometer

Although the index of refraction of the cornea is considered to be 1.376, most keratometers (including the Bausch and Lomb and American Optical instruments) are calibrated for an index of refraction of 1.3375. This value, originally used by Helmholtz, was selected because it takes into consideration the refraction at the *back* surface of the cornea (note that because of the small index change, amounting to (1.336 − 1.376), the refraction taking place at this interface is on the order of only about 5.00 D). Since it is difficult to measure the radius of curvature of the back surface of the cornea in routine clinical practice, the assumption is made that it is equal to that of the front surface of the cornea.

It is an interesting coincidence that the index of refraction used for the calibration of the keratometer is very close to the actual index of refraction of the *tear layer* (1.336): Thus, the keratometer comes very close to measuring the refraction taking place at the *air/tear interface.*

The initial refraction that we usually think of as taking place between air and the cornea actually takes place at the interface between air and the *tear layer*. When a "hard" contact lens is placed on the cornea, the front surface of the tear layer will tend to con-

form to the back surface curvature of the contact lens, with the result that the contact lens may alter the radius of curvature of the front surface of the tear layer. However, the contact lens would not be expected to alter the radius of curvature of the tear/cornea interface, or the refraction taking place at that interface. Hence the radius of curvature of the tear/cornea interface can be considered as a constant.

Using the actual values for the indices of the tears and the cornea, let us calculate the sum of the refracting power of the air/tear interface and the tear/cornea interface, for a tear layer having a radius of curvature of 7.5 mm (see Figure 13-4). At the air/tear interface,

$$F_{A/T} = \frac{1.336 - 1}{0.0075} = +44.80 \text{ D},$$

and at the tear/cornea interface,

$$F_{T/C} = \frac{1.376 - 1.336}{0.0075} = +5.33 \text{ D}.$$

$$F_{total} = +44.80 + 5.33 = +50.13 \text{ D}.$$

If these two interfaces are treated as a single surface, having a radius of curvature of 7.5 mm, the index of refraction of that surface may be found as follows:

$$\frac{n - 1}{0.0075} = +50.13$$

$$n - 1 = 0.376$$

$$n = 1.376.$$

Thus, from a practical point of view, in measuring the radius of curvature of the cornea we can ignore the tear layer: If we assume it to be infinitely thin, it will have zero power and will therefore act as a plano

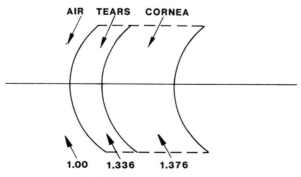

FIGURE 13-4. **Calculation of the refracting power at the air/tear interface and the tear/cornea interface, using actual values for indices of the tears and cornea.**

lens. We can, therefore, consider that all of the refraction takes place at the air/cornea interface:

$$F = \frac{n' - n}{r} = \frac{1.376 - 1}{0.0075} = +50.13 \text{ D}.$$

For the same reason, the tear layer *on the front surface of a contact lens* can also be ignored.

However, since the keratometer is calibrated for an index of refraction of 1.3375, the keratometer reading for a cornea having a radius of curvature of 7.5 mm. would be

$$F_K = \frac{1.3375 - 1}{0.0075} = +45.00 \text{ D}.$$

As mentioned earlier, this discrepancy in the apparent power of the cornea, as determined by the keratometer, takes into consideration the power of the back surface of the cornea. The front surface power of the cornea is actually

$$F_{A/C} = \frac{+1.376 - 1}{0.0075} = +50.13 \text{ D},$$

and the back surface power of the cornea is

$$F_{C/aq} = \frac{1.336 - 1.376}{0.0075} = -5.33 \text{ D}.$$

The resultant refracting power of the cornea is therefore

$$+50.13 + (-5.33) = +44.80 \text{ D},$$

or 0.20 D less than the keratometer reading of 45.00 D.

The keratometer underestimates the front surface power of the cornea by a factor of

$$\frac{5.33}{50.13} = 0.106, \text{ or approximately 10\%.}$$

The keratometer finding, therefore, measures approximately 9/10 of the total front surface power of the cornea. If corneal astigmatism is present, the keratometer measures only 9/10 of the actual amount, underestimating it by 1/10. However, this does not produce a 10% error, because the back surface of the cornea has a negative power (in each principal meridian) of approximately 1/9 of the total front surface power. In marked corneal astigmatism, the negative back surface is likely to be responsible for reducing a similar percentage of front surface astigmatism. On this basis, the refractive index for which the keratometer is calibrated works very well in practice.

The calibration index of 1.3375 not only provides keratometric readings that take into consideration the back corneal surface, but also provides, through an averaging process, an approximate power of the entire cornea as if the cornea were a homogeneous medium having an index of refraction of 1.3375. This index is so close to the actual index of the tears (1.336) that one can assume that the corneal astigmatism indicated by the keratometer will be essentially neutralized when a "hard" contact lens having a spherical back surface is placed on the eye, thereby rendering the front surface of the tear layer spherical. However, as we will see later, complications arise if the contact lens has a *toroidal*, rather than spherical, back surface.

For convenience in problem solving, the index of refraction of the tears is assumed to be equal to the index for which the keratometer has been calibrated: That is, the front surface power of the cornea is equal to the back surface power of the tear layer, but opposite in sign. This assumption involves only a very small constant error (0.20 D as shown in the above example) and greatly simplifies calculations.

In solving contact lens problems, it is helpful to make use of an important general principle of optics: If an optical system is composed of two media on either side of a spherical surface, the power of the system will be the same (*a*) if the two surfaces are considered to be in contact with each other, or (*b*) if the two surfaces are considered to be separated by a thin layer of air. This principle will be illustrated by an example:

### EXAMPLE

If two media, the first having an index of 1.60 and the second having an index of 1.50, having adjacent surfaces with radii of curvature of 20 cm, are in contact with each other (see Figure 13-5), the

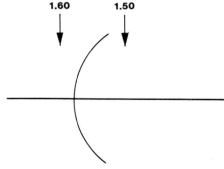

FIGURE 13-5. **Calculation of refracting power at the interface between media having indices of 1.60 and 1.50, with the two surfaces in contact.**

power at the interface would be

$$F_1 = \frac{1.50 - 1.60}{0.2} = -0.50 \text{ D.}$$

If the two surfaces are considered to be separated by an infinitely thin layer of air (Figure 13-6), the sum of the two surface powers would be

$$F_1 = \frac{1 - 1.60}{0.2} = -3.00 \text{ D.}$$

$$F_2 = \frac{1.50 - 1}{0.2} = +2.50 \text{ D.}$$

$$F_1 + F_2 = -3.00 \text{ D} + 2.50 \text{ D} = -0.50 \text{ D.}$$

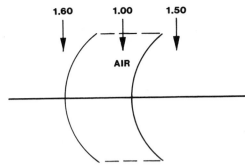

FIGURE 13-6. **Calculation of refracting power at the interface between media having indices of 1.60 and 1.50, with the two surfaces considered to be separated by an infinitely thin layer of air.**

# THE CONTACT LENS/EYE OPTICAL SYSTEM

## 13.4. The Contact Lens on the Eye

The optical system consisting of the contact lens and the eye can be considered by the use of either of two methods, based on the methods just described: the "in situ" method, or the "exploded" method. Using either method, the contact lens is assumed to be infinitely thin.

1. Using the "*in situ*" method, the lens is treated as if it were in place on the cornea (see Figure 13-7). Using this method, we are concerned with two refractions: one taking place at the air/contact-lens interface and the other taking place at the contact-lens/tear interface. Therefore, if the contact lens perfectly corrects the eye,

(power at air/lens interface)
+ (power at lens/tear interface)
    = (corneal refracting power) + (ocular refraction)
or
$$F_{A/C} + F_{C/T} = K + R_x. \tag{13.1}$$

**Important Note.** With the "in situ" method, power can be transferred back and forth between the air/contact-lens interface and the contact-lens/tear interface, as long as the power at the two interfaces adds up to the sum of the corneal refracting power (as determined by the keratometer) and the ocular refraction. Using this method, it is assumed that the tear/cornea interface will not be affected by the presence of the contact lens. Also, it should be understood that the term $R_x$ in Eq. (13.1) indicates the *ocular*

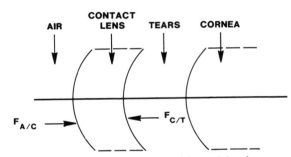

FIGURE 13-7. **The "in situ" method of determining the power of a contact lens on the eye. This method considers two refractions: one taking place at the air/contact-lens interface and the other at the contact-lens/tear interface.**

refraction, that is, refraction determined at the plane of the cornea. This is equal to the power of the contact lens in air only if the back surface of the contact lens has the same radius of curvature as the front surface of the cornea (fit on $K$).

2. Using the "*exploded*" method, we assume that there is a layer of air between the contact lens and the tear layer, and also between the tear layer and the cornea (see Figure 13-8). Using this method, we are concerned with the refraction by the contact lens in air and the refraction by the tear lens in air. Therefore, if the contact lens perfectly corrects the eye,

(power of contact lens in air)
    + (power of tear lens in air) = ocular refraction

or
$$F_C + F_T = R_x. \tag{13.2}$$

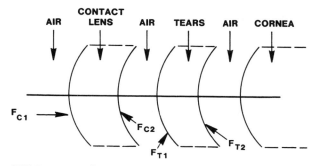

FIGURE 13-8. **The "exploded" method of determining the power of a contact lens on the eye. This method considers refractions by the contact lens and by the tear lens, with an infinitely thin layer of air considered to be between the contact lens and the tear layer and between the tear layer and the cornea.**

**Important Note.** With the exploded method, power can be transferred back and forth between the tear lens and the contact lens, as long as the two add up to the ocular refraction. Using this method, it should be understood that the back surface curvature of the tear layer is assumed to match that of the front surface of the cornea.

The two methods will be compared by the use of some examples. In these examples it will be assumed that the patient is being fitted with a "hard" contact lens (a hard lens being considered as a lens that is sufficiently rigid so that it will maintain its curvature while on the eye, but will not change the curvature of the tear/cornea interface).

*EXAMPLE 1*

Assume that we are providing a perfect optical correction for a patient using a contact lens having an index of refraction of 1.490, and having both front and back surface radii of curvature of 7.5 mm, to be fitted on a cornea having an apical radius of curvature (i.e., keratometer finding) of 7.5 mm. Neglecting the thickness of the contact lens, find the ocular refraction, using (*a*) the "in situ" method and (*b*) the "exploded" method.

(*a*) Using the "in situ" method (Figure 13-9),

$$F_{A/C} + F_{C/T} = K + R_x,$$

or

$$R_x = F_{A/C} + F_{C/T} - K$$

$$= \frac{1.490 - 1}{0.0075} + \frac{1.3375 - 1.490}{0.0075} - \frac{1.3375 - 1}{0.0075}$$

$$= +65.33 + (-20.33) + (-45.00) = +45.00 - 45.00$$

$$= 0.00 \text{ D.}$$

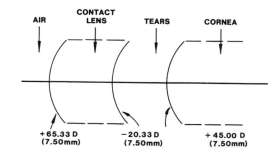

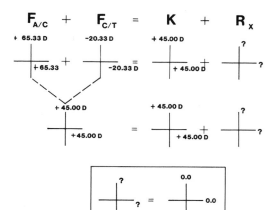

FIGURE 13-9. **Calculation of the refracting power of a contact lens on the eye, using the "in situ" method, for a contact lens having an index of refraction of 1.490, front and back radii of curvature of 7.5 mm, fitted on a cornea having an apical radius of curvature of 7.5 mm.**

(*b*) Using the "exploded" method (Figure 13-10),

$$R_x = F_C + F_T$$

$$= F_{C1} + F_{C2} + F_{T1} + F_{T2}$$

$$= \frac{1.490 - 1}{0.0075} + \frac{1 - 1.490}{0.0075} + \frac{1.3375 - 1}{0.0075}$$

$$+ \frac{1 - 1.3375}{0.0075}$$

$$= +65.33 + (-65.33) + 45.00 - 45.00$$

$$= 0.00 \text{ D.}$$

*EXAMPLE 2*

An eye has an apical radius of curvature of 7.5 mm, and the refraction in the plane of the corneal apex is −3.00 D. If the back surface (base curve) of the contact lens is to have the same radius of curvature of the corneal apex, what will be the power of the front surface of the required contact

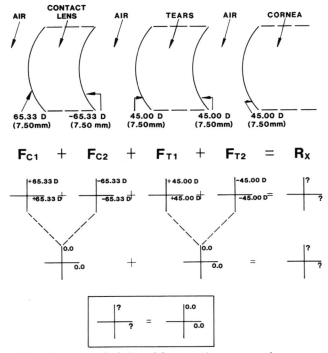

$$F_{C1} + F_{C2} + F_{T1} + F_{T2} = R_x$$

FIGURE 13-10. **Calculation of the refracting power of the example shown in Figure 13-9, using the "exploded" method.**

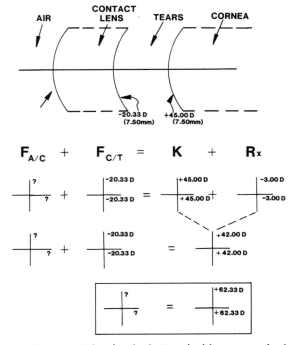

FIGURE 13-11. **Using the "in situ" method for a cornea having an apical radius of curvature of 7.5 mm, fitted with a contact lens having a back surface radius (base curve) of 7.5 mm and a refracting power of −3.00 D.**

lens, using (*a*) the "in situ" method and (*b*) the "exploded" method?

(*a*)  Using the "in situ" method (Figure 13-11).

$$F_{A/C} + F_{C/T} = K + R_x$$

$$F_{A/C} = K + R_x - F_{C/T}$$

$$= \frac{1.3375 - 1}{0.0075} + (-3.00)$$

$$- \frac{1.3375 - 1.490}{0.0075}$$

$$= 45.00 + (-3.00) - (-20.33)$$

$$= +62.33 \text{ D.}$$

(*b*)  Using the "exploded" method (Figure 13-12),

$$R_x = F_C + F_T$$

$$= F_{C1} + F_{C2} + F_{T1} + F_{T2}$$

$$F_{C1} = R_x - F_{C2} - F_{T1} - F_{T2}$$

$$= -3.00 - \frac{1 - 1.490}{0.0075}$$

$$- \frac{1.3375 - 1}{0.0075} - \frac{1 - 1.3375}{0.0075}$$

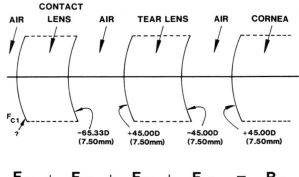

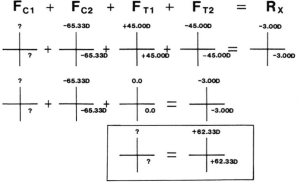

$$F_{C1} + F_{C2} + F_{T1} + F_{T2} = R_x$$

FIGURE 13-12. **Using the "exploded" method for the example shown in Figure 13-11.**

$$= -3.00 - (-65.33) + 45.00 - 45.00$$
$$= +62.33 \text{ D}.$$

### A Soft Contact Lens on the Eye

A "soft" contact lens behaves on the eye quite differently than a "hard" contact lens. Whereas a hard lens, as already stated, is assumed to maintain its shape while on the eye, a soft lens is assumed to conform completely to the corneal surface with the result that there is no "tear lens," as such. There is, of course, an extremely thin layer of tears between the soft lens and the cornea, but, except in an extreme case such as a very thick plus lens in aphakia, the contact lens is sufficiently flexible so that the "tear lens" has no power.

In spite of the fact that a soft lens tends to conform to the cornea whereas a hard lens tends to retain its shape while on the cornea, a soft lens can be considered to behave in the same way as a hard lens as long as:

1. The lens is fitted on a *spherical* cornea.
2. The back surface of the lens has the same radius of curvature as the front surface of the cornea.

Thus, for the two examples just discussed, the results would have been the same if a soft lens rather than a hard lens had been fitted, since the cornea was assumed to be spherical and the lens was fitted so that its back surface had the same radius of curvature as that found by the keratometer. However, in Sections 13.6 and 13.7, which deal with contact lenses fitted on astigmatic corneas and contact lenses fitted "steeper than K" and "flatter than K," it must be kept in mind that the "tear lens" concept applies only to the *hard lens*, and that a soft lens will fit very differently than a hard lens on an astigmatic cornea or if it is fitted steeper or flatter than the corneal radius. Parenthetically, in actual practice a soft contact lens is normally fitted somewhat *flatter* than the corneal apex, being parallel to the corneal periphery, with the result that the lens may be considered to "drape" the corneal apex.

## 13.5. Over-refraction

In the routine fitting of contact lenses, the optometrist often does an "over-refraction": a refraction with the patient wearing either a pair of trial contact lenses or his or her own contact lenses. Over-refraction through trial contact lenses has the advantage that, rather than depending on the *prediction* that a given contact lens will correct the eye's ametropia, the practitioner determines the actual refractive state while wearing the lens.

For calculating the results of a contact lens "in situ," for an eye whose refractive error is perfectly corrected by the contact lens, we can use formula (13.1):

$$F_{A/C} + F_{C/T} = K + R_x.$$

If the over-refraction (OR) demonstrates that the contact lens does not provide a complete correction, then the formula must read

$$F_{A/C} + F_{C/T} + F_{OR} = K + R_x. \qquad (13.3)$$

If the eye is perfectly corrected by the contact lens, the "exploded" system may be used in the following manner, using Eq. (13.2):

$$F_C + F_T = R_x.$$

However, if the contact lens does not provide a complete correction, the equation for the exploded system changes to

$$F_C + F_T + F_{OR} = R_x. \qquad (13.4)$$

### EXAMPLE

Given the following information, what is the ocular refraction?

> Keratometer:    44.00 @ 180/44.00 @ 90
> Trial lens:     base curve = 7.58 mm
>                 power = plano
> Over-refraction: −1.00 DS.

Using the in situ method (see Figure 13-13),

$$F_{A/C} + F_{C/T} + F_{OR} = K + R_x$$

$$\frac{1.49 - 1}{0.00758} + \frac{1.3375 - 1.49}{0.00758} + (-1) = 44.00 + R_x$$

$$64.64 + (-20.12) + (-1) = 44.00 + R_x$$

$$R_x = 64.64 - 21.12 - 44.00$$

$$= -0.48 \text{ D}.$$

Using the exploded method (see Figure 13-14),

$$F_C + F_T = F_{OR} = R_x, \qquad \text{but } F_C = \text{plano}.$$

The power of the front surface of the tear lens is equal to

$$F_{T1} = \frac{1.3375 - 1}{0.00758} = \frac{0.3375}{0.00758} = +44.52 \text{ D}.$$

The power of the back surface of the tear lens is equal to the power of the cornea but opposite in sign, or,

$$F_{T2} = -44.00 \text{ D},$$

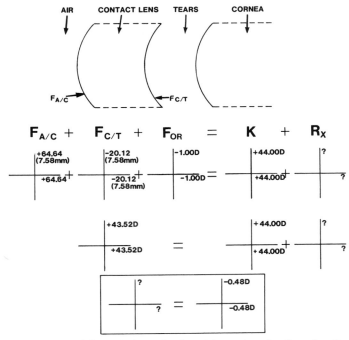

FIGURE 13-13. Schematic determination of the ocular refraction when the keratometer reading, base curve, contact lens power, and over-refraction are known, using the "in situ" method.

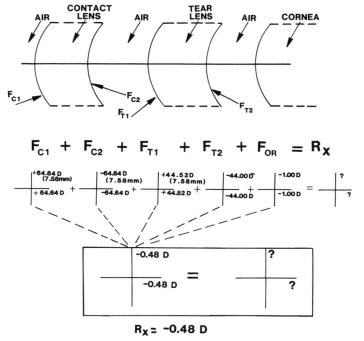

FIGURE 13-14. Schematic determination of the ocular refraction when the keratometer reading, base curve, contact lens power, and over-refraction are known, using the "exploded" method.

and

$$F_{T1} + F_{T2} = +44.52 + (-44.00) = +0.52 \text{ D.}$$

Since

$$F_C + F_T + F_{OR} = R_x,$$

therefore

$$0 + 0.52 + (-1) = R_x,$$

or

$$R_x = -0.48 \text{ D.}$$

## 13.6. Contact Lenses and Corneal Astigmatism

In order to understand the effect of a hard contact lens on corneal astigmatism it must be realized, as already pointed out, that a contact lens interfaces with (or fits on) the *tear layer*, rather than the cornea. Since a contact lens, unless fitted very badly, has no effect on the curvature of the cornea itself, we need be concerned only with the relationship between the contact lens and the tear layer. This being the case, when a spherical hard contact lens is fitted on a spherical cornea, the lens has no effect on the tear layer if its back surface is parallel to the front surface of the cornea.

Whereas in the fitting of spectacles a cylindrical or toric lens is routinely used for an astigmatic eye, in the fitting of hard contact lenses, toric lenses are seldom used. The reason for this is that as long as the hard lens maintains its curvature while on the eye, the interface between the lens and the tear layer will be *spherical* no matter how much corneal astigmatism existed before the lens was put on. A spherical hard lens, in effect, *eliminates* corneal astigmatism. Internal astigmatism (astigmatism due to all ocular factors other than the toricity of the front surface of the cornea) is obviously *not* corrected by the spherical contact lens. Internal astigmatism averages about $-0.50$ D axis 90°, and, using a retinoscope or subjective refraction, the practitioner will almost always find a small amount of against-the-rule astigmatism when doing an "over-refraction" on a patient wearing a hard contact lens.

### EXAMPLE

Given an eye having the following subjective refraction and keratometer readings:

Subjective refraction: plano $-1.00$ DC x 180
Keratometer reading: 44.00 @ 180/45.00 @ 90.

If the back surface radius of a spherical hard contact lens is fitted parallel to the flattest corneal meridian (commonly referred to as "on K" fitting), what will be the required front surface ra-

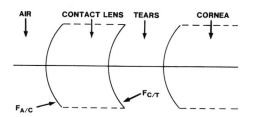

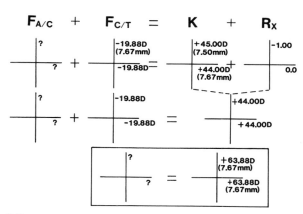

FIGURE 13-15. **Using the "in situ" method for an eye having a keratometer reading of 44.00 @ 180/45.00 @ 90 and a refraction of plano $-1.00$ DC x 180, fitted with a hard contact lens whose back surface radius is parallel to the flattest meridian of the cornea.**

dius of curvature, using (*a*) the "in situ" system and (*b*) the "exploded" system?

(*a*) Using the "in situ" method (Figure 13-15) the radius of curvature of the back surface of the contact lens, using the index of refraction for which the keratometer is calibrated, and the keratometer reading in the horizontal meridian,

$$r_2 = \frac{1.3375 - 1}{44.00} = 0.00767 \text{ m}$$

(normally, this would be done by the use of a table). The refraction taking place at the interface between the contact lens and the tear layer will be the same in both the horizontal and vertical meridians (since the contact lens renders the front surface of the tear layer spherical), and will be

$$F_{C/T} = \frac{1.3375 - 1.490}{0.00767} = -19.88 \text{ D,}$$

while at the front surface of the contact lens,

$$F_{A/C} + F_{C/T} = R_x + K$$

$$F_{A/C} = R_x + K - F_{C/T}$$

$$= 0.00 + 44.00 - (-19.88)$$

$$= +63.88 \text{ D.}$$

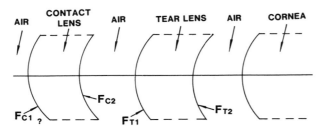

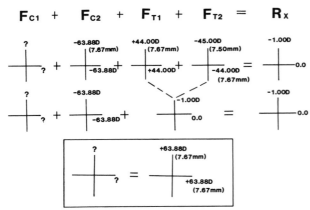

FIGURE 13-16. **Using the "exploded" method for the example shown in Figure 13-15.**

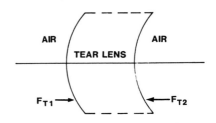

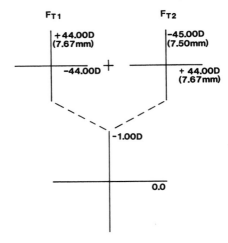

**PLANO –1.00 DC X 180**

FIGURE 13-17. **Power diagram used for determining the power of the contact lens illustrated in Figure 13-16.**

Therefore,

$$r_1 = \frac{1.490 - 1}{63.88} = 0.00767 \text{ m} = 7.67 \text{ mm}.$$

(*b*) Using the "exploded" system (Figure 13-16), the refracting power of the *tear lens* will be calculated, first for the horizontal meridian and then for the vertical meridian, using the subscripts T1 and T2 for the front and back tear layer surfaces.

In the *horizontal meridian*,

$$F_{T2} = \frac{1 - 1.3375}{0.00767} = -44.00 \text{ D}.$$

$$F_{T1} = \frac{1.3375 - 1}{0.00767} = +44.00 \text{ D}.$$

$$F_T = F_{T1} + F_{T2} = 44.00 + (-44.00) = 0.00 \text{ D}.$$

In the *vertical meridian*,

$$r_2 = \frac{1 - 1.3375}{-45.00} = 0.0075 \text{ m}.$$

Therefore,

$$F_{T2} = \frac{1 - 1.3375}{0.0075} = -45.00 \text{ D}.$$

$$F_{T1} = \frac{1.3375 - 1}{0.00767} = +44.00 \text{ D}.$$

$$F_T = F_{T1} + F_{T2} = +44.00 + (-45.00) = -1.00 \text{ D}.$$

Considering both meridians, the power of the tear lens (as shown in the power diagram in Figure 13-17) is

$$F_T = \text{plano} -1.00 \text{ DC axis } 180.$$

As for the *contact lens*, in the horizontal meridian,

$$F_{CL} + F_T = R_x = 0.00,$$

and since

$$F_T = 0.00 \text{ D},$$

$$F_{CL} + 0.00 = 0.00$$

$$F_{CL} = 0.00 \text{ D},$$

and in the vertical meridian,

$$F_{CL} + F_T = R_x = -1.00,$$

and since

$$F_T = -1.00 \text{ D},$$

$$F_{CL} + (-1.00) = -1.00$$

$$F_{CL} = 0.00 \text{ D}.$$

Since the back surface of the contact lens is spherical, having a radius of 7.67 mm, the back surface power of the contact lens *in both meridians* will be

$$F_{CL2} = \frac{1 - 1.490}{0.00767} = -63.88 \text{ D}.$$

Since the contact lens has zero refracting power in both meridians,

$$F_{CL1} + F_{CL2} = 0.00$$

$$F_{CL1} + (-63.88 \text{ D}) = 0$$

$$F_{CL1} = +63.88 \text{ D},$$

and therefore

$$r_1 = \frac{1.490 - 1}{+63.88} = 0.00767 \text{ m} = 7.67 \text{ mm}.$$

Therefore the contact lens is spherical on both surfaces with zero power. This example illustrates that corneal astigmatism is neutralized with a hard contact lens that has a spherical back surface.

### *Soft Contact Lenses and Corneal Astigmatism*

Unlike a hard contact lens, a soft lens is not fitted parallel to the cornea, but, as already described, is normally fitted considerably *flatter* than the optical zone of the cornea. Since the radius of curvature of the peripheral portion of the cornea is typically much flatter than the corneal optical zone, a soft contact

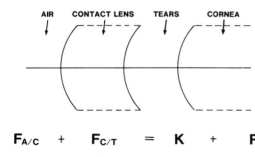

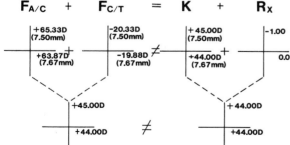

FIGURE 13-18. **The refracting power of a soft contact lens on the eye used in the example illustrated in Figure 13-15, using the "in situ" method.**

lens roughly parallels the corneal periphery, "draping" over the corneal apex. Moreover, whereas diagnostic lens sets for hard contact lenses are available in any desired radii of curvature (usually in steps of 0.05-mm radius), a diagnostic set for soft contact lenses may include no more than two or three radii (such as 8.0, 8.3, and 8.6 mm), which are expected to fit all corneas.

Turning now to the example of the eye having a subjective refraction of plano −1.00 DC x 180 and a keratometer finding of 44.00 @ 180/45.00 @ 90, Figure 13-18 shows that this lens will parallel the cornea in both principal meridians, with the result that it will correct no astigmatism at all. Thus, the refractive power of this lens system while on the eye, instead of being plano −1.00 x 180, would simply be *plano*. This example shows that if a lens parallels a toric cornea in both principal meridians (Figure 13-18), no corneal astigmatism will be corrected.

### 13.7. Fitting Steeper or Flatter than the Cornea

Some hard contact lens fitting philosophies require that the lens be fitted either steeper or flatter than the flattest corneal meridian, rather than the more conventional "on K" fit. When this is done, the effect of the contact lens on the curvature of the tear layer must be taken into consideration (for the contact-lens/tear interface in the "in situ" system and for the air/tear interface in the "exploded" system).

#### *EXAMPLE*
An emmetropic eye having a spherical keratometer reading of 45.00 D (therefore having a corneal radius of 7.5 mm) is fitted with a hard contact lens having a radius of curvature of 7.3 mm (0.2 mm steeper than the cornea). What is the required radius of the front surface of the contact lens?

(*a*) Using the "in situ" method (Figure 13-19), for the refraction at the interface between the contact lens and the tear layer,

$$F_{C/T} = \frac{1.3375 - 1.490}{0.0073} = -20.89 \text{ D},$$

$$F_{A/C} = R_x + K - F_{C/T}$$

$$= 0.00 + 45.00 - (-20.89)$$

$$= +65.89 \text{ D}.$$

$$r_1 = \frac{1.490 - 1}{65.89} = 0.00744 \text{ m}.$$

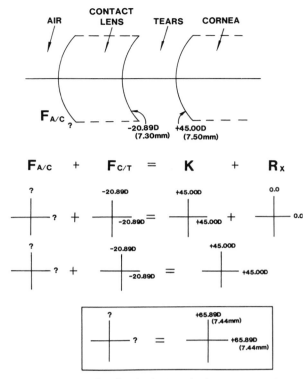

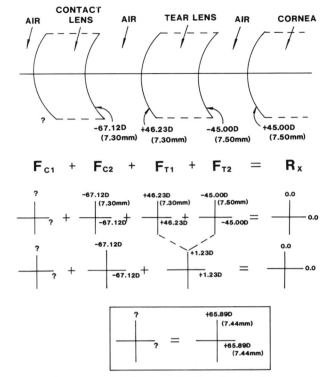

FIGURE 13-19. **Using the "in situ" method, an emmetropic eye is fitted with a hard contact lens 0.2 mm steeper than the cornea, which has a radius of curvature of 7.5 mm. The front surface of the contact lens must have a radius of curvature of 7.44 mm.**

FIGURE 13-20. **Using the "exploded" method for the example shown in Figure 13-19.**

Is there something wrong here? Even though the wearer of this contact lens is emmetropic, the two radii of curvature, $r_1$ and $r_2$, are not equal, indicating that the lens is not a plano lens. The required power of the lens will be found by the use of the "exploded" system.

(b) Using the "exploded" system (Figure 13-20), we first determine the refraction occurring at each surface of the tear lens:

$$F_{T2} = \frac{1 - 1.3375}{0.0075} = -45.00 \text{ D.}$$

$$F_{T1} = \frac{1.3375 - 1}{0.0073} = +46.23 \text{ D.}$$

$$F_T = +46.23 - (-45.00) = +1.23 \text{ D,}$$

while for the contact lens,

$$F_{CL} + F_T = R_x$$

$$F_{CL} = R_x - F_T = 0.00 - 1.23 = -1.23 \text{ D.}$$

$$F_{CL2} = \frac{1 - 1.490}{0.0073} = -67.12 \text{ D.}$$

$$F_{CL1} = F_{CL} - F_{CL2}$$

$$= -1.23 - (-67.12) = +65.89 \text{ D.}$$

$$r_{CL1} = \frac{1.490 - 1}{+65.89} = 0.00744 \text{ m.}$$

These results show that (1) when a lens is fitted steeper than the flattest corneal meridian, the front surface must be made *flatter*, in order to maintain the same refracting power while on the eye (which means that *more minus* power must be added to the lens), and (2) the amount of added minus power is approximately 0.50 D per 0.1 mm of lens steepening.

**Clinical Note.** Many optometrists think of steepening or flattening in terms of diopters of change in the keratometer finding. In the example just given, instead of steepening the back surface of the lens from 7.50 to 7.30 mm, the optometrist could think of steepening the curve from 45.00 to 46.23 D. This has the advantage that the optometrist knows immediately the change that must be made in the *back vertex power* of the contact lens, as compared to that required for an "on K" fit. In the example, the back vertex power of the lens must now be −1.23 D rather than plano as would be required if the lens were fitted "on K." It should be obvious that if the base curve were to be fitted *flatter* than the flattest corneal meridian, it would be necessary to *steepen* the front surface of the lens, thus adding *plus power* to the lens.

We therefore have the following rules of thumb:

1. If a hard contact lens is fitted steeper than the flattest corneal meridian, add more minus power to the lens.
2. If a hard contact lens is to be fitted flatter than the flattest corneal meridian, add plus power to the lens.
3. The amount of minus or plus power to be added is approximately 0.50 D for each 0.10 mm of steepening or flattening of the radius of the back surface of the lens. If the base curve is specified in "diopters" (that is, in diopters as determined by the keratometer), a straight "diopter-for-diopter" relationship exists, in which minus power is added to the lens when the back surface is steepened, and plus power is added when the back surface is flattened.

**Important Note.** Some practitioners (and even some textbooks), when explaining the need for altering the power of a contact lens when the base curve is steepened or flattened, make the mistake of attributing changes in power to the *tear lens*. A steep contact lens is said to result in a "plus tear lens" and a flat contact lens is said to result in a "minus tear lens." Even a casual inspection of the two "tear lenses" in Figure 13-21 shows why this mistake is continually made: The front surfaces of these two tear lenses cause them to *look like* plus and minus lenses, respectively. However, *when the lens is on the eye*, in both cases the tear lens will act as a *minus lens*, since there is a greater index change at the contact-lens/tear interface (a *minus* refraction) than at the tear/cornea interface (a *plus* refraction). The refracting power of the tear lens—and its change when the base curve is steepened—will be shown by an example.

### EXAMPLE

A cornea has a keratometer reading of 45.00 D. Assuming that the tear layer is infinitely thin, what will be the refracting power of the "tear lens" if the lens is fitted (a) parallel to the cornea or (b) 0.2 mm steeper than the cornea. This, of course, is the problem we just considered, but now we will calculate the "in situ" power of the lens.

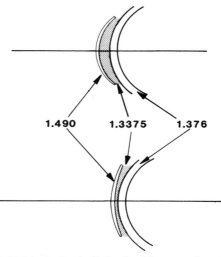

FIGURE 13-21. **A steeply fitting hard lens may be mistakenly considered to result in a tear lens having "plus power"; and a flat-fitting hard lens may be mistakenly considered to result in a tear lens having "minus power." However, in both cases the tear layer acts as a *minus lens*. (See text for explanation.)**

(a) The refraction taking place at the contact-lens/tear interface is

$$\frac{1.3375 - 1.490}{0.0075} = -20.33 \text{ D},$$

and the refraction taking place at the tear/cornea interface is

$$\frac{1.376 - 1.3375}{0.0075} = +5.13 \text{ D}.$$

The total refraction by the tear lens is therefore

$$-20.33 + 5.33 = -15.20 \text{ D}.$$

(b) With the lens fitted 0.2 mm steeper than the cornea, the refraction taking place at the contact-lens/tear interface will be

$$\frac{1.3375 - 1.490}{0.0073} = -20.89 \text{ D},$$

and since the refraction at the tear/cornea interface is still +5.13 D, the total refraction by the tear layer is now

$$-20.89 + 5.13 = -15.76 \text{ D}.$$

We see, therefore, that by fitting the lens 0.2 mm steeper than the corneal radius, the refraction by the "tear lens" is now 0.56 D *more minus* than it would have been for an "on K" fit. If a tear lens that looks like a plus lens is actually a *minus* (diverging) lens, and if fitting the contact lens steeper than the cornea *increases* the minus power of the tear lens, why do we have to add minus power to the contact lens to properly correct the eye?

What actually happens is that when we "steepen" the base curve of the contact lens *while keeping the same back vertex power*, we are steepening *both* the front and back surfaces of the lens, and, as shown in Part (*a*) of the solution to the original example, steepening of the front surface results in the front surface power increasing from its original value of

$$\frac{1.490 - 1}{0.0075} = +65.33 \text{ D}$$

to the value

$$\frac{1.490 - 1}{0.0073} = +67.12 \text{ D},$$

or an increase in refracting power of +1.79 D.

We see, therefore, that by steepening the lens 0.2 mm (both front and back surfaces), we bring about an increase in plus power at the air/contact-lens interface of 1.79 D, and an increase in the minus power at the contact-lens/tear interface (from −20.33 to −20.89 D) or −0.56 D, so the *net* change is equal to

$$+1.79 \text{ D} - 0.56 \text{ D} = +1.23 \text{ D},$$

which is the same as the original +1.23 D that we found would have to be compensated by reducing the plus power of the front surface.

It is of interest that the ratio between the power change at the front surface and the change at the back surface, brought about by steepening or flattening the lens, is

$$\frac{1.79}{0.56}, \quad \text{or approximately} \quad \frac{3.2}{1},$$

which is the same ratio as the index change at the front surface of the contact lens to the index change at its back surface, or

$$\frac{1.490 - 1}{1.3375 - 1.49} = \frac{3.2}{1}.$$

*Fitting Soft Contact Lenses Steeper or Flatter Than the Cornea*

As stated in Section 13.5, soft contact lenses are routinely fitted much flatter than the optical zone of the cornea, and since a soft lens conforms to the corneal surface so completely, not only is there no "tear lens," as such, but a flattening or steepening of the lens has no clinically significant effect on the ability of the lens to correct the eye's refractive error. An excep-

tion to this rule is a very *thick* lens, designed for an aphakic eye, which may fail to completely conform to the corneal curvature. In addition a *badly fitting* soft contact lens will sometimes "flex" or "bend" on the cornea in such a way that the refractive error will not be adequately corrected by the lens.

## 13.8. The Optics of Bifocal Contact Lenses

As with spectacle lenses, a bifocal contact lens may be made either by using two materials having different indices of refraction (a fused bifocal) or by using a lens made of a single material but having two different radii of curvature on either the front or the back surface (a one-piece bifocal). Although both methods are used in the manufacture of bifocal contact lenses, we will first consider one-piece "hard" bifocals, having two radii of curvature on the back surface.

*One-Piece Back Surface Hard Bifocals*

Referring to Figure 13-22, the power of the addition (on the eye) is equal to the difference in the back surface interface powers of the two portions of the lens:

$$F_{add} = F_{near} - F_{distance},$$

where $F_{near}$ = the back surface interface power of the near portion of the lens, and $F_{distance}$ = the back surface interface power of the distance portion of the lens.

For a material having an index of refraction of 1.490, a distance portion radius of curvature of 7.50 mm, and an addition of +2.00 D, we can find the

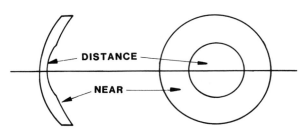

FIGURE 13-22. **For a one-piece back surface hard bifocal, the power of the addition is equal to the difference in the back surface interface powers in the two portions of the lens. Using a lensometer, the power of the addition is equal to the difference in back vertex powers in air (which is 3.2 times the power of the addition on the eye).**

radius of curvature of the back surface of the near portion $r_n$ of the lens as follows:

$$+2.00 = \frac{1.3375 - 1.490}{r_n} - \frac{1.3375 - 1.490}{0.0075}$$

$$+2.00 = \frac{0.1525}{r_n} - \left(\frac{-0.1525}{0.0075}\right)$$

$$+2.00 = \frac{0.1525}{r_n} - (-20.33)$$

$$-18.33 = \frac{0.1525}{r_n}$$

$$r_n = 0.00832.$$

What would we expect the power of the add to be if we verified the finished lens (above) by means of a lensometer? The refracting power of the distance and near portions of the back surface of the lens, *in air*, will be

$$F_{distance} = \frac{1 - 1.490}{0.0075} = -65.33 \text{ D.}$$

$$F_{near} = \frac{1 - 1.490}{0.00832} = -58.89 \text{ D.}$$

Since the refraction at the front surface of the lens will obviously be the same for the distance and near portions, the power of the add when measured on the lensometer will be

$$F_{near} - F_{distance} = -58.89 - (-65.33) = +6.44 \text{ D.}$$

Thus we see that the ratio of the power of the add as found by the lensometer (in air) as compared to the power of the add on the eye (the contact-lens/tear refraction) is

$$\frac{6.44}{2.00} = \frac{3.2}{1},$$

or the familiar 3.2/1 ratio discussed in Section 3.7.

## One-Piece Front Surface Hard Bifocals

Both hard and soft contact lens bifocals may be made in one-piece form, with the segment on the front surface. Since the power of the add is the difference between the powers of the distance and near portions of the front surface, when a front surface segment is verified on the lensometer we would expect to find the exact power of the add that was ordered.

## Fused Front Surface Hard Bifocals

The optical principles of fused front surface bifocal contact lenses (available in hard lenses only) are similar to the principles of fused bifocal spectacle lenses, as discussed in Section 8.12. The basic formula [Eq. (8.12)] is

$$\text{power of add} = \frac{F_1 - F_C}{\text{ratio}},$$

where $F_1$ = the front surface power of the major lens and $F_C$ = the surface power of the countersink surface. In this expression, the term "ratio" is equal to

$$\frac{n' - n}{n'' - n'},$$

where $n$ = the index of refraction of air, $n'$ = the index of the major lens, and $n''$ = the index of the segment.

When a fused front surface bifocal contact lens is verified by means of the lensometer, the power of the add is equal to the difference between the powers of the distance and near portions of the lens.

## Fused Back Surface Hard Bifocals

This lens is made by fusing a higher index material into a countersink curve on the back surface of the lens. The curvature of the back surface of the segment is the same as that of the base curve (i.e., the back surface of the major lens). Due to the strong concave curvature of the back surface of the lens, there is an increase in *minus* power at the back surface of the segment as compared to the back surface of the major lens. Therefore, the radius of curvature of the countersink surface must be much shorter for a fused back surface bifocal than for a fused front surface bifocal.

Suppose, for example, that we wish to make a fused back surface bifocal, using a major lens having an index of refraction of 1.490, a segment having an index of refraction of 1.570, with the lens having a base curve radius of 7.5 mm and an addition of +2.00 D. What is the radius of curvature of the countersink surface (see Figure 13-23)?

Using the "in situ" method, the power of the contact-lens/tear interface in the distance portion is

$$F_{C/T} = \frac{n_T - n_{C'}}{r_2} = \frac{1.3375 - 1.490}{0.0075} = -20.33 \text{ D.}$$

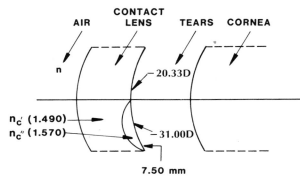

FIGURE 13-23. **For a fused back surface hard bifocal, the "in situ" method is used to find the curvature of the countersink curve for a lens having a base curve radius of 7.5 mm and an addition of +2.00 D, using a major lens having an index of 1.490 and a segment having an index of 1.570.**

The power of the contact-lens/tear interface in the near portion is

$$F_{C/T} = \frac{n_T - n_{C''}}{r} = \frac{1.3375 - 1.570}{0.0075} = -31.00 \text{ D}.$$

Therefore, the near contact-lens/tear interface is more minus in power than the distance contact-lens/tear interface by

$$-31.00 - (-20.33) = -10.67 \text{ D}.$$

In order to determine the radius of curvature of the countersink curve, we use the following relationship: The power of the addition is equal to the power of the front surface of the segment plus the power change between the distance and near contact-lens/tear interfaces, or

$$\text{power of add} = \frac{n_{C''} - n_{C'}}{r_C} + (-10.67)$$

$$+2.00 = \frac{1.570 - 1.490}{r_C} + (-10.67)$$

$$\begin{array}{c}\text{power of}\\\text{countersink} = 12.67 \text{ D} = \frac{1.570 - 1.490}{r_C}\\\text{interface}\end{array}$$

$$+12.67 = \frac{0.08}{r_C}$$

$$r_C = \frac{0.08}{12.67} = 0.00631 \text{ m} = 6.31 \text{ mm}.$$

If the back surface of the contact lens were in air, the power of the contact-lens/tear interface for the distance portion of the lens would be

$$F_{C/A} = \frac{n - n_{C'}}{r_2} = \frac{1.00 - 1.490}{0.0075} = -65.33 \text{ D},$$

and the power of the contact-lens/tear interface for the segment portion of the lens would be

$$F_{C/A} = \frac{n - n_{C''}}{r_2} = \frac{1.00 - 1.57}{0.0750} = -76.00.$$

The change in power between the distance and near contact-lens/air interfaces is therefore

$$-76.00 - (-65.00) = -10.67 \text{ D},$$

which is the same as we found when the back surface of the lens was placed against the tear layer. Thus, the effective power of the addition on the eye will be the same as the power of the add as determined by the use of a lensometer.

## 13.9. Residual Astigmatism and Its Correction

Residual astigmatism is the astigmatism that is present when a contact lens with spherical surfaces is placed on the cornea to correct the eye's ametropia. Residual astigmatism may be caused by any of a number of optical and anatomical components of the contact-lens/eye system. An anatomical factor which may cause residual astigmatism is the displacement of the fovea from the posterior pole of the eye, with the result that the visual axis and the optic axis fail to coincide: Thus we have astigmatism due to obliquity of incidence. Other sources of residual astigmatism include the back surface of the cornea, toroidal crystalline lens surfaces, and the tilting of these surfaces.

The relationship between the total astigmatism of the eye (also called *refractive* astigmatism), corneal astigmatism, and residual astigmatism is given by the relationship

$$\begin{array}{c}\text{total astigmatism} = \text{corneal astigmatism} + \text{residual}\\\text{astigmatism.} \qquad (13.5)\end{array}$$

According to Javal's rule, which attempts to predict the total astigmatic correction on the basis of keratometric findings,

$$\begin{array}{c}\text{total astigmatism} = (1.25) \text{ (corneal astigmatism)}\\+ (-0.50 \text{ DC x } 90).\end{array}$$

Although Javal's rule was determined strictly on an empirical (rather than a theoretical) basis, residual astigmatism can be thought of as averaging about $-0.50$ DC x 90. (Note that Javal's rule is in the form

$y = mx + b$, with the result that $b$ represents residual astigmatism.)

Residual astigmatism has a rather narrow distribution, normally ranging from zero to about 1.50 DC x 90, and very seldom having its axis at or near 180.

Since a hard contact lens (unless it is very thin and flexible) eliminates corneal astigmatism, a spherical hard lens would be expected to leave only *residual* astigmatism uncorrected. For this reason, if retinoscopy or subjective testing is done while a patient wears spherical hard contact lenses, the expected finding would be −0.50 D of astigmatism with the axis at or near 90 degrees. In order to predict the amount of uncorrected astigmatism a patient will have when fitted with hard contact lenses, we need only compare total astigmatism (as determined by retinoscopy or by subjective testing) to keratometric astigmatism. This process will be illustrated by three examples.

### EXAMPLE 1
A given eye has the following keratometric finding and subjective refraction:

> Keratometer:   42.50 @ 180/45.00 @ 90
> Subjective:    plano −2.00 DC x 180.

What will be the amount of astigmatism that is not corrected by a spherical hard contact lens (i.e., the residual astigmatism)?

Residual
astigmatism = refractive astigmatism
                − corneal astigmatism
         = (−2.00 DC x 180) − (−2.50 x 180)
         = +0.50 DC x 180, or −0.50 DC x 90.

If retinoscopy or subjective refraction were to be done while a spherical hard contact lens is worn on this eye, one would expect the astigmatic portion of the refractive finding to be −0.50 DC x 90.

### EXAMPLE 2
A given eye has the following keratometric finding and subjective refraction:

> Keratometer:   44.00 @ 180/43.50 @ 90
> Subjective:    plano −1.50 DC x 90.

What will be the amount of astigmatism that is not corrected by a spherical hard contact lens?

Residual
astigmatism = (−1.50 DC x 90) − (−0.50 DC x 90)
         = −1.00 DC x 90.

### EXAMPLE 3
Given the keratometric findings and subjective refraction:

> Keratometer:   42.50 @ 180/42.50 @ 90
> Subjective:    plano −0.75 DC × 90.

What will be the amount of astigmatism that is not corrected by a spherical hard contact lens?

Residual
astigmatism = (−0.75 DC x 90) − 0.00
         = −0.75 DC x 90.

In this case, since there is no corneal astigmatism, *all* of the refractive astigmatism is accounted for by residual astigmatism.

In routine hard contact lens fitting, once the lens has been placed on the eye and has been allowed to "settle down," the predicted residual astigmatism is verified by performing an over-refraction.

### The Correction of Residual Astigmatism

As a general rule a contact lens wearer can tolerate, without asthenopia or blurred vision, as much as 0.50 D of against-the-rule or oblique astigmatism, or as much as 0.75 D (or even 1.00 D) of with-the-rule astigmatism. Therefore, when it is estimated (by comparing keratometer readings and subjective findings) or found by direct measurement (i.e., by over-refraction) that a spherical contact lens will result in more than these amounts of uncorrected astigmatism, the use of a toric contact lens should be seriously considered.

Since spherical contact lenses (either hard or soft) tend to *rotate* while on the wearer's eye, some method must be found to *stabilize* a toric lens, to prevent it from rotating on the eye. The most common method of stabilizing a contact lens on the eye is what is known as *prism ballast* (Figure 13-24): base-down prism is placed on the lens, in the lathe cutting process, so that the bottom of the lens is thicker than the top. It is this increased thickness of the lower part of the lens that keeps the lens from rotating on the eye.

Using either a hard or soft contact lens, a toric surface may be put on either the front or the back of

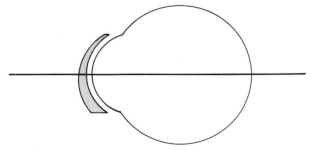

FIGURE 13-24. **A prism-ballasted lens used for the correction of astigmatism, having base-down prism so that the bottom of the lens is thicker than the top.**

the lens. In the great majority of cases, a *front* toric surface is used. However, both possibilities will be examined here.

### EXAMPLE 1

Given the keratometer readings and subjective refraction findings:

Keratometer:    44.00 @ 180/43.00 @ 90
Subjective:    plano −2.00 DC x 90.

(*a*) What will be the amount of uncorrected astigmatism if a *spherical* hard contact lens is fitted with its back surface radius parallel to the flattest corneal meridian? (*b*) What will be the required parameters of a *front surface toric* lens, fitted so that the astigmatism is corrected in both principal meridians? (*c*) What will be the back vertex power of the contact lens in (*b*) above?

(*a*) Since the corneal astigmatism can be corrected by a −1.00 D cylinder axis 090 and the refractive astigmatism can be corrected by a −2.00 D cylinder axis 090, the residual astigmatism with a spherical hard contact lens would be

(−2.00 DC x 090) − (−1.00 DC x 90)
= −1.00 DC x 90.

The parameters of a spherical hard lens for this eye can be determined with the aid of Figure 13-25, using the "in situ" method. For the 43.00 D keratometer reading in the *vertical meridian*, the

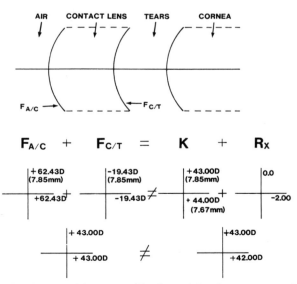

$$F_{A/C} + F_{C/T} = K + R_x$$

FIGURE 13-25. **Diagram used for determining the parameters of a spherical hard contact lens for an eye having a keratometer reading of 44.00 @ 180/43.00 @ 90 and a subjective refraction of plano −2.00 DC x 90. Note that the horizontal meridian is overcorrected by 1.00 D.**

use of a table (or calculation) results in a back surface radius of 7.85 mm. Therefore, for the refraction at the back surface of the contact lens (contact lens/tears),

$$F_{C/T} = \frac{1.3375 - 1.490}{0.00785} = -19.43 \text{ D},$$

and for the refraction at the front surface of the contact lens, if the lens is ordered in plano power (due to the spherical portion of the spectacle refraction being plano) and if the lens is assumed to be infinitely thin,

$$F_{A/C} = \frac{1.490 - 1}{0.00785} = +62.43 \text{ D}.$$

In order to be perfectly corrected in each principal meridian, the power of the air/contact-lens interface ($F_{A/C}$) plus the power of the contact-lens/tear interface ($F_{C/T}$) must be equal to the keratometer reading plus the ocular refraction. For the horizontal meridian, the keratometer reading plus the ocular refraction will be

$$44.00 + (-2.00) = 42.00 \text{ D},$$

and for the vertical meridian the sum of the keratometer reading and the ocular refraction will be

$$43.00 + 0.00 = 43.00 \text{ D}.$$

We may now determine how well this lens will correct the refractive error of the eye in each of the principal meridians:

Horizontal:  $F_{A/C} + F_{C/T} = K + R_x$
$+62.43 + (-19.43) = +44.00$
$+ (-2.00) + 43.00 \neq +42.00$ D.

Vertical:    $+62.43 + (-19.43) = 43.00 + 0.00$
$+43.00$ D $= +43.00$ D.

This result shows that whereas the refractive error is corrected in the vertical meridian, the horizontal meridian is overcorrected by 1.00 D. This, of course, is what we found out by comparing the keratometric and refractive astigmatism.

(*b*) Referring to Figure 13-26, the parameters of a *front surface toric* hard lens for this eye will be determined, using the "in situ" method. Since the correction in the vertical meridian is plano, the radius of curvature of the front surface in the vertical meridian will be the same as the radius of the back surface, that is, 7.85 mm. Therefore, we need concern ourselves only with the horizontal meridian. We'll first determine the refracting power of the front surface, $F_{A/C}$, in the horizontal meridian and then the radius of curvature, $r_{1H}$:

Horizontal: $F_{A/C} + F_{CT} = K + R_x$

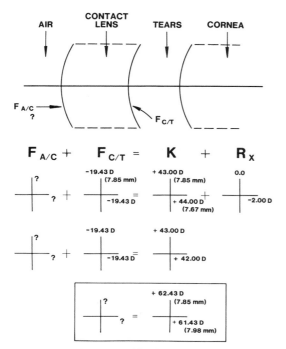

FIGURE 13-26. **Diagram used for determining the parameters of a front surface toric hard contact lens for the eye considered in Figure 13-25.**

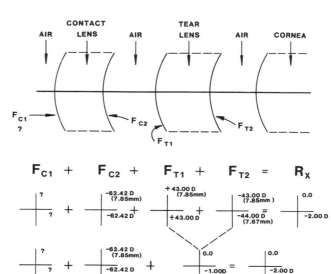

$F_{C1} + F_{C2} = $ PLANO $- 1.00$ DC $\times$ 90

FIGURE 13-27. **The use of schematic power crosses to illustrate the "exploded" method of determining the back vertex power of the front toric hard contact lens referred to in Figure 13-25.**

or

$$F_{A/C} = K + R_x - F_{C/T}$$
$$= -2.00 + 44.00 - (-19.43)$$
$$= +61.43 \text{ D},$$

and

$$r_{1H} = \frac{1.490 - 1}{61.43} = +0.00798 \text{ m}.$$

Hence, the radii of this lens would be

Back surface: 7.85 mm.
Front surface: vertical meridian = 7.85 mm.
horizontal meridian = 7.98 mm.

(c) The back vertex power is more easily seen if the "exploded" method is used. This method is more easily visualized if schematic power crosses are used (see Figure 13-27). The use of these power crosses shows that the radii of curvature of this lens are

Front surface:
Vertical meridian = 7.85 mm,
therefore $F_{A/C} = +62.43$ D.
Horizontal meridian = 7.98 mm
therefore $F_{A/C} = 61.43$ D.
Back surface = 7.85 mm
therefore $F_{C/A} = -62.43$ D.

Hence the back vertex power of this contact lens is

plano $-1.00$ DC x 90

### EXAMPLE 2

For the same eye, what will be the required parameters of a *back surface* toric hard lens, with one of the principal meridians of the back surface of the lens parallel to the flattest corneal meridian? What is the back vertex power of the contact lens?

Referring to Figure 13-28, and again using the "in situ" method, the radius of curvature in the *vertical meridian* of the back surface will parallel the cornea and hence will be 7.85 mm. This contact-lens/tear interface, therefore, generates the following power:

$$F_{C/T} = \frac{1.3375 - 1.490}{0.00785} = -19.43 \text{ D}.$$

Since the correction in the vertical meridian is plano, the power in the *vertical meridian* of the front surface would be

$$F_{A/C} + F_{C/T} = K + R_x$$
$$F_{A/C} = K + R_x - F_{C/T}$$
$$= +43.00 + 0.00 - (-19.43)$$
$$= +62.43 \text{ D}.$$

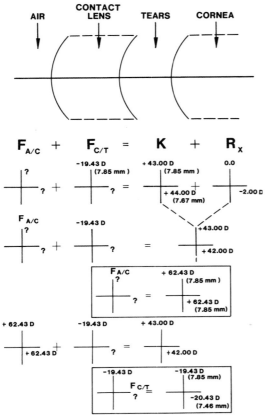

FIGURE 13-28. **Diagram used for determining the parameters of a back surface toric hard contact lens, using the "in situ" method, for the eye referred to in Figure 13-25.**

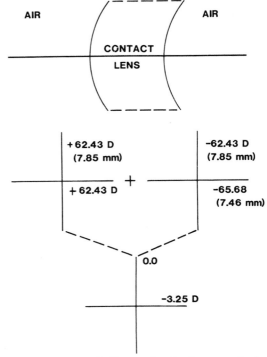

FIGURE 13-29. **Schematic diagram showing the power, in air, of the back surface toric hard contact lens illustrated in Figure 13-28.**

Hence the radius of curvature in the vertical meridian of the front surface is found by the relationship

$$r_{1V} = \frac{1.490 - 1}{62.43} = 0.00785 \text{ m.}$$

Since the front surface is spherical, the radius of curvature in the horizontal meridian of the front surface is 7.85 mm as well.

In the horizontal meridian,

$$F_{A/C} + F_{C/T} = K + R_x$$

$$F_{C/T} = K + R_x - F_{A/C}$$

$$= +44.00 + (-2.00) - 62.43$$

$$F_{C/T} = -20.43 \text{ D.}$$

$$r_{2H} = \frac{1.3375 - 1.490}{-20.43}.$$

Therefore, in the horizontal meridian,

$$r_{2H} = 0.00746 \text{ m, or } 7.46 \text{ mm.}$$

A schematic diagram of the power of this contact lens, in air, is given in Figure 13-29. This diagram shows that the radii of curvature of this lens are:

Front surface = 7.85 mm, therefore
$F_{A/C}$ = +62.43 D.
Back surface:
vertical meridian = 7.85 mm,
therefore $F_{C/A}$ = −62.43 D.
horizontal meridian = 7.46 mm,
therefore $F_{C/A}$ = −65.68 D.

Hence, the back vertex power of this lens would be

plano −3.25 DC x 90.

Note that a comparison of the power of the back surface cylinder to that of the front surface cylinder gives us the familiar ratio,

$$\frac{3.25}{1.00}, \quad \text{or approximately 3.2/1.}$$

*The Effect of Lens Flexure on Residual Astigmatism*

In the preceding discussion it has been assumed that a hard contact lens maintains its spherical curvature when fitted on a toric cornea. However, when a spherical polymethyl methacrylate (PMMA) contact lens is made with a center thickness of less than about 0.13 mm, the lens has a tendency to *flex* or "bend" on a toric cornea in such a manner that it assumes a part of the corneal toricity. Gas permeable hard lenses (for example those made of silicone acrylate) are even more flexible than PMMA lenses and must be made somewhat thicker in order to prevent excessive flexure. The amount of flexure may be measured by taking a keratometer reading while the spherical lens is worn: if the front surface of the lens is found to be spherical, no flexure is present; but if astigmatism is found, it will be due to flexure.

Flexure can be either an advantage or a disadvantage, in terms of its effect on residual astigmatism. If a lens flexes on a cornea having with-the-rule toricity in such a manner that it fails to eliminate 0.50 to 0.75 D of the corneal astigmatism, the uncorrected astigmatism may be just sufficieint to compensate for 0.50 to 0.75 D of against-the-rule *internal* astigmatism, with the result that little or no residual astigmatism will be found when an over-refraction is done. However if a lens flexes *more* than 0.50 to 0.75 D (which may occur with a thin lens on a highly toric cornea), a large amount of astigmatism may remain uncorrected. In such a case, a thicker lens (or one made of a less flexible material) may be used, or a toric contact lens may be fitted.

### Residual Astigmatism and Soft Contact Lenses

When the term "residual astigmatism" is used in connection with soft contact lenses, it has a different meaning than when used in connection with hard lenses. Rather than being used as a synonym for "internal astigmatism," the term is used to denote any astigmatism not corrected by the contact lens in question, which may better be termed "uncorrected astigmatism."

As stated earlier, a well-fitting soft contact lens completely conforms to the corneal surface with the result that it fails to correct any corneal astigmatism; and since we know that a contact lens cannot correct internal astigmatism, it follows that a spherical soft contact lens *corrects no astigmatism at all*. Thus, the astigmatism that is not corrected by a spherical soft contact lens (the "uncorrected" astigmatism) is the *refractive*, or *total*, astigmatism. In order to illustrate this principle, we'll recalculate each of the above examples for a spherical soft lens.

**EXAMPLE 1**

What is the amount of uncorrected astigmatism for a spherical soft contact lens, given the following findings?

> Keratometer:    42.50 @ 180/45.00 @ 90
>
> Subjective:    plano −2.00 DC x 180.

Since a spherical soft contact lens corrects no corneal astigmatism, the answer will be

$$-2.00 \text{ DC x } 180.$$

**EXAMPLE 2**

What is the amount of uncorrected astigmatism for a spherical soft contact lens, for the following findings?

> Keratometer:    44.00 @ 180/43.50 @ 90
>
> Subjective:    plano −1.50 DC x 90.

The uncorrected astigmatism would be the refractive astigmatism, or

$$-1.50 \text{ DC x } 90.$$

**EXAMPLE 3**

Given the following findings, what is the amount of uncorrected astigmatism for a spherical soft contact lens?

> Keratometer:    42.50 @ 180/42.50 @ 90
>
> Subjective:    plano −0.75 DC x 90.

The answer is, again, the refractive astigmatism, or

$$-0.75 \text{ DC x } 090.$$

It should be understood, on the basis of these examples, that if a prospective contact lens wearer has any amount of astigmatism, he or she will be likely to have *less* uncorrected astigmatism if a *hard* contact lens is worn than if a *soft* lens is worn. Furthermore, since internal astigmatism is almost always against the rule, and since corneal astigmatism and refractive astigmatism are more commonly with the rule, it follows that:

1. When a spherical hard lens is worn, any uncorrected astigmatism is most likely to be *against-the-rule* astigmatism.
2. When a spherical soft lens is worn, any uncorrected astigmatism is very likely to be *with-the-rule* astigmatism.

### Correction of Astigmatism with Soft Contact Lenses

When a patient is to be fitted with soft contact lenses, the use of a toric soft lens is normally considered whenever the total astigmatism of the eye exceeds 0.50 or 0.75 D of against-the-rule or oblique astigmatism, or 0.75 or 1.00 D of with-the-rule astigmatism. Although a toric soft contact lens can be made in either front toric or back toric form, our discussion will be limited to the front toric form. Since a well-fitting soft lens conforms completely to the corneal surface, it will be assumed that a front surface toric lens fits in such a way that the back surface radii of curvature of the lens are equal, in each of the principal meridians, to the radii of curvature of the cornea. Since soft contact lenses contain a high percentage of water, they have a lower index of refraction than hard contact lenses, usually considered to be approximately 1.430.

### EXAMPLE

Given the following keratometric and subjective findings:

Keratometer:    44.00 @ 180/43.00 @ 90
Subjective:     plano −2.00 DC x 90.

Using the index of refraction of 1.430, find the necessary front and back radii of curvature in each of the principal meridians of the contact lens.

We first calculate the radii of curvature of the cornea in each meridian:

$$r_H = \frac{1.3375 - 1}{44.00} = 0.00767 \text{ m.}$$

$$r_V = \frac{1.3375 - 1}{43.00} = 0.00785 \text{ m.}$$

Thus, the radii of curvature of the back surface of the lens will be 7.67 mm in the horizontal meridian and 7.85 mm in the vertical meridian.

The power of the contact-lens/tear interface in the *horizontal* meridian is

$$F_{C/T} = \frac{1.3375 - 1.430}{0.00767} = \frac{-0.0925}{0.00767} = -12.06 \text{ D,}$$

and the power of the contact-lens/tear interface in the *vertical* meridian is

$$F_{C/T} = \frac{1.3375 - 1.430}{0.00785} = -11.78 \text{ D.}$$

Using the "in situ" method for the refraction in the horizontal meridian at the *front surface* of the contact lens,

$$F_{A/C} + F_{C/T} = K + R_x$$

$$F_{A/C} = K + R_x - F_{C/T}$$

$$= +44.00 + (-2.00) - (-12.06)$$

$$= +54.06 \text{ D.}$$

The radius of curvature of the front surface of the lens in the horizontal meridian is

$$r_{1H} = \frac{1.430 - 1}{54.06} = 0.00795 \text{ m,}$$

while for the refraction in the vertical meridian of the front surface,

$$F_{A/C} = K + R_x - F_{C/T}$$

$$= +43.00 + 0.00 - (-11.78)$$

$$= +54.78 \text{ D.}$$

Hence, the radius of curvature of the front surface in the vertical meridian is

$$r_{1V} = \frac{1.430 - 1}{54.78} = 0.00785 \text{ m.}$$

The radii of curvature of this soft contact lens on the eye would therefore be:

Front surface:  horizontal = 7.95 mm
                vertical   = 7.85 mm

Back surface:   horizontal = 7.67 mm
                vertical   = 7.85 mm

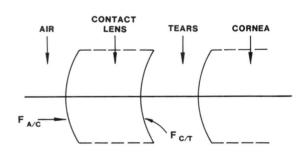

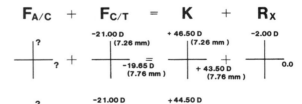

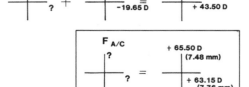

FIGURE 13-30. **Diagram used for determining the parameters of a back surface toric hard lens used to provide a comfortable physical fit on a highly toric cornea.**

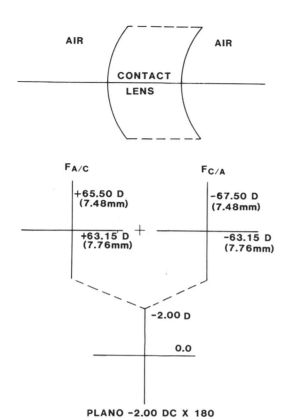

FIGURE 13-31. **Diagram showing the power, in air, of the back surface toric lens illustrated in Figure 13-30.**

*The Use of a Toric Back Surface Hard Lens for a Highly Toric Cornea*

For a highly toric cornea, a spherical hard contact lens will tend to fit very badly, "rocking" on the flat meridian of the cornea and causing discomfort due to pressure on the lid margins. In such a case, the use of a toric back surface lens will often provide a comfortable physical fit. It should be understood that the lens is fitted *only* for the purpose of providing a comfortable fit: From the *optical* point of view, such a lens fails to "mask" corneal astigmatism (as does a spherical lens) with the result that uncorrected astigmatism may be a problem unless a toric *front* surface is also used.

### EXAMPLE

Given the following findings:

    Keratometer:   43.50 @ 180/46.50 @ 90

    Subjective:      plano −2.00 DC x 180.

If a back surface toric lens is fitted so that it is

parallel to the cornea in each principal meridian, what would be the parameters of the resulting lens, designed in such a way as to completely correct the eye's astigmatism?

Using the "in situ" method, this problem can be solved by reference to the optical crosses shown in Figure 13-30. As shown in this figure, the lens radii would be as follows:

Back surface interface:
    horizontal = 7.76 mm ($F_{C/T}$ = −19.65 D).
    vertical    = 7.26 mm ($F_{C/T}$ = −21.00 D).

Front surface:
    horizontal = 7.76 mm ($F_{A/C}$ = +63.15 D).
    vertical    = 7.48 mm ($F_{A/C}$ = +65.15 D).

The back vertex power of the contact lens, in air (see Figure 13-31), would be

$$\text{plano} -2.00 \text{ DC x } 180.$$

It is obvious that this lens is *bi-toric*. When a toric back surface, matching the toricity of the cornea, is used, a front toric curve usually must be used to provide adequate correction of the eye's astigmatism.

## 13.10. Magnification Effects of Contact Lenses

As stated in Section 10.9, if we assume a correcting lens to be infinitely thin (so that the shape magnification factor is equal to unity), the size of the retinal image formed by a lens is inversely proportional to the power of the lens; and when this relationship is applied to the fitting of a contact lens as opposed to a spectacle lens on the same eye, we have the equation

$$\frac{\text{retinal image size with a contact lens}}{\text{retinal image size with a spectacle lens}} = \frac{F_S}{F_C},$$

and

$$\frac{\text{retinal image size with a spectacle lens}}{\text{retinal image size with a contact lens}} = \frac{F_C}{F_S},$$

where $F_S$ is the back vertex power of the spectacle lens and $F_C$ is the back vertex power of the contact lens. It will be recalled that, for a myopic eye, a contact lens will have a lower back vertex power than a spectacle lens for the same eye, with the result that the retinal image will be *larger* with a contact lens than with a spectacle lens, whereas for a hyperopic eye a contact lens will have a higher power than a spectacle lens with the result that the retinal image will be *smaller* with a contact lens than with a spectacle lens. These relationships will be illustrated by two examples.

*EXAMPLES*

Assuming the correcting lens to be infinitely thin in each case, what will be the change in retinal image size when switching from a spectacle lens to a contact lens for each of the following eyes: (*a*) an eye requiring a $-5.00$ D spectacle lens at a vertex distance of 12 mm, and (*b*) an eye requiring a $+13.00$ D spectacle lens at a vertex distance of 10 mm.

(*a*) The required power of a contact lens for this eye will be

$$\frac{-5.00}{1 - 0.012\,(-5.00)} = -4.72 \text{ D.}$$

$$\text{Magnification ratio} = \frac{-5.00}{-4.72} = 1.06,$$

and

$$\%\text{SM change} = (1.06 - 1.00)100 = +6\%.$$

(*b*) The required power of a contact lens for this eye will be

$$\frac{+13.00}{1 - 0.010\,(+13.00)} = +14.94 \text{ D.}$$

$$\text{Magnification ratio} = \frac{+13.00}{+14.94} = 0.87,$$

and

$$\%\text{SM change} = (0.87 - 1.00)100 = -13\%.$$

A more exact comparison can be made by using the entire spectacle magnification formula, taking into consideration the shape factor as well as the power factor. This will be done for the $+13.00$ D spectacle lens, assuming that the spectacle lens has a front curve of $+16.00$ D and a center thickness of 8.0 mm, and is made of a plastic having an index of refraction of 1.498, and that the contact lens has a front curve of $+77.00$ D and a center thickness of 0.4 mm, and is made of a plastic having an index of 1.490. For the spectacle lens,

$$\text{SM} = \left( \frac{1}{1 - \dfrac{0.008}{1.498}(+16)} \right)\left( \frac{1}{1 - 0.013(+13)} \right)$$

$$= (1.094)(1.203) = 1.316,$$

while for the contact lens,

$$\text{SM} = \left( \frac{1}{1 - \dfrac{0.0004}{1.490}(+77)} \right)\left( \frac{1}{1 - 0.003(+14.94)} \right)$$

$$= (1.021)(1.047) = 1.069.$$

$$\text{Magnification ratio} = \frac{1.069}{1.316} = 0.81.$$

$$\%\text{SM change} = (0.81 - 1.00)100 = -19\%.$$

The error of 6% (i.e., 19% $-$ 13%) due to considering an aphakic lens as an infinitely thin lens shows that this is an invalid assumption, and that an aphakic lens (whether a spectacle lens or a contact lens) must be treated, for purposes of determining spectacle magnification, as a thick lens.

*Relative Spectacle Magnification*

It will be recalled from the discussions in Sections 10.12, 10.13, and 10.20 that when a patient has anisometropia, if the difference in the spherical ametropia for the two eyes is assumed to be due to a difference in axial length, a minimum of aniseikonia will be induced if spectacles are prescribed; whereas if the difference in the spherical ametropia is assumed to be due to a difference in the refractive power of the two eyes (i.e., cornea, anterior chamber, or lens), a minimum of aniseikonia will be induced if contact lenses are prescribed. Since astigmatism is always refractive in origin, it follows that when a patient has significant astigmatism (particularly if the axes are oblique and symmetrical), a minimum of meridional aniseikonia will be induced if contact lenses are prescribed.

On the other hand, since most instances of spherical anisometropia tend to have both axial and refractive components, the above statements should not be interpreted as being hard and fast. For example, if an anisometropic patient desires contact lenses, he or she should not be told that contact lenses cannot be worn: The patient should be allowed to wear contact lenses for a trial period before the final decision is made.

# EFFECTS OF CONTACT LENSES
## ON BINOCULAR VISION

### 13.11. Accommodative Demand

Since accommodation, like refraction, takes place at the first principal plane of the eye (which is only slightly more than a millimeter behind the corneal apex), the accommodative demand can be significantly different for a contact lens wearer than for a spectacle wearer. This difference increases with increasing ametropia, and reaches significance in the pre-presbyopic and presbyopic years. As shown in Section 5.7, a spectacle-wearing 10.00 D hyperope must accommodate 3.29 D for a 40-cm reading distance, whereas a spectacle-wearing 10.00 D myope needs to accommodate only 1.83 D for the same distance.

Since the amount of accommodation while wearing contact lenses would be very close to 2.50 D, no matter what the refractive error, these examples show that the 10.00 D hyperope will have to accommodate about 0.75 D *less* while wearing contact lenses than while wearing glasses, whereas the 10.00 D myope will have to accommodate about 0.75 D *more* while wearing contact lenses than while wearing glasses. Even for patients having far less than 10.00 D of myopia, the increased accommodative demand required for contact lenses may constitute a problem. For example, if a 42-year-old 5.00 D myope, who has had no problem at all with near work while wearing glasses, is fitted with contact lenses, he or she may immediately find that close work is difficult. This is simply due to the fact that this myopic patient has lost the "advantage" of having to accommodate less than an emmetrope or a spectacle-wearing hyperope. On the other hand, it sometimes occurs that a spectacle-wearing hyperope, who has just begun to need an add while wearing spectacles, finds that if she switches to contact lenses, reading and other near work can be done comfortably without the need for an add.

### 13.12. Accommodative Convergence

The change in accommodative demand that occurs when a patient switches from glasses to contact lenses means that there will also be a change in the amount of accommodative convergence present for reading and other close work resulting in a change in the amount of fusional convergence required for fusion. The amount of change in accommodative convergence will depend upon the AC/A ratio.

For example, if a 10.00 D myope changes from glasses to contact lenses, the increased accommodative demand of almost 0.75 D will be accompanied by an increase in accommodative convergence. For a 6/1 AC/A ratio, the increased amount of accommodative convergence at 40 cm would be 6(0.75), or 4 prism diopters. This means that if the patient happened to be exophoric at 40 cm, the exophoria would be reduced by 4 prism diopters; but if the patient was esophoric at 40 cm, the esophoria would be increased by 4 prism diopters. In the former case, the patient would have to use 4 prism diopters less positive fusional vergence; while in the latter case he would have to use 4 prism diopters more negative fusional vergence.

Similarly, the decrease in accommodative demand for the high hyperope would result in a decreased amount of accommodative convergence, the change in accommodative convergence again depending upon the value of the AC/A ratio. For the 10.00 D hyperope having a 6/1 AC/A ratio, switching from glasses to contact lenses would mean that any exophoria at 40 cm would be increased and any esophoria at 40 cm would be decreased by 4 prism diopters.

Fortunately, the great majority of contact lens wearers have refractive errors in the ±1.00 to ±4.00 or ±5.00 D range, with the result that the change in accommodative convergence required with contact lenses is not often a problem.

### 13.13. Prismatic Effects

When spectacles are worn, whether primarily for distance vision or for both distance and near vision, they are normally centered in front of the wearer's pupils, thus avoiding prismatic effects during distance vision but allowing prismatic effects to be induced for near vision or for viewing through the lens periphery. When spectacle lenses of minus power are centered for the distance PD, *base-in* prism is induced for near vision; and when plus-powered spectacle lenses are centered for the distance PD, *base-out* prism is induced for near vision.

This induced prism power may be an advantage or a disadvantage to the wearer, depending upon the near phoria. As shown in Figure 13-32, if a myope has a significant exophoria at near, the presence of the induced base-in prism at the near point will be an advantage (reducing the amount of positive fusional vergence that must be used to compensate for the phoria); but a myope having esophoria at near would be at a disadvantage, since the prismatic effect at near would increase the amount of negative fusional vergence needed to compensate for the esophoria. On the other hand, for a hyperopic patient (see Figure 13-33), a *base-out* prismatic effect would be present at near, which would be an advantage if the wearer happened to be esophoric at near but a disadvantage if he or she were exophoric at near.

When contact lenses are worn rather than glasses, the contact lenses (if fitting correctly) are centered before the pupils for both distance and near vision, with the result that the induced prismatic effects for

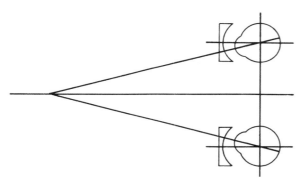

FIGURE 13-32. **If a myope's spectacle lenses are centered for distance vision, induced base-in prism is encountered for near work.**

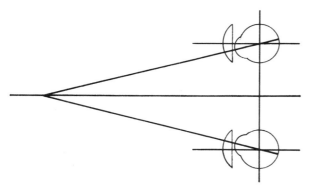

FIGURE 13-33. **If a hyperope's spectacle lenses are centered for distance vision, induced base-out prism is encountered for near work.**

near vision are no longer present. For relatively small refractive errors, this change in the near phoria situation does not often present a problem: However, it is possible that many symptoms reported by contact lens wearers that are considered by practitioners to be due to "adaptation" to the contact lenses are actually due to the change in prismatic effects for near vision.

### EXAMPLE 1

A 10.00 D myope who is orthophoric at 6 m but has 12 prism diopters of exophoria at 40 cm is accustomed to wearing spectacles with the lenses centered for distance vision, but switches to contact lenses which are centered in front of the pupils for both distance and near vision. What will be the necessary change in positive fusional vergence at 40 cm, with the contact lenses as compared to the glasses, if the distance PD is 64 mm and the near PD is 60 mm?

Using Prentice's rule, the prismatic effect induced at 40 cm while wearing the glasses, for each eye, will be

$P = dF = 0.2(-10) = 2$ prism diopters, base in,

or a total of 4 prism diopters base in for both eyes. This prismatic effect therefore reduces the exophoria by 4 prism diopters (from 12 to 8 prism diopters), reducing the positive fusional vergence demand by the same amount. However, when contact lenses are fitted, the near phoria would presumably be 12 prism diopters, requiring 4 prism diopters more positive fusional vergence than required when glasses were worn.

The word "presumably" was used because we have considered only the change in the near phoria due to the change in prismatic effect. We must also consider, however, the change in the near phoria brought about by the change in accommodative demand while wearing contact lenses as compared to glasses. As shown in Section 5.7 and referred to in Section 13.11, the increase in accommodative demand when a 10.00 D myope changes from glasses to contact lenses is approximately 0.75 D, and, as noted in the previous section, the change in accommodative convergence varies with the patient's AC/A ratio. In this case, the AC/A ratio was low: The calculated AC/A ratio for a patient having orthophoria at 6 m, and 12 prism diopters of exophoria at 40 cm, would be

$$\frac{AC}{A} = \frac{15 - 12}{2.5} = \frac{3}{2.5} = \frac{1.2}{1},$$

where 15 = total convergence demand at 40 cm, 2.5 = total accommodative demand at 40 cm, and 12 = (exophoria at 40 cm) − (exophoria at 6 m).

With an AC/A ratio of only 1.2/1, the increased amount of accommodative convergence while wearing contact lenses rather than glasses would be only

1.2(0.75) = 0.9 prism diopter,

with the result that the exophoria at 40 cm would be increased by only about 3 (rather than 4) prism diopters while wearing contact lenses.

For a patient having a higher AC/A ratio, the increased accommodative convergence available at 40 cm would more nearly offset the increased exophoria due to the lack of the base-in prismatic effect. For example, if the 10.00 D myope had a 6/1 AC/A ratio (which would mean that the phorias at 6 m and at 40 cm would be equal), the increased amount of accommodative convergence while wearing contact lenses would be

6(0.75) = 4 prism diopters,

decreasing the near exophoria by 4 prism diopters, with the result that the increased accommodative convergence at 40 cm while wearing contact lenses would be just enough to compensate for the lack of the induced base-in prismatic effect.

Thinking in practical terms, most people who have a clinically significant exophoria have a greater exophoria at 40 cm than at 6 m and therefore have a relatively *low* AC/A ratio. Therefore, the exophoria at 40 cm (for a myope) is likely to be increased more due to the lack of induced base-in prism than to be reduced due to increased accommodative convergence. Therefore, the practitioner should consider the possibility that a myope who is exophoric at near may experience adaptation symptoms when changing from glasses to contact lenses, due to the need to use a greater amount of positive fusional vergence at 40 cm.

Turning now to the myope who has esophoria at 40 cm, spectacle lenses that are centered for 6 m, inducing base-in prism at 40 cm will increase the amount of near esophoria.

### EXAMPLE 2
Let us consider a 10.00 D myope who is orthophoric at distance and has 6 prism diopters of esophoria at 40 cm, with a distance PD of 64 mm and a near PD of 60 mm. Using Prentice's rule, we find the induced prismatic effect at 40 cm, for each eye, to be

$P = dF = 0.2(10) = 2$ prism diopters, base in,

or 4 prism diopters base in for both eyes, with the result that the near esophoria would increase from 6 to 10 prism diopters, requiring the use of 4 prism diopters of additional negative fusional vergence. If this patient wears contact lenses

rather than glasses, he or she will have the advantage of a reduction in the near esophoria of 4 prism diopters (due to the lack of induced prismatic effect at near). But what about the change in the near phoria due to the increased accommodative demand?

Being orthophoric at distance and having a near esophoria of 6 prism diopters, the AC/A ratio would be

$$\frac{AC}{A} = \frac{15 - (-6)}{2.5} = \frac{21}{2.5} = \frac{8.4}{1}.$$

The increased accommodative convergence while wearing contact lenses would therefore be

8.4(0.75) = 6.3 prism diopters,

increasing the esophoria at near by 6.3 prism diopters, resulting in a predicted esophoria of 12.3 prism diopters.

Thus we see that if a patient who changes from glasses to contact lenses has high myopia accompanied by a high esophoria at 40 cm and a high AC/A ratio, the two effects on the near phoria tend to cancel each other out. Therefore, in most cases we need be less concerned about causing binocular vision problems when an esophoric myope changes from glasses to contact lenses, as opposed to an exophoric myope.

Since the great majority of contact lens wearers are myopes, we tend to ignore the problems *hyperopes* may have when changing from glasses to contact lenses. Since contact lens wearers requiring high plus lenses are usually aphakics, let us consider the change in the near phoria that is likely to occur when an aphakic wears contact lenses. Since aphakics tend to have high exophorias at near, only this condition will be considered.

### EXAMPLE 3
Suppose an aphakic patient wearing +10.00 D spectacle lenses, being orthophoric at distance and having 10 prism diopters of exophoria at near, switches from glasses to contact lenses. What will be the change in the near phoria due to the lack of induced base-out prismatic effect?

Assuming the difference between the distance and near PD to be 4 mm, the use of Prentice's rule shows that

$P = dF = 0.2(10) = 2$ prism diopters, base out, each eye,

or a total of 4 prism diopters base out for both eyes, therefore increasing the near exophoria from 10 to 14 prism diopters. With contact lenses, the lack of base-out prism would mean a reduction in the near exophoria of 4 prism diopters.

**Clinical note:** Since the aphakic patient is unable to accommodate, accommodative convergence is non-existent and therefore does not change the demand on fusional convergence at the near-point.

## 13.14. Prescribing Prism in a Contact Lens

The point has already been made that a contact lens may be provided with base-down prism ("prism ballast") to eliminate rotation of the lens on the eye. Vertical prism may also be prescribed to compensate for a vertical heterophoria. It has been stated in the literature that when a contact lens containing prism is placed in contact with the tear fluid, it retains only about one-fourth of its deviating power in air.[1,2] However, it has been shown by both Bailey[3] and Mandell[4] that placing a contact lens on the eye has no effect on the prism power. For a given amount of prism power, for *distance vision*, the deviation will be the same for a contact lens as for the spectacle lens. However, for *near vision*, a given amount of prism in a contact lens will produce somewhat *more* deviation than a prism in a spectacle lens because (as stated in Section 4.17) when a prism is used for near objects its deviating power is greater as the prism is moved closer to the center of rotation of the eye; that is, for a given amount of prism, the effective power of the prism for near objects will be greater for a contact lens than for a spectacle lens.

It should be understood that, whether or not a contact lens contains prism, the presence of *refractive power* in a contact lens brings about a vertical prismatic effect whenever the lens moves upward or downward on the eye with respect to the center of the pupil. This induced vertical prismatic effect would, of course, add to (or subtract from) the prismatic effect due to any vertical prism present in the lens.

## ————————ABERRATIONS AND FIELD OF VIEW————————

## 13.15. Aberrations

It will be recalled that all lenses are subject, to varying degrees, to chromatic aberration and to the five monochromatic aberrations. The effects of each of these aberrations on contact lenses will be briefly described.

### Chromatic Aberration

As discussed in Section 6.2, chromatic aberration may be considered as either (1) *longitudinal* chromatic aberration, defined as the separation along the optic axis of the focus for rays of light originating from a single axial point, or (2) *transverse* chromatic aberration, considered in terms of (a) the chromatic difference in magnification produced by a thick lens, which occurs in the image plane for rays from an extended object, or (b) the angular dispersion due to the presence of prismatic effects. The eye tolerates longitudinal chromatic aberration very well, probably because the eye itself is subject to about 1 D of longitudinal chromatic aberration. As for transverse chromatic aberration, the angular dispersion due to prismatic effects is present only when the eye turns to look through a point in or near the periphery of a high-powered lens: Since a contact lens wearer always looks through a point at or near the optical center of the lens, this aberration does not present a problem for a contact lens wearer.

### The Monochromatic Aberrations

The aberrations of *spherical aberration* and *coma* occur only for large-aperture optical systems. As with spectacle lenses, these aberrations do not present a problem for contact lenses of moderate power, because the eye looks through only a small portion of the lens (the apical portion, in the case of a contact lens) at any time.

The aberrations of *oblique astigmatism* and *curvature of image* occur when a narrow pencil of light from an object passes obliquely through a spherical surface. For a spectacle lens, these aberrations occur only when the eye rotates away from the primary position so that the foveal line of sight passes through a peripheral point in the lens. Since a contact lens fits the eye in such a way that the foveal line of sight always passes through a point at or near the optical center of the lens, these aberrations do not present a problem for the contact lens wearer.

The aberration of *distortion*, although a problem

for wearers of spectacle lenses of high power, applies only to extended objects, occurring as a result of the gradual change in magnification brought about by a lens from the center toward the periphery. It increases in magnitude with the distance between the correcting lens and the pupil of the eye. Because the eye always fixates through the center of a contact lens, and because the distance between the contact lens and the pupil of the eye is very small, distortion is not apt to be a problem for wearers of contact lenses.

In summary, neither chromatic aberration nor any of the five monochromatic aberrations is likely to present a problem to the contact lens wearer.

## 13.16. Field of View

As discussed in Section 11.4, field of view may be considered either in terms of the *peripheral* field of view, subtended at the entrance pupil of a steadily fixating eye, or in terms of the *macular* field of view of the moving eye (also known as field of fixation), subtended at the center of rotation of the eye.

As given in Eqs. (11.4) and (11.5) and illustrated in Figures 11-5 and 11-6, the extent of the peripheral field of view can be determined by the relationship

$$\tan \phi = y(E - F),$$

while the extent of the macular field of view can be determined by the relationship

$$\tan \theta = y(S - F),$$

where $\phi$ = one-half the peripheral field of view, $\theta$ = one-half the macular field of view, $y$ = one-half the lens aperture, in meters, $E$ = the vergence of light at the entrance pupil of the eye, $S$ = the vergence of light at the center of rotation of the eye, and $F$ = the power of the correcting lens.

The difference between the peripheral fields of view for a spectacle lens and for a contact lens will be illustrated by the following examples.

### EXAMPLE 1

How does the size of the peripheral field of view compare for a 3.00 D myope (a) for a spectacle lens having an aperture of 50 mm, located 15 mm from the entrance pupil of the eye, and (b) for a contact lens having an aperture of 7 mm, located 3 mm from the entrance pupil?

For the spectacle lens (a),

$$\tan \phi = 0.025(66.67 + 3) = 1.742$$
$$\phi = 60.14°$$
$$2\phi = 120.28°,$$

while for the contact lens (b),

$$\tan \phi = 0.0035(333.33 + 3) = 1.1777$$
$$\phi = 49.65°$$
$$2\phi = 99.31°.$$

### EXAMPLE 2

How does the size of the peripheral field of view compare for an aphakic patient (a) wearing a +12.50 D spectacle lens having an aperture of 50 mm, located 15 mm from the entrance pupil of the eye, and (b) wearing (for the same eye) a +14.50 D contact lens having an aperture of 7 mm, located 3 mm from the entrance pupil?

For the spectacle lens (a),

$$\tan \phi = 0.025(66.67 - 12.50) = 1.3543$$
$$\phi = 53.56°$$
$$2\phi = 107.12°,$$

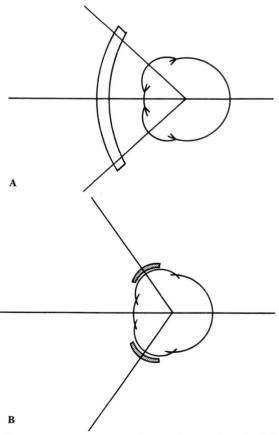

**A**

**B**

FIGURE 13-34. **For a spectacle lens (A) the macular field of view is limited by the rim, or edge, of the lens; for a contact lens (B) the macular field of view is limited only by the ability of the eye to make rotational movements.**

and for the contact lens (b),

$$\tan \phi = 0.0035(333.33 - 14.50) = 1.116$$
$$\phi = 48.135°$$
$$2\phi = 96.27°.$$

In view of the fact that many people (including many vision practitioners) believe that "the field of view is bigger" with a contact lens than with a spectacle lens, it is of interest that, in both of the above examples, the peripheral field of view is about 20 degrees *smaller* with a contact lens than with a spectacle lens. Although these results could vary somewhat, depending on the vertex distance of the spectacle lens and the aperture sizes of the spectacle lens and the contact lens, it is apparent that a contact lens does not necessarily provide a *larger* peripheral field of view than a spectacle lens. Why, then, does the idea persist that a contact lens provides a larger field of view than that provided by a spectacle lens?

The answer to this question is that the field of view that is larger for a contact lens wearer is not the peripheral field of view, but the *macular* field of view. The macular field of view can be shown to be larger for a minus lens than for a plus lens (as in the case of the peripheral field of view). However, the important consideration is that for a spectacle wearer the macular field of view (as shown in Figure 13-34) is limited by the presence of the rim, or edge, of the lens, whereas for a contact lens, the lens moves with the eye with the result that the macular field of view is limited only by the eye's ability to make rotational movements.

Therefore, when the statement is made that "a contact lens provides a larger field of view than a spectacle lens," the statement is true if it is applied to the *macular* field of view, but not if it is applied to the *peripheral* field of view.

## APHAKIA

### 13.17. Aphakic Contact Lenses

As discussed in Section 11.4, aphakic spectacle lenses present a number of inherent problems. Although most of these problems can be controlled to a great extent by careful selection of aphakic lenses and frames, it is fortunate that these problems can, for all practical purposes, be *eliminated* by prescribing contact lenses rather than glasses, allowing the aphakic patient to function in very much the same manner as in the pre-aphakic condition. Problems due to aphakic spectacles include the following:

1. Increased retinal image size
2. Decreased field of view
3. Presence of a ring scotoma
4. Increase in ocular rotations
5. Increased lens aberrations
6. Motion of objects in the field of view
7. Demands on convergence
8. Appearance of the wearer

*Increased Retinal Image Size*

In Section 11.4, it was found that for the aphakic schematic eye, corrected by a +10.34 D lens at a vertex distance of 12 mm, spectacle magnification (including both the shape factor and the power factor) would be 28%. What would be the spectacle magnification (or, rather, the contact lens magnification) if the aphakic schematic eye were fitted with a contact lens rather than a spectacle lens?

First we must consider the fact that the power of the aphakic contact lens would differ from that of the spectacle lens, because of the change in vertex distance. Using the effective power formula,

$$F_{CL} = \frac{F_S}{1 - dF_S}$$

$$= \frac{+10.34}{1 - .012(+10.34)}$$

$$= +11.80 \text{ D.}$$

In order to apply the spectacle magnification formula, we must supply values for the index of refraction, the center thickness, and the front surface power of the contact lens. Assuming that these values are 1.490, 0.50 mm, and +74.00 D, respectively, the magnification brought about by the contact lens would be

$$SM = \left(\frac{1}{1 - \dfrac{0.0005}{1.490}74}\right)\left(\frac{1}{1 - 0.003(11.80)}\right)$$

$$= (1.025)(1.037) = 1.063$$

$$\%SM = 6.3.$$

Even though 6.3% is a large amount of magnification, it is sufficiently small (when compared to the 28% magnification brought about by spectacle lenses) that the bilateral aphakic is able to adapt relatively easily to the aphakic world. Even this amount of magnification can be a problem for the *unilateral* aphakic: The reader is referred to Section 11.8 for a discussion of unilateral aphakia.

### Decreased Field of View and Presence of a Ring Scotoma

As shown in Section 11.4, when the pre-aphakic and aphakic fields of view are compared for a typical spectacle-wearing aphakic patient, the peripheral field of view decreases by about 7°, whereas the macular field of view decreases by about 17°. The decreases in both the peripheral and macular fields of view are responsible for the *ring scotoma* that is encountered when the eye rotates toward the periphery of the lens. However, since (as shown in the previous section) the macular field of view is virtually *unlimited* when a contact lens is worn, an aphakic wearing contact lenses is not aware of any limitation in the field of view.

As for the problem of the ring scotoma and the accompanying "jack-in-the-box phenomenon," it will be recalled that the ring scotoma occurs because of the fact that the spectacle-wearing aphakic has peripheral vision *beyond* the edge of the correcting lens. This, obviously, is not possible with a contact lens, with the result that the ring scotoma and the "jack-in-the-box phenomenon" are not present.

### Increase in Ocular Rotations

The increase in magnitude of ocular rotation when viewing an object in the periphery, for the spectacle-wearing aphakic eye as compared to the pre-aphakic state, is due to the *prismatic effect* that is present when the visual axis passes through a peripheral point in the lens. Since a contact lens rotates with the eye, the visual axis of an aphakic wearing contact lenses always passes through the center of the lens, with the result that increased ocular rotations are not required.

### Increased Lens Aberrations

As concluded in Section 13.14, neither chromatic aberration nor the five monochromatic aberrations are likely to present problems for the contact lens wearer. This statement applies equally to the *aphakic* contact lens wearer.

### Motion of Objects in the Field of View

This problem, sometimes referred to as "swim," is an "against motion" of the field of view that occurs when a wearer of aphakic spectacles moves the head to look at a peripheral object while the eyes are held stationary (but moving with the movement of the head). It occurs because of the fact that, as the head is moved, the visual axes pass through progressively more peripheral portions of the spectacles lenses, thus bringing about progressively increasing prismatic effects (with the prism base corresponding to the direction of the head movement). When an aphakic wearing contact lenses moves his or her head (and eyes) to look at a peripheral object, all objects in the field of view are seen (as the head moves) through the central or para-central portion of each contact lens, with the result that "swim" does not occur.

### Demands on Convergence

As discussed in Section 13.13, when spectacle lenses are centered for the wearer's distance PD, a horizontal prismatic effect is encountered during reading and other close work, requiring either an increase or a decrease in the demand on fusional vergence. For an aphakic lens, the base-out effect occurring with any plus lens is exaggerated, and can be easily determined by the use of Prentice's rule. Since a contact lens wearer looks through the centers of his or her lenses for reading as well as for distance vision, no base-out prismatic effect is present at near, with the result that the demand on fusional vergence is only that required by the patient's near phoria.

### Appearance

Even before the patient has undergone surgery (and is therefore relatively unaware of the perceptual problems brought about by aphakic spectacles), many patients will choose contact lenses rather than spectacles strictly because of what they consider to be the unsightly appearance of aphakic spectacles.

# INSTRUMENTATION

Just as the optics of a spectacle lens can be assessed by the use of the lensometer, the lens measure, and the thickness calipers, the optics of a contact lens can be assessed by means of the lensometer, the radiuscope, and the thickness gauge.

The optical principles of the *lensometer* have been described in Chapter 3. The back vertex power of a contact lens is measured by placing the back surface of the lens against the lensometer stop. As described in Section 13.1 and illustrated in Figure 13-1, modern lensometers are equipped with a smaller "stop" than that used for a spectacle lens, in order to provide for the steeper back surface curvature of a contact lens. A *thickness gauge* is a mechanical instrument (similar to that used by a machinist) having a stage upon which the contact lens is placed, and provided with a gear assembly which supplies a direct reading to the nearest 0.01 mm.

The *Radiuscope*, whose optical principles are described below, was developed by American Optical Corporation, specifically for the measurement of the back surface radius of curvature of a hard contact lens. (Note that as with other instrumentation including the refractor, the keratometer and the lensometer, the term "radiuscope" is now used in a generic sense.)

## 13.18. The Radiuscope

The radiuscope is an instrument designed to measure the radius of curvature of the back surface (the base curve) of a contact lens. The instrument is based on a principle described by a Drysdale[5] in 1900. The Drysdale principle is incorporated in all radiuscopes, although these instruments differ from one another in design and in the method of displaying the reading.

If a collimated beam of light is focused on a concave reflecting surface, the light will be reflected along the same path as the incident light; and if the focus of the collimated beam is then moved to the center of curvature of the concave surface, it will again be reflected back along the same path as the incident light. Because of the fact that the center of a concave reflecting surface and the center of curvature of the surface both cause incident light to be reflected back along its original path, these points are sometimes referred to as *self-reflecting points*.

The radiuscope incorporates a compound microscope in which an internally illuminated target (a radial-line target similar to a "clock dial") is projected along the axis of the instrument. An image of the target will be seen clearly by an observer, through the eyepiece, when the microscope objective focuses the image of the target either on the concave reflecting surface or at its center of curvature. The instrument is fitted with a half-silvered mirror, set at an angle of 45°, above the microscope objective.

The measurement consists of placing a contact lens (with its concave surface upward) on the stage of the instrument, with the convex surface in contact with a small pool of water. The water reduces the intensity of the image reflected from the lower (convex) surface of the contact lens so that it cannot be readily seen when the reflection from the upper (concave) surface is scrutinized. The stage of the instrument is then adjusted so that the illuminated target is aligned with the mirror, with the result that light from the mirror is reflected through the objective lens and forms a real image in the working plane of the objective lens.

Once the instrument has been aligned, the microscope objective is moved downward, toward the surface of the contact lens, until the working plane of the objective lens coincides with the plane of the back surface of the contact lens. When this occurs, the light is reflected back along its incident path, passing through the half-silvered mirror and forming an image in the focal plane of the eyepiece, where it can be seen clearly by the observer, as shown in Figure 13-35A. This is the "zero" reading, and the operator places a fiduciary line on the zero point on the instrument's scale.

The microscope objective is then raised until a second position is found where the image of the target is in optimum focus. This occurs when the working plane of the objective lens is positioned at the center of curvature of the back surface of the contact lens, as shown in Figure 13-35B. In this position, the concave surface will again return the light along its own path, which is normal to the back surface of the contact lens.

The distance through which the microscope has been moved between the first and second clearly focused images is the actual (physical) length of the radius of curvature of the concave surface (the base curve radius) of the contact lens.

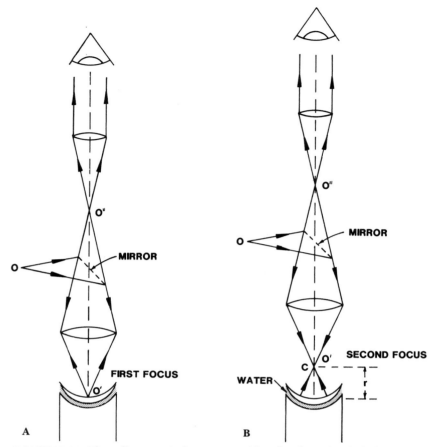

FIGURE 13-35. **The radiuscope: A, the "zero" position (first focus) in which the working plane of the microscope objective coincides with the plane of the back surface of the contact lens; B, the final position (second focus), in which the working plane of the microscope objective coincides with the center of curvature of the back surface of the contact lens.**

## References

1. Girard, L.J., Soper, J.W., and Sampson, W.G. *Corneal Contact Lenses*, p. 290. C.V. Mosby, St. Louis, 1964.
2. Filderman, I.P., and White, P.F. The Effect of Surrounding Media on Prismatic Contact Lenses. *Optom. Weekly*, Vol. 56, No. 26, pp. 19–24, July 1, 1965.
3. Bailey, N.J. Prism in a Contact Lens. *J. Amer. Optom. Assoc.*, Vol. 37, pp. 44–45, 1966.
4. Mandell, R.B. The Prism Controversy. *J. Amer. Optom. Assoc.*, Vol. 38, pp. 188–189, 1967.
5. Drysdale, C.V. On a Simple and Direct Method of Obtaining the Curvature of Small Lenses. *Trans. Ophthalmol. Soc. U.K.*, Vol. 2, pp. 1–12, 1900.

## Questions

In working these problems, it is assumed that the keratometer is calibrated for an index of refraction of 1.3375 and that both the cornea and tears have an index of refraction of 1.3375. The index of refraction of rigid contact materials is assumed to be 1.49.

1. A patient's refraction using a phoropter at a vertex distance of 13 mm is +14.75 DS OU. If hard contact lenses are prescribed, what will be the approximate power of the contact lenses for this patient?

2. Given:
   K readings: 43.00 D @ 180/44.25 D @ 090
   Spectacle $R_x$: $-1.00$ DS $-1.25$ DC x 090
   Base curve of contact lens: 43.00 D

   What is the residual astigmatism predicted with a hard contact lens in place?

3. A hard contact lens with a refractive index of 1.49 has a center thickness of 0.19 mm and a vertex power of $-2.00$ D. If the radius of curvature of the back surface is 7.4 mm, what is the radius of curvature of the front surface?

4. A 7.50-mm (45.00 D) base curve, $-3.00$ D hard contact lens is placed on a patient with a refraction of $-3.50$ DS $-0.25$ DC x 090, central "K" readings of 44.50 D in the 180 meridian and 45.00 D in the 90 meridian. What is the probable over-refraction?

5. A patient had K readings of 42.50 D at 180 and 44.00 D at 90 and a spectacle refraction of $-1.50$ DS $-1.50$ DC x 180. What standard rigid contact lens power is needed if a 7.67 D base curve (44.00 D) lens is fitted on this patient?

6. Given the following data:
   Spectacle refraction: $-1.00$ DS $-1.00$ DC x 180
   Hard contact lens specifications:
      Base curve = 7.80 mm
      Power = plano DS
      Refraction, contact lens, in situ = $-0.25$ DS $-0.25$ DC x 090

   What are the corneal curvatures?

7. A $+6.50$ D hyperopic patient who wears her spectacle prescription at a 16-mm vertex distance is to be fitted with a hard contact lens (of refractive index 1.49). Her cornea is spherical and measures 44.25 D. If the contact lens is fitted 0.50 D flatter than "K," what is the expected contact lens prescription with consideration given to lens effectivity changes?

8. Consider the following clinical information:
      Spectacle Rx: $-3.00$ DS $-0.50$ DC x    180
      K readings: 44.00 @ 180/45.00 @ 090
      Base curve of the contact lens: 7.58 mm

   What is the contact lens power necessary for complete correction?

9. If a patient has a spectacle prescription of $-1.50$ DS $-1.00$ DC x 180 with a 7.50-mm base curve radius, $-2.00$ D rigid diagnostic lens on the eye, and the refractive error completely corrected, what is the patient's keratometer reading?

10. A patient with a spectacle correction in one eye of $-1.00$ DS $-0.50$ DC x 180 is to be fitted with a rigid contact lens. A diagnostic lens with a radius of curvature of 8.04 mm and power $-3.00$ D is placed on the eye. Over-refraction to best acuity is $+1.50$ DS. The decision is to order a lens with a base curve radius of 7.94 mm for the patient. What power should be ordered for the contact lens?

11. A patient having a spectacle correction of $-3.00$ DS $-0.50$ DC x 180 is fitted with a $-3.50$ DS contact lens. If keratometer readings are 42.75 D at 180 and 43.25 D at 90, and the over-refraction is $+1.00$ DS, what is the power of the base curve of the contact lens measured on a keratometer?

12. A patient has K readings of 43.50 D at 180 and 46.50 D at 090. The spectacle prescription is plano. The patient is fit with a toric back surface that matches perfectly with the cornea. What is the radius of curvature of the vertical meridian of the front surface of the contact lens?

13. A hard contact lens with a spherical front curve is analyzed on a lensometer to be $-0.50$ DS $-4.37$ DC x 180. The radius of curvature is 8.49 mm in the horizontal meridian. What is the radius of curvature in the vertical meridian on the back side?

14. The power of a toric-base spherical front surface corneal contact lens is measured with a lensometer and is found to be plano $-3.25$ DC x 180. What is the approximate amount of over-refraction when properly placed on an emmetropic eye with a spherical cornea?

15. A front surface toric contact lens has a power of plano $-3.00$ DC x 020. The contact lens is placed on a patient's eye, but it is suspected that the lens is not oriented properly on the eye. An over-refraction of $+0.50$ DS $-1.00$ DC x 060 gives the best acuity. Assume the orientation is the only reason for the nonzero over-refraction. What is the appropriate minus axis cylinder orientation in the eye?

16. A patient is fitted with the following rigid contact lens: base curve radius = 7.67 mm, power = $-2.00$ D. At the time of the fitting the over-refraction was plano and the patient's K readings were 44.00 D at 180 and 44.50 D at 090. Three months later the K readings were found to be 44.75 D at 180 and 45.25 D at 090. The

patient wishes to wear spectacle lenses immediately after removal of the contact lenses. What is the spectacle prescription found immediately upon removal?

17. A patient's refractive error is fully corrected with a spherical corneal contact lens. The base curve of the lens is 43.00 D and the power is −3.00 D. The patient is refitted with a bi-toric lens having a base curve of 43.50 D at 180 and 45.50 D at 90. What is the power ordered for this lens?

18. A fused bifocal is made on the back surface of a hard contact lens with the back surface of the segment having the same radius of curvature as the base curve, that is, 8.0 mm. The refractive index of the segment is 1.57 and the radius of the

countersink surface is 6.67 mm. What is the power of the add on the eye?

19. A fused bifocal is made on the back surface of a hard contact lens (of refractive index 1.49) with the back surface of the segment having the same radius of curvature as the base curve, that is, 8.0 mm. The index of the segment is 1.56 and the power of the add on a lensometer is +3.00 D. What is the radius of curvature of the countersink surface?

20. A rigid one-piece back surface bifocal contact lens is ordered with a distance power of −3.00 D with a +2.50 D add. When verified on a lensometer, what is the difference in the readings between the near and distance portions of the lens?

# Answers to Questions

No answers are provided for essay or discussion questions.

## Chapter 2

1. 2.99 nm
2. 1.698
3. +5.07 D
4. (a) $F_1$ of major portion = +12.83 D
   (b) $F_1$ of segment = +16.92 D
5. +2.65 D
6. 1.625
7. 1.70
8. 1.60
9. +70.00 D
10. +3.50 DS +2.50 DC x 90
11. (a) −1.50 D
    (b) −4.50 D
12. (a) +1.00, +2.00, +1.73, +1.00, −1.00, −2.00, −1.00, +1.00 D
    (b) −1.50, +0.50, +1.87, +3.00, +3.50, +1.50, −1.00, −1.50 D
13. +1.31 D
14.

Horizontal line

| (a) | (b) | (c) |
|-----|-----|-----|
| +10 cm | +20 cm | +20 cm |
| (d) | (e) | |
| −20 cm | +33.3 cm | |

Vertical line

| (a) | (b) | (c) |
|-----|-----|-----|
| +20 | −20 | ∞ |
| (d) | (e) | |
| −10 | −33.3 | |

Circle

| (a) | (b) | (c) |
|-----|-----|-----|
| +13.3 | ∞ | +40 |
| (d) | (e) | |
| −13.3 | ∞ | |

15. +2.00 DS +1.00 DC x 90
16. 1−7
17. −2.00 DS +8.00 DC x 180
18. (a) −1.80 DS +4.85 DC x 2° 44′ or +3.05 DS −4.85 DC x 92° 44′
    (b) +0.85 DS +3.30 DC x 51° 42′ or +4.15 DS −3.30 DC x 141° 42′
    (c) −0.705 DS +0.916 DC x 8° 46′ or +0.211 DS −0.916 DC x 98° 46′
    (d) −1.52 DS +5.05 DC x 69′03″ or +3.53 DS −5.05 DC x 159′03″
    (e) −3.90 DS +6.05 DC x 30° 21′ or +2.15 DS −6.05 DC x 120° 21′
19. +3.00 DC x 180/+1.00 DC x 90
20. +10.00 D
21. 1.622
22. −10.10 DS −0.31 DC x 180
23. +3.50 D

## Chapter 3

1. −5.00 D
2. +11.27 D
3. +9.80 D
4. (a) −12.00 D
   (b) −13.03 D
   (c) −13.03 D
   (d) −3.03 D
5. −9.26 D
6. +13.90 DS −3.38 DC x 180
7. 1.33 cm
8. −2.25 DS +1.75 DC x 15
9. +9.14 D
10. +16.67 D
11. +25.00 D
12. +2.00 DS −6.00 DC x 180
13. −4.00 D
14. +2.00 DS +1.25 DC x 15, add +2.00 D
15. +2.50 DS −7.50 DC x 30
16. 2.3 mm
17. 4.4 mm
18. −10.46 D
19. +0.39 D increase
20. 4.5 mm

## Chapter 4

1. (a) 1°22′
   (b) 1°5′
2. 2.6 prism diopters
3. Base up

4. (a) 4.47 prism diopters base at 243°29'
   (b) 2.2 prism diopters base at 116°34'

5. (a) 2.59 prism diopters base in/1.5 prism diopters base up
   (b) 1 prism diopter base in/1.73 prism diopters base down

6. 12 prism diopters

7. 4.4 degrees

8. 11.39 prism diopters

9. (a) 0.6 prism diopter base out/3.6 prism diopters base up
   (b) 0.4 prism diopter base in/none vertical
   (c) none horizontal/2.4 prism diopters base up

10. (a) 3.8 BU
       0.2 BI
    (b) 4.5 BU
       3.0 BI
    (c) 2.68 BU
       2.18 BI

11. 1.40 prism diopters BD in the OD
    1.64 prism diopters BO in the OS

12. +3.00 DS −7.00 DC x 180

13. −5.00 DS +8.33 DC x 180

14. 13.33 prism diopters

15. 3 mm in from the geometrical center

16. 2.688 mm

17. 2 prism diopters

18. 3 mm temporally

19. 65 mm

20. 5.0 mm in; 1 mm out

21. 2.4 mm

22. 1.2 mm less

## Chapter 5

1. 0.77 D

2. OD +1.55 D
   OS +1.36 D

3. (a) +2.36 D add
   (b) 0.62 D

4. 3.29 D

5. 2.01 D

6. +2.97 DS −7.97 DC x 90

7. +2.40 DS +5.60 DC x 180

8. 0.33 meter in front of the eye

9. 10 cm

10. 10.00 D

11. 20 cm

12. 2.00 D hyperopia

13. 1.50 D

14. 1.00 D myopia

15. 6.00 D

## Chapter 6

1. 0.1

2. 10.0

3. 0.092 D

4. 0.7 prism diopter

5. +0.19 D

6. +11.66 D and −6.67 D

7. Crown = 8.18 prism diopters
   Flint = 5.18 prism diopters

8. −72.00 cm

9. +4.00 D

10. 0.18 D

11. 0.26 D

12. −508 mm

13. +360 mm

14. −20.00 D

## Chapter 7

1. 1.3

2. 116 nm

3. 5.6%

4. 87.0%

5. 95.6%

6. 6.25%

7. 2.1

8. 25

9. 1.4

10. 10.0%

11. 30 mm

12. 20.17%

13. 4.74%

14. 68.1%

15. 10

16. 0.1%

17. 4.48 mm

## Chapter 8

1. 3.3 prism diopters base down

2. 1.1 cm

3. 1.2 prism diopters base down

4. 4.0 prism diopters base up

5. 4.4 prism diopters base down

6. 30 mm

7. 7 mm below the top of the segment

8. −6.00 D

9. plano

10. +3.00 D

11. +15.00 D

12. Near segment = 1.60
    Intermediate segment = 1.54

13. 0.756 mm

14. $F_1 = +22.00$ D
    $F_2 = -31.00$ D

15. 1.58 mm

16. 10 mm

17. +1.00 DS −1.50 DC x 90;
    add = +2.50 D.

18. −4.00 D horizontal
    −6.00 D vertical

19. 1.18 D

20. 0.21 D

21. 41.76%

## Chapter 9

1. 5.0%

2. 63.2 mm

3. 62.05 mm

4. 2 mm in; 2 mm in

5. 4 mm in; 4 mm in

6. 3.5 mm each

7. 3 mm below the geometrical center

8. +2.00 D

9. 18.0 mm

## Chapter 10

1. 2.5 prism diopters OD

2. 9 prism diopters BU

3. OD R-5; OS R-9 or
   OD R-6; OS R-10

4. (e) OD Executive; OS Kryptok

5. 1.08 mm

6. +11.50 D

7. 2.5 mm above the reading level

8. 1.0 prism diopter left hyperphoria

9. orthophoria

10. 2.0%

11. −29.50 D

12. 6.75 mm

13. +11.00 D

14. 20.2% increase

15. 12.0% decrease

16. Increase by 3%
17. 1.0% decrease
18. 2.0% increase
19. +18.55 D

## Chapter 11

1. +13.00 D
2. +10.00 D
3. 73.15 mm behind the corneal apex
4. +11.88 D
5. 31.25 prism diopters
6. +14.02 D
7. 2.07 mm

## Chapter 12

1. 4x
2. 2.67x
3. (a) −2.5x
   (b) 28 cm

4. Objective lens = +10.00 D
   Eyepiece lens = −40.00 D
5. 16 cm
6. −3.5x
7. 1.30 or 30%
8. 8x
9. 8 D
10. 16 D
11. 25 cm
12. 5x
13. 2.5x telescope with a +4.0 D reading cap
14. +7.8 D
15. 40 mm
16. +15.00 D
17. 1.36x
18. 1.60x

## Chapter 13

1. +18.25 D

2. −2.50 DC x 90
3. 7.70 mm
4. −0.50 DS −0.75 DC x 90
5. −3.00 D
6. 43.75 D at 180; 45.00 D at 90
7. +7.75 D
8. −3.00 DS −0.50 DC x 90
9. 44.50 D at 180; 45.50 D at 90
10. −2.00 D
11. 42.25 D
12. 7.258 mm
13. 7.89 mm
14. +1.00 DS −1.00 DC x 90
15. 10 degrees
16. −2.75 DS −0.50 DX x 180
17. −3.50 DS −2.00 DC x 180
18. +2.00 D
19. 5.96 mm
20. 8.03 D

# Index

457

DISP formula ⎫ Pg 93
REL DISP    "  ⎭